Receptor Tyrosine Kinases: Structure, Functions and Role in Human Disease

Deric L. Wheeler • Yosef Yarden
Editors

Receptor Tyrosine Kinases: Structure, Functions and Role in Human Disease

Humana Press

Editors
Deric L. Wheeler
Department of Human Oncology
University of Wisconsin
Madison, WI, USA

Yosef Yarden
Department of Biological Regulation
Weizmann Institute of Science
Rehovot, Israel

ISBN 978-1-4939-2052-5 ISBN 978-1-4939-2053-2 (eBook)
DOI 10.1007/978-1-4939-2053-2
Springer New York Heidelberg Dordrecht London

Library of Congress Control Number: 2014954081

Printed on acid-free paper

Humana Press is a brand of Springer
Springer is part of Springer Science+Business Media (www.springer.com)

This book is dedicated to the memory of Anthony (Tony) Pawson, who passed away in August 2013 (age 60). Tony was born and educated in England and lived in Canada. His research has revolutionized our understanding of tyrosine phosphorylation networks and signal transduction in general. Many of the principles and examples discussed in these handbooks were laid down by Tony's pioneering work uncovering the phosphotyrosine-binding Src homology 2 (SH2) domain, a prototype module of phosphorylation-based protein-protein interaction domains.

Preface

Multicellular organisms have evolved complex cellular communication pathways that enable cells to function and communicate in specific ways. These processes are mediated through a diverse group of cell surface receptors, including a group called receptor tyrosine kinases (RTKs). Receptor tyrosine kinases have intrinsic ability to transfer the gamma phosphate group from adenosine triphosphate to a tyrosine residue on a substrate protein, serving as an important signal for maintenance of cellular homeostasis or changes in cellular function. Accordingly, RTKs regulate a diverse array of cellular functions including cellular proliferation, survival, differentiation, migration, and metabolism.

The human genome encodes 58 RTKs that fall into 20 subfamilies based on the ligands they bind, their sequence homology, and structure. The architecture of all RTKs is highly conserved from the nematode *Caenorhabditis elegans,* with an extracellular ligand-binding domain, a single transmembrane α helix, an intracellular tyrosine kinase domain, and a tyrosine-rich C-terminal tail. Receptor tyrosine kinases are activated upon binding to cognate ligands present in the extracellular milieu, leading to receptor homo- or hetero-dimerization, kinase domain activation, and subsequent phosphorylation of tyrosine residues located within the cytoplasmic tail. Phosphorylated tyrosines serve as docking sites for a variety of intracellular adaptors and effector kinases that transmit signals to the nucleus, resulting in changes in cell function or fate. Termination of RTK activation is tightly controlled through the activation of a variety of tyrosine phosphatases, receptor-mediated endocytosis, and subsequent receptor degradation. Indeed, the importance of many RTKs in mammalian development has been displayed through the study of genetically altered mice, resulting in either severe developmental abnormalities or embryonic lethality. Collectively, RTK activation of signaling networks provides an essential mechanism by which cells communicate to regulate a multitude of different cellular responses.

Since RTKs regulate both developmental and regulatory cellular processes, it is not surprising that abnormalities in RTK structure or activity can result in human diseases. Several diseases result from RTK mutation, constitutive activation, and/or overexpression such as cancer, diabetes, inflammation, arteriosclerosis, angiogenesis, autoimmune disorders, and skeletal diseases. Due to this causal relationship, therapeutic targeting of RTK activity is clinically approved for treatment of several human diseases.

The focus of this book is to provide a comprehensive outline for the role of RTKs in mammalian cell development, with a focus on RTK evolution over time, cell-signaling networks, with a focus on RTK structure and function, and finally how RTKs play fundamental roles in human disease. In the opening chapter, **Tony Hunter and Gerard Manning** provide a comprehensive overview of the eukaryotic protein kinase superfamily and the emergence of RTKs. Further, they discuss the human kinome and the role it plays in pathology.

In the next chapter**, Manfred Schartl, Jean-Nicolas Volff, and Frederic Brunet** dovetail and expand the discussion of the evolution of RTKs. This chapter provides a useful reconstruction of the evolutionary history of the mammalian RTKs and the interrelationship of RTK families and their individual members, and how the emergence of RTKs grew in parallel with the increasing complexity of cell types and their individual functions in development and disease. After a thorough discussion on the emergence and evolution of RTKs **Ben-Zion Shilo** describes the role of RTKs in invertebrates. Specifically, the chapter highlights lessons learned from invertebrates and primitive signal transduction pathways and how they relate to higher level signaling in vertebrate animals.

Building on RTK signaling in invertebrates, **Pierre De Meyts** advances this discussion on classical signaling mechanisms emanating from RTKs localized to the plasma membrane. Specifically, he highlights what is known about ligand requirements and dimerization events resulting in the activation of the kinase domain and phosphorylation of C-terminal tyrosines. Further, he provides an in-depth analysis of downstream signaling cascades activated by RTKs and how these events regulate cellular homeostasis, cell cycle, and apoptosis. Extending the conversation of RTK signaling**, Yi Du, Jennifer L. Hsu, Ying-Nai Wang, and Mien-Chie Hung** highlight an emerging novel signaling role of RTKs localized from within the nucleus. Here they highlight families of RTKs that have been identified in the nucleus, namely the HER/ERBB, FGFR, VEGFR, insulin and insulin-like growth factor receptor, HGFR, ROR, and EPH receptor families. Further, they provide a comprehensive analysis, notably centered on the epidermal growth factor (EGFR), and the events associated with EGFR translocation from the cell membrane to the nucleus. They further discuss novel functions of nuclear RTKs, such as their ability to function as co-transcription factors and nuclear kinases, and how these functions lead to poor prognosis of cancer patients and resistance to cancer therapeutics.

Yehoshua Enuka, Morris E. Feldman, and Yosef Yarden build upon these two chapters and describe in-depth computational modeling of RTK signaling networks. Specifically they discuss the complexity of signaling and how a single ligand-receptor interaction can lead to a complex layered signaling network with both positive- and negative-feedback loops controlling this signaling network, ultimately feeding the output layer. Closing this section on RTK signaling, **Alexander Sorkin and Arola Fortian** provide an in-depth review of the events occurring after RTK activation, highlighting the EGFR as the prototypic experimental model. Specifically, they highlight the process of receptor-mediated endocytosis, a critical cascade necessary for RTK attenuation. In conclusion they discuss other RTK systems and deviations from lessons learned from EGFR trafficking.

To better understand the structural features necessary for RTK activation and signaling, **Michael C. Lawrence and Colin W. Ward** discuss in depth the structural features shared among the ectodomains of RTKs. In this chapter they describe the modular structure of the ectodomains of each RTK and how these structural features define the 20 phylogenetic families of the RTK superfamily. As important as the ectodomains for ligand binding, the tyrosine kinase domain is just as critical for coupling extracellular signals across the membrane resulting in a signaling output. **Dániel Süveges and Natalia Jura** present a thorough review of the kinase domain and the architecture of the ATP-binding site. Further, they describe shared mechanisms for the activation of the kinase domain and provide discussion on the structural basis for regulation of kinase activity, including receptor oligomerization, the intracellular juxtamembrane segment, and regulation by the C-terminal tail.

In the penultimate chapter **Wolfgang J. Köstler and Christoph C. Zielinski** advance our knowledge of how RTKs are being identified and pursued as molecular targets in the clinical arena with a thorough discussion on how RTKs drive tumor progression, approaches to target RTKs in the clinic, and the development of resistance to advanced therapeutics targeting RTKs. Further, this chapter elaborates on the approaches taken in the clinic to inhibit RTK activation, highlighting the next generation of RTK inhibitors. In the closing chapter of this book, **Maria Sibilia and colleagues** systematically review the contribution of genetic mouse models, including transgenic, knockout, and knock-in mice for several RTKs, used to elucidate the functions of individual receptors and their roles in mammalian development, embryogenesis, immune function, and pathophysiological processes. This chapter represents the most comprehensive summary of mouse models leading to the advancement of our understating of RTKs. Collectively, the chapters presented in this book provide the reader with a cumulative summary of how RTKs function and behave in mammalian systems, with hopes of fostering new endeavors by the next generation of RTK researchers.

Madison, WI, USA — Deric L. Wheeler
Rehovot, Israel — Yosef Yarden

About the Editors

Associate Professor
Deric L. Wheeler

Born in Washington State, Deric Wheeler received his Bachelor of Science degrees in Botany and Zoology from the University of Idaho (Moscow, ID), followed by a Masters in Pathology from the University of Iowa (Iowa City, IA) and a Ph.D. in Cancer Biology from the University of Wisconsin (Madison, WI). After his graduate training Deric Wheeler started his postdoctoral training at the Fred Hutchinson Cancer Research Center in Seattle, WA (c/o Larry R. Rohrschneider), and at the University of Wisconsin School of Medicine and Public Health in Madison (c/o Paul M. Harari). He completed his postdoctoral training and transitioned to faculty, at the level of Assistant Professor, at the University of Wisconsin School of Medicine and Public Health in the Department of Human Oncology. Dr. Wheeler's lab research centers on the role of subcellular localization of receptor tyrosine kinases and their roles in oncogenesis and therapeutic resistance. Currently, Dr. Wheeler's research is supported by the American Cancer Society, Mary Kay Foundation, the Department of Defense Lung Cancer Research Program, and the Clinical and Translational Science Award (CTSA) program through the NIH National Center for Advancing Translational Sciences (NCATS).

Professor Yosef Yarden

Born in Israel, Yosef Yarden received his B.Sc. in Biological and Geological Sciences from the Hebrew University of Jerusalem, and a Ph.D. in Molecular Biology from the Weizmann Institute of Science. His postdoctoral training was undertaken at Genentech, Inc. (c/o Axel Ullrich) in San Francisco, and at the Massachusetts Institute of Technology (c/o Robert A. Weinberg). He returned to the Weizmann Institute of Science as an Assistant Professor, eventually becoming a Full Professor there. His past administrative responsibilities at the Weizmann Institute include Dean of the Faculty of Biology, Vice President for Academic Affairs, Director of the M.D. Moross Institute for Cancer Research, and Dean of the Feinberg Graduate School. Dr. Yarden's lab research centers on the roles played by growth factors and receptor tyrosine kinases in tumor progression. Currently, Yarden's research is supported by the US National Cancer Institute (MERIT Award), a grant from the European Research Council (ERC-AdG), the German Research Foundation (DIP), and a professorship from the Israel Cancer Research Fund (ICRF).

Contents

Contributors

Nicole Amberg Department of Medicine I, Comprehensive Cancer Center, Institute of Cancer Research, Medical University of Vienna, Vienna, Austria

Frederic Brunet Institut de Génomique Fonctionnelle, University of Lyon I/Ecole Normale Supérieure, Lyon, France

Pierre De Meyts De Meyts R&D Consulting, Kraainem, Belgium

Barbara Drobits Department of Medicine I, Comprehensive Cancer Center, Institute of Cancer Research, Medical University of Vienna, Vienna, Austria

Yi Du Department of Molecular and Cellular Oncology, The University of Texas MD Anderson Cancer Center, Houston, TX, USA

Yehoshua Enuka Department of Biological Regulation, Weizmann Institute of Science, Rehovot, Israel

Morris E. Feldman Department of Biological Regulation, Weizmann Institute of Science, Rehovot, Israel

Arola Fortian Department of Cell Biology, University of Pittsburgh School of Medicine, Pittsburgh, PA, USA

Elisabeth Glitzner Department of Medicine I, Comprehensive Cancer Center, Institute of Cancer Research, Medical University of Vienna, Vienna, Austria

Martin Holcmann Department of Medicine I, Comprehensive Cancer Center, Institute of Cancer Research, Medical University of Vienna, Vienna, Austria

Mien-Chie Hung Department of Molecular and Cellular Oncology, The University of Texas MD Anderson Cancer Center, Houston, TX, USA

Center for Molecular Medicine, China Medical University, Taichung, Taiwan

Graduate Institute of Cancer Biology, China Medical University, Taichung, Taiwan

Tony Hunter Salk Institute for Biological Studies, La Jolla, CA, USA

Jennifer L. Hsu Department of Molecular and Cellular Oncology, The University of Texas MD Anderson Cancer Center, Houston, TX, USA

Center for Molecular Medicine, China Medical University, Taichung, Taiwan

Graduate Institute of Cancer Biology, China Medical University, Taichung, Taiwan

Natalia Jura Department of Cellular and Molecular Pharmacology, Cardiovascular Research Institute, University of California, San Francisco, San Francisco, CA, USA

Karin Komposch Department of Medicine I, Comprehensive Cancer Center, Institute of Cancer Research, Medical University of Vienna, Vienna, Austria

Wolfgang J. Köstler Clinical Division of Oncology, Department of Medicine I, Medical University of Vienna, Vienna, Austria

Comprehensive Cancer Center, Medical University of Vienna, Vienna, Austria

Michael C. Lawrence Walter + Eliza Hall Institute of Medical Research, Parkville, VIC, Australia

Department of Medical Biology, University of Melbourne, Parkville, VIC, Australia

Gerard Manning Salk Institute for Biological Studies, La Jolla, CA, USA

Department of Bioinformatics and Computational Biology, Genentech Inc., South San Francisco, CA, USA

Jonathan Robson Department of Medicine I, Comprehensive Cancer Center, Institute of Cancer Research, Medical University of Vienna, Vienna, Austria

Fabio Savarese Department of Medicine I, Comprehensive Cancer Center, Institute of Cancer Research, Medical University of Vienna, Vienna, Austria

Manfred Schartl Department of Physiological Chemistry, Biocenter, University of Würzburg, Würzburg, Germany

Ben-Zion Shilo Department of Molecular Genetics, Weizmann Institute of Science, Rehovot, Israel

Maria Sibilia Department of Medicine I, Comprehensive Cancer Center, Institute of Cancer Research, Medical University of Vienna, Vienna, Austria

Alexander Sorkin Department of Cell Biology, University of Pittsburgh School of Medicine, Pittsburgh, PA, USA

Sriram Srivatsa Department of Medicine I, Comprehensive Cancer Center, Institute of Cancer Research, Medical University of Vienna, Vienna, Austria

Gabriel Stulnig Department of Medicine I, Comprehensive Cancer Center, Institute of Cancer Research, Medical University of Vienna, Vienna, Austria

Dániel Süveges Department of Cellular and Molecular Pharmacology, Cardiovascular Research Institute, University of California, San Francisco, San Francisco, CA, USA

Jean-Nicolas Volff Institut de Génomique Fonctionnelle, University of Lyon I/Ecole Normale Supérieure, Lyon, France

Ying-Nai Wang Department of Molecular and Cellular Oncology, The University of Texas MD Anderson Cancer Center, Houston, TX, USA

Center for Molecular Medicine, China Medical University, Taichung, Taiwan

Graduate Institute of Cancer Biology, China Medical University, Taichung, Taiwan

Colin W. Ward Walter+Eliza Hall Institute of Medical Research, Parkville, VIC, Australia

Yosef Yarden Department of Biological Regulation, Weizmann Institute of Science, Rehovot, Israel

Christoph C. Zielinski Clinical Division of Oncology, Department of Medicine I, Medical University of Vienna, Vienna, Austria

Comprehensive Cancer Center, Medical University of Vienna, Vienna, Austria

Chapter 1
The Eukaryotic Protein Kinase Superfamily and the Emergence of Receptor Tyrosine Kinases

Tony Hunter and Gerard Manning

The regulation of protein function by reversible phosphorylation was discovered over 50 years ago [1], and the first protein kinase activity capable of phosphorylating proteins was reported in 1954 [2]. Since those early days, there has been intense interest in the role of protein phosphorylation in regulating protein function. At the outset, a few protein kinases were purified by conventional means, and this ultimately led to a sequence of the catalytic subunit of the cAMP-dependent protein kinase (PKA) catalytic domain (C), which was determined by classical protein sequencing [3]! The PKA C subunit sequence revealed a surprising relationship to the sequences of two retroviral transforming proteins, one of which had just been reported to have tyrosine kinase activity [4, 5]. This immediately suggested that there might be a superfamily of protein kinases. With the advent of facile DNA cloning and sequencing in the 1970s and 1980s, it quickly became apparent that a large family of related eukaryotic protein kinases (ePKs) exists, and the burgeoning numbers of cloned protein kinase sequences led to the speculation that a mammalian genome might encode as many as 1,001 protein kinases [6]. Since then, the importance of protein phosphorylation as a regulatory mechanism has continued to grow, and recent phosphoproteomic analyses indicate that the majority of intracellular proteins can be phosphorylated at one or more sites under an appropriate

T. Hunter, Ph.D. (✉)
Salk Institute for Biological Studies, 10010 North Torrey Pines Road,
La Jolla, CA 92037-1002, USA
e-mail: Hunter@salk.edu

G. Manning, Ph.D.
Salk Institute for Biological Studies, 10010 North Torrey Pines Road,
La Jolla, CA 92037-1002, USA

Department of Bioinformatics and Computational Biology, Genentech Inc.,
South San Francisco, CA 94080, USA
e-mail: manning.gerard@gene.com

D.L. Wheeler, Y. Yarden (eds.), *Receptor Tyrosine Kinases: Structure, Functions and Role in Human Disease*, DOI 10.1007/978-1-4939-2053-2_1

condition [7, 8]. The phosphorylation of a protein at a specific site can regulate enzymatic activity by inducing conformational changes or through direct steric effects and can also modulate the activity of structural proteins through conformational and charge effects. In addition, a major function of protein-linked phosphates is to provide docking sites for other proteins, thus promoting inducible protein-protein association.

The catalytic domains of the Ser/Thr- and Tyr-specific protein kinases in the ePK superfamily are related in sequence to each other and to the larger class of protein kinase-like (PKL) kinases; PKLs all share a common fold and catalytic mechanism [9] and are present in both eukaryotes and prokaryotes. Distinct from ePKs, most PKLs phosphorylate nonprotein substrates, an example being the choline and aminoglycoside kinases in bacteria. PKLs appear to have evolved before the ePKs, but whether the first ePK originated from a PKL is unclear. Some protein kinases are PKLs but not ePKs. These include the phosphatidylinositol kinase-like kinases (PIKKs), a small family that includes ATM/ATR and DNAPK, which are closely related to and may have evolved from the phosphatidylinositol kinases.

A few proteins structurally unrelated to the ePKs and PKLs also reportedly have protein kinase activity and as a group are known as atypical kinases (see below). Some of these phosphorylate Ser, Thr, or Tyr, but in addition to these three hydroxyamino acids, several other amino acids can be phosphorylated, including Lys, Arg, and His. In most cases, the provenance of the responsible protein kinases remains unclear, but NME2 (NDPK-B/Nm23) has been reported to be a bona fide mammalian histidine kinase [10], and the other nine nucleoside diphosphate kinase (NDK) family members might also have this specificity. The bacterial two-component protein kinases, commonly known as "histidine kinases," which primarily transmit signals across the cell membrane, form yet another distinct family. These kinases are really misnamed, because although they autophosphorylate on a catalytic histidine residue, they transfer this phosphate onto an aspartate in a substrate protein and therefore are more properly known as aspartate kinases. Such kinases are also found in plants and many protists, but are absent from metazoa, apart from the unusual mitochondrial PDHK family members, which phosphorylate Ser/Thr. Finally, bacteria also possess a family of tyrosine-specific kinases, the BY kinases, but these are unrelated in sequence or structure to any other class of protein kinase, including the tyrosine kinase branch of the ePK superfamily [11].

The ~270 AA ePK catalytic domain is characterized by a series of conserved short sequence motifs along its length, which define 11 subdomains and serve as key catalytic elements of the kinase domain, being primarily positioned around the catalytic cleft of the bilobate kinase domain [12, 13]. In combination with the overall catalytic domain sequence, these motifs can be used to identify other protein kinases through pairwise and HMM profile sequence searches; likewise, other PKLs and atypical protein kinases can be found using these approaches. Using this strategy, genomes from diverse eukaryotic species, ranging from protists to vertebrates, have been surveyed to define the protein kinase complement (kinome) of each organism [14–23]. These compiled kinomes have served as a basis to explore the evolution and global functions of all protein kinases.

1.1 The Human Kinome

In the original protein kinase survey completed in 2002 [16], the human genome was predicted to contain 518 protein kinase genes (2.5 % of the ~21,000 genes). Of these, 478 encode ePKs, with the others being divided between 9 small families, which include the PIKK, the PDHK (pyruvate dehydrogenase kinase), and alpha kinase (E2F kinase) families. Based on catalytic domain sequence similarity, most of the 478 ePKs can be divided into seven major branches: AGC, CAMK, CMGC, CK1, STE, TKL, and TK [13]. The TK branch contains the 90 tyrosine kinase (TK) genes (16 % of all protein kinases). The complexity of the expressed kinome is increased by the existence of splice isoforms of over half of all protein kinases [24], the existence of regulatory subunits, and as a result of differential targeting within the cell through association with scaffolding proteins. Other eukaryotic kinomes are significantly larger than the human kinome; to date, the *Paramecium* kinome with 2,921 kinases and one species of eucalyptus with 2,532 kinases are the largest kinomes reported [25, 26]. The original prediction that a single genome could encode as many as 1,001 protein kinases has even proven to be an underestimate!

Since the publication of the 2002 human kinome catalogue, 52 new remote and atypical putative kinases have been added [27]. The new remote kinases include 6 four-jointed family kinases, named after the *Drosophila* four-jointed kinase, as well as NME2 and nine homologues. One member of the four-jointed family, FAM20C, is localized in the lumen of the ER and appears to be the long-sought casein kinase that phosphorylates casein and other secreted proteins, usually on Ser.Xxx.Glu/P.Ser motifs [28, 29]. No new classical TKs have been found; however, WSTF/BAZ1B, which is structurally unrelated to any protein kinase, has been reported to phosphorylate histone H2AX on Tyr142 [30]. As of 2012, the human kinome totals 566 protein kinases. However, the list may yet grow, as humans have members of other classes of PKLs that are usually thought of as small-molecule kinases but could have protein kinase activity, and other proteins that have another primary function can also moonlight as protein kinases. An example here is pyruvate kinase M2 (PKM2 is the embryonic and cancer-specific splice isoform of PKM), which normally utilizes PEP to phosphorylate ADP to generate ATP in the final step in glycolysis and can also act as a protein kinase, using PEP to phosphorylate Tyr705 in STAT3 [31] and Thr11 in histone H3 [32]. Certain ATPases, such as the MinD family of bacterial BY tyrosine kinases [11], or even structurally quite distinct proteins, may also prove to have genuine protein kinase activity. Note that a few protein kinases currently in the original kinome catalogue, such as SgK424 and PRKY, are dubious and may be relegated to pseudogenes, and the twinfilin/A6 and H11 atypical tyrosine kinases are no longer thought to possess intrinsic kinase activity.

In the initial human kinome [16], 48 catalytic domains out of a total of 492 ePK catalytic domains were predicted to lack catalytic activity, because they were missing one or more of the conserved catalytic residues (Lys72/Asp166/Asp184 in PKA) known to be essential for catalytic activity based on mutagenesis studies with classical kinases. Such proteins were termed pseudokinases; many have been conserved through evolution and because of this were presumed to serve key noncatalytic functions. Such "pseudokinase domains" can serve as docking platforms or scaffolds (e.g., ILK), structural elements (receptor guanylyl cyclase kinase homology domains), and/or regulatory domains, which might bind and sense ATP levels [33, 34]. Alternatively, they can act as regulators of protein kinases, mimicking mechanisms used by active protein kinases. Most human pseudokinase domains are conserved in all vertebrates, and several are even more ancient; CCK4 is inactive in all metazoans, and the inactive second kinase domain of GCN2 is found in almost all eukaryotes, suggesting that these domains play vital biological roles [16]. However, it is important to note that, despite the original prediction, subsequent analyses have shown that a number of protein kinases classified as pseudokinases do in fact have reported phosphotransfer activity; these include CASK [35], ErbB3 [36], KSR [37], and the JAK2 N-terminal kinase domain [38]. In contrast, crystal structure studies indicate that other pseudokinases, such as STRAD [39], VRK3 [40], and ILK [41], are very unlikely to possess catalytic activity, although this is still debated in the case of ILK [42]. The lesson learned is that each pseudokinase has to be analyzed individually, especially since some, like CASK, prove to have activity only under very unusual conditions.

Protein kinase catalytic function is often dependent on additional domains within the protein, which can regulate catalytic activity through binding second messengers, localize the kinase through intrinsic motifs or interaction with scaffolding proteins, or recruit regulatory proteins and substrates. About half the protein kinases are predicted to have additional domains, many of which are implicated in signaling. Of the tyrosine kinases (TKs), 25 have P.Tyr-binding SH2 domains that play a cardinal role in establishing tyrosine phosphorylation-based signaling networks. In contrast, perhaps surprisingly, only one serine kinase contains a P.Ser/Thr-binding domain (the FHA domain in CHK2). In addition, 46 protein kinases have domains that interact with other proteins (e.g., SH3); 55 Tyr and Ser/Thr kinases have lipid interaction domains (e.g., PH); 38 kinases have domains linked to small GTPase signaling; and 29 Ser/Thr kinases have domains linked to calcium signaling. Generally, most members of a protein kinase family have the same set of ancillary domains, but there are some exceptions, and alternative splicing is often used to generate distinct domain combinations from a single gene. A listing of additional domains found in human protein kinases is given at http://kinase.com/.

1.2 Protein Kinases and Human Disease

The major control functions of protein kinases are reflected in their involvement in human disease. Approximately 35 % of protein kinases (175 genes) are directly implicated in human disease, as a result of mutation, misexpression, or copy number changes [43]. Of these, 121 protein kinases are implicated in cancer, including 51 of the 90 TKs. Many more protein kinases are weakly implicated, and their possible involvement in cancer is emerging from genome-scale studies through sequence, copy number, and expression analyses. Historically, cancer-causing mutations in protein kinases were found through the analysis of retroviral transforming proteins and tumor oncogenes present in either hereditary cancer predisposition syndromes or sporadic cancers and also chromosomal translocations that lead to the expression of a fusion protein in which the protein kinase domain is constitutively active, because the fusion partner domain often has the propensity to dimerize and thereby juxtapose the two catalytic domains leading to autoactivation.

More recently, however, potential cancer driver mutations in protein kinases have been pinpointed by targeted re-sequencing of the exons of all kinase genes or sequencing of all coding exons in a wide variety of human tumor cell lines and primary tumor tissue. These efforts have revealed a large number of cancer-associated non-synonymous mutations in protein kinases. About 65 % of such putative driver mutations lie within the catalytic domain, whereas the rest of the driver mutations fall outside, especially in the extracellular domain in the case of the receptor tyrosine kinases (RTKs) and in unstructured interdomain linker regions. Within the catalytic domain, mutations within the P-loop and activation loop segment are more common than in the remainder of the catalytic domain [44]. Cancer mutations in highly conserved catalytic domain residues tend to be loss-of-function mutations, but in some cases such mutations have the paradoxical effect of being able to activate the wild-type kinase through a (hetero)dimerization mechanism (e.g., RAF family kinases). Among the cancer-associated kinase mutations shown to cause loss of activity are mutations in TKs, such as those in the c-FES nonreceptor TK [45] and the EPHA3 RTK [46], and in Ser/Thr kinases, e.g., MAP2K4, DAPK3 [47], and various PKC isoforms. Such kinases may have a role as tumor suppressor kinases, analogous to well-studied examples of tumor suppressors with protein kinase activity, such as ATM and LKB1, for which loss-of-function mutations are found in both hereditary and sporadic cancers. Many other catalytic domain mutations, however, are gain-of-function activating mutations. Activating mutations in oncogenic protein kinases, such as the ABL nonreceptor TK and the HER2/ERBB2 RTK, were identified early on, but more have emerged from cancer genome sequencing, including mutations in the JAK2 nonreceptor TK, the EGFR and ALK RTKs, and the B-RAF Ser/Thr MAP3K.

The first effort to develop selective protein kinase inhibitors began in the 1980s with the goal of using these to study protein kinase function in the cell and ultimately to develop them as therapeutics. The importance of elevated tyrosine phosphorylation in cancer triggered efforts to develop selective inhibitors against individual TKs known to be activated by mutation or overexpression in different types of cancer. In 1998, the first drug antagonizing a Tyr kinase was approved for cancer therapy; trastuzumab (Herceptin™), a monoclonal antibody directed against the extracellular domain of HER2, is used for therapy of HER2-positive breast cancer. This was quickly followed by the approval of a small-molecule TK inhibitor (TKI), imatinib (Gleevec™), an inhibitor of the ABL TK and BCR-ABL, the activated chromosomal fusion protein TK that drives chronic myelogenous leukemia (CML). Several additional small-molecule TKIs and protein drugs have been approved for cancer therapy since that time. Notably, eight out of the 13 approved TKIs target activated RTKs, and five protein drugs are directed against RTK extracellular domains or their ligands.

1.3 Kinome Evolution

Kinome catalogues allow the comprehensive identification of related kinases in the same organism and equivalent (orthologous) kinases in others, and this in turn allows one to draw inferences about the evolution of protein kinases in eukaryotes and deduce their evolutionary gains and losses. To enable these comparisons, a hierarchical classification of all protein kinases into groups, families, and subfamilies has been developed. In general, groups share overall substrate site preference (e.g., the "proline-directed" CMGC group), families share similar biochemical functions (e.g., the MAPK family), and subfamilies are evolutionarily conserved variations within the families, including the ERK, JNK, p38, ERK5, and ERK7 MAP kinase subfamilies. The full classification scheme and several kinomes are available at the KinBase database (http://kinase.com) and extend the original Hanks and Hunter classification [13]. Family and subfamily sizes can vary from a single, deeply conserved member to expansions of ten or more members. Although most protein kinases fall into major groups, many remain outliers ("Other" group), either singly or as small families.

The comparison of several kinomes at the subfamily level shows a rich primordial diversity of protein kinases and their dynamic gain and loss throughout evolution. The common ancestor of all sequenced eukaryotes appears to have had at least 68 distinct protein kinase subfamilies, of which 31 are still present in all nonparasitic genomes surveyed. By contrast, only three protein kinases are shared with archaea, and while bacteria have ePK-like kinases, none are evolutionarily widespread, indicating that an elaborate kinase apparatus emerged early in eukaryotic evolution. The pan-eukaryotic protein kinases include well-known kinases involved in the cell cycle, DNA repair and replication, lipid signaling, the MAPK cascade, and splicing, as well as several other kinases whose functions are still quite obscure, despite being essential for eukaryotic life (http://kinase.com/evolution/) [23].

The size of the kinome grows with genomic and organismal complexity and usually corresponds to ~2 % of the full gene count. Exceptions include the large expansion of innate immunity-related TKL group kinases in higher plants, accounting for over 60 % of the ~1,100 *Arabidopsis* protein kinases (4 % of the proteome) [48], the recent expansions in *C. elegans* [17], and the record 2,921 kinases of *Paramecium* (7 % of all genes) [49], largely due to recent whole genome duplications. Curiously, the atypical protein kinase families are mostly pan-eukaryotic but are rarely expanded. This is most obvious in the case of the members of the PIKK family, such as ATM. Thus, it is the ePK superfamily that is most dynamic by far, perhaps due to its specific adaptations for regulated activity and diversity of substrates [9].

A large increase in kinase diversity comes with the metazoan lineage of the evolutionary tree, which includes the holozoans (holozoa are defined as a group of organisms that includes animals and their closest single-celled relatives) near the base of this split from fungi (see below), which gave rise to multicellular animals. In this branch, 80 new subfamilies emerge, including the canonical tyrosine kinases (TK group), the TGFβ-receptor-like Ser/Thr kinases, and a wide variety of cytoplasmic kinases [22]. In particular, the TKs are largely cell surface receptors or receptor associated, and their existence correlates with the need for intercellular communication in the switch to multicellularity. Metazoans have a concomitant expansion of the P.Tyr-binding domains (SH2 and PTB) that transduce TK signals [50] and the tyrosine-specific protein-tyrosine phosphatases (PTPs), all of which are needed for a functional tyrosine phosphorylation-based signaling system.

Most vertebrate kinomes are highly similar to that of human. For instance, the mouse kinome has single orthologues for 510 of the original 518 human kinases, and most of the differences are due to individual cases of gene loss and duplication by retrotransposition [19]. The divergence increases in more distant vertebrates, but our current estimate is that all vertebrates share a core of 531 protein kinases. The analysis of the sea urchin kinome [18] and other invertebrate genomes shows that the major advance in vertebrate kinomes was due to the proposed fourfold genome duplication leading to increases in the number of kinases in most subfamilies [51]. By contrast, the only vertebrate-specific subfamilies are the SRM and PKR ePKs, and the ChaK and FASTK atypical kinases.

1.4 The Genesis of Receptor Tyrosine Kinases

Because of their key role in intercellular communication, receptor tyrosine kinases (RTKs) were originally proposed to have evolved in parallel with multicellularity in animals, and it was suggested that the development of tyrosine phosphorylation-based signaling may have played a vital part in the emergence of metazoans by providing the essential means of coordinating function between different cells in a multicellular organism [52]. The lack of tyrosine phosphorylation in single-celled organisms, such as the yeasts, reinforced this idea. However, an unexpectedly complex repertoire of TKs has recently been found in the genomes of two unicellular organisms that are close relatives of metazoans, namely, the choanoflagellate *Monosiga brevicollis* [53] and the filasterian *Capsaspora owczarzaki* [54]. These two genomes each encode over 100 TKs, more than the total human count, although both organisms are predominantly or completely unicellular. Almost nothing is known about the functions of these TKs, but their sequences do shed light on TK evolution and the diversity of domain contexts in which TKs can operate. The majority of TKs in both species are predicted to be membrane-spanning cell surface receptors (RTKs) (88/128 in *Monosiga* and 92/103 in *Capsaspora*), but for the most part, these RTKs do not appear to be orthologous to metazoan RTKs. In contrast, a subset of their nonreceptor TKs (CTKs) are recognizable—all four Src subgroup kinases (Src, Csk, Abl, Tec) are found in both species, whereas Fak is found in *Capsaspora*, and Fer is found in another holozoan, *Ministeria vibrans*. Possible Syk or Jak homologues may also be present in choanoflagellates, although they are too remote to be certain. Thus, six or more of the ten human CTK families predate animals. By contrast, the RTKs are far more labile. Possible homologues of Eph and IGF1R may be present in choanoflagellates [55], but the evidence is weak and no other examples of human RTK families can be seen. Interestingly, no other RTK families are shared between *Monosiga* and *Capsaspora*, suggesting that the large numbers of RTKs in both clades evolved independently. The recent sequencing of a second choanoflagellate genome, that of *Salpingoeca rosetta* [56], shows that even within the choanoflagellates, the CTKs are very stable, whereas the RTKs are much more dynamic in evolution, perhaps in response to the diversity of environmental ligands.

Not only is their number of TKs greater than that of a typical vertebrate, but choanoflagellates possess all the machinery needed for tyrosine phosphorylation-based signaling, including an abundance of PTPs and SH2 domain proteins. These include homologues of several specific metazoan PTP and SH2 proteins as well as many novel proteins with unique domain arrangements. These findings suggest that an early holozoan had a mature set of 6–8 CTKs, several fast-evolving RTKs, and an extensive network of SH2 and PTB P.Tyr-binding domains and PTP domain proteins to reverse the action of TKs, and presumably downstream TK target substrates [53].

At what point in the emergence of tyrosine phosphorylation-based signaling did RTKs arise? To consider this, we need to know when and how tyrosine phosphorylation and TKs themselves evolved [57]. Since conventional TKs are not found in bacteria, it is reasonable to suppose that the first eukaryotic TK evolved from an ePK with Ser/Thr specificity. Tyrosine phosphorylation and PTPs are widespread in eukaryotes, but canonical TKs are almost exclusively found in animals and related holozoans. However, the TKL group is present in almost all eukaryotes other than some fungi [23] and is intermediate in sequence between the main branches of Ser/Thr kinases and the TK branch. Indeed, some extant TKLs possess the capacity to phosphorylate both Tyr and Ser/Thr in vitro [58, 59], and several plant and *Dictyostelium* TKLs are receptor-like and are implicated in tyrosine phosphorylation. It is therefore plausible that a dual-specificity TKL was the progenitor to the TK branch, having acquired additional mutations that eliminated the Ser/Thr specificity without affecting the Tyr specificity, which is characterized by the unique HRDLAARN and WMAPE catalytic domain motifs [54]. In this scenario, a receptor TKL could have given rise directly to an RTK, which in turn engendered a CTK. Alternatively, a soluble CTK may have preceded the first RTK. In this regard, it is worth noting that each of the conserved CTK families has a conserved membrane interaction domain or lipid modification that allows them to interact with the plasma membrane and thereby transduce signals from outside the cell when they interact with a membrane-spanning protein. In this second scenario, a cytoplasmic kinase with dual specificity for Ser/Thr and Tyr was the progenitor of the catalytic domain of the first CTKs, which in turn gave rise to the RTKs. Finally, it should be remembered that while Tyr phosphorylation, like Ser/Thr phosphorylation, could serve to regulate protein activity through allosteric and electrostatic effects, the true power of tyrosine phosphorylation as a regulatory mechanism could not be realized without the evolution of a P.Tyr-binding domain to transmit signals through induced protein-protein interactions, such as the SH2 domain.

Since *Monosiga* and *Capsaspora*, the simplest eukaryotes we know of with well-defined TKs, have both RTKs and CTKs, we cannot deduce whether a CTK or RTK evolved first. However, once a catalytic domain with TK activity evolved, the stage was set for the emergence of Src family TKs, which are represented in all holozoan tyrosine kinomes. This step required not only a catalytic domain with TK activity but also SH3 and SH2 domains. The SH3 protein interaction domain was apparently an early eukaryotic invention, being found in all eukaryotes, and an SH3 domain could be linked together with a P.Tyr-binding SH2 domain, whose evolutionary origins are less clear, and then joined to the TK catalytic domain to form a Src family TK. Although the first CTK may have been an unadorned catalytic domain, no well-established extant examples of isolated TK catalytic domains exist in choanoflagellates or metazoa, and at least one other domain is always found linked to the catalytic domain. Indeed, in contrast to the Ser/Thr kinases, where there are several examples of isolated catalytic domains that are regulated through association with another protein (e.g., PKA, CK2, and the cyclin-dependent kinases), the TKs instead rely on regulation through intramolecular interactions between the catalytic domain and another domain or through dimerization-mediated phosphorylation of regulatory Tyr phosphorylation sites. This difference may reflect the need to maintain very tight negative regulation of TK activity, which will be more efficiently accomplished with intramolecular interactions.

It seems likely that Tyr phosphorylation was first used for transmembrane signaling in response to external factors. This could have been mediated through non-covalent interactions between a CTK and a surface receptor, a principle that is still evident in modern metazoans in several systems, such as antigen receptors, which use Src family CTKs, and cytokine receptors, which utilize JAK family CTKs. This type of bimolecular receptor system provides versatility and economy by allowing a single CTK to be used in conjunction with multiple receptors. Consistent with this idea, the Src and Tec CTK families have a conserved lipid modification and membrane interaction domain that allows them to interact with the plasma membrane and transduce signals from outside the cell. However, it is easy to imagine how through gene duplication and transposition a gene for a CTK catalytic domain could be fused to the C-terminus of a receptor to form the first unimolecular receptor TK, which would provide a dedicated Tyr phosphorylation response to a single extracellular ligand.

1.5 Evolution of the Receptor Tyrosine Kinases

Once the first RTKs appeared, what does comparative kinomics tell us about the subsequent evolution of RTKs? As discussed above, the choanoflagellate RTKs are largely unrelated to metazoan RTKs and even vary significantly between the two choanoflagellate species that have been sequenced, with <20 % orthology among their RTKs [55]. Seventy-three of the 88 *Monosiga* RTKs belong to 15 families, but none of these have clear-cut metazoan relatives with the possible exception of divergent FGFR and EPH RTK orthologues [53]. Strikingly, many choanoflagellate RTKs have unique structures; most of the RTKs are type 1 transmembrane proteins, but at least two are predicted to have multiple transmembrane domains. No functions have been ascribed to any of the choanoflagellate RTKs, but their diversity may have been driven by the diversity of the extracellular ligands that these organisms have to respond to. This ligand diversity is in contrast to the conserved nature of ligands that activate RTKs in multicellular organisms, which will presumably have constrained their evolution. Thus, so far, the study of holozoan RTKs has not afforded significant insights into the evolution of the major metazoan RTK families.

If we examine RTKs through animal evolution, most RTK families are already evident in the parazoa, such as the sponges (Porifera), which are the simplest metazoans. Sponges have EGFR, MET, DDR, ROR/TRK/MUSK-like, EPH, CCK4, LMR, RET, RYK, and Sevenless family RTKs (170 total) [22]. The EGFR and some EPH extracellular domain architectures are similar to those in their eumetazoan counterparts, but other sponge RTKs have unique extracellular domains, with the DDRs having Ig repeats, and some members of the expanded EPH and MET families having sushi domains. This indicates that the activating ligands, presumably found largely in the external ocean environment, may be quite distinct. Within the bilateria, deuterostomes, such as sea urchins, have 17 of the 19 vertebrate RTK families; 14 of these RTKs, ALK, CCK4, DDR, EGFR, EPH, LMR, MET/RON, MUSK, RET, ROR, ROS, RYK, TIE, and TRK, are present as single copy genes, whereas VEGFR, FGFR, and INSR are represented by pairs of related genes [18]. The *C. elegans* kinome has 37 RTKs, with members of 11 conserved RTK families, and two unique RTK families, Kin6 and Kin16 [17]. The *Drosophila* kinome has 19 RTKs, all of which are in recognizable families [15]. Ultimately, as a result of genome duplications, mammals have ended up with a repertoire of 58 RTKs distributed in 20 families [60].

All modern RTKs can be joined to the main TK branch of the kinome tree, with the main families each forming a separate branch (INSR/TAM/MET/TRK, FGFR/PDGFR/VEGFR, and HER). Although the progenitor RTKs for these families could have arisen independently, it seems plausible that they all arose from a common ancestral RTK. In addition, each of the RTK pseudokinases probably arose subsequent to an active RTK in the same family. This is clearly the case for ERBB3 (N.B. this RTK has residual kinase activity), EPHB6, and EPHA10; for CCK4 and RYK, the progenitor RTK is less obvious, although both lie on the INSR/TAM/MET/TRK RTK branch.

1.6 The Development of Ligand-Regulated RTK Activity

Apart from the acquisition of a TK catalytic domain, what were the other important principles of RTK evolution? Extracellular domains (ECDs) of RTKs are usually built up of a collection of recognizable domains, such as FNIII and Ig domains, although there are a few exceptions [60]. A major function of these domains, ECD, is to recognize a specific ligand, and it seems plausible that the ability of these domains to bind cognate ligands had been established at some level before the RTK itself was created by fusing them with a CTK or the cytoplasmic domain of another RTK. Following generation of an RTK gene encoding a type I transmembrane protein that could be trafficked to the cell surface, the next step would be to evolve a mechanism for coordinating ligand binding to the ECD with thc activation of the cytoplasmic catalytic domain.

The binding of a secreted ligand to the ECD of a single RTK molecule with a type I topology would not by itself lead to the activation of the cytoplasmic TK catalytic domain, because ligand binding would be unlikely to result in the propagation of any conformational change through the plasma membrane via the single transmembrane domain to the catalytic domain. In consequence, a specific kinase activation mechanism would have been required to make ligand binding into a signal transduction event. It is formally possible that early RTKs were activated in trans by a CTK in response to ligand binding, perhaps as a result of ligand-induced clustering, and there are extant examples where this still occurs [61]. However, the major principle of modern RTK activation is through ligand-mediated dimerization (or oligomerization) of receptor molecules in the plasma membrane, and it seems likely that this activation principle was used early on in RTK evolution. The dimerization of two RTK molecules, which can either be identical or different members of the same family, juxtaposes the catalytic domains. Catalytic activation within an RTK dimer requires specific interactions between the two catalytic domains. Such interactions can promote phosphorylation between the two domains in trans leading to activation but can also drive allosteric interaction of one catalytic domain with the other leading to the activation of one of the two domains in a manner that does not involve transphosphorylation of the activation loop segment.

Over time, various mechanisms of ligand-dependent RTK activation through dimerization arose. There are at least four principles. Monomeric ligands can bind and induce dimer formation; dimeric ligands can bind and stabilize RTK-RTK dimers; monomeric ligands can bind to a multimeric coreceptor, such as a heparin sulfate proteoglycan, which then in conjunction with the ligand binds two RTKs and induces dimerization; and, finally, in some cases the RTK is a preformed disulfide-bonded dimer [60], and ligand binding alters the relative conformation of the two RTK molecules in the dimer. In some RTK dimers, the two transmembrane domains (TMD) must interact, and this may be important in establishing the correct rotational coupling of the two catalytic domains for a productive interaction. Whatever dimerization mechanism was used initially during the evolution of RTKs, effective ligand activation ultimately required coordinated changes in the surfaces of ECDs and intracellular domains (ICDs), in order to promote the specific, ligand-dependent molecular interactions that allow a productive association between the two kinase domains.

Finally, individual RTK and ligand pairs evolved to trigger the activation of distinct cytoplasmic signaling pathways through the acquisition of different autophosphorylation sites lying in the membrane juxtamembrane domain and in the C-terminal tail. This allowed the binding of distinct sets of SH2 domain proteins, which in turn each coupled to distinct pathways. Given that SH2 domains in general selectively recognize a P.Tyr residue and the four downstream amino acids, this step of RTK evolution may have been relatively straightforward, since it only required mutation of few neighboring residues, in an unstructured region to acquire a new autophosphorylation site and coupling to a new SH2 domain protein. In addition, the FGFR/PDGFR/VEGFR RTK families acquired a long unstructured insert in the catalytic domain, which also contains crucial autophosphorylation sites used for signaling. Another important step was the acquisition of tissue- and cell-type-specific expression of RTKs. In higher metazoans, this was facilitated by RTK gene duplications, which permitted the acquisition of cell-specific enhancers. In this way, the 90-member vertebrate RTK kinome evolved to transduce distinct patterns of signals dependent on the repertoire of SH2 proteins coexpressed with the RTK.

1.7 Coda

The elucidation of the kinomes of both unicellular and multicellular eukaryotes provides us with rich insights into the evolutionary history of protein kinases in general and RTKs in particular. Once the first RTKs evolved, they elaborated and generated new families. During eukaryotic evolution, RTKs became one of the major transmembrane signaling mechanisms used to respond to extracellular signals. Through parallel acquisition of distinct autophosphorylation sites and evolution of the SH2 domain family, RTKs became coupled to many of the major cytoplasmic signaling pathways, including the RAS/MAPK and the PI-3K/AKT pathways. RTK gene duplications allowed the specialization of different RTK families, which over time became one of the most versatile and diversified types of transmembrane receptor signaling system.

References

1. Fischer EH, Krebs EG. Conversion of phosphorylase b to phosphorylase a in muscle extracts. J Biol Chem. 1955;216:121–32.
2. Burnett G, Kennedy EP. The enzymatic phosphorylation of proteins. J Biol Chem. 1954;211:969–80.
3. Shoji S, Parmelee DC, Wade RD, Kumar S, Ericsson LH, Walsh KA, et al. Complete amino acid sequence of the catalytic subunit of bovine cardiac muscle cyclic AMP-dependent protein kinase. Proc Natl Acad Sci U S A. 1981;78:848–51.

4. Barker WC, Dayhoff MO. Viral src gene products are related to the catalytic chain of mammalian cAMP-dependent protein kinase. Proc Natl Acad Sci U S A. 1982;79:2836–9.
5. Hunter T, Sefton BM. Transforming gene product of Rous sarcoma virus phosphorylates tyrosine. Proc Natl Acad Sci U S A. 1980;77:1311–5.
6. Hunter T. A thousand and one protein kinases. Cell. 1987;50:823–9.
7. Olsen JV, Blagoev B, Gnad F, Macek B, Kumar C, Mortensen P, et al. Global, in vivo, and site-specific phosphorylation dynamics in signaling networks. Cell. 2006;127:635–48.
8. Huttlin EL, Jedrychowski MP, Elias JE, Goswami T, Rad R, Beausoleil SA, et al. A tissue-specific atlas of mouse protein phosphorylation and expression. Cell. 2010;143:1174–89.
9. Kannan N, Taylor SS, Zhai Y, Venter JC, Manning G. Structural and functional diversity of the microbial kinome. PLoS Biol. 2007;5:e17.
10. Srivastava S, Li Z, Ko K, Choudhury P, Albaqumi M, Johnson AK, et al. Histidine phosphorylation of the potassium channel KCa3.1 by nucleoside diphosphate kinase B is required for activation of KCa3.1 and CD4 T cells. Mol Cell. 2006;24:665–75.
11. Grangeasse C, Nessler S, Mijakovic I. Bacterial tyrosine kinases: evolution, biological function and structural insights. Phil Trans Roy Soc B. 2012;367:2640–55.
12. Hanks SK, Quinn AM, Hunter T. The protein kinase family: conserved features and deduced phylogeny of the catalytic domains. Science. 1988;241:42–52.
13. Hanks SK, Hunter T. Protein kinases 6. The eukaryotic protein kinase superfamily: kinase (catalytic) domain structure and classification. FASEB J. 1995;9:576–96.
14. Hunter T, Plowman GD. The protein kinases of budding yeast: six score and more. Trends Biochem Sci. 1997;22:18–22.
15. Manning G, Plowman GD, Hunter T, Sudarsanam S. Evolution of protein kinase signaling from yeast to man. Trends Biochem Sci. 2002;27:514–20.
16. Manning G, Whyte DB, Martinez R, Hunter T, Sudarsanam S. The protein kinase complement of the human genome. Science. 2002;298:1912–34.
17. Plowman GD, Sudarsanam S, Bingham J, Whyte D, Hunter T. The protein kinases of Caenorhabditis elegans: a model for signal transduction in multicellular organisms. Proc Natl Acad Sci U S A. 1999;96:13603–10.
18. Bradham CA, Foltz KR, Beane WS, Arnone MI, Rizzo F, Coffman JA, et al. The sea urchin kinome: a first look. Dev Biol. 2006;300:180–93.
19. Caenepeel S, Charydczak G, Sudarsanam S, Hunter T, Manning G. The mouse kinome: discovery and comparative genomics of all mouse protein kinases. Proc Natl Acad Sci U S A. 2004;101:11707–12.
20. Eisen JA, Coyne RS, Wu M, Wu D, Thiagarajan M, Wortman JR, et al. Macronuclear genome sequence of the ciliate Tetrahymena thermophila, a model eukaryote. PLoS Biol. 2006;4:e286.
21. Goldberg JM, Manning G, Liu A, Fey P, Pilcher KE, Xu Y, et al. The dictyostelium kinome – analysis of the protein kinases from a simple model organism. PLoS Genet. 2006;2:e38.
22. Srivastava M, Simakov O, Chapman J, Fahey B, Gauthier ME, Mitros T, et al. The Amphimedon queenslandica genome and the evolution of animal complexity. Nature. 2010;466:720–6.
23. Manning G, Reiner D, Lauwaet T, Dacre M, Smith A, Zhai Y, et al. The minimal kinome of Giardia lamblia illuminates early kinase evolution and unique parasite biology. Genome Biol. 2011;12:R66.
24. Forrest AR, Taylor DF, Crowe ML, Chalk AM, Waddell NJ, Kolle G, et al. Genome-wide review of transcriptional complexity in mouse protein kinases and phosphatases. Genome Biol. 2006;7:R5.
25. Talevich E, Tobin AB, Kannan N, Doerig C. An evolutionary perspective on the kinome of malaria parasites. Phil Trans Roy Soc B. 2012;367:2607–18.
26. Lehti-Shiu MD, Shiu SH. Diversity, classification and function of the plant protein kinase superfamily. Phil Trans Roy Soc B. 2012;367:2619–39.
27. Legler A, Manning G. in preparation [unpublished].
28. Tagliabracci VS, Engel JL, Wen J, Wiley SE, Worby CA, Kinch LN, et al. Secreted kinase phosphorylates extracellular proteins that regulate biomineralization. Science. 2012;336: 1150–3.

29. Ishikawa HO, Xu A, Ogura E, Manning G, Irvine KD. The Raine syndrome protein FAM20C is a Golgi kinase that phosphorylates bio-mineralization proteins. PLoS One. 2012;7:e42988.
30. Xiao A, Li H, Shechter D, Ahn SH, Fabrizio LA, Erdjument-Bromage H, et al. WSTF regulates the H2A.X DNA damage response via a novel tyrosine kinase activity. Nature. 2009;457:57–62.
31. Gao X, Wang H, Yang JJ, Liu X, Liu ZR. Pyruvate kinase M2 regulates gene transcription by acting as a protein kinase. Mol Cell. 2012;45:598–609.
32. Yang W, Xia Y, Hawke D, Li X, Liang J, Xing D, et al. PKM2 phosphorylates histone H3 and promotes gene transcription and tumorigenesis. Cell. 2012;150:685–96.
33. Boudeau J, Miranda-Saavedra D, Barton GJ, Alessi DR. Emerging roles of pseudokinases. Trends Cell Biol. 2006;16:443–52.
34. Zeqiraj E, van Aalten DM. Pseudokinases-remnants of evolution or key allosteric regulators? Curr Opin Struct Biol. 2010;20:772–81.
35. Mukherjee K, Sharma M, Urlaub H, Bourenkov GP, Jahn R, Sudhof TC, et al. CASK Functions as a Mg2+-independent neurexin kinase. Cell. 2008;133:328–39.
36. Shi F, Telesco SE, Liu Y, Radhakrishnan R, Lemmon MA. ErbB3/HER3 intracellular domain is competent to bind ATP and catalyze autophosphorylation. Proc Natl Acad Sci U S A. 2010;107:7692–7.
37. Brennan DF, Dar AC, Hertz NT, Chao WC, Burlingame AL, Shokat KM, et al. A Raf-induced allosteric transition of KSR stimulates phosphorylation of MEK. Nature. 2011;472:366–9.
38. Ungureanu D, Wu J, Pekkala T, Niranjan Y, Young C, Jensen ON, et al. The pseudokinase domain of JAK2 is a dual-specificity protein kinase that negatively regulates cytokine signaling. Nat Struct Mol Biol. 2011;18:971–6.
39. Zeqiraj E, Filippi BM, Deak M, Alessi DR, van Aalten DM. Structure of the LKB1-STRAD-MO25 complex reveals an allosteric mechanism of kinase activation. Science. 2009;326:1707–11.
40. Scheeff ED, Eswaran J, Bunkoczi G, Knapp S, Manning G. Structure of the pseudokinase VRK3 reveals a degraded catalytic site, a highly conserved kinase fold, and a putative regulatory binding site. Structure. 2009;17:128–38.
41. Fukuda K, Gupta S, Chen K, Wu C, Qin J. The pseudoactive site of ILK is essential for its binding to alpha-parvin and localization to focal adhesions. Mol Cell. 2009;36:819–30.
42. Maydan M, McDonald PC, Sanghera J, Yan J, Rallis C, Pinchin S, et al. Integrin-linked kinase is a functional Mn^{2+}-dependent protein kinase that regulates glycogen synthase kinase-3beta (GSK-3beta) phosphorylation. PLoS One. 2010;5:e12356.
43. Manning G. Protein kinases in human disease. 2005-06 Catalog and technical reference. Beverly, MA: Cell Signaling Technologies; 2005. p. 402–9.
44. Torkamani A, Schork NJ. Prediction of cancer driver mutations in protein kinases. Cancer Res. 2008;68:1675–82.
45. Delfino FJ, Stevenson H, Smithgall TE. A growth-suppressive function for the c-fes protein-tyrosine kinase in colorectal cancer. J Biol Chem. 2006;281:8829–35.
46. Lisabeth EM, Fernandez C, Pasquale EB. Cancer somatic mutations disrupt functions of the EphA3 receptor tyrosine kinase through multiple mechanisms. Biochemistry. 2012;51:1464–75.
47. Brognard J, Zhang YW, Puto LA, Hunter T. Cancer-associated loss-of-function mutations implicate DAPK3 as a tumor-suppressing kinase. Cancer Res. 2011;71:3152–61.
48. Initiative TAG. Analysis of the genome sequence of the flowering plant Arabidopsis thaliana. Nature. 2000;408:796–815.
49. Bemm F, Schwarz R, Forster F, Schultz J. A kinome of 2600 in the ciliate Paramecium tetraurelia. FEBS Lett. 2009;583:3589–92.
50. Liu BA, Nash PD. Evolution of SH2 domains and phosphotyrosine signalling networks. Phil Trans Roy Soc B. 2012;367:2556–73.
51. Manning G, Scheeff E. How the vertebrates were made: selective pruning of a double-duplicated genome. BMC Biol. 2010;8:144.

52. Miller WT. Tyrosine kinase signaling and the emergence of multicellularity. Biochim Biophys Acta. 1823;2012:1053–7.
53. Manning G, Young SL, Miller WT, Zhai Y. The protist, Monosiga brevicollis, has a tyrosine kinase signaling network more elaborate and diverse than found in any known metazoan. Proc Natl Acad Sci U S A. 2008;105:9674–9.
54. Suga H, Dacre M, de Mendoza A, Shalchian-Tabrizi K, Manning G, Ruiz-Trillo I. Genomic survey of premetazoans shows deep conservation of cytoplasmic tyrosine kinases and multiple radiations of receptor tyrosine kinases. Sci Signal. 2012;5:ra35.
55. Kramer E, Manning G. unpublished.
56. Nichols SA, Roberts BW, Richter DJ, Fairclough SR, King N. Origin of metazoan cadherin diversity and the antiquity of the classical cadherin/beta-catenin complex. Proc Natl Acad Sci U S A. 2012;109:13046–51.
57. Lim WA, Pawson T. Phosphotyrosine signaling: evolving a new cellular communication system. Cell. 2010;142:661–7.
58. Lawler S, Feng XH, Chen RH, Maruoka EM, Turck CW, Griswold-Prenner I, et al. The type II transforming growth factor-beta receptor autophosphorylates not only on serine and threonine but also on tyrosine residues. J Biol Chem. 1997;272:14850–9.
59. Oh MH, Wang X, Kota U, Goshe MB, Clouse SD, Huber SC. Tyrosine phosphorylation of the BRI1 receptor kinase emerges as a component of brassinosteroid signaling in Arabidopsis. Proc Natl Acad Sci U S A. 2009;106:658–63.
60. Lemmon MA, Schlessinger J. Cell signaling by receptor tyrosine kinases. Cell. 2010;141:1117–34.
61. Tice DA, Biscardi JS, Nickles AL, Parsons SJ. Mechanism of biological synergy between cellular Src and epidermal growth factor receptor. Proc Natl Acad Sci U S A. 1999;96:1415–20.

Chapter 2
Evolution of Receptor Tyrosine Kinases

Manfred Schartl, Jean-Nicolas Volff, and Frederic Brunet

2.1 Introduction

Receptor tyrosine kinases (RTKs) are important molecules for communication of cells in multicellular animals. It is therefore not surprising that the basal metazoans, like *Hydra* and sponges, already have a well-developed repertoire of RTKs. Even the choanoflagellate *Monosiga brevicollis*, a unicellular eukaryote from a group of protozoa that are thought to represent the ancestors of metazoa or be at least closely related to those, has up to seven genes coding for proteins that present clear features of receptor tyrosine kinases: an extracellular domain composed of EGF and Ig domains, a transmembrane domain, and an intracellular tyrosine kinase domain with a long carboxy-terminus that has several tyrosine residues which may act as substrate docking site after transphosphorylation [1]. It is unclear whether these genes are representing indeed the common ancestors of all RTKs in metazoa, and biochemical studies on signaling properties of the encoded proteins are still missing. The basal metazoans like sponges or *Hydra* already have a well-developed repertoire of RTKs [2–5]. Several of the *Hydra* genes show sequence similarity to the RTKs of *Drosophila* and vertebrates and might represent ancestors of the RTK families that we know today. These include insulin receptors; MUSK, ROR, PTK7, and FGF receptors; and VEGF-/PDGF-receptor-like RTKs [3]. How the progenitors of the RTKs may have evolved in the first place is unknown and probably difficult to infer from the genomes of the today existing organism. It can be hypothesized that in the metazoan ancestors,

M. Schartl, Ph.D. (✉)
Department of Physiological Chemistry, Biocenter, University of Würzburg, Am Hubland, 97074 Würzburg, Germany
e-mail: phch1@biozentrum.uni-wuerzburg.de

J.-N. Volff • F. Brunet
Institut de Génomique Fonctionnelle, UMR5242 CNRS/INRA/University of Lyon I/Ecole Normale Supérieure, 46 allée d'Italie, Lyon, France
e-mail: Jean-Nicolas.Volff@ens-lyon.fr; frederic.brunet@ens-lyon.fr

D.L. Wheeler, Y. Yarden (eds.), *Receptor Tyrosine Kinases: Structure, Functions and Role in Human Disease*, DOI 10.1007/978-1-4939-2053-2_2

preexisting cytoplasmic tyrosine kinases were fused by domain shuffling to surface-associated extracellular proteins that functioned in cell-cell interactions or binding of soluble extracellular proteins [6].

With the increase in complexity of cellular interactions and their regulations, RTKs became more numerous and diversified. This has reached a highest level in vertebrates and created the extensive RTK inventory of the human genome.

A major driving force for the evolution of gene diversity is gene duplication. This can happen on the single gene level and is then called local or tandem gene duplication. It can also happen that the whole genome is duplicated (whole genome duplication (WGD)). Such ideas go back to several visionary evolutionary biologists, most importantly S. Ohno, who has pointed out, already at times where only scarce genomic information was available, the importance of gene duplications because of the creation of new gene loci with *primarily* nonexistent functions. As even more advantageous, he considered complete genome duplications because they increase gene number without upsetting gene dosage and regulatory interactions [7]. Today, we know that such WGD events have indeed happened several times during evolution [8]. In particular, the evolution of vertebrate genomes has been shaped by such events. There were two rounds of such WGDs (called the 1R and 2R event) during early vertebrate evolution when the ancestors of the extant fish groups evolved (the "head-bearing vertebrates," Craniata) but before the fish lineages that are existing today and land-living vertebrates diverged. In the lineage leading to the modern bony fishes, the teleosts, another WGD (3R) occurred, the fish-specific whole genome duplication (FSGD).

The evolutionary significance of gene duplications becomes evident when one considers the possible consequences. Gene duplications create a situation where two identical copies exist for a function that was previously sufficiently fulfilled by a single one. Consequently, in the first place, only one of the copies is necessary and the second one is dispensable for this function. Now one of the redundant genes can take one of several evolutionary routes. Firstly, it can acquire mutations that will impair its function, and the gene will degenerate and eventually disappear without leaving any detectable trace in the genome. This is the fate of most genes, and in general, the majority of gene duplicates are rapidly lost. Secondly, the dispensable copy can evolve a new somehow related or even unrelated function (neofunctionalization) and will be maintained as a paralogue of the original gene. For instance, a duplicated RTK could be expressed in a different cell type, change its ligand interaction, or can signal through a related pathway or adaptor molecule. Such neofunctionalizations have happened for the vertebrate RTK duplicates from the 1R/2R WGDs. The third possibility is that both duplicates share the functions of the common ancestral gene (gene function partition or subfunctionalization). For example, if an RTK progenitor gene has functioned in the central and the peripheral nervous system, one of the duplicates can exert the CNS function and the other duplicate can take over the function in the periphery. Subfunctionalization appears to be a fate that is very common for the RTK duplicates from the 3R WGD in fish.

Paralogous RTK genes, due to their common origin, share a mostly very obvious sequence similarity and are thus grouped into gene families. The families are named (often arbitrarily) after their prototypic member, e.g., the EGFR/ERBB family, the insulin receptor family, or the PDGF receptor family. Due to the strong biochemical constraints, the kinase domains show the highest sequence similarity within and between families, and basically, two types of kinase domains are found throughout all RTKs: the uninterrupted and the interrupted enzymatic domains. The extracellular ligand-binding domains are much more divergent in sequence, but within an RTK family, they are a diagnostic common feature. While some families share related domains, like Ig domains or fibronectin type III domains, in general, it is the ligand-binding domain difference that separates the RTK families.

The evolutionary relationships of the RTK families to each other are with few exceptions difficult to resolve, and some might have arisen independently by fusing (even the same) kinase domains to different extracellular domains. To decipher the evolutionary relationships and trajectories inside the RTK gene families, one has to look into the genomes of related organisms, in particular those which diverged in parallel from a common ancestor. In such studies, it is often difficult to identify which is the corresponding family member in the other other organism that derived from the same ancestral gene (the true orthologue) and what are paralogous genes that were already separated in the common ancestor. These doubts can be solved by looking at the genomic neighborhood. When a WGD has occurred, whole chromosomes were duplicated and at least some neighboring genes should also be represented on the same chromosome where the duplicated RTK copy is found. This phenomenon is called conserved synteny and is a strong evolutionary tracefinder tool. Gene synteny is reasonably well conserved throughout the vertebrate genomes and has even allowed the reconstruction of the genomic organization of the vertebrate ancestor that has lived more than 500 million years ago [9–12].

This chapter is designed to give an overview of what we know about the phylogenetic history of the RTK gene superfamily and how the repertoire of RTKs in the human genome evolved. A glance at the evolution of some ligands—often occurring also in corresponding families to their receptors—will also be taken, although the evolutionary relationships of this extraordinarily large group of genes are much less clear. We will use the HUGO gene nomenclature of the annotated human genome primarily but will mention the most common and sometimes even better known synonyms.

2.2 The EGF Receptor Family

The EGF receptor (or ERBB) family is represented in the human genome by four members: the prototypic *EGF receptor*, *ERBB2* (*HER2/NEU*), *ERBB3*, and *ERBB4* genes. This situation is a clear result of the WGDs in the vertebrate lineage. In the well-studied invertebrate model species, there is a single bona fide homologue, the *Drosophila* Egf receptor (DER) (also known as the torpedo gene product [13]), and in *C. elegans*, the Let-23 receptor [14]. The sea squirt as the protochordate outgroup has two *egfr* genes, which are derived from a lineage-specific duplication of the common ancestral gene. In the vertebrate lineage, the 1R event has created two genes, one of which generated by the 2R WGD, the *EGF receptor* and *erbb2*, and the other one, the progenitor of *erbb3* and *erbb4*.

While these four family members appear to represent the default situation for all tetrapods, teleost fish have generally seven *erbb*'s as a result of the FSGD (3R), with two copies of *egfr*, *erbb3* and *erbb4*. *Erbb2* does not show any duplication in the fish genomes analyzed so far; hence, the second copy may have been lost again very early on. From all the RTK families, the EGFR family has retained the most of its FSGD duplicates.

The unifying characteristic feature of this family is the structure of the extracellular domain, being composed of two cysteine-rich and two L domains. This is reflected in the common, exclusively "receptor-mediated" [15] activation mechanism, where first the bivalent ligand monomer contacts two distinct sites within a single receptor molecule. The resulting conformational change allows then the dimerization arm of domain II to interact with a second ligand-bound receptor molecule.

Interestingly, two of the three amino acid substitutions (H740Q and D815N) in the Erbb3 catalytic domain that are made responsible for the loss of tyrosine kinase activity of this RTK in mammals are also seen in the fish homologues [16]. Thus, these mutations would also have happened early on in the evolutionary history of this gene family and have shaped the functional interactions of Erbb3 with the other family members.

An enigmatic situation is found in a certain lineage of fish, the platyfishes and swordtails. Here a tandem duplication of one of the fish *egfr* genes produced another version, so that these fish have three EGF receptors [17]. However, this third gene, which became known as *xmrk* (*Xiphophorus* melanoma receptor kinase), acquired two mutations in the extracellular domain. These allow the formation of intermolecular cysteine bridges. Hence, the receptor is present as constitutively active dimer and leads to ligand-independent signaling [18], a mechanism of oncogenic RTK activation that has also been found in several human cancers. The expression of the *xmrk* oncogene is normally repressed, so that the increased Egfr signal becomes apparent only by the development of large pigment spots. In hybrid crosses, this suppression is lost and the pigment spots grow out to highly malignant melanoma [19] (Fig. 2.1).

2.3 The Insulin Receptor Family

There are three different genes found in the insulin receptor family: the insulin receptor (*insr*) itself, insulin receptor-related protein (*insrr*), and insulin-like growth factor 1 receptor (*igf1r*). The phylogenetic analyses of these three proteins revealed that they can be rooted by one single sequence found in *Ciona* or

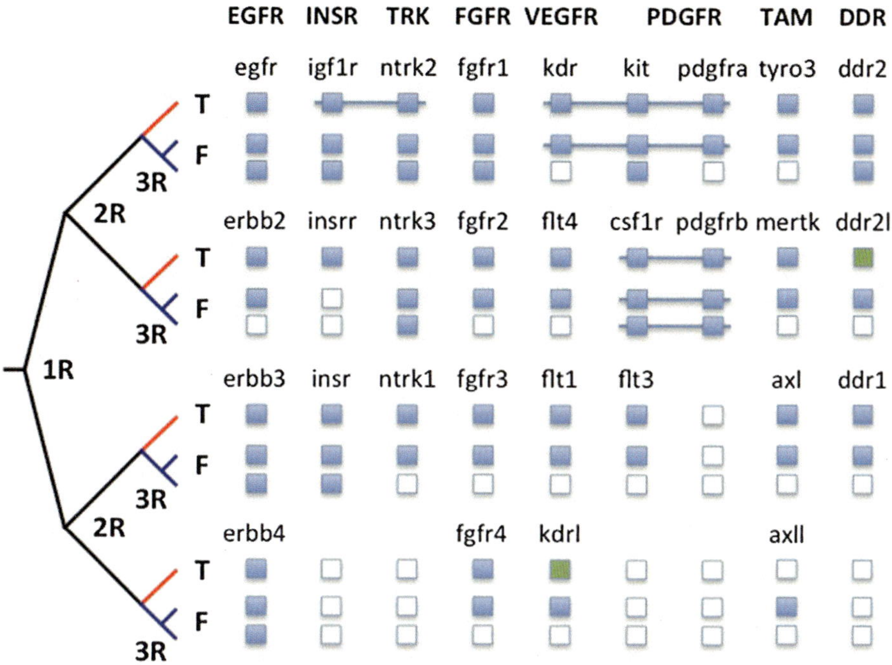

Fig. 2.1 Evolution of the vertebrate EGFR, insulin receptor, FGF receptor, VEGF receptor, PDGF receptor, TRK, TAM, and DDR families, which show a clear 1R/2R/3R pattern. *Filled boxes* represent the occurrence of an RTK gene, in *blue* whenever a gene is found in all main lineages of tetrapods (T) or fish (F) and in *green* for those that have not been found in mammals. *Ddr2l* is also not found in therian mammals and *krdl* in eutherians. A *white box* indicates that no RTK of this paralogy type has been found among the group of species considered. The *blue lines* indicate genes that are linked on the same chromosome

Amphioxus. The insulin signaling pathway is well conserved in *Drosophila* and *C. elegans*, and consequently, an insulin (or insulin-like) receptor is found there [20]. Even in sponges [2] and *Hydra* [3], insulin receptor-like RTKs are found. The WGD events have generated the *igfr1* and *insrr* paralogues from 2R, while the corresponding paralogue of the *insr* has obviously been lost very early on. Surprisingly, the *insrr* gene has also been lost in the fish lineage. Nevertheless, fish have in general four members of the insulin receptor family, because the FSGD duplicates of *insr* and *ifgr1* have been retained.

The insulin receptor family is, besides its sequence similarity, also characterized by a unique quaternary structure. The precursor polypeptide of the receptor monomer is cleaved into an α-fragment and a β-fragment, which are covalently linked in the extracellular part. The mature receptor monomer is further on bound to another monomer. Therefore, the paradigm of RTK receptor proteins existing as monomers in the inactive state is a feature absent from all members of this family.

2.4 The FGF Receptor Family

Alike the EGFR family, the mammalian FGF receptor family is composed of four members, *fgfr1–4*. In the fruit fly, two FGF receptors have been characterized, Btl (breathless) and Htl (heartless), referring to their important developmental function [21, 22]. Heartless, a *Drosophila* FGF receptor homologue, is essential for cell migration and establishment of several mesodermal lineages. They are the result of a lineage-specific duplication, like in planarians, which also have two FGFRs. But, although these invertebrate genes are sometimes named *fgfr1* and *fgfr2* like the mammalian ones, they are clearly not orthologues of the vertebrate receptors, but the result of a lineage-specific local gene duplication. Other invertebrates, like *C. elegans*, have only one FGF receptor [23]. *Fgfr*-like RTKs were also found in *Hydra* [3]. There is a single gene in the protochordate out-group. During the 1R WGD, the progenitors of *fgfr1/2* and *fgfr3/4* were generated, which then during 2R produced the four receptor paralogues that we know today. In fish, the family is less expanded than the EGFR family. Only the 3R duplicate of *fgfr1* has been retained.

The members of this family are united by the split tyrosine kinase domain and three Ig domains in the extracellular ligand-binding part with a small acid box between the first and second Ig domains.

2.5 The VEGF Receptor Family

In placental mammals, the family of receptors for the vascular endothelial growth factors consists of three members: *flt1/vegfr1*, *kdr/vegfr2*, and *flt4/vegfr3*. In marsupials, platypus, chicken, and several fish genomes, a fourth member is found, which is called *kdrl* (for kinase insert domain receptor-like) or consequently *vegfr4*. The family thus shows a clear 1R/2R pattern (Brunet et al. in preparation), which generated at first four duplicates, of which obviously the *kdrl/vegfr4* gene has been lost in the placental mammals. *Vegfr2* and *vegfr3* are the product of the second WGD, and obviously, *vegfr4* arose together with *vegfr1*. Why one member of this family after hundreds of million years of existence became dispensable in mammals is unclear. Either it was accidentally lost and its function was taken over by another RTK or its function is related to features that are not present anymore in the higher mammals. Interestingly, all four fish-specific duplicates of the FSGD have been lost. The common vertebrate ancestor had a single *vegfr* gene. The situation in invertebrates is intriguing as there are receptors, which bind a ligand that is obviously the common precursor of the PDGFs and the VEGFs. In fact, this receptor by sequence similarity could be the ancestor of both receptor families. In *Drosophila*, a single Pdgf/Vegf receptor (PVR) is present, while in *Hydra*, three of these RTKs are found [3].

The structural features of the VEGF receptor family are the interrupted kinase domain and seven Ig domains that form the ligand-binding domain. This part of the receptors is most similar to the PGDF receptor family, which has five instead of seven Ig domains. Such structural similarity hints to a common origin. Indeed, this has been proposed on the basis of the common exon/intron structure [24] and is further supported by the molecular phylogeny on the basis of the kinase domains (Brunet et al., in preparation). Considering also a common origin of VEGFR and PDGFR ligands, it becomes reasonable that these receptors have a common origin. The evolutionary scenario for the origin of these receptors therefore is that in the protochordate ancestor, a single receptor gene was first duplicated to produce the ancestors of the VEGFR family and of the PDGFR family. The latter one duplicated once more to give the progenitors of the PDGFR (sensu stricto) subfamily and the KIT/CSF1R/FLT3 subfamily (see below). Then during the vertebrate WGDs, the whole repertoire of this group of RTKs was generated (Fig. 2.2).

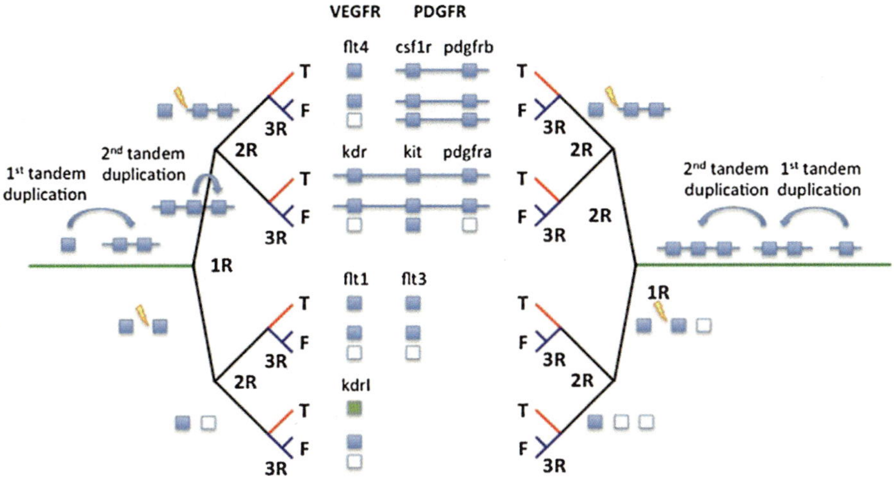

Fig. 2.2 Hypothetical scenarios for the evolution of the PDGF and VEGF receptor families. Due to the fact that *pdgfra*, *kit*, and *kdr* are found in tandem, as well as *pdgfrb* and *csf1r* in tetrapods (T) and in fish (F), two evolutionary scenarios can be proposed that explain this feature by considering two tandem duplications, either with a tandem duplication that generated the ancestors of the VEGFR and PDGFR families before the 1R WGD (*right*) or a tandem duplication between the 1R and 2R events (*left*). The *left* scenario is more parsimonious because it requires one event less of gene loss to fit with the actual distribution of the two RTK families. Synteny disruptions are shown with *yellow lightnings*. *Filled boxes* represent the occurrence of an RTK gene, in *blue* whenever a gene is found in all main lineages of tetrapods (T) or fish (F). *Krdl* is in *green* because it is not found in eutherians. *White boxes* indicate that no RTK is found among the group of species considered

2.6 The PDGF Receptor Family

The platelet-derived growth factor family is composed of five members in the human genome: *PDGFRa* and *b*, *KIT/SCFR*, *FMS/CSF1R*, and *FLT3*. Phylogenetically, they are placed in two subfamilies, the PDGFRs and the remaining three RTKs. *PDGFRa* and *b* originate from the 2R event (Brunet et al. in preparation). There is a third PDGF receptor in the teleost fishes, which is called *pdgfrb2* to distinguish it from the *pdgfrb1*. Both are the result of the FSGD of *pdgfrb*, while the duplicate of *pdgfra* was lost early on during the evolution of the teleost fishes.

The KIT/FMS/FLT3 subfamily is also again the product of the early WGDs. *Kit* and *csf1r* were generated in the 2R event from a common precursor, while the 2R paralogue of *flt3* was obviously lost. *Kit* and *csf1r* in fish have their FSGD paralogues, but not *flt3*. Thus, the whole PDGFR superfamily is represented by eight members in the teleosts compared to only five in mammals.

Intriguingly, *kit* and *pdgfra* as well as *csf1r* and *pdgfrb* are closely located side by side on the same chromosomes in tetrapod and some fish genomes. Thus, it can be inferred with fidelity that before the 2R event, a local gene duplication generated the precursors of the genes that then gave rise to *kit* and *csf1r*, on the one hand, and *pdgfra* and *b*, on the other hand (Fig. 2.2) (Brunet et al. in preparation).

Moreover, *kdr* is linked to *kit* and *pdgfra*, suggesting another evolutionary relationship within the RTKs with Ig domain containing ligand-binding domains. In birds and lizards, *flt4* is also only 2–4 genes away from the *csfr1/pdgfrb* gene pair. Thus, a scenario becomes likely that a single RTK gene in the protochordate ancestor with seven Ig domains was first locally duplicated on one chromosome to become the progenitor of the seven Ig and five Ig domain RTKs. One of the genes became the progenitor of the VEGFR family, which has seven Ig domains. The other one lost two Ig domains and then underwent another local gene duplication. These two genes then constituted the progenitors of the five Ig domain RTK gene families, the PDGFR, and FMS/CSF1R/FLT3 subfamilies. The resulting cluster of three RTK genes then was expanded through the 1R/2R WGDs with subsequent loss of some genes [23] (Fig. 2.2).

2.7 The NTRK Family

The human gene family is composed of *NTRK1/TRKA*, *NTRK2/TRKB*, and *NTRK3/TRKC*. The phylogenetic tree of this RTK family again proposes an evolutionary history from the 1R/2R WGDs, indicating that one of the four genes has been lost early on, most probably the duplicate of *ntrk1*. *Ntrk2* and *3* group together by sequence similarity but also by a common genomic structure (large size of the gene), which is different in *ntrk1* [25]. In fish, *ntrk2* and *3* have 3R WGD duplicates, but an NTRK1 duplicate is not present. In the vertebrate ancestors, a gene homologous to the NTRKs has been found in *Amphioxus* and sea urchin, but not in the sea squirt [26, 27]. *Drosophila* and *C. elegans* do not have *ntrk*'s, but the pond snail *Lymnaea stagnalis* has a Trk-like RTK [28]. Hence, it is possible that this RTK family might also have its origin in a common ancestor of the deuterostomes and the protostomes, which is supported by a similar evolutionary scenario for the neurotrophin ligands [25].

NTRKs share with many other RTK families the uninterrupted kinase domain and extracellular Ig domains but uniquely have an N-terminal solenoid-like leucine-rich domain. Moreover, all three NTRKs function together with a common neurotrophin receptor, the p75NTR. This co-receptor is structurally and functionally unrelated to the NTRKs but apparently has coevolved with this RTK family. p75NTRs are present in all vertebrates and in the protochordates.

A possible relationship of NTRKs with other RTK families could be inferred from the fact that in tetrapods, *ntrk2* is located side by side with *igf1r*. However, the total different structure of the extracellular domains of both families makes a common origin from an ancestral gene duplication difficult to explain without involving a domain shuffling event. On the other hand, *ntrk2* and *ror2* are only 25 genes away (~5 Mb) on human chromosome 9 and linked on chrZ in birds and chr2 in the lizard as well. Sequence similarity also indicates a closer relationship and thus possibly a common origin of the NTRK and ROR RTK families. We hypothesize that the progenitors of both families were generated by a local gene duplication of an RTK gene with affiliation to the *Drosophila* sevenless receptor in the chordate ancestors before the 1R WGD.

2.8 The TAM Receptor Family

The three TAM family members *TYRO3*, *AXL/UFO*, and *MERTK* are the results of the WGDs. *TYRO3* and *MERTK* evolved from a common ancestor during the 2R event. The corresponding 2R duplicate of *AXL/UFO* is most likely represented by the *axl-like* gene in the teleost fishes, but it has been lost in the tetrapod lineage. Strikingly, in the bird genomes, no homologue of *axl/ufo* gene is found.

The origin of this family is obscure. There are homologues in the sea squirts, which are now considered to be more closely related to the vertebrates. No homologue was found in sea urchin or *Drosophila*.

2.9 The DDR Family

The two *DDR* RTK genes in humans originated during the early WGDs (Brunet et al. in preparation). The nonmammalian vertebrates have a third family member, *ddr2-like*, which is clearly the 2R ohnolog of *ddr2*. Thus, the corresponding 2R duplicate of *ddr1* was lost without leaving a trace in any of the known vertebrate genomes. While *ddr1* was lost in reptiles and birds but kept in the therian mammals, the opposite happened for *ddr2-like*, which was lost in mammals but kept in reptiles and birds. In the *Xenopus* genome, all three DDRs can be found. Fish have a 3R duplicate in the *ddr2* branch; consequently, this group has four family members compared to only two in mammals. *Ddr* is present as a single gene in the protochordates. In the *Drosophila* genome, also a clear homologue of *ddr* is present. A ddr gene was also identified in *Hydra* [3]. The DDRs are characterized by their unique extracellular domain, which consists of two discoidin domains.

2.10 The EPH Receptor Family

The ephrin receptors are the largest family of RTKs. In humans, 15 members can be identified: *EPHa1–8*, *10*, and *EPHb1–6*. EPHa and EPHb receptors are members of two phylogenetically distinct clades and are distinguished also by their different ligands: A-type receptors bind GPI-linked ephrin-A ligands (ephrin-A1–6) and B-type receptors bind to ephrin-B ligands (ephrin-B1–3), with the exception of EPHa4 and EPHb2, which can bind a ligand of the other type in addition. *Drosophila* and sea urchin have a single ephrin receptor, but in the sea squirts, a lineage-specific expansion of the ephrin receptors occurred. In *Hydra magnipapillata*, three different ephrin receptors were identified [3], which also were generated specifically during *Hydra* evolution. Ephrin receptors are also present in sponges [29]. It has been proposed that until the appearance of the protochordates, a single EPH receptor was present, which was a common receptor for the ancestral ephrin-A and ephrin-B ligands [30]. How the components of the two subfamilies were generated during vertebrate evolution by local gene duplications, WGDs, and gene losses is unclear at present and remains a challenging task for the future. In fish, clearly, the 3R WGD has generated additional genes for *epha2*, *4*, *6*, and *ephb1–5* (Brunet et al. in preparation). The EPHa4-like receptor is only found in fish and *Xenopus* and appears to be lost in the "higher" vertebrates. Conversely, *epha1* is absent from these organisms but present in reptiles, birds, and mammals. *Eph5* exists only in chicken, where it appears to be lineage-specific duplicate of *eph6* (Fig. 2.3).

2.11 The ROR, TIE, MET, and ALK Receptor Families

The ROR, TIE, MET, and ALK RTK families have only two members each (*ror1* and *2*; *tie1* and *2*; *met* and *mst1r*; *alk* and *ltk*) in the mammalian genomes and all other vertebrates, except for the *mst1 receptor*, which in fish is represented by two paralogues resulting from the 3R WGDs. All go back to a single protochordate ancestor; however, it is unclear so far how the two family members were generated during the early vertebrate WGDs. In *Drosophila*, RTK homologues of *alk* and *tie*'s from the vertebrate families are present. *Met* and *ron* appear to be absent from insects and other protostomes, but *ror* has a homologue in *Drosophila*, at least what

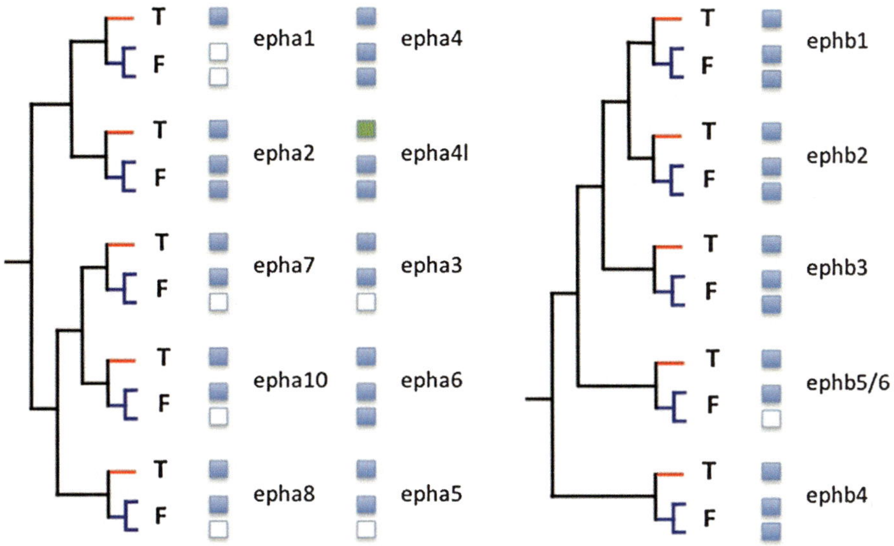

Fig. 2.3 Phylogenetic distribution and evolutionary relationship of the *epha* and *ephb* genes. The nature of the duplication events (WGDs and local gene duplications) that occurred among these genes could not be traced down by molecular phylogenetic analyses and conserved synteny. *Filled boxes* represent the occurrence of an RTK gene, in *blue* whenever a gene is found in all main lineages of tetrapods (T) or fish (F). *Epha4l* is in *green* because it is only found in *Xenopus*. *White boxes* indicate that no RTK is found among the group of species considered

can be said from the sequence similarity of the kinase domain, while the other portions are less obviously related. In *Hydra*, a Ror1-like RTK was identified [3].

All four families have an uninterrupted kinase domain. The ALK family is most peculiar because the extracellular domain is composed of two Mam domains and one Ldla domain, which are not found in any other RTK. The three other families share with many other RTK families the presence of Ig domains. Ror1 and 2 are somehow similar to MuSK by the additional Fz domain but uniquely have a kringle domain. Tie1 and Tie2/Tek show similarity to the AXL family having three Ig and fibronectin domains instead of two. They are unique by the presence of three EGF domains. Met and Ron have in front of their four Ig domains a Sema/Psi domain which is absent from all other RTKs [15] (Fig. 2.4).

2.12 MUSK, PTK7, RET, ROS, and RYK

This group of RTKs is peculiar, because they present themselves as "sibling-less" without a clear paralogue in the genomes of vertebrates.

Musk has in its extracellular portion a Fz domain like the Ror's and an Ig domain, which is present in Ror's and many other RTKs, but the phylogenetic

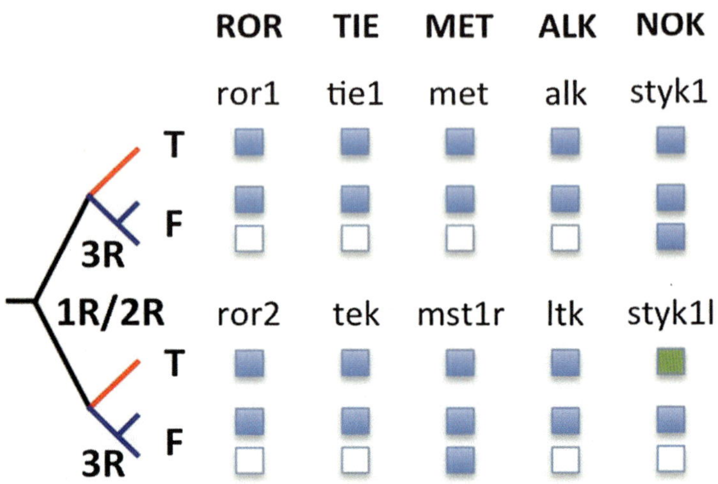

Fig. 2.4 Schematic view of the ROR, TIE, MET, and ALK RTK families that expanded either in the 1R or 2R whole genome duplications, where, however, the exact history cannot be inferred. The gene duplication of *mst1r* and *styk1* resulting from the 3R is also shown by two boxes in fish. *Filled boxes* represent the occurrence of an RTK gene, in *blue* whenever a gene is found in all main lineages of tetrapods (T) or fish (F). A *white box* indicates that no RTK has been found among the group of species considered. *Styk1l* is in *green* because it is only found in *Xenopus* within the tetrapods

analyses did not group it together with any of the others. The "earliest" homologue of *musk* was found in the ray *Torpedo californica* from the basal vertebrate group of cartilaginous fish. Neither in fish nor in any other vertebrates a paralogue could be identified. On the other hand, a Musk-like RTK has been reported from sea urchin [27] and from *Hydra* [3]. Thus, its evolutionary origin remains obscure. From sequence similarities, a relationship to the ROR family can be drawn.

For *ptk7*, *ret*, *ros*, and *ryk* each a distantly related but clear protochordate outgroup orthologue could be identified in *Ciona*, *Branchiostoma* (Brunet et al. in preparation), and sea urchin [27]. In *Hydra*, a putative orthologue of *ptk7* was described [3]. No remnants of the three WGDs were detected in any of the vertebrate genomes. The "orphan" character or *ryk* is reflected by the unique presence of a WIF extracellular domain. Besides an ephrin-binding domain that is shared with the many members of the EPH receptor family, Ret harbors uniquely a WIF domain. Similarly, Ros has besides the widespread fibronectin type III domains exclusively the YWTD propeller domain, which supports the status of singularity. Ptk7, also known as Cck4, has seven Ig domains and no other type as the extracellular part alike the VEGFR family members (and five or three for the PDGFR and FGFR families, respectively) and an uninterrupted kinase domain as many other RTKs (Fig. 2.5).

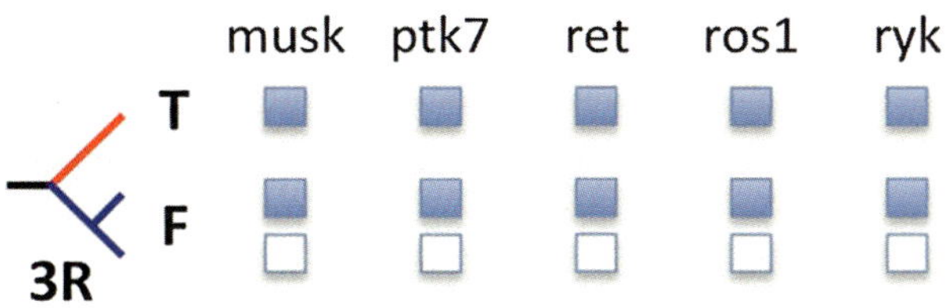

Fig. 2.5 Common origin of the *musk*, *ptk7*, *ret*, ros, and *ryk* RTKs in tetrapods (T) and fish (F). These RTKs have not build up a gene family even though the ancestral gene surely went through the two whole 1R and 2R genome duplications. Also none of the FSGD genes has been retained

2.13 The NOK Family

The NOK family is represented in vertebrates by *styk1*, which is found in all vertebrates. In teleost fishes, two genes are present, *styk1a* and *styk1b*, which are clearly duplicates from the 3R WGD. Interestingly, there is another *styk1-like* gene (*styk1l*), which is only present in *Xenopus* and in fishes (without a 3R paralogue). No homologue could be found in protochordates or invertebrates. Styk1 and Styk1l possess no extracellular domain and are therefore here discussed outside the RTK families, but their kinase domains show clear sequence similarity to the kinase domains of the PDGFRs and FGFRs. It appears that *styk1* emerged in a vertebrate ancestor from a duplicate of an RTK gene of the seven Ig domain/split kinase domain family, which subsequently lost the extracellular part (Fig. 2.6).

2.14 Evolution of RTK Ligands

The evolutionary history of the RTK ligands is much more difficult to elucidate. First, the number of known and potential ligands is very high and exceeds by far the number of RTKs. Second, they are much less conserved on the amino acid level, making interferences from protein sequence comparisons more difficult or often impossible. In many cases, only a few amino acids, which are important for the common three-dimensional structure of a certain growth factor family, are conserved, like cysteines involved in cystin bridges. Only a few studies have approached therefore a comprehensive evolutionary history of groups of ligands, and thus, only some prototypic examples can be discussed here.

2.15 EGFR Ligands

In mammals, there are seven recognized ligands that bind and activate Egfr, namely, epidermal growth factor (Egf), transforming growth factor alpha (Tgfa), amphiregulin (Areg), betacellulin (Btc), epiregulin (Ereg), heparin-binding EGF-like growth

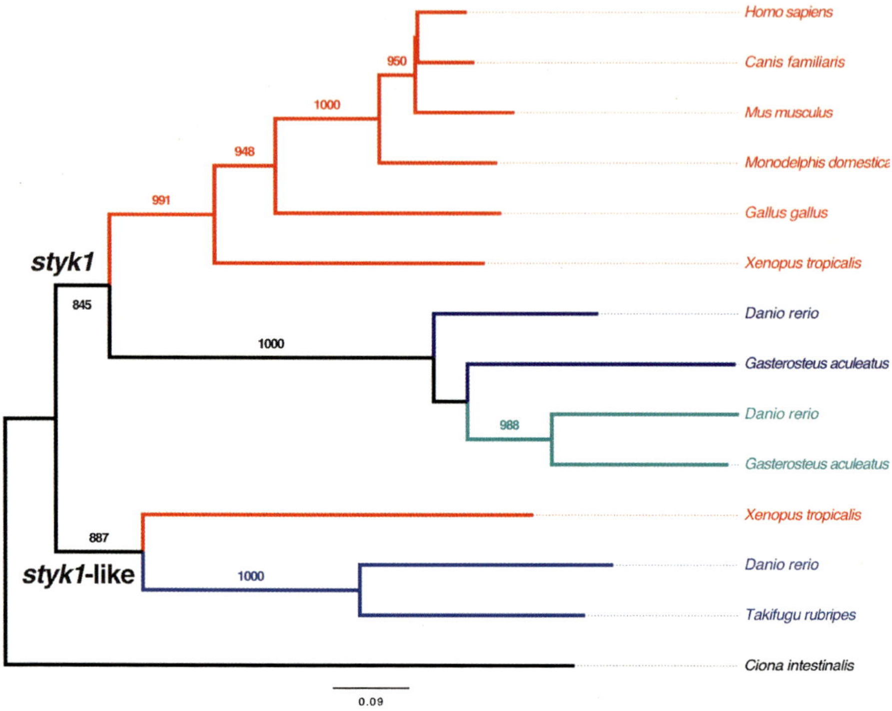

Fig. 2.6 Phylogenetic tree reconstruction of the NOK family using maximum likelihood (PhyML). This family is rooted by the protochordate *Ciona intestinalis* ENSCINP00000025821. *Styk1* is shown for human, *Homo sapiens* ENSP00000075503; dog, *Canis familiaris* ENSCAFP00000019768; mouse, *Mus musculus* ENSMUSP00000044098; opossum, *Monodelphis domestica* ENSMODP 00000022763; chicken, *Gallus gallus* ENSGALP00000023762; and *Xenopus tropicalis* ENSXETP 00000035460, and in the fish, *styk1a* is shown for zebrafish, *Danio rerio* ENSDARP00000095367, and stickleback, *Gasterosteus aculeatus* ENSGACP00000002580, and *styk1b* for the same species ENSDARP00000101865 and ENSGACP00000013787; *styk1*-like is shown for *Xenopus tropicalis* ENSXETP00000020968; zebrafish, *Danio rerio* ENSDARP00000115394; and fugu, *Takifugu rubripes* ENSTRUP00000012210

factor (Hbegf), and epigen (Epgn). Although a lot of genomic information exists for the other vertebrate lineages, there is for most organisms no clear answer how many ligands the Egfr and any of the other family members would be exposed to. The problem is that only the mature part of the ligands with the Egfr-interacting motif, which covers not more than approximately 140 bases, is suitable for phylogenetic tree reconstruction. This small region gives in most cases not enough phylogenetic information, so only the synteny analyses can give final cues to define true orthology vs. paralogy and to trace the phylogenetic relationships. Such analyses revealed the evolution of the Egfr ligands in vertebrates [31] (Fig. 2.7).

All ligands of the Egf receptor can be traced back to a single ancestral gene on protochordate chromosome C. This gene was duplicated twice during the 1R and 2R whole genome duplication events of early vertebrate evolutionary history in

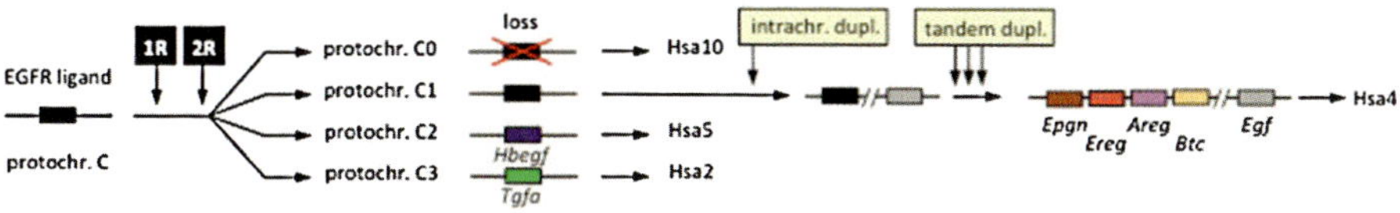

Fig. 2.7 Evolution of the Egfr ligands from a single common ancestor. Hsa, human chromosome

organisms that preceded the appearance of sharks and fish. One of the four progenitor genes was lost, one became *tgfα*, and the other one the *hbegf* gene. The remaining gene on the protochordate chromosome C1 underwent a local gene duplication. One of the resulting duplicates evolved into the *egf* gene, while the other one was again duplicated twice giving rise to the Btc, Areg, Ereg, and Epgn encoding genes. This cluster is, however, not always fully intact in all vertebrates; for instance, in birds, *btc* has been moved to another chromosome. A similar situation is seen in some fish, and some organisms have obviously even lost one or the other gene family member or the cluster broke further apart.

Of special evolutionary interest is the situation in fish. Here most species have as a result of the 3R FSGD two *egfr* genes (*egfra* and *egfrb*), but there is no indication that there are also more ligands. Obviously, immediately after the FWGD, the second version of each of the ligand genes was lost. Moreover, there is evidence that the two *egfr* genes have partitioned the tasks of the ancestral *egfr* gene by being differentially expressed in separate tissues and organs and they are working in concert with specific subsets of the Egfr ligands [31].

It is intuitive that other ligand genes with EGF-like domains, i.e., neuregulins, tomoregulins, and others, share an evolutionary history with the seven Egfr ligands due to their striking sequence similarity [32, 33]. However, the tomoregulin genes do not show any obvious synteny with the gene regions that harbor the Egfr ligands, so their evolutionary history remains unclear for the moment. The neuregulin gene family members, NRG 1, 2, 3, are also located on mammalian chromosomes that are derived from the ancestral vertebrate chromosome C and therefore could have evolved along with the canonical Egfr ligands [31].

2.16 The FGF Family

The mammalian FGF family (reviewed in [34]) comprises 22 members (*fgf1–23*). *Fgf15* is not present in humans, while in mouse and rat, *fgf19* is missing. The protein sequence similarity and syntenic relationships define seven subfamilies, which mirrors also the classification by mode of action: the intracrine FGFs (Fgf11–14), the paracrine FGFs (subfamilies 1/2/5, 3/4/6, 7/10/22, 8/17/18, 9/16/20), and the endocrine FGFs (FGF15/19/21/23). Notably, the intracrine FGF subfamily members are not ligands for the FGFRs but interact with the intracellular domains of voltage-gated sodium channels and with a neuronal MAPK scaffold protein, islet-brain-2 [35, 36]. The subfamily of endocrine FGFs is also united by their very low affinity to FGFRs and heparin/heparin sulfate.

An evolutionary scenario has been proposed [34] with one heparin binding but secreted signal peptide lacking ancestral gene of the intracrine subfamily. This gene was duplicated in the early metazoan history and a secreted, receptor-binding gene evolved. This further duplicated in the protochordate ancestor to generate the ancestor of the endocrine FGFs. Although in the genome of *Amphioxus* no homologue of the intracrine FGFs was found [37], one of the six FGF genes present in the *Ciona* genome is firmly nested within the *fgf11–14* branch of the phylogenetic tree [38]. The expansion of the family that we see today then was the result of the 1R/2R WGDs. Remarkably, the intracrine and endocrine subfamilies consist of four members, being in line with this hypothesis. All paracrine subfamilies have only three members, which could be interpreted as if one of the four 2R duplicates got lost very early after the WGD event.

2.17 The PDGF/VEGF Family of Growth Factors

Vegfa–d, Plgf, and Pdgfa–d form a family of RTK ligands with a well-conserved growth factor domain, the so-called Pdgf/Vegf homology domain. This domain forms a tight cysteine knot structure and in most cases is involved in (homo- or hetero-) dimerization by disulfide bonding. Pdgfc and d have a CUB domain in addition to the classical Pdgfs (a and b).

On the basis of their amino acid sequence, the evolutionary history of these growth factors is difficult to decipher. *Pdgfa* and *b*, *c*, and *d*, *vegfc* and *d*, and *vegfa* and *plgf* are more closely related to each other, but there is no further clear rooting [39].

In *Drosophila*, three PDGF-/VEGF-like factors (PVFs) have been identified, which act through a single receptor, and a single PVF homologue is present in *C. elegans* [40]. Although a lineage-specific expansion of the corresponding receptors has occurred in the worm, the entire PDGF/VEGF family and all their receptors go back to a single growth factor/receptor pair in the early bilaterian ancestor. The PDGF/VEGF family appears to be related as a superfamily by the cysteine knot feature to the glycoprotein hormone- and mucin-like proteins and more distantly to the transforming growth factor ß (TGFß) family [40].

2.18 Conclusions

Tracing the evolutionary history of the whole complement of 58 RTKs in the human genome uncovers that they developed from a set of progenitor RTK genes in the ancestor of vertebrates. In general, the picture emerges that all RTK families and superfamilies go back to a single common ancestral gene in the early metazoan lineage and in most cases in extant invertebrates only a single representative of each family is found. Representatives of the EGF receptor, insulin receptor, FGF receptor, the PDGFR/VEGFR superfamily, EPH receptor, DDR and ROR families, and MuSK and PTK7 receptors have been found. This basic set of RTKs has diverged and expanded in certain branches of the protostomian invertebrates, but it has been this basic repertoire of RTK that was used again and again for creating evolutionary and developmental diversity.

The set of 9 early metazoan RTKs was expanded to 16 in the protochordate ancestor, and representatives can be found in the genomes of sea urchin, *Ciona*, and *Amphioxus*. They are the ancestors of the mammalian gene families. How these ancestors are related is difficult to say. The sequence similarity in the kinase domain suggests a close relationship of the insulin receptor family to Ptk7 and Ros and also to the DDR family, MUSK with the ALK family, the AXL family with the MET family and RYK, the TIE family with RET and the FLT/CSF1R/KIT subfamily, and the FGFR family with the PDGFR subfamily. The full complement of receptors of the human RTK families is the result of the early vertebrate WGDs. The average retention rate for gene duplicates from the 1R/2R events is 20–30 % [41]. Without considering the EPH receptor family, for which the evolutionary relationships could not be fully solved, the duplicate retention rate for RTKs is much higher: 55–58 %. This high tendency for duplicates to be kept—rather than being lost in the process of reinstalling the normal 2 N gene dosage of the genome—in this class of proteins is easily explained by the important cellular function of these receptors. With the increasing complexity of cell types, cell functions, and the intracrine, autocrine, juxtacrine, paracrine, and endocrine cellular interactions, the RTKs were targets to implement evolutionary innovations that required novel signaling pathways.

It appears that each RTK family is mirrored by a family of corresponding growth factors that also expanded during the evolution of the vertebrates. The need for specific ligand/receptor interactions has the immediate consequence that both partners coevolve. A receptor mutation in an amino acid that is involved in ligand binding and which will change the intermolecular interaction forces will only be fixed if a compensating mutation occurs in the corresponding growth factor. On the other hand, the need for creating evolutionary novel interactions to make an increased complexity possible has led to new ligand/receptor interactions. The ability for receptors and ligand to dimerize as heteromers has probably also driven the retention of many duplicates after local and whole genome duplications and created many new ligand/receptor interactions. But obviously, the coevolution paradigm has restricted this to occur only within families. At least to our knowledge, there is no case that a ligand of a certain gene family has switched its binding affinity to another family of RTKs outside its original affiliation.

Coevolution would also predict that following a WGD, each set of duplicated RTKs should have kept a corresponding set of duplicated ligands, but this is obviously not the case as can be seen, for instance, for the lack of any 3R duplicates in fish for the EGFR ligands, despite two paralogues of the receptor have been retained in evolution.

With the same argument, it is predicted that the intracellular signal transduction network downstream of RTKs should have expanded together with the RTK families. To trace the evolutionary history and origins of this extremely complex system where indeed many components are paralogues is a challenging task for evolutionary systems biology [41]. An intriguing hypothesis has been put forward in this field. It is proposed that WGDs would have initially duplicated RTKs with a single cognate ligand and a linear private signal transduction machinery. Over time, these uncoupled cascades helped by the structural similarity of their paralogous members started to horizontally interact, leading finally to the cross-talking pathways [6].

Acknowledgments We gratefully acknowledge the support by the Deutsche Forschungsgemeinschaft (DFG) to MS and Agence Nationale de la Recherche (ANR) to JNV. Due to restrictions in space and in light of the plethora of publications on RTKs, only a selection of references could be considered. We apologize to all whose work has not been cited or discussed and declare that such omissions do not reflect any validation of the relevance and quality of the research reported there.

References

1. King N, Carroll SB. A receptor tyrosine kinase from choanoflagellates: molecular insights into early animal evolution. Proc Natl Acad Sci U S A. 2001;98(26):15032–7 [Research Support, U.S. Gov't, P.H.S.].
2. Skorokhod A, Gamulin V, Gundacker D, Kavsan V, Muller IM, Muller WE. Origin of insulin receptor-like tyrosine kinases in marine sponges. Biol Bull. 1999;197(2):198–206 [Research Support, Non-U.S. Gov't].
3. Reddy PC, Bidaye SS, Ghaskadbi S. Genome-wide screening reveals the emergence and divergence of RTK homologues in basal Metazoan Hydra magnipapillata. J Biosci. 2011;36(2):289–96 [Research Support, Non-U.S. Gov't].
4. Müller WE, Skorokhod A, Müller IM. Receptor tyrosine kinase, an autapomorphic character of metazoa: identification in marine sponges. Acta Biol Hung. 1999;50(4):395–411 [Research Support, Non-U.S. Gov't Review].
5. Müller WE, Kruse M, Blumbach B, Skorokhod A, Muller IM. Gene structure and function of tyrosine kinases in the marine sponge Geodia cydonium: autapomorphic characters in Metazoa. Gene. 1999;238(1):179–93 [Research Support, Non-U.S. Gov't].
6. Amit I, Wides R, Yarden Y. Evolvable signaling networks of receptor tyrosine kinases: relevance of robustness to malignancy and to cancer therapy. Mol Syst Biol. 2007;3:151 [Research Support, Non-U.S. Gov't Review].
7. Ohno S. Evolution by gene duplication. New York, NY: Springer; 1970.
8. Van de Peer Y, Maere S, Meyer A. The evolutionary significance of ancient genome duplications. Nat Rev Genet. 2009;10(10):725–32 [Research Support, Non-U.S. Gov't Review].
9. Nakatani Y, Takeda H, Kohara Y, Morishita S. Reconstruction of the vertebrate ancestral genome reveals dynamic genome reorganization in early vertebrates. Genome Res. 2007;17(9):1254–65 [Research Support, Non-U.S. Gov't].
10. Jaillon O, Aury JM, Brunet F, Petit JL, Stange-Thomann N, Mauceli E, et al. Genome duplication in the teleost fish Tetraodon nigroviridis reveals the early vertebrate proto-karyotype. Nature. 2004;431(7011):946–57 [Research Support, Non-U.S. Gov't].

11. Kasahara M, Naruse K, Sasaki S, Nakatani Y, Qu W, Ahsan B, et al. The medaka draft genome and insights into vertebrate genome evolution. Nature. 2007;447(7145):714–9 [Research Support, Non-U.S. Gov't].
12. Muffato M, Crollius RH. Paleogenomics in vertebrates, or the recovery of lost genomes from the mist of time. Bioessays. 2008;30(2):122–34.
13. Livneh E, Glazer L, Segal D, Schlessinger J, Shilo BZ. The Drosophila EGF receptor gene homolog: conservation of both hormone binding and kinase domains. Cell. 1985;40(3):599–607 [Comparative Study, Research Support, Non-U.S. Gov't, Research Support, U.S. Gov't, P.H.S.].
14. Aroian RV, Koga M, Mendel JE, Ohshima Y, Sternberg PW. The let-23 gene necessary for Caenorhabditis elegans vulval induction encodes a tyrosine kinase of the EGF receptor subfamily. Nature. 1990;348(6303):693–9 [Research Support, Non-U.S. Gov't Research Support, U.S. Gov't, P.H.S.].
15. Lemmon MA, Schlessinger J. Cell signaling by receptor tyrosine kinases. Cell. 2010;141(7):1117–34 [Research Support, N.I.H., Extramural Review].
16. Volff JN, Schartl M. Evolution of signal transduction by gene and genome duplication in fish. J Struct Funct Genomics. 2003;3(1–4):139–50.
17. Gomez A, Volff JN, Hornung U, Schartl M, Wellbrock C. Identification of a second egfr gene in *Xiphophorus* uncovers an expansion of the epidermal growth factor receptor family in fish. Mol Biol Evol. 2004;21(2):266–75.
18. Gomez A, Wellbrock C, Gutbrod H, Dimitrijevic N, Schartl M. Ligand-independent dimerization and activation of the oncogenic Xmrk receptor by two mutations in the extracellular domain. J Biol Chem. 2001;276(5):3333–40.
19. Meierjohann S, Schartl M. From Mendelian to molecular genetics: the *Xiphophorus* melanoma model. Trends Genet. 2006;22(12):654–61.
20. Broughton S, Partridge L. Insulin/IGF-like signalling, the central nervous system and aging. Biochem J. 2009;418(1):1–12 [Research Support, N.I.H., Extramural Review].
21. Klambt C, Glazer L, Shilo BZ. Breathless, a Drosophila FGF receptor homolog, is essential for migration of tracheal and specific midline glial cells. Genes Dev. 1992;6(9):1668–78 [Research Support, Non-U.S. Gov't Research Support, U.S. Gov't, P.H.S.].
22. Beiman M, Shilo BZ, Volk T. Heartless, a Drosophila FGF receptor homolog, is essential for cell migration and establishment of several mesodermal lineages. Genes Dev. 1996;10(23):2993–3002 [Research Support, Non-U.S. Gov't].
23. Grassot J, Gouy M, Perriere G, Mouchiroud G. Origin and molecular evolution of receptor tyrosine kinases with immunoglobulin-like domains. Mol Biol Evol. 2006;23(6):1232–41 [Research Support, Non-U.S. Gov't].
24. Kondo K, Hiratsuka S, Subbalakshmi E, Matsushime H, Shibuya M. Genomic organization of the flt-1 gene encoding for vascular endothelial growth factor (VEGF) receptor-1 suggests an intimate evolutionary relationship between the 7-Ig and the 5-Ig tyrosine kinase receptors. Gene. 1998;208(2):297–305 [Research Support, Non-U.S. Gov't].
25. Hallböök F, Wilson K, Thorndyke M, Olinski RP. Formation and evolution of the chordate neurotrophin and Trk receptor genes. Brain Behav Evol. 2006;68(3):133–44 [Research Support, Non-U.S. Gov't Review].
26. Benito-Gutierrez E, Nake C, Llovera M, Comella JX, Garcia-Fernandez J. The single AmphiTrk receptor highlights increased complexity of neurotrophin signalling in vertebrates and suggests an early role in developing sensory neuroepidermal cells. Development. 2005;132(9):2191–202 [Research Support, Non-U.S. Gov't].
27. Lapraz F, Röttinger E, Duboc V, Range R, Duloquin L, Walton K, et al. RTK and TGF-beta signaling pathways genes in the sea urchin genome. Dev Biol. 2006;300(1):132–52 [Research Support, Non-U.S. Gov't].
28. Beck G, Munno DW, Levy Z, Dissel HM, Van-Minnen J, Syed NI, et al. Neurotrophic activities of trk receptors conserved over 600 million years of evolution. J Neurobiol. 2004;60(1):12–20 [Comparative Study Research Support, Non-U.S. Gov't].

29. Suga H, Hoshiyama D, Kuraku S, Katoh K, Kubokawa K, Miyata T. Protein tyrosine kinase cDNAs from amphioxus, hagfish, and lamprey: isoform duplications around the divergence of cyclostomes and gnathostomes. J Mol Evol. 1999;49(5):601–8 [Comparative Study Research Support, Non-U.S. Gov't].
30. Mellott DO, Burke RD. The molecular phylogeny of eph receptors and ephrin ligands. BMC Cell Biol. 2008;9:27 [Research Support, Non-U.S. Gov't].
31. Laisney JA, Braasch I, Walter RB, Meierjohann S, Schartl M. Lineage-specific co-evolution of the Egf receptor/ligand signaling system. BMC Evol Biol. 2010;10:27.
32. Stein RA, Staros JV. Insights into the evolution of the ErbB receptor family and their ligands from sequence analysis. BMC Evol Biol. 2006;6:79 [Research Support, N.I.H., Extramural].
33. Schneider MR, Wolf E. The epidermal growth factor receptor ligands at a glance. J Cell Physiol. 2009;218(3):460–6 [Research Support, Non-U.S. Gov't Review].
34. Itoh N, Ornitz DM. Functional evolutionary history of the mouse Fgf gene family. Dev Dyn. 2008;237(1):18–27 [Research Support, N.I.H., Extramural, Research Support, Non-U.S. Gov't, Review].
35. Goldfarb M, Schoorlemmer J, Williams A, Diwakar S, Wang Q, Huang X, et al. Fibroblast growth factor homologous factors control neuronal excitability through modulation of voltage-gated sodium channels. Neuron. 2007;55(3):449–63 [In Vitro Research Support, N.I.H., Extramural, Research Support, Non-U.S. Gov't].
36. Schoorlemmer J, Goldfarb M. Fibroblast growth factor homologous factors and the islet brain-2 scaffold protein regulate activation of a stress-activated protein kinase. J Biol Chem. 2002;277(51):49111–9.
37. Bertrand S, Camasses A, Somorjai I, Belgacem MR, Chabrol O, Escande ML, et al. Amphioxus FGF signaling predicts the acquisition of vertebrate morphological traits. Proc Natl Acad Sci U S A. 2011;108(22):9160–5 [Research Support, Non-U.S. Gov't].
38. Satou Y, Imai KS, Satoh N. Fgf genes in the basal chordate Ciona intestinalis. Dev Genes Evol. 2002;212(9):432–8 [Research Support, Non-U.S. Gov't].
39. Li X, Eriksson U. Novel PDGF family members: PDGF-C and PDGF-D. Cytokine Growth Factor Rev. 2003;14(2):91–8 [Review].
40. Holmes DI, Zachary I. The vascular endothelial growth factor (VEGF) family: angiogenic factors in health and disease. Genome Biol. 2005;6(2):209 [Research Support, Non-U.S. Gov't Review].
41. Huminiecki L, Heldin CH. 2R and remodeling of vertebrate signal transduction engine. BMC Biol. 2010;8:146 [Research Support, Non-U.S. Gov't].

Chapter 3
RTKs in Invertebrates: Lessons in Signal Transduction

Ben-Zion Shilo

3.1 Families of RTKs in *Drosophila*

The major families of RTKs are represented in the *Drosophila* genome and some also in the *C. elegans* genome. This is an indication for the early divergence of RTK families and the essential roles each of them plays during multicellular development. These include the EGF receptor [1, 2], the FGF receptors (Breathless and Heartless) [3–6], the insulin receptor [7, 8], Torso [9], Alk [10], an Eph receptor [11], a PDGF/VEGF receptor (PVR) [12], Sevenless [13, 14], Derailed [15], and Stitcher [16]. Interestingly, Trk receptors were not found in the *Drosophila* or *C. elegans* genomes [17].

For the *Drosophila* EGF receptor, a detailed structural analysis was carried out. It provided novel insights regarding a mechanism by which ligand triggers the activation of this receptor in a biphasic manner [18]. The first ligand-binding event induces the formation of an asymmetric dimer with only one bound ligand. The unoccupied site in this dimer is structurally restrained, leading to reduced affinity for binding of the second ligand and thus negative cooperativity. Individual EGFR ligands might thus stabilize distinct dimeric species with different signaling properties.

Interestingly, except for the FGF receptor family which has two members, all other receptors represent a single member per family. This proved to be an enormous experimental advantage, as the phenotypes observed from mutations in a given receptor provide a direct and comprehensive indication for the role of the signaling pathway they trigger.

B.-Z. Shilo (✉)
Department of Molecular Genetics, Weizmann Institute of Science, Rehovot 76100, Israel
e-mail: benny.shilo@weizmann.ac.il

D.L. Wheeler, Y. Yarden (eds.), *Receptor Tyrosine Kinases: Structure, Functions and Role in Human Disease*, DOI 10.1007/978-1-4939-2053-2_3

3.2 Linear vs. Branched RTK Signaling Pathways

In vertebrates, RTKs trigger several parallel intracellular signaling pathways. First and foremost is the Ras/MAPK pathway, but additional branches involving PLC gamma, PI3K/AKT, and others also play prominent roles. In sharp contrast, most of the *Drosophila* RTKs trigger only the Ras/MAPK branch. This was demonstrated in a variety of ways. First, mutations in the downstream components of the signaling pathway give identical phenotypes to mutations in the receptor or the ligands triggering the pathway. When a given signaling component is activated by several receptors, its loss will lead to a composite phenotype of the different receptors. Conversely, activated versions of downstream components are epistatic to mutations in the receptor [19, 20].

It follows that there is no signaling signature for a given receptor, all giving rise to activated MAPK. Indeed, early experiments which exchanged the kinase domains between receptors showed no obvious alterations in signaling outcomes [21]. The pathway triggered by the two FGF receptors requires a unique intracellular adaptor termed Dof [22]. However, even this pathway could be replaced by cytoplasmic kinase domains which trigger MAPK, but do not require Dof.

The lack of specificity at the level of the intracellular signal suggested that pathways which dictate different developmental switches cannot be activated simultaneously in a given cell, since they generate the same output. An overlap would lead to an additive effect on a single output. This is true for the induction of R7 photoreceptor cell fate in the eye disc, where the combined additive input of EGF receptor and Sevenless signaling is required [23]. Similarly, for the migrating border cells in the ovary, the activation of the EGF receptor and PVR is redundant. Their respective ligands are produced at the target for migration, i.e., the oocyte, and thus lead to graded activation of their receptors [12].

This notion of limited overlap could indeed be tested by immunohistochemical staining, with an antibody directed against the activated (double phosphorylated) form of MAPK (also termed dpERK). A dynamic picture of activation is observed during embryonic and postembryonic stages [24, 25] (Fig. 3.1). At the outset, according to the existing knowledge about sites where each receptor is active, the patterns were tentatively attributed to distinct RTKs. These notions were verified by systematically examining mutants for the different RTKs and identifying the dpERK patterns that were missing in each case. Importantly, in the absence of a given receptor, complete elimination of the dpERK pattern was observed at the relevant tissue, indicating that a single RTK is operating at each case.

The MAPK atlas was also informative in distinguishing between RTKs that function only once, like Torso, and RTKs that display a highly dynamic activation pattern, like EGFR. For receptors that are activated multiple times utilizing the same signaling cassette, activation may take place in very different biological circumstances. The question arises whether the regulation of the pathway can be modulated, to accommodate distinct signaling requirements, e.g., autocrine vs. paracrine activation. We will provide below several examples for mechanisms that allow such variations.

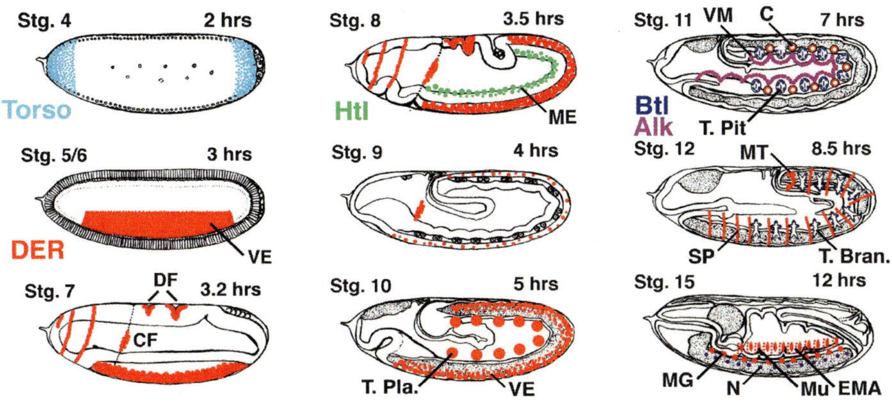

Fig. 3.1 Atlas of dpERK during embryogenesis. The distribution of dpERK during embryogenesis is presented schematically, based on [24]. The RTKs inducing each pattern are color-coded (Torso, *light blue*; DER (EGFR), *red*; Heartless, *green*; Breathless, *dark blue*; Alk, *magenta*; Unidentified receptors, *purple*). Note that there is no spatial overlap between the different RTKs, i.e., any given pattern can be eliminated by the removal of a distinct RTK. Some RTKs are activated only once (e.g., Torso), while others, most notably EGFR (DER), are activated numerous times, in a highly dynamic pattern. *VE* ventral ectoderm, *CF* cephalic furrow, *DF* dorsal anterior and posterior folds, *M* mesoderm, *T. Pla.* tracheal placodes, *T. pit* tracheal pits, *VM* visceral mesoderm, *C* chordotonal organs, *MT* Malpighian tubules, *SP* segmental pattern, *T. Bran.* tracheal branches, *MG* midline glial cells, *EMA* ectodermal muscle-attachment cells, *MU* muscles; *N* unidentified neurons

In sharp contrast to the developmental pathways where no spatial and temporal overlap is the rule, RTK pathways that are involved in adjusting or maintaining the metabolic state were shown to converge within the same cell. For example, InR (insulin receptor) and PVR activation can trigger MAPK in an overlapping manner. The difference between the developmental and metabolic regulation may be explained by the type of decisions that are triggered. For developmental processes, the decision is irreversible, i.e., a cell fate switch. The cues from the environment that activate a given receptor lead to a sharp switch. In the case of metabolic pathways, the cells are also monitoring the environment, but the signals they receive dictate reversible transitions or adjustments. The shared elements in the pathway thus serve as a "token" to monitor the physiological state of the cell.

3.3 Spatial and Temporal Features of RTK Signaling

The atlas of dpERK signaling in *Drosophila* development was extremely illuminating, since it marked not only the tissues which are activated at a given stage but also the level of activation and its duration. Moreover, the spatial distribution of activation in relation to the source of the signal, i.e., the ligand, could be visualized [24, 25]. As a rule, the signaling ranges observed did not exceed 4–5 cells from the ligand source (e.g., production of active EGF in the midline cells to trigger EGFR in the ventral ectoderm). In other cases, the signal was exclusively autocrine, as is the case for EGFR in the wing veins. The determination of the correct range of RTK activation is critical during development, and we will describe below some of the mechanisms responsible for restricting the range of activation and response.

It is interesting to consider the possibility that the responsiveness of different tissues could be predetermined, thus facilitating longer-range responses in some cases. A protein termed cytohesin, representing a guanine-nucleotide exchange factor, was recently shown to enhance the capacity of dimerized EGF receptors to carry out transphosphorylation in vertebrate cells, by associating directly with the cytoplasmic domains [26]. The expression of this protein or others in a given tissue could provide a means to make this tissue more responsive to the signal. This could generate a useful variation in the range of biological responses to the same signaling cascade.

3.4 Restricting the Range of RTK Signaling

As a rule, the expression of RTKs was shown to be fairly broad, encompassing all cells or at least all cells within a given tissue. Thus, the spatial response to the signal is usually not dictated by the distribution of the receptor or its downstream components. Mechanisms to restrict signaling range by controlling the amount of ligand that is secreted, the distribution of the secreted ligand, or the capacity of the receiving cells to respond to the signal were identified.

The range of EGFR signaling in *Drosophila* is regulated primarily by the amount of secreted ligand that is provided to the receiving cells. Three out of the four EGFR ligands are initially produced as an inactive transmembrane precursor that is sequestered in the endoplasmic reticulum (ER). The trafficking of ligand to a secretory compartment where processing takes place is facilitated by a transmembrane chaperone termed Star [27, 28]. Within the secretory compartment, cleavage of the precursor is carried out by intramembrane proteases of the Rhomboid family, and the cleaved extracellular ligand portion is subsequently secreted [29]. The chaperone Star is also cleaved by Rhomboids, but in contrast to the ligands, this cleavage generates an inactive molecule [30].

Some of the Rhomboid proteins are localized not only to the secretory compartment but also to the ER. When Star encounters Rhomboid in the ER, it is inactivated before it can traffic the ligand precursors to the secretory compartment, where productive ligand cleavage should take place [31]. Thus, only a fraction of the chaperone molecules that escaped cleavage in the ER is functional, and hence the level of ligand precursor that is trafficked is significantly compromised. This leads to a reduction in the amount of active ligand that is subsequently secreted and hence to a corresponding decline in the range of signaling that is elicited. In tissues where a restricted range of EGFR activation is required, such as the eye disc or the germ line, Rhomboid proteins with the dual intracellular distribution are utilized.

Once the ligand is secreted, another tier of regulation is employed. High levels of EGFR activation induce the expression of the target gene *argos*, encoding a secreted molecule that traps the secreted ligand [32, 33]. The induction of Argos thus functions as a sieve which reduces the levels of active ligand that can diffuse from the source and hence the range of signaling.

At the level of the responding cells, additional mechanisms are employed to restrict the signaling range, functioning in a cell-autonomous manner. In parallel to the induction of *argos* by the signal-receiving cells, two additional genes are induced, *kekkon* [34] and *sprouty* [35–37]. The former encodes a transmembrane protein that generates inactive heterodimers with EGFR, while the latter represents an intracellular inhibitor of the MAPK signaling cascade. In both cases, the induction of the inhibitors in a fairly broad range of the receiving cells results in productive signaling only in the cells that are closer to the ligand source and hence receive enough input to counteract the inhibitory effects.

An interesting variation is found in the context of the Breathless (Btl) FGF receptor, whereby the receptor itself restricts the diffusion of the ligand Branchless (Bnl). The role of Btl is to guide the migration of the tracheal cells towards the ligand source [6]. In order to increase the sharpness of the attractive ligand gradient, the expression of the receptor is induced by high levels of signaling, thus generating a trap that restricts the diffusion of the Bnl ligand to more distant cells [38].

3.5 Concept of "Ratchet-Like" Signal Expansion

In most tissues, a given RTK will signal only once, to trigger a particular developmental switch. However, in complex organs, there are interesting examples where the same pathway is triggered sequentially in adjacent cells, in a ratchet-like mechanism (reviewed in [39]). In *C. elegans*, the anchor cell expresses Lin-3, the EGF ligand, to trigger primary cell fates in the underlying vulval cells. It also induces the expression of a Rhomboid within the vulval cells [40]. Presumably, Rhomboid can then process ligand that is expressed within the vulval cells, thus amplifying the activation of the EGFR within the vulval cells, by autocrine signaling.

In *Drosophila* oogenesis, the EGF ligand Gurken, expressed in the egg, triggers EGFR on the adjacent follicle cells. In this case, a target gene for EGFR is *rhomboid-1*. The expression of Rhomboid-1 by the follicle cells coincides with the time when they become insulated from the egg, due to the secretion of the vitelline membrane. Once Rhomboid is expressed, the cells continue to trigger EGFR signaling by virtue of their capacity to process the ligand Spitz, which they express [41, 42]. This wave of EGFR activation triggers the formation of structures called dorsal appendages.

Perhaps the most complicated scenario of EGFR activation takes place in the developing eye. The primary cell for each of the 800 ommatidia is defined in an EGFR-independent manner as the R8 cell and expresses Rhomboid. The production of ligand by the R8 cell induces EGFR activation in the neighboring cells.

These cells then express Rhomboid to induce the next round of differentiation [31, 43]. Different photoreceptor cell fates are induced in each round, despite the common EGFR signal, due to a distinct combination of transcription factors within the receiving cells at each stage. In this type of sequential activation, the restricted range of the signal at each round is critical. It is determined by the expression of rhomboid-3 which has a dual ER and late compartment localization, thus restricting the amount of ligand that is trafficked and cleaved [31].

Restricting the amount of ligand that is secreted, and defining a transient time window of signaling by EGFR during eye development, poses a new challenge. Will the cells receiving the transient signal indeed be able of inducing the expression of relevant target genes in a sustained and reliable manner? Pointed is an ETS protein that represents the cardinal transcriptional response to EGFR signaling [44, 45]. Two protein forms, sharing the same ETS DNA-binding domain, are generated by the gene: PointedP2 is broadly expressed but is inactive. Phosphorylation by MAPK triggers its activity. PointedP1 is constitutively active, and its expression is triggered by EGFR activation. In some tissues, only one of the two forms is utilized, e.g., Pointed P1 in the embryonic ectoderm and PointedP2 in the embryonic mesoderm [44, 45]. Interestingly, in the developing eye, both forms are utilized and were shown to be essential for the recruitment of photoreceptor cells. It turns out that the two forms are utilized in a sequential manner. PointedP2 is transiently activated by EGFR/MAPK signaling. It then leads to the induction of PointedP1 expression. Once expressed, PointedP1 activity no longer relies on sustained activation of EGFR and is capable of inducing target genes as long as the protein persists [46]. PointedP1 does not induce its own expression such that its transcriptional activity eventually terminates.

3.6 Integral vs. Auxiliary Elements in the Signaling Pathway of RTKs

The genetic screens that identified the RTK signaling pathways indicated that at the organismal level, the pathways are linear. The fact that different mutations in the pathway gave rise to similar developmental phenotypes suggested that the core pathway is indeed operating as a linear module. However, the overlap between these linear pathways and general processes within the cell remained open. It is important to keep in mind that the screens uncovered mutations that gave rise to severe phenotypes and in this respect focused on elements that are integral components of the pathway. In addition, the genetic screens focused on discrete phenotypes that are characteristic for a given pathway. Genes that impinge on signaling of more than one pathway may give a phenotype that is too complex or severe to place within the known sets of embryonic developmental phenotypes.

The ability to carry out genome-wide RNAi screens in cell culture provided an opportunity to address these open issues. Different reporters for a signaling pathway, at the cell culture level, were devised and used to screen for genes that would alter the response. For RTK pathways, these reporters include intracellular localization of the EGF ligand, phosphorylation of MAPK, and induction of target genes [47, 48]. These screens are inherently different from the analysis of phenotypes in the whole organism, because they follow a cell-based output. They are quantitative, in that partial effects can also be scored. In addition, because the effect of RNAi is not complete, essential genes that impinge on several signaling processes can still be identified, by virtue of the effect they exert on the distinct pathway being scored, even when it is only partial.

The observation from a variety of comprehensive RNAi screens, carried out not only for RTK pathways but also for other developmental signaling pathways, is that a continuous range of effects is observed, with the number of genes identified being at least tenfold higher than the genes isolated by the genetic screens [49]. One category of genes to consider is genes that may function only in the context of a given pathway and comprise an important layer of quantitative modulation, such as positive or negative feedback loops. These genes are expected to display distinct phenotypes at the level of the whole organism, which would be compatible with the enhancement or suppression of activity of a distinct pathway. Elements like Argos or Sprouty, which comprise feedback loops, were indeed genetically identified. One example of a dedicated regulator that was identified is Small wing (PLC gamma), which has a scaffold activity that is independent of its enzymatic activity. In cases where the EGF ligand is cleaved in the ER, Small wing facilitates the retention of the cleaved ligand in the ER. Accordingly, in such tissues, the hyperactivation of EGFR signaling is observed in *small wing* mutants [48].

For the most part, the RNAi screens have not identified new integral elements in the pathways that were missed by the genetic screens. Rather, they have uncovered elements that impinge on the pathway and are tangential to it. They may modify the efficiency of pathway output in a given cellular context, but are not absolutely necessary for its activity. Since the signaling process is deeply embedded within the cell-biological features of the cell, we would expect signaling to be affected by modulations of the state of the cell. For example, the translocation of signaling components between cellular compartments may be influenced by accessibility of recycled membrane, availability of nuclear localization machinery, or efficiency of translation. Thus, the RNAi screens provide a glimpse into the way in which a developmental signaling pathway is embedded within the most basic properties of the cell.

3.7 Cell Biology of RTK Signaling During Development

Several examples have emerged for the way in which the basic features of the cell, its compartments and the cross talk between them, are utilized to coordinate RTK signaling at the level of the whole organism.

Ligand processing of EGF ligands in invertebrates follows a different strategy to the one used in vertebrate organisms and relies heavily on intracellular compartmentalization. As mentioned above, trapping the ligand precursor in the ER prevents it from reaching the plasma membrane and being cleaved by non-specific metalloproteases [27, 28]. It follows that the restricted expression of the ligand is not necessary, and in fact it is broadly expressed. The trafficking of the ligand precursor out of the ER relies on a chaperone protein, Star. Modulating the levels of active Star by proteolytic cleavage provides another tier of regulation [30]. Finally, only when the protease Rhomboid and the ligand precursor meet in a late secretory compartment do they yield productive cleavage. Variations in the intracellular localization of Rhomboid play an important role in this regulation [31].

An important role for intracellular trafficking was identified in a situation where spatial asymmetry in the activation is essential for guided migration of a group of cells, termed border cells. In the *Drosophila* female ovary, a cluster of eight cells migrate from the anterior part of the egg chamber, through the nurse cells, to a contact site with the oocyte. These cells are guided posteriorly by virtue of a restricted source for the ligands emanating from the egg, triggering the PVR and EGFR. However, since these ligands diffuse over long distances, the gradient at any given point is shallow, and amplification of the spatial cues is essential. Such an amplification is provided by endocytosis of PVR and EGFR and their selective reinsertion into the membrane at sites where a small bias in signaling is detected [50]. The E3 ubiquitin ligase Cbl, which plays an important role in tagging receptors for endocytosis, was thus shown to play an important role in generating spatial restriction of signaling.

3.8 Transcriptional Output of RTK Pathways

The transcriptional output of RTK signaling is mediated by proteins of the ETS family. Most prominent is the ETS-domain protein termed Pointed which has two isoforms generated by alternative splicing [45, 51]. Activation by MAP kinase activates each of the two forms in a different manner. The phosphorylation of PointedP2 converts an inactive protein to the active form. The mechanistic basis for activation by phosphorylation is not known and may involve stabilization, nuclear translocation, and exposure of the transcription activation or DNA-binding domains. The second isoform, PointedP1, is constitutively active even in the absence of MAPK. However, its expression is dependent on MAPK activity [44]. The transcription factor which responds to MAPK activity, to trigger PointedP1 expression, is not known.

Another twist in the regulation of ETS target genes was described. The ETS protein Yan contains an ETS DNA-binding domain, but is devoid of the transcription activation domain. Yan is also a target for MAPK phosphorylation, but in this case MAPK leads to its inactivation, by promoting nuclear exit and degradation [52]. The dual and opposite activity of MAPK on the activators and inhibitor may make the induction of ETS target genes more robust. A mechanism for generating transcriptional thresholds termed "zero-order hypersensitivity" has been proposed. In cases where the transcription factor, or the transcriptional repressor, is undergoing reversible phosphorylation, and the protein undergoing modification is in excess, an interesting paradigm stemming from the zero-order reaction takes place. Essentially, even small differences between the reversible phosphorylation and dephosphorylation reactions will lead to the complete accumulation of the protein in one form or another, generating a sharp threshold response [53].

3.9 Overlap Between RTKs and Other Signaling Pathways

As a rule, most of the integration between RTK pathways and other developmental signaling pathways takes place at the level of the regulatory regions of target genes. Thus, every relevant promoter is regulated according to a different set of rules that is applicable for the tissue and developmental timing. Typically, a responsive region will contain binding sites for tissue-specific transcription factors that will define and restrict the tissue context in which this promoter is activated. In addition, it will contain ETS binding sites, which define the cardinal transcriptional output of activated EGFR and FGFR. Finally, it will contain binding sites for transcription factors activated by other signaling pathways, such as Notch, BMP, or Wnt [54–56]. The organization of these sites within the context of a given gene will define whether two pathways are required simultaneously for activation or whether they function in an antagonistic manner. One way to achieve antagonism is by generating overlapping binding sites for the two pathways. The integration of signals at the level of target genes provides an enormous flexibility, in the sense that every gene can be designed to integrate the limited number of developmental signaling pathways, in a fashion that would be relevant for its activity. A large number of combinations are provided by this design.

In contrast to the convergence of regulation at the enhancer level, relatively few examples for horizontal cross talk between RTKs and other signaling pathways have been documented. Such interactions are not tailor-made for a given gene or setting and may represent a more "hardwired" solution. The complex of proteins comprised of Capicua, which contains a DNA-binding domain, and Groucho, which provides a repressor function, provides one such example. The phosphorylation of Capicua by MAPK gives rise to its nuclear exit and degradation [57, 58]. The phosphorylation of Groucho by MAPK reduces its biological activity in the repression of target gene expression [59]. In the case of the Torso pathway activated at the termini of the early *Drosophila* embryo, the neutralization of these two proteins in regions of high levels of RTK activation gives rise to the relief of the repression and induction of the terminal target genes. Groucho plays a prominent role as a general transcriptional repressor that can be recruited to the DNA by many proteins and was originally identified as the critical element in mediating Notch-dependent repression. The attenuation of Groucho activity by MAPK signaling will thus also impinge on signaling by the Notch pathway [60].

In conclusion, the study of RTKs and their signaling pathways in invertebrate organisms provided many novel insights into the architecture of the cascades and their integration with the cellular machinery of the cells in which they operate. The ability to follow these pathways and the diverse processes they regulate at the level of the whole organism has uncovered the logic and execution of these developmental switches.

Acknowledgments I am grateful to my lab members that have contributed throughout the years to deciphering many of the paradigms discussed here and to my colleagues for discussions. This work was funded by an ISF converging technologies grant. B.S. is an incumbent of the Hilda and Cecil Lewis chair in Molecular Genetics.

References

1. Livneh E, Glazer L, Segal D, Schlessinger J, Shilo BZ. The Drosophila EGF receptor gene homolog: conservation of both hormone binding and kinase domains. Cell. 1985;40(3):599–607. PubMed PMID: 2982499.
2. Wadsworth SC, Vincent 3rd WS, Bilodeau-Wentworth D. A Drosophila genomic sequence with homology to human epidermal growth factor receptor. Nature. 1985;314(6007):178–80. PubMed PMID: 2983232.
3. Beiman M, Shilo BZ, Volk T. Heartless, a Drosophila FGF receptor homolog, is essential for cell migration and establishment of several mesodermal lineages. Genes Dev. 1996;10(23):2993–3002. PubMed PMID: 8957000.
4. Gisselbrecht S, Skeath JB, Doe CQ, Michelson AM. Heartless encodes a fibroblast growth factor receptor (DFR1/DFGF-R2) involved in the directional migration of early mesodermal cells in the Drosophila embryo. Genes Dev. 1996;10(23):3003–17. PubMed PMID: 8957001.
5. Glazer L, Shilo BZ. The Drosophila FGF-R homolog is expressed in the embryonic tracheal system and appears to be required for directed tracheal cell extension. Genes Dev. 1991;5(4):697–705. PubMed PMID: 1849109.
6. Klambt C, Glazer L, Shilo BZ. Breathless, a Drosophila FGF receptor homolog, is essential for migration of tracheal and specific midline glial cells. Genes Dev. 1992;6(9):1668–78. PubMed PMID: 1325393.
7. Nishida Y, Hata M, Nishizuka Y, Rutter WJ, Ebina Y. Cloning of a Drosophila cDNA encoding a polypeptide similar to the human insulin receptor precursor. Biochem Biophys Res Commun. 1986;141(2):474–81. PubMed PMID: 3099787.
8. Petruzzelli L, Herrera R, Arenas-Garcia R, Fernandez R, Birnbaum MJ, Rosen OM. Isolation of a Drosophila genomic sequence homologous to the kinase domain of the human insulin receptor and detection of the phosphorylated Drosophila receptor with an anti-peptide antibody. Proc Natl Acad Sci U S A. 1986;83(13):4710–4. PubMed PMID: 3014506.
9. Sprenger F, Stevens LM, Nusslein-Volhard C. The Drosophila gene torso encodes a putative receptor tyrosine kinase. Nature. 1989;338(6215):478–83. PubMed PMID: 2927509.
10. Loren CE, Scully A, Grabbe C, Edeen PT, Thomas J, McKeown M, et al. Identification and characterization of DAlk: a novel Drosophila melanogaster RTK which drives ERK activation in vivo. Genes Cells. 2001;6(6):531–44. PubMed PMID: 11442633.
11. Boyle M, Nighorn A, Thomas JB. Drosophila Eph receptor guides specific axon branches of mushroom body neurons. Development. 2006;133(9):1845–54. PubMed PMID: 16613832.
12. Duchek P, Somogyi K, Jekely G, Beccari S, Rorth P. Guidance of cell migration by the Drosophila PDGF/VEGF receptor. Cell. 2001;107(1):17–26. PubMed PMID: 11595182.

13. Banerjee U, Renfranz PJ, Pollock JA, Benzer S. Molecular characterization and expression of sevenless, a gene involved in neuronal pattern formation in the Drosophila eye. Cell. 1987;49(2):281–91. PubMed PMID: 2882857.
14. Hafen E, Basler K, Edstroem JE, Rubin GM. Sevenless, a cell-specific homeotic gene of Drosophila, encodes a putative transmembrane receptor with a tyrosine kinase domain. Science. 1987;236(4797):55–63. PubMed PMID: 2882603.
15. Yoshikawa S, McKinnon RD, Kokel M, Thomas JB. Wnt-mediated axon guidance via the Drosophila Derailed receptor. Nature. 2003;422(6932):583–8. PubMed PMID: 12660735.
16. Wang S, Tsarouhas V, Xylourgidis N, Sabri N, Tiklova K, Nautiyal N, et al. The tyrosine kinase Stitcher activates Grainy head and epidermal wound healing in Drosophila. Nat Cell Biol. 2009;11(7):890–5. PubMed PMID: 19525935.
17. Beck G, Munno DW, Levy Z, Dissel HM, Van-Minnen J, Syed NI, et al. Neurotrophic activities of trk receptors conserved over 600 million years of evolution. J Neurobiol. 2004;60(1):12–20. PubMed PMID: 15188268.
18. Alvarado D, Klein DE, Lemmon MA. Structural basis for negative cooperativity in growth factor binding to an EGF receptor. Cell. 2010;142(4):568–79. PubMed PMID: 20723758.
19. Ambrosio L, Mahowald AP, Perrimon N. Requirement of the Drosophila raf homologue for torso function. Nature. 1989;342(6247):288–91. PubMed PMID: 2554148.
20. Brand AH, Perrimon N. Raf acts downstream of the EGF receptor to determine dorsoventral polarity during Drosophila oogenesis. Genes Dev. 1994;8(5):629–39. PubMed PMID: 7926754.
21. Dossenbach C, Rock S, Affolter M. Specificity of FGF signaling in cell migration in Drosophila. Development. 2001;128(22):4563–72. PubMed PMID: 11714681.
22. Vincent S, Wilson R, Coelho C, Affolter M, Leptin M. The Drosophila protein Dof is specifically required for FGF signaling. Mol Cell. 1998;2(4):515–25. PubMed PMID: 9809073.
23. Nagaraj R, Banerjee U. The little R cell that could. Int J Dev Biol. 2004;48(8–9):755–60. PubMed PMID: 15558468.
24. Gabay L, Seger R, Shilo BZ. MAP kinase in situ activation atlas during Drosophila embryogenesis. Development. 1997;124(18):3535–41. PubMed PMID: 9342046.
25. Gabay L, Seger R, Shilo BZ. In situ activation pattern of Drosophila EGF receptor pathway during development. Science. 1997;277(5329):1103–6. PubMed PMID: 9262480.
26. Bill A, Schmitz A, Albertoni B, Song JN, Heukamp LC, Walrafen D, et al. Cytohesins are cytoplasmic ErbB receptor activators. Cell. 2010;143(2):201–11. PubMed PMID: 20946980.
27. Lee JR, Urban S, Garvey CF, Freeman M. Regulated intracellular ligand transport and proteolysis control EGF signal activation in Drosophila. Cell. 2001;107(2):161–71. PubMed PMID: 11672524.
28. Tsruya R, Schlesinger A, Reich A, Gabay L, Sapir A, Shilo BZ. Intracellular trafficking by Star regulates cleavage of the Drosophila EGF receptor ligand Spitz. Genes Dev. 2002;16(2):222–34. PubMed PMID: 11799065.
29. Urban S, Lee JR, Freeman M. Drosophila rhomboid-1 defines a family of putative intramembrane serine proteases. Cell. 2001;107(2):173–82. PubMed PMID: 11672525.
30. Tsruya R, Wojtalla A, Carmon S, Yogev S, Reich A, Bibi E, et al. Rhomboid cleaves Star to regulate the levels of secreted Spitz. EMBO J. 2007;26(5):1211–20. PubMed PMID: 17304216.
31. Yogev S, Schejter ED, Shilo BZ. Drosophila EGFR signalling is modulated by differential compartmentalization of Rhomboid intramembrane proteases. EMBO J. 2008;27(8):1219–30. PubMed PMID: 18369317.
32. Golembo M, Schweitzer R, Freeman M, Shilo BZ. Argos transcription is induced by the Drosophila EGF receptor pathway to form an inhibitory feedback loop. Development. 1996;122(1):223–30. PubMed PMID: 8565833.
33. Klein DE, Nappi VM, Reeves GT, Shvartsman SY, Lemmon MA. Argos inhibits epidermal growth factor receptor signalling by ligand sequestration. Nature. 2004;430(7003):1040–4. PubMed PMID: 15329724.
34. Ghiglione C, Carraway 3rd KL, Amundadottir LT, Boswell RE, Perrimon N, Duffy JB. The transmembrane molecule kekkon 1 acts in a feedback loop to negatively regulate the activity

of the Drosophila EGF receptor during oogenesis. Cell. 1999;96(6):847–56. PubMed PMID: 10102272.
35. Casci T, Vinos J, Freeman M. Sprouty, an intracellular inhibitor of Ras signaling. Cell. 1999;96(5):655–65. PubMed PMID: 10089881.
36. Kramer S, Okabe M, Hacohen N, Krasnow MA, Hiromi Y. Sprouty: a common antagonist of FGF and EGF signaling pathways in Drosophila. Development. 1999;126(11):2515–25. PubMed PMID: 10226010.
37. Reich A, Sapir A, Shilo B. Sprouty is a general inhibitor of receptor tyrosine kinase signaling. Development. 1999;126(18):4139–47. PubMed PMID: 10457022.
38. Ohshiro T, Emori Y, Saigo K. Ligand-dependent activation of breathless FGF receptor gene in Drosophila developing trachea. Mech Dev. 2002;114(1–2):3–11. PubMed PMID: 12175485.
39. Shilo BZ. Regulating the dynamics of EGF receptor signaling in space and time. Development. 2005;132(18):4017–27. PubMed PMID: 16123311.
40. Dutt A, Canevascini S, Froehli-Hoier E, Hajnal A. EGF signal propagation during C. elegans vulval development mediated by ROM-1 rhomboid. PLoS Biol. 2004;2(11):e334.
41. Sapir A, Schweitzer R, Shilo BZ. Sequential activation of the EGF receptor pathway during Drosophila oogenesis establishes the dorsoventral axis. Development. 1998;125(2):191–200. PubMed PMID: 9486793.
42. Wasserman JD, Freeman M. An autoregulatory cascade of EGF receptor signaling patterns the Drosophila egg. Cell. 1998;95(3):355–64. PubMed PMID: 9814706.
43. Baonza A, Casci T, Freeman M. A primary role for the epidermal growth factor receptor in ommatidial spacing in the Drosophila eye. Curr Biol. 2001;11(6):396–404. PubMed PMID: 11301250.
44. Gabay L, Scholz H, Golembo M, Klaes A, Shilo BZ, Klambt C. EGF receptor signaling induces pointed P1 transcription and inactivates Yan protein in the Drosophila embryonic ventral ectoderm. Development. 1996;122(11):3355–62. PubMed PMID: 8951052.
45. Klambt C. The Drosophila gene pointed encodes two ETS-like proteins which are involved in the development of the midline glial cells. Development. 1993;117(1):163–76. PubMed PMID: 8223245.
46. Shwartz A, Yogev S, Schejter ED, Shilo BZ. Sequential activation of ETS proteins provides a sustained transcriptional response to EGFR signaling. Development. 2013;140(13):2746–54. PubMed PMID: 23757412.
47. Friedman A, Perrimon N. A functional RNAi screen for regulators of receptor tyrosine kinase and ERK signalling. Nature. 2006;444(7116):230–4. PubMed PMID: 17086199.
48. Schlesinger A, Kiger A, Perrimon N, Shilo BZ. Small wing PLCgamma is required for ER retention of cleaved Spitz during eye development in Drosophila. Dev Cell. 2004;7(4):535–45. PubMed PMID: 15469842.
49. Friedman A, Perrimon N. Genetic screening for signal transduction in the era of network biology. Cell. 2007;128(2):225–31. PubMed PMID: 17254958.
50. Jekely G, Sung HH, Luque CM, Rorth P. Regulators of endocytosis maintain localized receptor tyrosine kinase signaling in guided migration. Dev Cell. 2005;9(2):197–207. PubMed PMID: 16054027.
51. O'Neill EM, Rebay I, Tjian R, Rubin GM. The activities of two Ets-related transcription factors required for Drosophila eye development are modulated by the Ras/MAPK pathway. Cell. 1994;78(1):137–47. PubMed PMID: 8033205.
52. Rebay I, Rubin GM. Yan functions as a general inhibitor of differentiation and is negatively regulated by activation of the Ras1/MAPK pathway. Cell. 1995;81(6):857–66. PubMed PMID: 7781063.
53. Melen GJ, Levy S, Barkai N, Shilo BZ. Threshold responses to morphogen gradients by zero-order ultrasensitivity. Mol Syst Biol. 2005;1:2005.0028. PubMed PMID: 16729063.
54. Flores GV, Duan H, Yan H, Nagaraj R, Fu W, Zou Y, et al. Combinatorial signaling in the specification of unique cell fates. Cell. 2000;103(1):75–85. PubMed PMID: 11051549.
55. Halfon MS, Carmena A, Gisselbrecht S, Sackerson CM, Jimenez F, Baylies MK, et al. Ras pathway specificity is determined by the integration of multiple signal-activated and tissue-restricted transcription factors. Cell. 2000;103(1):63–74. PubMed PMID: 11051548.

56. Xu C, Kauffmann RC, Zhang J, Kladny S, Carthew RW. Overlapping activators and repressors delimit transcriptional response to receptor tyrosine kinase signals in the Drosophila eye. Cell. 2000;103(1):87–97. PubMed PMID: 11051550.
57. Jimenez G, Guichet A, Ephrussi A, Casanova J. Relief of gene repression by torso RTK signaling: role of capicua in Drosophila terminal and dorsoventral patterning. Genes Dev. 2000;14(2):224–31. PubMed PMID: 10652276.
58. Roch F, Jimenez G, Casanova J. EGFR signalling inhibits Capicua-dependent repression during specification of Drosophila wing veins. Development. 2002;129(4):993–1002. PubMed PMID: 11861482.
59. Hasson P, Egoz N, Winkler C, Volohonsky G, Jia S, Dinur T, et al. EGFR signaling attenuates Groucho-dependent repression to antagonize Notch transcriptional output. Nat Genet. 2005;37(1):101–5. PubMed PMID: 15592470.
60. Hasson P, Paroush Z. Crosstalk between the EGFR and other signalling pathways at the level of the global transcriptional corepressor Groucho/TLE. Br J Cancer. 2007;96(Suppl):R21–5. PubMed PMID: 17393581.

Chapter 4
Receptor Tyrosine Kinase Signal Transduction and the Molecular Basis of Signalling Specificity

Pierre De Meyts

4.1 Introduction

The field of growth factor research was arguably kicked off by the discovery of nerve growth factor (NGF) in 1953 by Rita Levi-Montalcini and Stanley Cohen [1] and epidermal growth factor (EGF) in 1962 by Stanley Cohen [2], winning both investigators the Nobel Prize in Physiology or Medicine in 1986. While biochemical evidence had demonstrated that protein tyrosine kinase activity was associated with the cell surface receptors for EGF and insulin, the cloning and determination of the cDNA sequences of the EGF receptor in 1984 [3], shortly followed by those of the insulin and IGF-I receptors [4–6], CSF-1 receptor [7] and PDGF receptors [8], showed that the tyrosine kinase activity was an intrinsic part of these receptor intracellular domains and that they formed a family of related proteins. These receptors were therefore named receptor tyrosine kinases (RTKs). All members of the RTK superfamily were subsequently cloned and characterised. In humans, they comprise 59 members distributed into 20 subfamilies according to the diverse architecture of their extracellular domains which determines ligand specificity (see Chap. 1).

RTKs control critical cellular processes such as proliferation and differentiation, survival and longevity and metabolism and migration. The structural biology and cellular signalling mechanisms of the RTKs have been the subject of numerous exhaustive reviews by prominent leaders in the field over the past 30 years [9–23]. In this brief review, I will focus on general organisational principles found in RTK signalling, using most often insulin, IGF or EGF receptors as prototypical examples, since the details specific to each subfamily members will be reviewed in separate chapters.

P. De Meyts, M.D., Ph.D. (✉)
De Meyts R&D Consulting, Avenue Reine Astrid 42, 1950 Kraainem, Belgium
e-mail: pierre.demeyts@gmail.com

D.L. Wheeler, Y. Yarden (eds.), *Receptor Tyrosine Kinases: Structure, Functions and Role in Human Disease*, DOI 10.1007/978-1-4939-2053-2_4

4.2 RTKs Assemble into Dimeric Allosteric Enzymes

RTKs are made of a single polypeptide chain with a single transmembrane (TM) domain. The RTK monomers' extracellular domain of several hundred amino acids, which contain the ligand-binding sites, is made of an arrangement of various structural motifs and features a distinctive pattern of Cys residues, as well as N-linked and O-linked sugars (see Chap. 8). The TM domain consists of a stretch of hydrophobic residues that are followed by several basic residues [12]. The intracellular part contains a juxtamembrane domain, the catalytic domain which is about 250 residues in length, excluding inserts present in some RTKs, and a C-terminal tail that varies from a few residues up to 200 residues [12].

It was initially proposed that RTKs are activated by ligand-induced dimerisation, allowing intermolecular cross-phosphorylation [24]. Ligand-induced dimerisation was also proposed for the growth hormone receptor [25] (Fig. 4.1a), a member of the cytokine receptor superfamily that does not have intrinsic tyrosine kinase activity but couples to JAK kinases. The simplest formulation of this concept is stated in

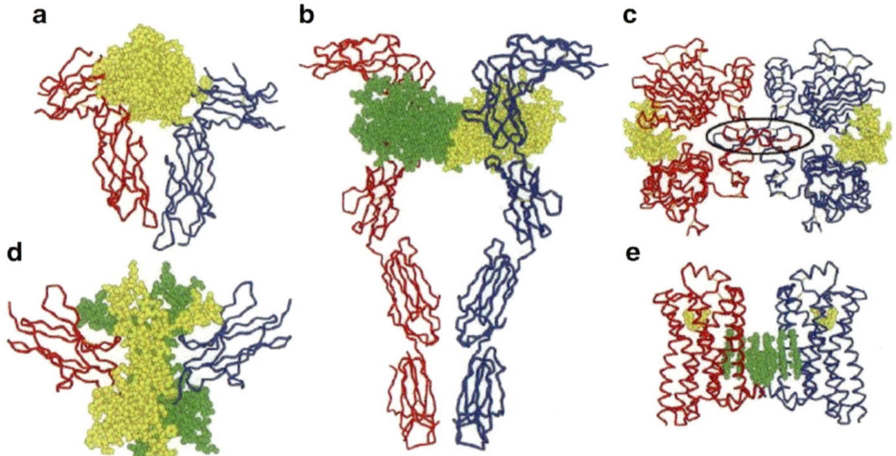

Fig. 4.1 Examples of dimeric membrane-receptor complexes. A sample of the various architectural modes of dimerization found in the crystallographic structures of liganded extracellular domains of three classes of membrane receptors: cytokine receptors (**a**), RTKs (**b–d**) and GPCRs (**e**). Receptor structures are shown as tubes (one monomer *red*, one *blue*), and the ligands are shown as CPK spheres (which are *yellow* when monomeric and *yellow* and *green* when dimeric) (**a**). Structure of human growth hormone bound to two growth-hormone receptor ectodomains (PDB code: 3HHR) [25]. (**b**) Structure of a dimeric SCF bound to two KIT ectodomains (PDB code: 2E9W) [27]. (**c**) Structure of two monomeric EGFs bound to two EGF-receptor ectodomain molecules (PDB code: 1 IVO), The black ellipse denotes the contact zone between the two autoinhibitory hairpin loops of the EGFR [28]. (**d**) Structure of a dimeric NGF bound to two TrkA receptor ligand-binding domains (PDB code: 1www) [26]. (**e**) Structure of the beta2-adrenergic receptor embedded in a lipid membrane (not shown) and bound to a diffusible ligand (*yellow*), with cholesterol and palmitic acid (*green*) between the two receptor molecules (PDB code: 2RH1) [29]. From ref. [18], used with permission

ref. [12]: "*After ligand binding*, the receptor dimerizes. This is presumed to occur by a ligand-induced conformational change in the external domain that results in dimerization of receptors". The fact that the insulin and IGF-I receptors occur naturally as covalently bound dimers in the absence of ligand [18] provides strong evidence that the "ligand-induced dimerisation" paradigm is not universal and that at least in some cases, the receptor may dimerise *before ligand binding* and that the ligand may stabilise the dimer by shifting a pre-existing monomer-dimer equilibrium. This issue is addressed in more detail in ref. [20].

Crystal structures of several liganded RTKs have revealed various arrangements explaining the stabilisation of the dimeric structure [20]. In some cases, the ligands are itself a dimer: NGF (Fig. 4.1d), stem cell factor (SCF), ligand of the c-KIT receptor (Fig. 4.1b), PDGF, VEGF and the ligands for Axl, Tie2 and Eph receptors (reviewed in ref. [20]). In the case of monomeric ligands, like EGF, insulin and IGF-I, the ligand is postulated to have two distinct binding sites for the receptor that cross-link two separate sites on the receptor, either in the same subunit (EGFR, Fig. 4.1c) or between the two distinct subunits (insulin and IGF-I receptors) ([18, 19], see Chap. 8) (Fig. 4.2). In the case of the FGF receptor, an accessory molecule (heparin) stabilises the FGF-receptor complex, for which two different X-ray structures have been observed (see ref. [20] for review).

It should be noted that RTKs also often form heterodimers within the same subfamilies.

Whatever the various individual modalities, it has become clear that a liganded RTK functions as an allosteric enzyme, with the dimeric structure providing a structural basis for cooperative ligand interactions as well as kinase activation by intermolecular transphosphorylation. Much structural work remains to be done to fully understand the molecular basis for transmembrane allosteric activation, since so far the extracellular and intracellular domains exist only as separate crystal structures.

4.3 Cooperative Interactions in Ligand Binding to RTKs and Other Surface Receptors

The dimeric nature of liganded RTKs provides a structural basis for cooperative interactions. Multimeric enzymes (and other proteins) almost always display either positive cooperativity, where binding of one ligand molecule enhances the affinity for further binding, or negative cooperativity, where the affinity for further binding decreases after the first ligand molecule is bound. The celebrated Monod, Wyman and Changeux concerted model of allosteric interactions published nearly 50 years ago [30] only allows positive cooperativity due to the conservation of symmetry (as observed in oxygen binding to haemoglobin), while the "induced fit" models of Koshand, Némethy and Filmer [31] allow for both positive and negative cooperativity (due to ligand-induced asymmetry). Many enzymes were subsequently found to display negative cooperativity [32, 33].

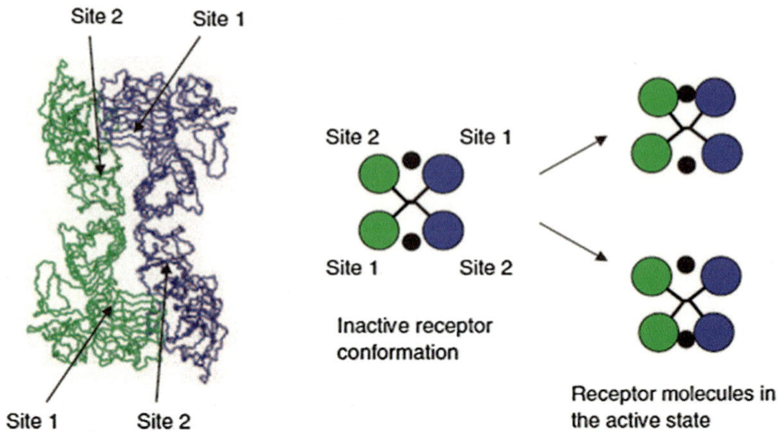

Fig. 4.2 The harmonic oscillator model of the insulin receptor. The crystal structure of the unliganded insulin receptor dimer viewed from the top is shown on the *left*, with the approximate locations of the two pairs of binding sites 1 and 2 shown. See refs. [55] and [56]. PDB code: 2DTG. This conformation is shown in a schematic way on the *right*, as oscillating between two possible conformations stabilised by bivalent binding of an insulin molecule depicted as a *black dot*. From ref. [54], used with permission

The occurrence of negative cooperativity in cell surface receptors was first described by De Meyts et al. over 40 years ago [34] for the insulin receptor (now known to be an RTK) using radioligand-binding assays, before any information on the structure of receptors was available. These authors demonstrated that the curvilinear Scatchard plots for insulin binding were explained by negatively cooperative site-site interactions rather than by pre-existing binding sites heterogeneity, by using an original experimental design where dissociation of a pre-bound tracer concentration of radioiodinated ligand was studied in an "infinite dilution" both in the absence and presence of a saturating concentration of unlabelled ligand. The observation that unlabelled insulin markedly accelerated the dissociation of the labelled insulin became the hallmark of negative cooperativity and was subsequently applied to many other receptors, including the related IGF-I receptor [35].

Shortly after the insulin receptor, the NGF receptor (now also known to be an RTK: Trk A, B and C) was also found to display accelerated dissociation by unlabelled ligand in an infinite dilution and proposed to show negative cooperativity [36]. The structural basis of this property remains unexplained in the context of the interactions of NGF and the three Trks with the common p75 receptor [37] and deserves a closer look.

Another RTK, the EGF receptor, was shown many years ago to display curvilinear Scatchard plots and to exhibit ligand-induced accelerated dissociation in the "infinite dilution" set-up of De Meyts et al. [38–40], attributed in the last two papers to negative cooperativity. In a review published in 1976, I proposed that negative cooperativity was a rather common property of cell surface receptors including the EGF receptor [41]. Yet this possibility was largely ignored by the EGF receptor community until 2008, when Macdonald and Pike rediscovered that EGF accelerates the dissociation of ^{125}-I EGF in an infinite dilution and attributed it again to negative cooperativity [42]. Moreover, by studying the EGF-binding properties in cells expressing varying levels of EGF receptors, they showed that EGF induces negative cooperativity in a system where EGF receptors bind to a pre-existing monomer-dimer equilibrium, as discussed in the previous section and as proposed earlier by Wofsy et al. [43].

A structural basis for negative cooperativity in the EGF receptor was not obvious from the crystal structure of the human EGF receptor ectodomain which showed a symmetrical dimeric arrangement with one EGF molecule bivalently bound to each subunit (Fig. 4.1 c) (reviewed in ref. [23]). It should be noted that the purified human insulin receptor ectodomain also lost its negative cooperativity [44]. Recent studies of the *D. melanogaster* EGF receptor provided a structural view as to how negative cooperativity arises in a member of the ErbB family [45, 46]. Firstly this EGF receptor crystallises as a dimer in the unliganded form, again supporting the pre-existing monomer-dimer equilibrium model. Secondly, it retains negative cooperativity in ligand binding to the soluble extracellular domain. Finally, it can be crystallised as an asymmetric singly ligated extracellular dimer, showing a classical half-of-the-sites negative cooperativity (reviewed in ref. [23]).

Dimerisation and negative cooperativity are not the apanage of RTKs [19]. Beta-adrenergic receptors (later identified as G protein-coupled receptors or GPCRs) were shown shortly after the insulin receptor to exhibit negative cooperativity as evidenced by ligand-accelerated dissociation kinetics [47] and were later shown to be dimeric (Fig. 4.1e). TSH receptors were also shown in 1976 by Len Kohn and colleagues at the NIH to show ligand-accelerated dissociation and negative cooperativity (quoted in ref. [41]), as rediscovered nearly 30 years later [48], also for FSH and LH receptors. More GPCRs were shown to have negative cooperativity in the context of GPCR dimers [19, 49–52].

The physiological relevance of negative cooperativity has been postulated to result from the spreading of the binding curve over several log units, thus providing exquisite sensitivity of the system to low ligand concentrations while preserving further stimulation at higher concentrations. Lemmon et al. [23] have suggested that negative cooperativity provides a potential mechanism for graded activation of EGF receptor in response to a concentration gradient of activating ligand (as morphogen), e.g. in *D. melanogaster*. It is also possible that the conformational change that results in ligand-induced asymmetry is somewhat linked to the transmission of signal from the extracellular part of the receptor to the transmembrane and intracellular domains, as shown in the glycoprotein receptors, where a direct relation was demonstrated between the conformational changes associated with the activation of the transmembrane domain and the allosteric behaviour of the receptor dimers, by studying activating receptor mutations [53].

A structure-based mathematical model of the allosteric behaviour or the insulin and IGF-I receptors, based on the alternative bivalent cross-linking model proposed by De Meyts in 1994 [44], was developed by Kiselyov et al. [54] by assuming that the insulin and IGF-I receptors behave as a harmonic oscillator (Fig. 4.2).

The interactions among all possible receptor intermediaries can be modelled using only 5 parameters and 33 differential equations (plus two for the small amount of endocytosis and exocytosis) and the model accurately reproduces for the first time all the complex kinetic properties of the receptor and provide unique and robust estimates of the kinetic parameters. Other RTKs like the EGF receptor should be amenable to a similar treatment.

Interestingly, while the receptors of the growth hormone/erythropoietin/cytokine superfamily also assemble as liganded dimers (Fig. 4.1a), negative cooperativity has not been reported in this class of receptors, as shown for the growth hormone receptor by De Meyts et al. in 1973 [34], likely due to the binding mechanism that involves a single ligand molecule.

4.4 Allosteric Activation of RTKs Intracellular Kinase Domains

How the intracellular tyrosine kinase domains (TKD) of dimerised ligand-bound RTKs are activated following ligand binding has been resolved for several RTKs, namely, the insulin and EGF receptors as well as KIT and FGF receptors, and the mechanisms involved are surprisingly different (for review see refs. [17, 20]). All TKDs have a small N-terminal lobe made of beta sheets and a larger C-terminal lobe made of alpha helices (Fig. 4.3), with the active site in the cleft in between. The crystal structures of all activated TKDs are very similar [57] (Fig. 4.3a). By contrast, the structures of inactive TKDs show considerable plasticity reflecting the diversity in their regulatory mechanisms [20, 57]. Each TKD is *cis*-autoinhibited by a unique set of intramolecular interactions specific for a given RTK. Ligand binding releases this *cis*-autoinhibition and triggers RTK activation [20]. The fact that the inactive insulin (and IGF-I) receptor is a constitutive dimer suggests that lateral juxtaposition of the TKDs in the dimer is not sufficient for full kinase activation. The unliganded insulin receptor has basal kinase activity, but ligand binding causes a 36-fold increase in the rate constant for the chemical step of the enzyme-catalysed reaction, a fivefold increase in the affinity for MgATP and an eightfold increase in the affinity for peptide substrate [58]. Cobb et al. [59] have reported a 200-fold increase in enzyme activity upon ligand binding. This clearly shows that kinase activation requires more than the lateral effect of dimerisation and that ligand binding must produce an additional "vertical (allosteric) kick", for which the structural details are missing. Ward et al. [56] have suggested a "yo-yo-like" vertical motion of the TKDs of the insulin receptor.

The mechanism of *cis*-autoinhibition differs among RTKs. In the case of the insulin and FGF receptors, the TKD is inhibited by its own activation loop which moves out of the cleft when phosphorylated, adopting the typical "active" configuration found in all TKDs (Fig. 4.3a) [20]. In the case of other RTKs such as KIT (Fig. 4.3a), MuSK, Flt3 and Eph family RTKs, there is a "juxtamembrane autoinhibition" in which sequences of the juxtamembrane region make extensive

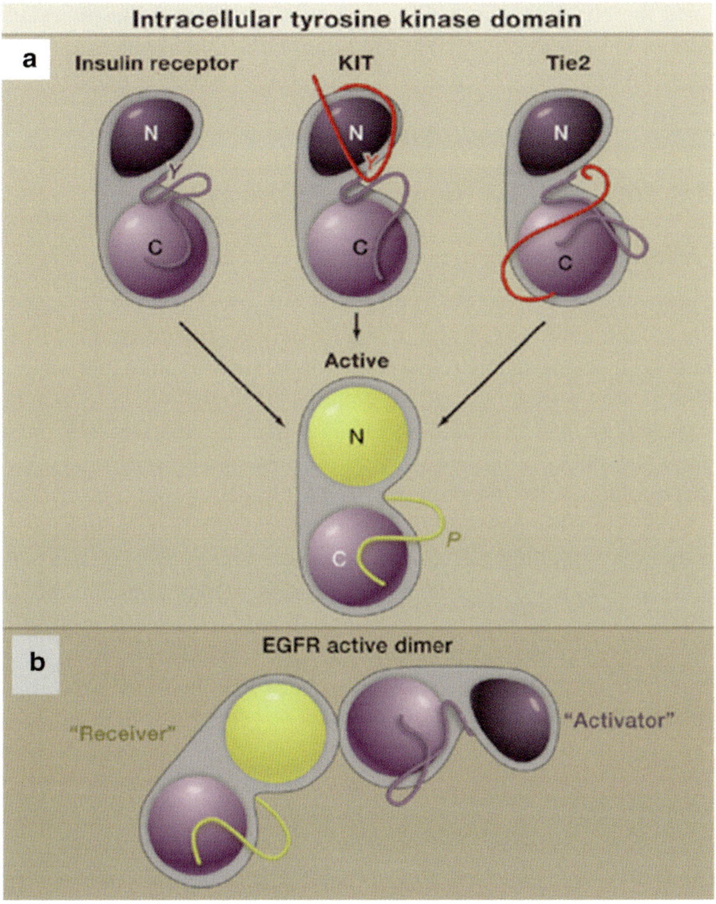

Fig. 4.3 Activation of the intracellular TKD of RTKs. Activation mechanism of the insulin receptor, KIT receptor and Tie2 receptor (**a**) and of the EGF receptor (**b**). See ref. [20] for detailed explanation. From ref. [20], used with permission

contacts with several parts of the TKD including the activation loop (Fig. 4.3a). In the case of Tie2 (Fig. 4.3a), it is a region of the C-terminal tail that contains tyrosine autophosphorylation sites that blocks substrate access to the active site. For a more detailed description of these inhibitory and activation mechanisms, see refs. [20] and [57].

While for most RTKs, activation requires *trans*phosphorylation of tyrosines in the activation loop, juxtamembrane segment and/or the C-terminal region, the EGF receptor/ErbB family [60] and Ret [61] are notable exceptions and do not require transphosphorylation to be activated. The EGF receptors form upon ligand binding an asymmetric dimer (Fig. 4.3b) in which the C-lobe of one TKD, called the "activator", makes intimate contacts with the N-lobe of the second TKD, called the "receiver" which adopts the characteristic active configuration without phosphorylation of its activation loop. This allosteric mechanism is similar to that observed for activation of cyclin-dependent kinases by cyclins [60]. See ref. [20] for more detailed discussion.

4.5 RTK Intracellular Signalling Pathways

Following ligand binding, the RTK's TKD has been enabled by the mechanisms described above to phosphorylate intracellular substrates. In fact the first such substrate is the receptor itself, with the phosphorylation of additional tyrosyl residues in other parts of the cytoplasmic region rather than the activation loop, which creates binding sites for downstream signalling proteins that contain SH2 (Src-homology 2) and PTB (phosphotyrosine-binding) domains that are recruited to the receptor, allowing their phosphorylation. SH2-containing proteins comprise enzymes, adaptors and docking proteins, scaffold proteins, signal regulators and transcription factors [62, 63]. Some signalling proteins also use a membrane-targeting site, the PH (pleckstrin homology) domain. This recruitment results in the assembly of a signal transduction particle that makes the activated RTK a node in a complex signalling network that transmits information from the exterior to the interior of the cell [20]. Most RTKs bind signalling substrates directly to tyrosine-phosphorylated residues in the cytoplasmic domain of the RTK, while a few like the insulin and IGF-I receptors recruit and phosphorylate docking proteins like IRS (insulin receptor substrate) 1-4, Shc and Gab (Grb2-associated binder) 1, while the FGF and NGF receptors recruit FRS2 [20]. Gab1 is recruited by multiple RTKs while IRS and FRS2 are more restricted to particular subsets of RTKs [20].

I will not attempt here an encyclopedic description of the complex signalling networks that are appended to the various RTK subfamilies. With many specific variations, most utilise overlapping pathways that are centred around the canonical Grb2/Sos/Ras/Raf/MEK/ERK or MAP kinase [64] and the PI 3K/PDK/Akt [65] pathways. A "generic" skeleton of the RTK signalling network is shown in Fig. 4.4. Other key downstream signalling proteins are the enzymes PLC (phospholipase C) γ and the isoforms of PKC (protein kinase C), p90RSK, mTOR (target of rapamycin) and GSK (glycogen synthase kinase) 3. Protein phosphatases (PTP) like PTEN and PP2A are constitutively active and play a major role in signal control and termination. FoxO is a transcription factor that is a substrate of Akt and plays a major role in cellular metabolism, differentiation and transformation and cell cycle regulation [66]. Phosphorylation by Akt results in nuclear exclusion and inhibition of function. Since Akt is activated by many RTKs, one would expect FoxO to be widely regulated by RTKs, but its function has mostly been studied to my knowledge in the context of insulin and IGF-I signalling, with the exception of a study showing inhibition by PDGF and FGF [67].

The simplified diagram shown in Fig. 4.4 gives the impression of independent linear signalling cascades as suggested in early work [20], but the RTK signalling networks proved to be much more complex and intertwined. As an example, the

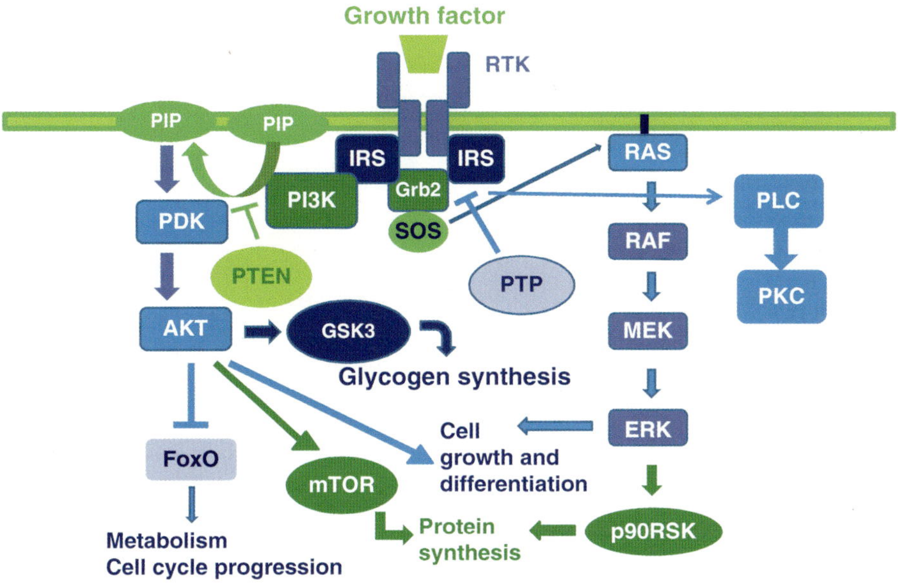

Fig. 4.4 A "generic" RTK signal transduction network. This is a very schematic representation of the activation of the canonical PI3K and ERK/MAPK pathways by RTKs, showing some of the main downstream targets. See text for explanation. IRS (insulin receptor substrate) is a docking protein used by the insulin and IGF-I receptors; other RTKs are able to bind Grb2/Sos and PI3K directly to phosphotyrosyl residues on the cytoplasmic part of the receptor. Negative feedback by protein tyrosine phosphatases (PTP) and the lipid phosphatase PTEN are shown

insulin receptor signalling network is shown in Fig. 4.5; for detailed description, see refs. [68–70]. A remarkably detailed annotated map of the EGF receptor signalling pathways from Hiroaki Kitano's group can be found in ref. [71].

4.6 The Paradox of Signalling Specificity

An important conundrum is how different RTKs elicit distinct physiological cellular responses (e.g. proliferation versus differentiation, metabolic versus mitogenic effects, etc.) while engaging overlapping and similar complements of receptor-proximal components in the cellular signalling network [20]. This is illustrated for the insulin versus IGF-I RTKs in the cartoon of Fig. 4.6.

The insulin receptor (IR) and the IGF-I receptor (IGF-IR) utilise common signalling pathways (Fig. 4.5) to mediate a broad spectrum of "metabolic" and "mitogenic" responses [68–70, 72]. This signalling network is also engaged by other receptor tyrosine kinases (RTKs).

That insulin and IGF-I receptors have distinct physiological functions is exemplified by the distinct phenotypes of IR versus IGF-IR knockout (KO) mice [73].

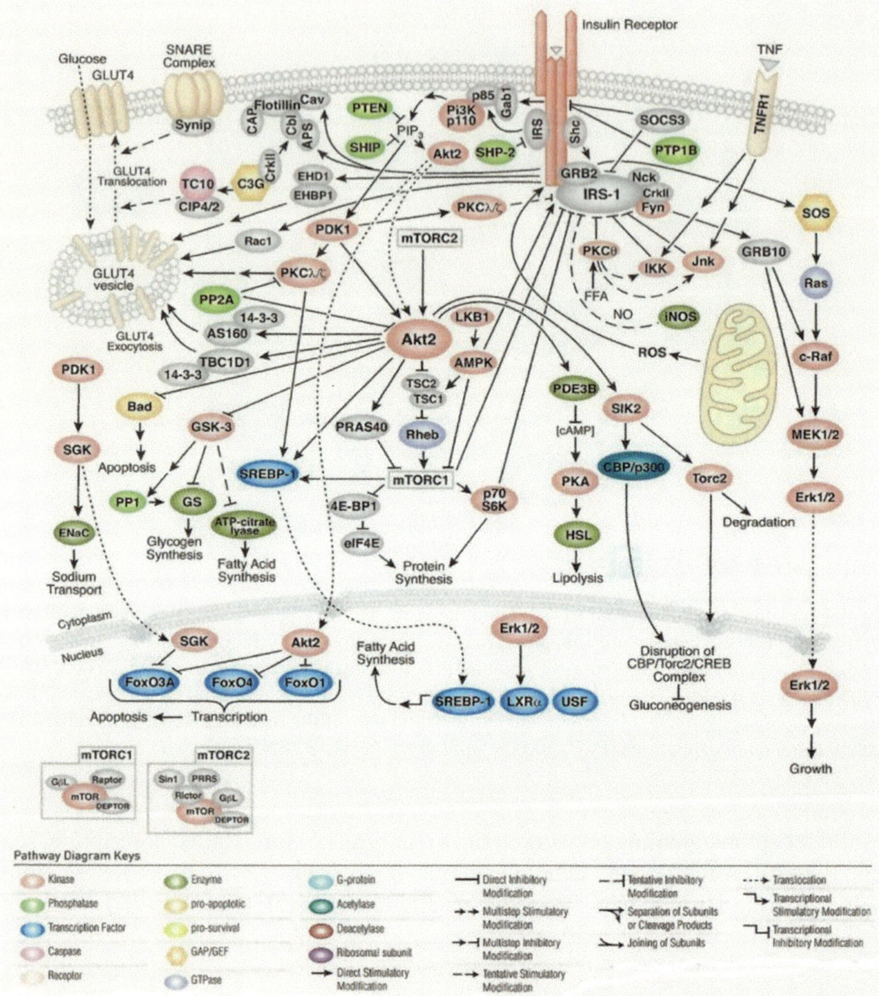

Fig. 4.5 The insulin receptor signal transduction network. Detailed representation of the insulin RTK signal transduction network. See refs. [68–70] for detailed explanation. One of the ultimate metabolic outcomes of this signalling network, the translocation of GLUT4 glucose transporters to the cell membrane of target cells, is shown in the *upper left corner*. Illustration reproduced courtesy of Cell Signaling Technology, Inc. (www.cellsignal.com)

IR KO mice are born with almost normal size but die within a few days from acute ketoacidosis, while IGF-IR KO mice are born small and die quickly from asphyxia due to underdevelopment of thoracic muscles. This downstream specificity is not inherent to the receptor structures since it has been shown that the IR and the IGF-IR can exert each other's function in cases where the other receptor is absent. Thus, the IGF-IR can induce stimulation of glucose transport and glycogen syn-

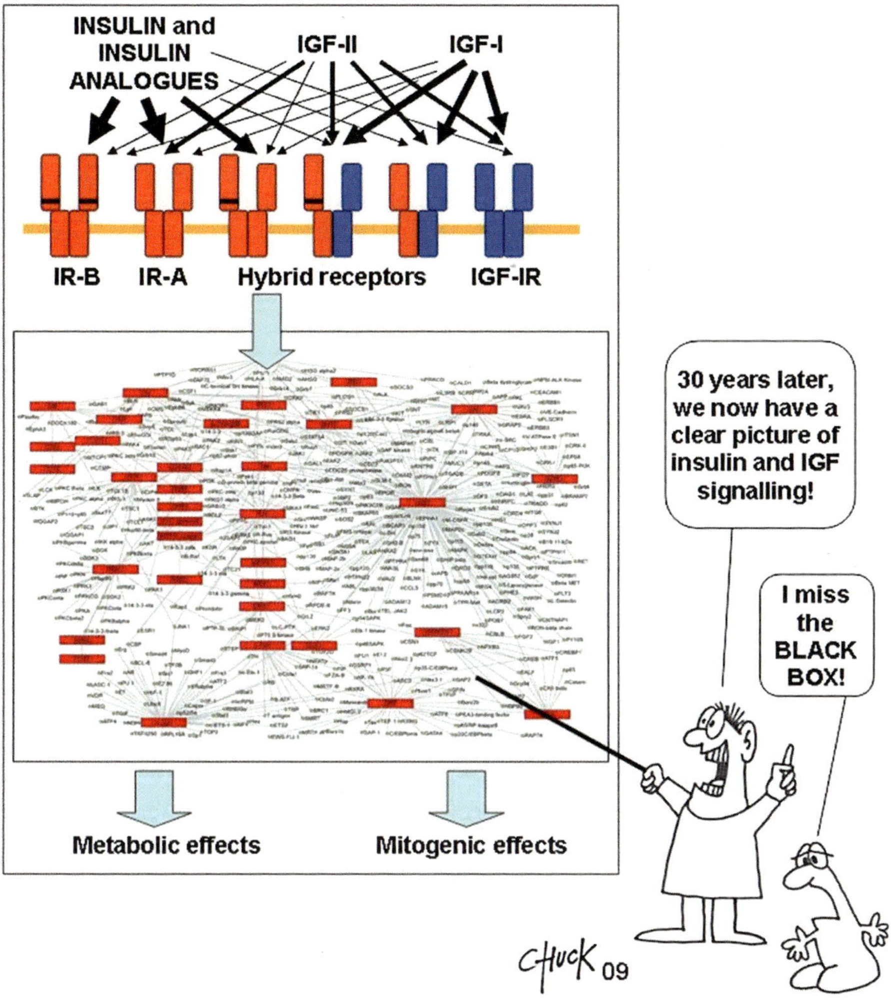

Fig. 4.6 The paradox of insulin versus IGF-I signalling specificity. This cartoon illustrates the conundrum of understanding how the complex array of ligands and receptors (and receptor isoforms and hybrids) for insulin, IGF-I and IGF-II generate specific mitogenic and metabolic responses through a shared signalling network

thesis, typical insulin metabolic effects, in fibroblasts from IR KO mice [74]. Reciprocally, the IR stimulates thymidine incorporation and cell growth in a T-cell lymphoma line devoid of IGF-I receptors [75].

There is therefore a paradox in having specific biological endpoints through ligand-specific receptors, while the downstream signalling networks appear to be largely overlapping [72, 76, 77]. Unravelling this complexity is complicated by the fact that most cells contain both IGF-I and insulin receptors (two isoforms, A and B) as well as hybrid receptors (Fig. 4.6) that bind insulin, IGF-I and IGF-II with differing affinities. In addition, they contain the IGF-II/mannose-6-phosphate (M6P) receptor that binds IGF-II and has a scavenger function (no signalling) [73]. In order to examine if different ligands acting through a single RTK may exhibit differential signalling, Versteyhe et al. [78] used fibroblasts from mice where both the IGF-II/M6P and IGF-II had been knocked out [79] and naturally devoid of insulin receptors. The cells were treated with equipotent IGF-I receptor saturating concentrations of insulin, IGF-I and IGF-II and submitted to Affymetrix gene microarray profiling. The results clearly showed that albeit signalling was exclusively mediated through the IGF-I receptor, each ligand specifically regulated a group of transcripts that was not regulated by the other two ligands. Thus, not only the affinity of a ligand determines its biological response but also its nature, even through the same receptor. This may be due to the fact that although the receptor is on average occupied the same fraction of time by equipotent ligand concentrations, the frequency of on and off binding will differ between ligands of different affinities.

In a different system, the importance of the "time factor" in signalling specificity [80] was demonstrated by the fact that the stimulation of MAP kinase by EGF and NGF in PC12 cells has different outcomes: proliferation with EGF acting through the EGF receptor or insulin through the insulin receptor, versus differentiation and neurite outgrowth with NGF acting through the NGF receptor. The major difference appears to be that EGF or insulin induces only transient MAP kinase activation, while NGF induces sustained activation allowing nuclear translocation [81]. The outcome could be reversed by overexpressing EGF or insulin receptors, which changed MAP kinase activation from transient to sustained [81]. This shift is partly explained by the differential engagement of positive and negative feedback mechanisms ([82], reviewed in [20]).

There is currently no precise solution to the conundrum of signalling specificity through common networks, except to state that the cellular context and the differential "combinatorial" use of the components of a common network, rather than intrinsic receptor activity, appear to be major determinants of differential biological endpoints such as mitogenic versus metabolic signalling or proliferation versus differentiation. A system biology approach to the modelling of complex signalling networks appears to be the way of the future in order to unravel this complexity.

4.7 Systems Biology of Combinatorial Signalling

Major efforts have been undertaken to analyse quantitatively signalling networks at a systems level, in particular for the EGF receptor/ERBB and MAP kinase signalling system [20, 71, 83–86]. The development of large-scale new technologies such as microarray gene profiling or quantitative phosphoproteomics has been helpful, but the whole field is still in its infancy pending the collection of much needed quantitative information on the multiple nodes and modules involved in a complex

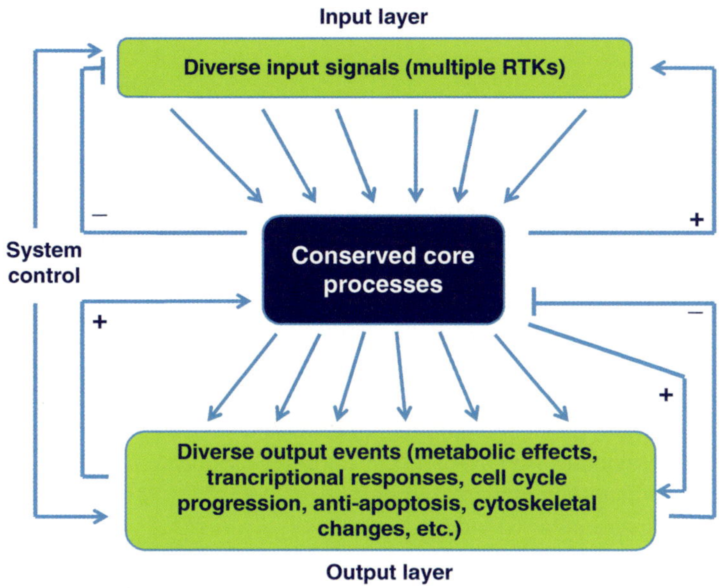

Fig. 4.7 A "bow tie" or "hourglass" network, as originally described by Kitano [84]. See refs. [20], [84] and [86] for detailed explanation. Adapted from ref. [20]

network. A useful concept that has emerged is that of a "bow tie" architecture of robust networks [20, 84, 86] (Fig. 4.7) linking the input layer of ligands and receptors to the output layer of biological outcomes, in which modularity and redundancy are important components [86].

The systems biology of RTK signalling will be discussed in more detail in Chap. 6, and I will just stress here some important elements that together may contribute to signalling specificity:

- Distribution and concentration of RTKs and subtypes in the input layer
- Subcellular localisation of signalling: surface versus endocytosed receptors ([20], see Chap. 7). Nuclear translocation of receptors and transcription factors (see Chap. 5)
- Positive and negative (including phosphatases, ubiquitylation) feedback loops [20, 86]
- Feedforward mechanisms [20]
- Scaffolds [87]
- Coincidence detection [20, 63, 88]

As pointed out in ref. [86], improved data sharing and a better interaction between experimentalists and theoreticians in this field should result in exponential advances, with important consequences for human health.

4.8 Cell Cycle Regulation by RTKs

One of the ultimate outcomes of growth factor stimulation of RTKs is naturally cell growth. An increase in the number of cells is the result of two opposing effects: an increase in the number of cells that traverse the cell cycle and divide into two daughter cells (mitosis), or a decrease in the number of cells that die, principally through programmed cell death (apoptosis), or both. Many RTKs are both capable of facilitating cell progression through the cell cycle and of inhibiting apoptotic mechanisms, although specific details are missing for most of them. In this section we will review the mechanisms that control the cell cycle, while apoptosis will be discussed in the next section of this chapter.

Growing cells go through four successive phases [89, 90] (Fig. 4.8):

- G_1 (Gap_1), where they get committed to grow and ribosomal biogenesis occurs
- S, where they synthesise DNA and chromosomes replicate
- G_2 (Gap_2), where they prepare to divide
- M (mitosis), where they divide

The phases before M are collectively known as "interphase". Cells not actively dividing are in a resting or quiescent state called G_0, from which they enter into G_1 under appropriate growth stimuli. The transitions most tightly regulated by growth factors are G_0/G_1 and G_1/S.

Much of what we know about the workings of the cell cycle is through the work of Sir Paul M. Nurse, Sir Timothy ("Tim") Hunt and Leland H. Hartwell. They won the Nobel Prize in Physiology or Medicine in 2001.

They showed that cellular processes required for cells to successfully replicate and divide are driven by the sequential activation and inactivation of a family of cyclin-dependent kinases (CDKs). These form bipartite complexes with different cyclins [89, 90]. Activation is driven predominately by the periodic expression of the cyclin subunit and requires activating phosphorylation of the kinase subunit.

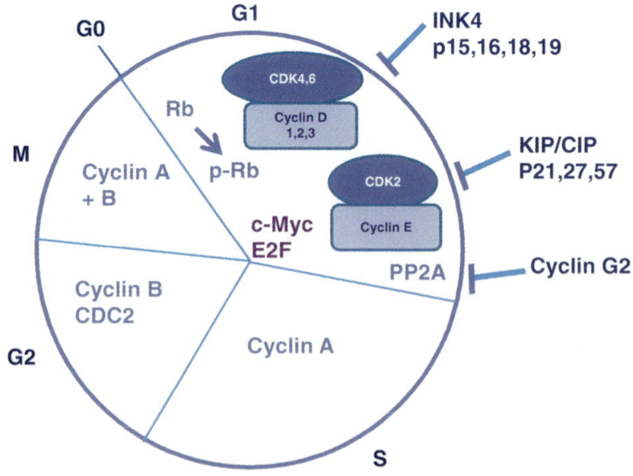

Fig. 4.8 The cell cycle. This graph depicts the successive phases of the cell cycle, as well as the various proteins involved in its regulation (progression or inhibition)

Inactivation is controlled by inhibitory phosphorylation of the kinase subunit, by ubiquitin-mediated degradation of the cyclin subunit and by interaction of the complex with small inhibitory proteins [89–91].

A detailed description of the complex regulation of the cell cycle is beyond the scope of this brief review. A clever minimisation of this complex machinery has shown that basic control of the eukaryotic cell cycle is a CDK oscillator that acts as the primary organiser of the cell cycle, imposing timing and directionality to a system of two CDK activity thresholds that define independent cell cycle phases [92].

The cell cycle is also negatively regulated by several CDK inhibitors that are in fact tumour suppressors [93]: retinoblastoma (Rb) protein, the INK4 family of proteins, $p21^{cip1}$ and $p27^{kip1}$ (Fig. 4.8).

The way the IGF-I receptor signalling system is involved in the regulation of the cell cycle has been thoroughly reviewed by Dupont et al. [93]. The PI3K/Akt pathway and the MAPK/ERK1/2 pathways are involved to variable degrees depending on the cell type [93].

Early work on chick embryo fibroblasts compared the effect of insulin with that of serum on confluent chick embryo fibroblasts [94]. The authors found that the total cell population that is stimulated by insulin or serum to enter S and G_2 is comparable initially (within 15 h). However, serum-stimulated cells entered S later than insulin-stimulated cells and had a shorter residence time in S and G_2. They therefore divided earlier than insulin-stimulated cells. Insulin stimulated only part of the cell population to divide. These observations together with others suggested that insulin has a "permissive" role that is additive to other growth factors.

Further work from the lab of W. Jackson Pledger at the University of North Carolina showed that human serum contains two sets of growth factors that function synergistically to promote cell growth through the G_0-G_1 traverse [95–97]. "Competence factors" such as platelet-derived growth factor or fibroblast growth factor commit cells to enter the division cycle and render them competent to respond to "progression factors" contained in platelet-poor plasma such as IGF-I (or its substitute insulin at high concentrations) or epidermal growth factor (EGF), up to a growth arrest point in mid-Go/G_1 termed the V point. In addition, IGF-I mediated the traverse from the V point to the S phase [97].

More recent studies have worked out some of the details of the components of the cell cycle affected by IGFs (for review see ref. [93]). These include increase in D-type cyclins, hyperphosphorylation of Rb, release of the transcription factor E2F and generation of cyclin E which then binds to CDK2 and leads to G1/S phase progression.

IGFs also regulate the CDK inhibitors (CDKIs). P27 transcription is downregulated through phosphorylation-dependent inhibition of FOXO-1 as well as post-transcriptionally [93].

In addition to its actions at the G1/S transition, IGF-I also affects the duration of G2 and the timely progression through G2/M phases, in synergy with estrogens [93].

More recently, a role for IGFs and insulin in regulating the cell cycle inhibitor cyclin G2 has been unravelled by the De Meyts group in collaboration with Mary

Horne at UC. Davis, the discoverer of cyclin G2 [98, 99]. Cyclin G2 is an unconventional cyclin highly expressed in postmitotic cells [100, 101]. Unlike classical cyclins that promote cell cycle progression, cyclin G2 causes cell cycle arrest at the G_1/S transition, not by associating with a CDK, but by associating with the phosphatase PP2A, thereby altering the activity of PP2A ([102], Fig. 4.8). Cyclin G2 is upregulated by FoxO which is inhibited by Akt. Cyclin G2 was found to be markedly downregulated by insulin, insulin analogues and IGF-I in L6 myoblasts overexpressing the insulin receptor, as well as in several other cell types [98, 99]. Overexpression of cyclin G2 inhibited cell proliferation induced by insulin, insulin analogues and IGF-I [99]. The use of specific inhibitors indicated that both the MAPK (mitogen-activated protein kinase) and the PI3K (phosphoinositide 3-kinase) pathways mediate the downregulation of cyclin G2. These results indicate that regulation of cyclin G2 is a key mechanism whereby insulin, insulin analogues and IGF-I stimulate cell proliferation.

The downregulation of cyclin G2 has also been suggested to play a role downstream of ERBB2 through the PI3K pathway in human breast cancer cells [103].

Continuous stimulation of cells with growth factors enables them to cross the restriction point in the G_1-phase after which cell cycle progression becomes growth factor independent. It has been shown that this continuous stimulation can be substituted by two pulses of growth factor stimulation separated by a fixed-length time interval [104]. A recent elegant study in human mammary epithelial cells [105] by the group of Yossi Yarden at the Weizmann Institute, using comprehensive proteomic and transcriptomic analyses of each pulse, indicated that the first pulse of growth factor (in this case EGF or IGF-I) primes cells by promoting lipid metabolism but also activates p53 which results in expression of antiproliferative genes. The second pulse of EGF suppresses p53 through PI3K/Akt and permits crossing of the restriction point if ERK activation and the consequent induction of the transcription factor EGR1 exceed a threshold (Fig. 4.9). IGF-I can substitute for EGF in the first but not the second pulse. This landmark study sets up a high standard for further investigation of the role of the various RTKs in controlling the cell cycle.

Interestingly, Jensen et al. have shown in microarray gene profiling of L6 myoblasts overexpressing the insulin receptor that while cyclin G2 was the gene most downregulated by insulin (but much less by the insulin mimetic peptide S597 which is a poor mitogen), the most upregulated gene was EGR1 (also much less by S597). This was associated with a weaker stimulation of MAP kinase by S597 [98, 106].

4.9 Regulation of Apoptosis by RTKs

Although there were earlier descriptions of some of the cell morphological changes associated with programmed cell death, the term apoptosis was first used in the seminal paper published in the British Journal of Cancer in 1972 [107], by John Foxton Ross Kerr, Alastair A. Currie and Andrew H. Wyllie. In 2002, the Nobel Prize in Physiology or Medicine was given to Sydney Brenner, John E. Sulston and

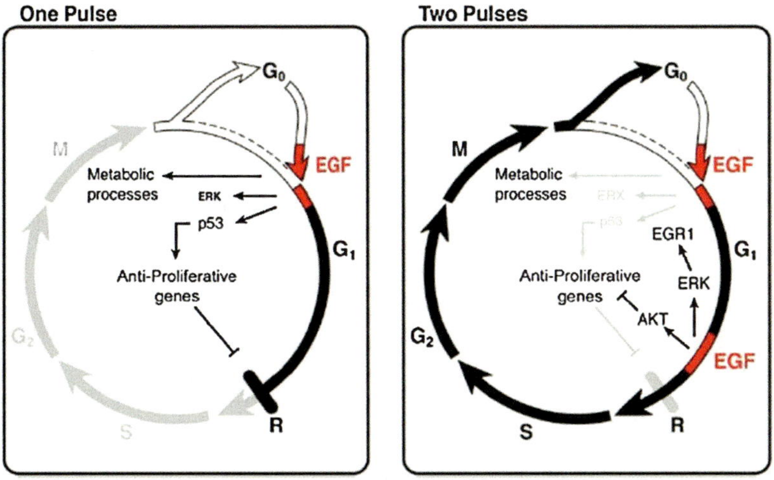

Fig. 4.9 Biochemical events elicited by a single and a dual EGF pulse. The first pulse of EGF induces expression of lipid biosynthesis-associated genes, along with activation of p53. The latter propels expression of antiproliferative genes, such as BTG2 and SESN1. When cells are treated with a second pulse of EGF, enhanced activation of ERK and subsequent induction of EGR1 exceed a critical threshold. In parallel, signalling through PI3K and resulting suppressed expression of antiproliferative genes permit cells to cross the restriction point (R) and enter the S phase. From ref. [105], used with permission

H. Robert Horwitz for their work on the genetic regulation of organ development and programmed cell death in the worm *Caenorhabditis elegans*.

This mechanism of controlled cell deletion is common to all multicellular organisms. Cells undergo marked morphological changes: blebbing, cell shrinkage, nuclear fragmentation, chromatin condensation and chromosomal DNA fragmentation. Cells eventually split into small closed entities (apoptotic bodies) that can be phagocytosed by other cells, preventing the release of toxic cellular substances that may cause necrosis. Apoptosis is triggered by a variety of environmental stimuli, both physiological and pathological including stress. The intracellular mechanisms involved in apoptosis are very complex (Fig. 4.10); while most proteins involved in the apoptotic network are known, much work remains to be done to unravel their interactions. Here, we will only briefly summarise the salient features of the apoptotic process: see ref. [108] for more detailed review.

Some cells apoptose through extrinsic pathways that involve death receptors such as Fas (fatty acid synthetase) or TNF (tumour necrosis factor) receptors; others have a default death pathway that must be blocked by a survival factor such as a hormone or a growth factor acting through an RTK [108]. Serum withdrawal is a classic way to initiate this pathway. In the end, apoptosis is an energy-dependent process that involves the activation of a group of cystine proteases called "caspases" and involves a complex cascade of events that link the initiating stimuli to the death of the cell [108]. The two main regulatory mechanisms used by extracellular signals

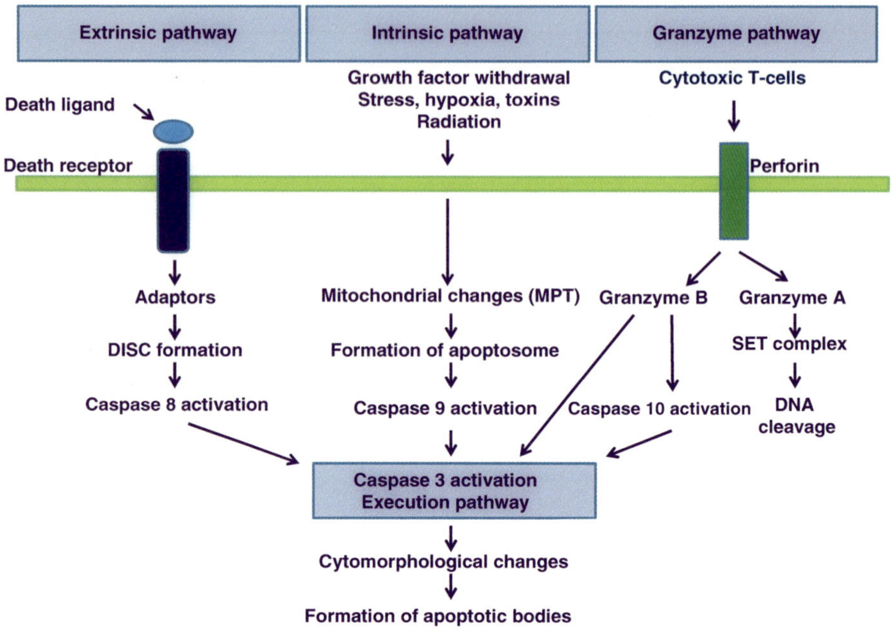

Fig. 4.10 Apoptotic pathways. The two main pathways of apoptosis are extrinsic and intrinsic as well as a perforin/granzyme pathway, each triggered by specific signals. Each pathway activates its own initiator caspase which in turn will activate the executioner caspase-3, except granzyme A which works in a caspase-independent fashion. The execution pathway results in characteristic cytomorphological features including cell shrinkage, chromatin condensation, formation of cytoplasmic blebs and apoptotic bodies and finally phagocytosis of the apoptotic bodies by adjacent parenchymal cells, neoplastic cells or macrophages [108]. Adapted from ref. [108]

are either by targeting mitochondrial functionality (intrinsic pathway) or by directly transducing the signal via adaptor proteins to the apoptotic mechanisms (extrinsic pathway). There is an additional pathway mediating T-cell-induced cytotoxicity and perforin-granzyme (a serine protease) A- or B-dependent killing of the cell (Fig. 4.10).

The mitochondrial pathway involves an increased mitochondrial permeability resulting in release in the cytosol of cytochrome c and SMACS (small mitochondria-derived activators of caspases) that bind to and deactivate IAPs (inhibitors of apoptosis proteins). IAPs repress the caspases. Mitochondrial permeability is regulated positively or negatively by 25 members of the Bcl-2 family of proteins [109], under the control of the tumour suppressor protein *p53*.

Cytochrome c and ATP released from the mitochondrial intermembrane space form the apoptosome consisting of ATP, apoptosis protease-activating factor (APAF)-1, cytochrome c and caspase-9, which becomes activated by autoproteolytic cleavage and activates the execution caspase-3, caspase-6 and caspase-7, which leads to the collapse of cellular infrastructure [108].

The extrinsic pathway involves binding of trimeric ligands to their receptors which cluster (FasL to the FasR or TNFα to the TNR1). Binding of FasL to FasR recruits the adapter protein FADD (Fas-associated death domain), while binding of TNFα to the TNR1 recruits the adapter protein TRADD (TNF receptor-associated death domain). TRADD then recruits FADD and RIP (receptor-interacting protein). FADD forms a death-inducing signalling complex (DISC) with procaspase-8 resulting in its autocatalytic activation [108] and triggering of the "execution phase".

The extrinsic, intrinsic and granzyme B pathways converge on the same terminal "execution" pathway that is initiated by the cleavage of caspase-3 by caspases 8, 9 or 10. The granzyme A pathway activates a parallel, caspase-independent pathway via single-stranded DNA damage [110] (Fig. 4.10).

There are many more players in the apoptosis cascades; see Tables 1–4 in ref. [108].

IGF-I is a potent antiapoptotic growth factor at low concentrations in multiple cell types [111, 112]. This antiapoptotic effect involves both the PI3K and MAPK/ERK1/2 pathways (Fig. 4.11) as shown by the use of specific inhibitors in serum-deprived PC12 cells [111], but other pathways also appear to be involved (for review, see ref. [112]). Many of the antiapoptotic actions of IGF-I appear to be through the regulation of mitochondrial membrane permeability via Bcl-2 proteins like Bad [112] via Akt. It also appears capable of inhibiting the extrinsic apoptotic pathway via regulation of death-inducing receptors [112]. These properties may

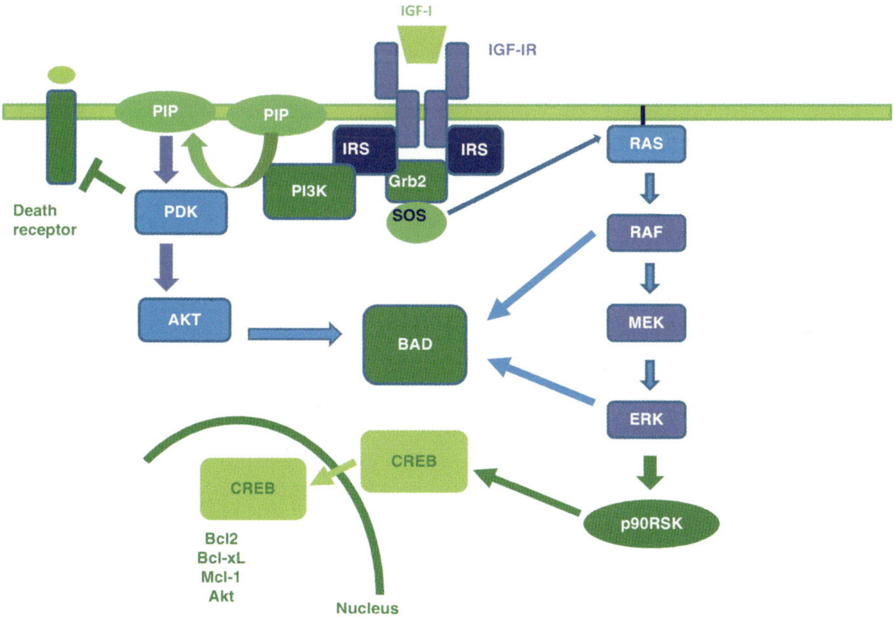

Fig. 4.11 The antiapoptotic pathways of insulin-like growth factor-I. A much simplified scheme for IGF-I antiapoptotic signalling. See ref. [112] for the full picture

hold therapeutic potential for hypoxic/ischemic brain injury, amyotrophic lateral sclerosis, Huntington disease, Alzheimer disease and cancer [112].

Numerous papers indicate that insulin exerts also a potent antiapoptotic role in a variety of cell types [113–122]. The experimental conditions (e.g. using high concentrations of insulin and IGF-I and no dose–response curve) do not always allow to conclude whether the insulin effect is through the insulin receptor or the IGF-I receptor. In a few carefully controlled studies using native cells or insulin receptor-transfected cells (e.g. [117, 118, 122]), the authors established unequivocally that insulin at physiological concentrations has antiapoptotic effects through the insulin receptor.

The PI3K pathway appears to play a prominent role in this effect.

In a recent intriguing paper [113], the group of C. Ronald Kahn at the Joslin Diabetes Center in Boston showed that brown adipocytes from mice with both the insulin receptor and IGF-I receptor knocked out were resistant to apoptosis induced by serum deprivation. Sensitivity to apoptosis was restored by knocking in either receptor, which was then prevented by the appropriate ligand. These data suggested that the unoccupied insulin and IGF-I receptors are proapoptotic (in a kinase-independent fashion), while the ligand-occupied receptors are antiapoptotic.

Finally we should mention that autophagy or autophagocytosis (a process involving the lysosomal degradation of a cell's own components), if dysregulated, can also lead to cell death [123]. A recent study suggests that this process is also inhibited by IGF-I [124].

In summary, both IGF-I and insulin have antiapoptotic effects on multiple cell lines. Both the IGF-I and insulin receptors are capable of mediating this effect. Multiple signalling pathways are implicated, but the PI3K pathway appears to play a prominent role.

I have not attempted to review the possible role of all known RTKs in regulating apoptosis; the role of the EGF receptor has been reviewed recently [125] with involvement of mechanisms similar to those described above for IGF-I and insulin.

4.10 Conclusion and Perspectives

RTKs play critical roles in many essential cellular processes, and their dysregulation by mutations or other pathogenic processes have been implicated in cancers and metabolic disease states. They represent important therapeutic targets with quite a number of success stories ([20], see Chap. 10). Despite considerable progress in the genetics, structural biology, biochemistry and cell biology of the RTKs and their signalling networks over the last 30 years, much remains to be done; as pointed out by Lemmon and Schlessinger [20], we only have detailed information for only half of the RTK families. The very complexity but at the same time commonality of their signalling networks makes it a challenge to understand the molecular basis of their signalling specificity. Large-scale proteomic and transcriptomic approaches as well as progress in mathematical modelling and the various facets of what is referred to as systems biology hold the promise to reach a new level of quantitative understanding of the behaviour of these complex systems and to be able to predict emergent behaviours and to make qualitative predictions with important potential therapeutic outcomes.

Acknowledgements I would like to acknowledge many enlightening discussions on RTKs over many years with Yossi Schlessinger, Yossi Yarden, Axel Ullrich, Ralph Bradshaw, Boris Kholodenko, Steve Wiley, Colin Ward and Mike Lawrence, as well as the many friends and colleagues who participated in the 37th European Symposium on Hormones and Cell Regulation on "RTKs from structural biology to systems biology" that I organised in Mont Sainte-Odile in October 2012.

Note I discovered after completion of this chapter an excellent review by Ichiro N. Maruyama [126] that discusses several issues also discussed in my chapter, and contains many additional references.

References

1. Cohen S, Levi-Montalcini R, Hamburger V. A nerve growth-stimulating factor isolated from Sarcomas 37 and 180. Proc Natl Acad Sci. 1954;40:1014–8.
2. Cohen S. Isolation of a mouse submaxillary gland protein accelerating incisor eruption and eyelid opening in the new-born animal. J Biol Chem. 1962;237:1555–62.
3. Ullrich A, Coussens L, Hayflick JS, Dull TJ, Gray A, Tam AW, Lee Y, Yarden Y, Libermann TA, Schlessinger J, Downward J, Mayes ELV, Whittle N, Waterfield M, Seeburg P. Human epidermal growth factor receptor cDNA sequence and aberrant expression of the amplified gene in A431 epidermoid carcinoma cells. Nature. 1984;309:418–25.
4. Ullrich A, Bell JR, Chen EY, Herrera R, Petruzelli LM, Dull TJ, Grey A, Coussens L, Liao Y-C, Tsubokawa M, Grunfeld C, Rosen OM, Ramachandran J. Human insulin receptor and its relationship to the tyrosine kinase family of oncogenes. Nature. 1985;313:756–61.
5. Ebina Y, Ellis L, Jarnagin K, Edery M, Graf L, Clauser E, Ou J-H, Masiarz F, Kan YW, Goldfine ID, Roth RA, Rutter WJ. The human Insulin receptor cDNA: the structural basis for hormone - activated transmembrane signalling. Cell. 1985;40:747–58.
6. Ullrich A, Gray A, Tam AW, Yang-Feng T, Tsubokawa M, Collins C, Henzel W, Le Bon T, Kathuria S, Chen E, Jacobs S, Francke U, Ramachandran J, Fujita-Yamaguchi Y. Insulin-like growth factor-I receptor primary structure: comparison with insulin receptor suggests structural determinants that define functional specificity. EMBO J. 1986;5:2503–12.
7. Coussens L, Van Beveren C, Smith D, Chen E, Mitchell RL, Isacke CM, Verma IM, Ullrich A. Structural alteration of viral homologue of receptor proto-oncogene *fms* at carboxyl terminus. Nature. 1986;321:277–81.
8. Yarden Y, Escobedo JA, Kwang W-J, Yang-Feng TL, Daniel TO, Tremble PM, Chen EY, Ando ME, Harkins NR, Francke U, Fried UA, Ullrich A, Williams LT. Structure of the receptor for platelet-derived growth factor helps define a family of closely related growth-factor receptors. Nature. 1986;323:226–32.
9. Yarden Y, Ullrich A. Growth-factor receptor tyrosine kinases. Ann Rev Biochem. 1988;57:443–78.
10. Ullrich A, Schlessinger J. Signal transduction by receptors with tyrosine kinase activity. Cell. 1990;61:203–12.
11. Schlessinger J, Ullrich A. Growth factor signalling by receptor tyrosine kinases. Neuron. 1992;9:383–91.
12. Van der Geer P, Hunter T, Lindberg RA. Receptor protein-tyrosine kinases and their signal transduction pathways. Ann Rev Cell Bio. 1994;10:251–337.

13. Heldin CH. Dimerization of cell surface receptors in signal transduction. Cell. 1995;80:213–23.
14. Schlessinger J. Cell signalling by receptor tyrosine kinases. Cell. 2000;103:211–25.
15. Hunter T. Signaling – 2000 and beyond. Cell. 2000;100:113–27.
16. Blume-Jensen P, Hunter T. Oncogenic kinase signalling. Nature. 2001;411:355–65.
17. Hubbard SR, Miller WT. Receptor tyrosine kinases: mechanisms of activation and signaling. Curr Opin Cell Biol. 2007;19:117–23.
18. De Meyts P. The insulin receptor: a prototype for dimeric, allosteric membrane receptors? Trends Biochem Sci. 2008;33:376–84.
19. De Meyts P, Gauguin L, Svendsen AM, Sarhan M, Knudsen L, Nøhr J, Kiselyov VV. Structural basis of allosteric interactions in the insulin/relaxin peptide family. Implications for other receptor tyrosine kinases and G protein-coupled receptors. Ann NY Acad Sci. 2009;1160:45–53.
20. Lemmon MA, Schlessinger J. Cell signalling by receptor tyrosine kinases. Cell. 2010;141: 1117–34.
21. Schlessinger J. Receptor tyrosine kinases: legacy of the first two decades. Cold Spring Harb Perspect Biol 2014; 6(3). pii: a008912, doi:10.1101/cshperspect.a008912.
22. Hunter T. The genesis of tyrosine phosphorylation. Cold Spring Harb Perspect Biol 2014; 6(5):a020644, doi:10.1101/cshperspecta620644.
23. Lemmon MA, Schlessinger J and Ferguson KM. The EGFR family: not so prototypical receptor tyrosine kinases. Cold Spring Harb Perspect Biol 2014; 6(4):a020768, doi:10.1101/cshperspecta020768.
24. Honegger AM, Kris RM, Ullrich A, Schlessinger J. Evidence that autophosphorylation of solubilized receptors for epidermal growth factor is mediated by intermolecular cross-phosphorylation. Proc Natl Acad Sci. 1989;86:925–9.
25. de Vos AM, Ultsch M, Kossiakoff AA. Human growth hormone and extracellular domain of its receptor: crystal structure of the complex. Science. 1992;255:306–12.
26. Wiesmann C, Ultsch MH, Bass CH, de Vos AM. Crystal structure of nerve growth factor in complex with the ligand-binding domain of the TrkA receptor. Nature. 1999;401:184–8.
27. Yuzawa S, Opatowsky Y, Zhang Z, Mandiyan V, Lax I, Schlessinger J. Structural basis for the activation of the receptor tyrosine kinase KIT by stem cell factor. Cell. 2007;130:323–34.
28. Ogiso H, Ishitani R, Nureki O, Fukai S, Yamanaka M, Kim JH, Saito K, Sakamoto A, Inoue M, Shirozou M, Yokoyama S. Crystal structure of the complex of human epidermal growth factor and receptor extracellular domains. Cell. 2002;110:770–87.
29. Cherezov V, Rosenbaum DM, Hanson MA, Rasmussen CG, Thian FS, Kobilka TS, Choi HJ, Kuhn P, Weis WI, Kobilka BK, Stevens RC. High-resolution crystal structure of an engineered human beta(2)-adrenergic G protein-coupled receptor. Science. 2007;318:1258–65.
30. Monod J, Wyman J, Changeux JP. On the structure of allosteric transitions. a plausible model. J Mol Biol. 1965;12:88–118.
31. Koshland DE, Nemethy G, Filmer D. Comparison of experimental binding data and theoretical models in proteins containing subunits. Biochemistry. 1966;5:365–85.
32. Levitzki A, Koshland DE. Negative cooperativity in regulatory enzymes. Proc Natl Acad Sci. 1969;62:1121–8.
33. Schlessinger J, Levitzki A. Molecular basis for negative cooperativity in rabbit muscle glyceraldehyde-3-phosphate dehydrogenase. J Mol Biol. 1974;82:547–61.
34. De Meyts P, Roth J, Neville Jr DM, Gavin 3rd JR, Lesniak MA. Insulin interactions with its receptors: experimental evidence for negative cooperativity. Biochem Biophys Res Commun. 1973;55:154–61.
35. Christoffersen CT, Bornfeldt KE, Rotella CM, Gonzales N, Vissing H, Shymko RM, ten Hoeve J, Groffen J, Heisterkamp N, De Meyts P. Negative cooperativity in the insulin-like growth factor-I (IGF-I) receptor and a chimeric IGF-I/insulin receptor. Endocrinology. 1994;135:472–5.
36. Frazier WA, Boyd LF, Pulliam MW, Szutowski A, Bradshaw RA. Properties and specificity of binding sites for ^{125}-I nerve growth factor in embryonic heart and brain. J Biol Chem. 1974;249:5918–23.

37. Zampieri N, Chao MV. The p75 NGF receptor exposed. Science. 2004;304:833–4.
38. Cuatrecasas P, Hollenberg MD. Binding of insulin and other hormones to non-receptor materials: saturability, specificity and apparent "negative cooperativity". Biochem Biophys Res Commun. 1975;62:31–41.
39. Rao CV, Carman FR, Chegini N, Schultz GS. Binding sites for epidermal growth factor in human fetal membranes. J Clin Endo Metab. 1984;58:1034–42.
40. Prahl M, Nederman T, Carlsson J, Sjödin L. Binding of epidermal growth factor (EGF) to a cultured human glioma cell line. J Recept Res. 1991;11:791–812.
41. De Meyts P. Cooperative properties of hormone receptors in cell membranes. J Supramol Struct. 1976;4:241–58.
42. Macdonald JL, Pike LJ. Heterogeneity in EGF-binding affinities arise from negative cooperativity in an aggregating system. Proc Natl Acad Sci U S A. 2008;105:112–7.
43. Wofsy C, Goldstein B, Lund K, Wiley HS. Implications of epidermal growth factor (EGF) induced EGF receptor aggregation. Biophys J. 1992;63:98–110.
44. De Meyts P. The structural basis of insulin and insulin-like growth factor-I (IGF-I) receptor binding and negative cooperativity, and its relevance to mitogenic versus metabolic signaling. Diabetologia. 1994;37 suppl 2:S135–48.
45. Alvarado D, Klein DE, Lemmon MA. Structural basis for negative cooperativity in growth factor binding to an EGF receptor. Cell. 2010;142:568–79.
46. Leahy DJ. The ins and outs of EGFR asymmetry. Cell. 2010;142:513–5.
47. Limbird L, De Meyts P, Lefkowitz RJ. Beta-adrenergic receptors: evidence for negative cooperativity. Biochem Biophys Res Commun. 1975;64:1160–8.
48. Urizar E, Montanelli L, Loy T, Bonomi M, Swillens S, Gales C, Bouvier M, Smits G, Vassart G, Costagliola S. Glycoprotein hormone receptors: link between receptor homodimerization and negative cooperativity. EMBO J. 2005;24:1954–64.
49. Christopoulos A, Kenakin T. G protein-coupled receptor allosterism and complexing. Pharmacol Rev. 2002;54:323–74.
50. Springael JY, Urizar E, Costagliola S, Vassart G, Parmentier M. Allosteric properties of G protein-coupled oligomers. Pharmacol Ther. 2007;115:410–8.
51. Svendsen AM, Vrecl M, Knudsen L, Heding A, Wade JD, Bathgate RA, De Meyts P, Nøhr J. Dimerization and negative cooperativity in the relaxin family peptide receptors. Ann NY Acad Sci. 2009;1160:54–9.
52. Roed SN, Orgaard A, Jorgensen R, De Meyts P. Receptor oligomerization in family B1 of G protein-coupled receptors: focus on BRET investigations and the link between GPCR oligomerization and binding cooperativity. Front Endocrin. 2012;3:62. doi:10.3389/fendo.2012.00062. eCollection.
53. Zoenen M, Urizar E, Swillens S, Vassart G, Costagliola S. Evidence for activity-regulated hormone-binding cooperativity across glycoprotein hormone receptors homomers. Nat Commun. 2012;3:1007. doi:10.1038/ncomms.
54. Kiselyov VV, Versteyhe S, Gauguin L, De Meyts P. Harmonic oscillator model of the insulin and IGF1 receptors' allosteric binding and activation. Mol Sys Biol. 2009;5:243. doi:10.1038/msb.208.78.
55. Ward CW, Lawrence MC, Streltsov VA, Adams TE, McKern NM. The insulin and EGF receptor structures: new insights into ligand-induced receptor activation. Trends Biochem Sci. 2007;32:129–37.
56. Ward CW, Menting JG, Lawrence MC. The insulin receptor changes conformation in unforeseen ways on ligand binding: sharpening the picture of insulin receptor activation. Bioessays. 2013;35:945–54.
57. Huse M, Kuriyan J. The conformational plasticity of protein kinases. Cell. 2002;109:275–82.
58. Ablooglu AJ, Kohanski RA. Activation of the insulin receptor kinase domain changes the rate-determining step of substrate phosphorylation. Biochemistry. 2001;40:504–13.
59. Cobb MH, Sang BC, Gonzales R, Golsmith E, Ellis L. Autophosphorylation activates the soluble cytoplasmic domain of the insulin receptor in an intermolecular reaction. J Biol Chem. 1989;264:18701–6.

60. Zhang X, Gureasko J, Shen K, Cole PA, Kuriyan J. An allosteric mechanism for activation of the kinase domain of epidermal growth factor receptor. Cell. 2006;125:1137–49.
61. Knowles PP, Murray-Rust J, Kjaer S, Scott RP, Hanrahan S, Santoro M, Ibáñez CF, McDonald NQ. Structure and chemical inhibition of the RET tyrosine kinase domain. J Biol Chem. 2006;281:33577–87.
62. Schlessinger J, Lemmon MA. SH2 and PTB domains in tyrosine kinase signaling. Sci STKE. 2003;191:RE12.
63. Pawson T. Specificity in signal transduction: from phosphotyrosine-SH2 domain interactions to complex cellular systems. Cell. 2004;116:191–203.
64. Avruch J. MAP kinase pathways: the first twenty years. Biochem Biophys Acta. 2007;1773:1150–60.
65. Cantley LC. The phosphoinositide 3-kinase pathway. Science. 2002;296:1655–7.
66. Accili D, Arden KC. FoxOs at the crossroads of cellular metabolism, differentiation and transformation. Cell. 2004;117:421–6.
67. Essaghir A, Dif N, Marbehant CY, Coffer PJ, Demoulin JB. The transcription of FoxO genes is stimulated by FoxO3 and repressed by growth factors. J Biol Chem. 2009;284:10334–42.
68. Taniguchi CM, Emmanuelli B, Kahn CR. Critical nodes in signaling pathways: insights into insulin action. Nat Rev Mol Cell Biol. 2006;7:85–96.
69. Taguchi A, White MF. Insulin-like signaling, nutrient homeostasis and lifespan. Ann Rev Physiol. 2008;70:191–212.
70. Boucher J, Kleinridders A, Kahn CR. Insulin receptor signaling in normal and insulin-resistant states. Cold Spring Harb Perspect Biol. 2014;6:a009191.
71. Oda K, Matsuoka Y, Fumahashi A, Kitano H. A comprehensive pathway map of epidermal growth factor receptor signaling. Mol Sys Biol. 2005;1:2005.0010.
72. Siddle K. Molecular basis of signalling specificity of insulin and IGF receptors: neglected corners and recent advances. Front Endocrin. 2012;3:34. doi:10.3389/fendo.2012.00034.
73. Efstratiadis A. Genetics of mouse growth. Int J Dev Biol. 1998;42:955–76.
74. Lamothe B, Baudry A, Christoffersen CT, De Meyts P, Jami J, Bucchini D, Joshi RL. Insulin receptor-deficient cells as a new tool for dissecting complex interplay in insulin and insulin-like growth factors. FEBS Lett. 1998;426:381–5.
75. Ish-Shalom D, Christoffersen CT, Vorwerk P, Sacerdotti-Sierra N, Shymko RM, De Meyts P. Mitogenic properties of insulin and insulin analogues mediated by the insulin receptor. Diabetologia. 1997;40:S25–31.
76. De Meyts P. Insulin and insulin-like growth factors: the paradox of signaling specificity. Growth Horm IGF Res. 2002;12:81–3.
77. Kim JJ, Accili D. Signaling through IGF-I and insulin receptors: where is the specificity? Growth Horm IGF Res. 2002;12:84–90.
78. Versteyhe S, Klaproth B, Borup R, Palsgaard J, Jensen M, Gray SG, De Meyts P. IGF-I, IGF-II and insulin stimulate different gene expression responses through binding to the IGF-I receptor. Front Endocrin. 2013;4:98. doi:10.3389/fendo.2013.00098. eCollection.
79. Dittmer F, Ulbrich EJ, Hafner A, Schmahl W, Meister T, Pohlmann R, Von Figura K. Alternative mechanisms for trafficking of lysosomal enzymes in mannose-6-phosphate receptor-deficient mice are cell-type specific. J Cell Sci. 1999;112:1591–7.
80. De Meyts P, Christoffersen CT, Ursø B, Wallach B, Grønskov K, Yakushiji F, Shymko RM. Role of the time factor in signaling specificity: application to mitogenic and metabolic signaling by the insulin and insulin-like growth factor-I receptor tyrosine kinases. Metabolism. 1995;44(10 suppl 4):2–11.
81. Marshall CJ. Specificity of receptor tyrosine kinase signaling: transient versus sustained extracellular signal-regulated kinase activation. Cell. 1995;80:179–85.
82. Santos SD, Verveer PJ, Bastiaens PJ. Growth factor-induced MAPK network topology shapes ERK response determining PC-12 cell fate. Nat Cell Biol. 2007;9:324–30.
83. Schoeberl B, Eichler-Johnsson C, Gilles ED, Mueller G. Computational modeling of the dynamics of the MAP kinase cascade activated by surface and internalized EGF receptors. Nat Biotechnol. 2002;20:370–1.

84. Kitano H. Biological robustness. Nat Rev Genet. 2004;5:826–37.
85. Kholodenko BN. Cell signaling dynamics in time and space. Nat Rev Mol Cell Biol. 2006;17:165–76.
86. Citri A, Yarden Y. EGF-ERBB signalling: towards the systems level. Nat Rev Mol Cell Biol. 2006;7:505–16.
87. Borisov NM, Markevich NI, Hoek JB, Kholodenko BN. Signaling through receptors and scaffolds: independent interactions reduce combinatorial complexity. Biophys J. 2005;89:951–66.
88. Shymko RM, Dumont E, De Meyts P, Dumont J. Timing dependence of insulin-receptor mitogenic versus metabolic signaling: a plausible model based on coincidence of hormone and effector binding. Biochem J. 1999;339:675–83.
89. McGowan CH. Regulation of the eukaryotic cell cycle. Progr Cell Cycle Res. 2003;5:1–4.
90. Morgan DO. Cyclin-dependent kinases: enzymes, clocks and microprocessors. Annu Rev Cell Dev Biol. 1997;13:261–91.
91. Olashaw N, Bagui TK, Pledger WJ. Cell cycle control. A complex issue. Cell Cycle. 2004;3:263–4.
92. Coudreuse D, Nurse P. Driving the cell cycle with a minimal CDK control network. Nature. 2010;468:1075–9.
93. Dupont J, Pierre A, Froment P, Moreau C. The insulin-like growth factor axis in cell cycle progression. Horm Metab Res. 2003;35:740–50.
94. Teng M-H, Bartholomew JC, Bissel MJ. Insulin effects on the cell cycle: analysis of the kinetics of growth parameters in confluent chick cells. Proc Natl Acad Sci U S A. 1976;73:3173–7.
95. Stiles CD, Capone GT, Scher CD, Antoniades HN, Van Wyk JJ, Pledger WJ. Dual control of cell growth by somatomedins and platelet-derived growth factor. Proc Natl Acad Sci U S A. 1979;76:1279–83.
96. O'Keefe EJ, Pledger WJ. A model of cell cycle control: sequential events regulated by growth factors. Mol Cell Endocrinol. 1983;31:167–86.
97. Olashaw NE, Van Wyk JJ, Pledger WJ. Control of late G_0/G_1 progression and protein modification by SmC/IGF-I. Am J Physiol. 1987;253:C575–9.
98. Jensen M, Palsgaard J, Borup R, De Meyts P, Schäffer L. Activation of the insulin receptor (IR) by insulin and a synthetic peptide has different effects on gene expression in IR-transfected L6 myoblasts. Biochem J. 2008;412:435–45.
99. Svendsen AM, Winge SB, Zimmermann M, Lindvig AB, Warzecha CB, Sajid W, Horne MC, De Meyts P. Down-regulation of cyclin G2 by insulin, IGF-I (insulin like growt factor-I), and X10 (AspB10 insulin): role in mitogenesis. Biochem J. 2014;457:69–77.
100. Horne MC, Goolsby GL, Donaldson KL, Tran D, Neubauer M, Wahl AF. Cyclin G1 and cyclin G2 comprise a new family of cyclins with contrasting tissue-specific and cell-cycle regulated expression. J Biol Chem. 1996;271:6050–61.
101. Horne MC, Donaldson KL, Goolsby GL, Tran D, Mulhelsen M, Hell JW, Wahl AF. Cyclin G2 is upregulated during growth inhibition and B cell antigen receptor-mediated cell cycle arrest. J Biol Chem. 1997;272:12650–61.
102. Bennin DA, Don AS, Brake T, Mckenzie JL, Rosenbaum H, Ortiz L, DePaoli-Roach AA, Horne MC. Cyclin G2 associates with protein phosphatase 2A catalytic and regulatory B subunits in active complexes and induces nuclear aberrations and a G1/S cell cycle arrest. J Biol Chem. 2002;277:27449–67.
103. Le XF, Arachchige-Don AS, Mao W, Horne MC, Bast Jr RC. Roles of human epidermal growth factor receptor 2, c-jun NH2-terminal kinase, phosphoinositide 3-kinase and p70S6 kinase in regulation of cyclin G2 expression in human breast cancer cells. Mol Cancer Ther. 2007;6:2843–57.
104. Jones SM, Kazlaukas A. Growth-factor-dependent mitogenesis requires two distinct phases of signalling. Nat Cell Biol. 2001;3:165–72.
105. Zwang Y, Sas-Chen A, Drier Y, Shay T, Avraham R, Lauriola M, Shema E, Lidor-Nili E, Jacob-Hirsch J, Amariglio N, Lu Y, Mills GB, Rechavi G, Oren M, Domany E, Yarden

Y. Two phases of mitogenic signaling unveil roles for p53 and EGR1 in elimination of inconsistent growth signals. Mol Cell. 2011;42:524–35.
106. Jensen M, De Meyts P. Molecular mechanisms of differential signaling from the insulin receptor. Vitam Horm. 2009;80:51–75.
107. Kerr JFR, Wyllie AH, Currie AR. Apoptosis: a basic biological phenomenon with wide-ranging implications in tissue kinetics. Br J Cancer. 1972;26:239–57.
108. Elmore S. Apoptosis: a review of programmed cell death. Toxicol Pathol. 2007;35:495–516.
109. Cory S, Adam JM. The Bcl2 family: regulators of the cellular life-or-death switch. Nat Rev Cancer. 2002;2:647–56.
110. Martinvalet D, Zhu P, Lieberman J. Granzyme A induces caspase-independent mitochondrial damage, a required first step for apoptosis. Immunity. 2005;22:355–70.
111. Párrizas M, Saltiel AR, LeRoith D. Insulin-like growth factor-I inhibits apoptosis using the phosphatidylinositol 3′-kinase and mitogen-activated protein kinase pathways. J Biol Chem. 1997;272:154–61.
112. Kooijman R. Regulation of apoptosis by insulin-like growth factor (IGF)-I. Cyt Growth Fact Rev. 2006;17:305–23.
113. Boucher J, Marcotella Y, Bezy O, Mori MA, Kriauciunas K, Kahn CR. A kinase-independent role for unoccupied insulin and IGF-I receptors in the control of apoptosis. Sci Signal. 2010;3(151):ra87. doi:10.1126/scisignal.2001173.
114. Tseng YH, Ueki K, Kriauciunas KM, Kahn CR. Differential roles of insulin receptor substrates in the anti-apoptotic function of insulin-like growth factor-1 and insulin. J Biol Chem. 2002;277:31601–11.
115. Diaz B, Pimentel B, De Pablo F, De La Rosa EJ. Apoptotic cell death of proliferating neuroepithelial cells in the embryonic retina is prevented by insulin. Eur J Neurosci. 1999;11:1624–32.
116. Bertrand E, Atti A, Cadoret A, L'Allemain G, Robin H, Lascols O, et al. A role for nuclear factor Kappa B in the antiapoptotic function of insulin. J Biol Chem. 1998;273:2931–8.
117. Lee-Kwon W, Park D, Baskar PV, Kole S, Bernier M. Antiapoptotic signalling by the insulin receptor in Chinese hamster ovary cells. Biochemistry. 1998;37:15747–57.
118. Rampalli AM, Zelinka PS. Insulin regulates expression of C-fos and c-jun and suppresses apoptosis of lens epithelial cells. Cell Growth Diff. 1995;6:945–53.
119. Kang S, Song J, Kang H, Kim S, Lee Y, Park D. Insulin can block apoptosis by decreasing oxidative stress via phosphatidylinositol 3-kinase and extracellular signal-regulated protein kinase-dependent signalling pathways in HepG2 cells. Eur J Endocrinol. 2003;148:147–55.
120. Tanaka M, Sawada M, Yoshida S, Hanaoka F, Manurouchi T. Insulin prevents apoptosis of external granular layer neurons in rat cerebellar slice cultures. Neurosci Lett. 1995;199:37–40.
121. Johnson JD, Bernal-Mizzachi E, Alejandro EU, Han Z, Kalynyak TB, Li H, Beith JL, Gross J, Warnock GL, Townsend RR, Permutt MA, Polonski KS. Insulin protects islets from apoptosis via Pdx1 and specific changes in the human islet proteome. Proc Natl Acad Sci U S A. 2006;103:19575–80.
122. Iida KT, Suzuki H, Sone H, Shimano H, Toyoshima H, Yatoh S, Asano T, Okuda Y, Yamada N. Insulin inhibits apoptosis of macrophage cell line, THP-1 cells, via phsphatidylinositol 3-kinase-dependent pathway. Artherioscler Thromb Vasc Biol. 2002;22:380–6.
123. Corcelle EA, Puustinen P, Jäättelä M. Apoptosis and autophagy: targeting autophagy signalling in cancer cells – "trick or treats"? FEBS J. 2009;276:6084–96.
124. Troncoso R, Vicenzio JM, Parra V, Nemchenko A, Kawashima Y, Del Campo A, Toro B, Battiprolu PK, Aranquiz P, Chiong M, Yakar S, Gillette TG, Hill JA, Abel ED, LeRoith D, Levandero S. Energy-preserving effects of IGF-I antagonize starvation-induced cardiac autophagy. Cardiovasc Res. 2012;93:320–9.
125. Danielsen AJ, Mailhe NJ. The EGF/ErbB receptor family and apoptosis. Growth Factors. 2002;20:1–15.
126. Maruyama IN. Mechanisms of activation of receptor tyrosine kinases: monomers or dimers. Cells. 2014;3:304-30. doi:10.3390/cells3020304.

Chapter 5
Nuclear Functions of Receptor Tyrosine Kinases

Yi Du, Jennifer L. Hsu, Ying-Nai Wang, and Mien-Chie Hung

5.1 Introduction

Receptor tyrosine kinases (RTKs) are well-known cell surface receptors for many growth factors and cytokines and play important roles in normal cellular processes [1, 2]. Abnormal expression and mutations of these RTKs are implicated in the development of several diseases, including many types of cancers [3, 4]. Activation of RTKs is initiated by ligand binding, and downregulation of ligand-activated RTKs is mediated by endosomal trafficking from early endosome to the lysosome [5, 6]. More recently, accumulated studies have shown that 11 of total 20 RTK subfamilies are detected in the nucleus, including the epidermal growth factor receptor (EGFR) family, the vascular endothelial growth factor receptor (VEGFR) family, the fibroblast growth factor receptor (FGFR) family, the insulin receptor family, the erythropoietin-producing hepatoma (Eph) family, the hepatocyte growth factor (HGF) receptor family, and the receptor tyrosine kinase-like orphan receptor (ROR) family as well as tropomyosin receptor kinase (Trk), receptor-like tyrosine kinase,

Y. Du, Ph.D.
Department of Molecular and Cellular Oncology, Unit 0079, The University of Texas MD Anderson Cancer Center, 1515 Holcombe Boulevard, Houston, TX 77030, USA

J.L. Hsu • Y.-N. Wang • M.-C. Hung, Ph.D. (✉)
Department of Molecular and Cellular Oncology, Unit 0079, The University of Texas MD Anderson Cancer Center, 1515 Holcombe Boulevard, Houston, TX 77030, USA

Center for Molecular Medicine, China Medical University, Taichung, Taiwan

Graduate Institute of Cancer Biology, China Medical University, Taichung, Taiwan
e-mail: mhung@mdanderson.org

D.L. Wheeler, Y. Yarden (eds.), *Receptor Tyrosine Kinases: Structure, Functions and Role in Human Disease*, DOI 10.1007/978-1-4939-2053-2_5

and Mer receptor tyrosine kinase. These novel observations of nuclear localization of RTKs and related functional studies demonstrate that RTKs function not only on the cell surface but also in the nucleus. Thus, it has become increasingly important to understand how the cell surface receptors can be trafficked to the nucleus and how these receptors function in the nucleus. A membrane-associated trafficking mechanism has been reported and provides a comprehensive pathway for nuclear translocation of EGFR [7–11]. This trafficking route may be a general mechanism by which other cell surface receptors also follow for their nuclear transport [12]. In this chapter, we will summarize the physiological and pathological functions of RTKs in the nucleus (Table 5.1).

5.2 EGFR Family

EGFR family includes EGFR/ErbB-1, ErbB-2/HER-2/neu, ErbB-3/HER-3, and ErbB-4/HER-4 [13]. All members of the EGFR family have tyrosine kinase activity except for ErbB-3. Upon ligand binding, the tyrosine kinase activity of EGFR family proteins is activated through either homo- or heterodimerization, triggering a multitude of downstream signaling pathways, including phosphatidylinositol-3 kinase (PI3K), mitogen-activated protein (MAP) kinases, signal transducer and activator of transcription (STAT), and phospholipase C (PLC), among others [14–18].

Both EGFR and ErbB-2 are overexpressed and/or constitutively activated in human tumors of epithelial origin, which cause aggressive tumor behaviors, such as cancer initiation, increased tumor growth/progression, poor patient outcome, metastasis, and chemoresistance [3, 19–21]. Many anticancer therapies targeting EGFR and ErbB-2 have been developed and approved by the US Food and Drug Administration such as monoclonal antibodies, cetuximab and trastuzumab, and small-molecule tyrosine kinase inhibitors (TKIs), lapatinib, erlotinib, and gefitinib. Although ErbB-3 does not have kinase activity, it can form heterodimer with other EGFR family members and lead to the activation of pathways related to cell proliferation or differentiation [22]. The activation of ErbB-4 requires ligand binding and dimerization with either EGFR or ErbB-3. ErbB-4 shares similar ligand binding such as HB-EGF, epiregulin, and β-cellulin binding with EGFR and neuregulins (NRG1 or NRG2) with ErbB-3 [23, 24]. Neuregulin/ErbB-4 signaling has been demonstrated to be critical for the regulation of cardiovascular system and nervous development [25].

5.2.1 Discovery of Nuclear EGFR Family

In addition to their roles in traditional signaling cascades, accumulating evidence shows that EGFR family translocates to the nucleus with unique biological functions [26–30]. Nuclear EGFR was first observed in the ducts of salivary glands more than 20 years ago [31]. Then, nuclear EGFR staining was observed in hepatocytes

Table 5.1 Summary of nuclear functions of RTKs

	Form of receptors in the nucleus				
	Full length	Truncated	Tissues or cell types	Functions	Clinical correlation
EGFR	Yes		Salivary gland [31], hepatocyte [33], keratinocyte [26], uterus [26], thyroids [35], mouth mucosa [26], placenta [205], adrenocortical carcinoma [32], epidermoid carcinoma [26], breast cancer [18, 26], gallbladder carcinoma [206], cervix cancer, follicular adenoma and carcinoma [35], head and neck cancer [207], pancreatic cancer [55], colorectal carcinomas [208], glioblastoma [44], ovarian cancer [209], lung cancer [28, 65]	Transcriptional regulation of target genes: cyclin D1 [26], iNos [18], b-myb [53], c-Myc [55], COX-2 [44], Aurora-A [54], BCRP [57], and thymidylate synthase [56]	Ductal invasive breast cancer [205], colorectal carcinomas [208], oropharyngeal squamous cell cancer [38], oral squamous carcinoma and breast cancer [37], ovarian cancer [209], esophageal squamous cell carcinoma [210]
				Kinase activity: PCNA [61]	
				Response to the DNA damage [28, 42, 64, 65]	
		EGFRvIII	Astrocyte [43], breast cancer [66], prostate cancer [67], glioblastoma [44]	Transcriptional regulation of target gene: COX-2 [44]	Hormone-refractory prostate cancer [67]
		mLEEK	Glioblastoma [45]	Transcriptional regulation of target gene: GRP78 [45]	None
ErbB-2 (HER-2)	Yes		Breast cancer, small bowel, esophagus, kidney, and oral tumors [27], glial cells [81], and ovarian cancer cells [69]	Transcriptional regulation of target genes: COX-2 [27], cyclin D1 [68], thymidylate synthase [56], and ErbB-2 [69]	Breast cancer [74]
				Translational regulation: ribosomal RNA [70]	
		p95, truncated 55, 39, and 30 kDa	Breast cancer [47, 72]	None	None
ErbB-3	Yes		Human epithelial cells and breast cancer [48], canine mammary tumor [211], mouse lung epithelial cell [76], glial cells [81], prostate cancer [77]		High frequency of nuclear ErbB-3 in hormone-refractory tissues [77]

(continued)

Table 5.1 (continued)

	Form of receptors in the nucleus				
	Full length	Truncated	Tissues or cell types	Functions	Clinical correlation
		Truncated 80, 55, 21 kDa, nuclear variant 50 kDa	Rhesus monkeys primate brain [49], rat Schwann cells [82]	Transcriptional regulation of target genes: ezrin and HMGB1 [82] and cyclin D1 [83]	
ErbB-4	Yes		Invasive breast cancer [50], rhesus monkeys primate brain [49], ventricular myocytes [90]	Doxorubicin-induced DNA damage response in cardiac myocytes [43]	A better differentiated phenotype [50]
		80 kDa ICD	Brain [86], secretory epithelium [84], osteosarcoma [87], Madin-Darby canine kidney II cells [89], and breast cancer cells [91]	Transcriptional regulation of target genes: β-casein [84], astrocytic genes in developing brain [86]	
FGFR1	Yes		Subpial astrocytes [93], mouse fibroblast cell [96], adrenal medullae [94], liver cells [212], glial cells [213], lung [120]	Transcriptional regulation of target genes: FGF-2 [97], tyrosine hydroxylase [106]	Never smoking lung cancer patients [120]
		55 kDa ICD	Breast cancer [109]	Transcriptional activity [109]	
FGFR2	Yes	None	Breast cancer [116], mammary gland [112], Sertoli cells [110], endometrial carcinoma [119], lung cancer [120]	FGFR2 colocalizes with STAT5 on DNA elements responsive to progesterone [116], mammary gland development [112], male sex determination [110]	Never smoking lung cancer patients [120], poor prognosis of breast cancer [114]
FGFR3	Yes		Breast epithelial cells [121], bladder carcinoma [124]		Bladder cancer [60]
		110 kDa FGFRIIIb, ICD	Breast tissue [122], monkey fibroblast cells [123]		
FGFR4	N/A		Glial progenitor cells [126]		
VEGFR1 (Flt-1)	Yes		Mouse hemangioma [134], breast cancer [135], gastrointestinal tissues [137], myxoid liposarcomas, fibrosarcoma, normal adipocytes, myxoid liposarcomas, ovarian cancer, pancreatic cancer, adipose tissue, smooth muscle, umbilical cord [136]	Angiogenesis [139]	

VEGFR2 (Flk-1/ KDR)	Yes	None	Endothelial cells [140], breast cancer, leukemia cancer, cervical cancer cells, and lung carcinoma sample [145]	Transcriptional regulation of target genes: VEGFR2 [144], wound healing [143], shear-stress response of vascular endothelial cells [142], angiogenesis [146]	
IR	Yes	None	Hepatocytes [150], osteosarcoma cells [154]		
IGF-1R	Yes	None	Hamster kidney cells [152], kidney cancer, prostate cancer, ductal carcinoma of the breast, ductal carcinoma in situ (DCIS), non-small cell lung cancer, pancreatic adenocarcinoma, colon cancer, lymphoma, malignant mixed mesodermal tumor and ovarian adenocarcinoma [155], hepatocellular carcinoma [156], orbital fibroblasts [161]	Transcriptional activity [155, 158, 160]	Survival of renal cell carcinoma [155], pathogenesis of immune disease [161]
				Target gene: CD133 [156]	
				Drug resistance [156]	
HGFR (Met)	Yes		Mouse hepatocytes [164], epithelial cells [171]	Calcium signals [166]	
		60 kDa CTF	Normal and cancerous samples of colon, testis, skin, and lymph nodes [165], breast carcinoma cells [167], biliary carcinoma [169], lung cancer [170]	Transcriptional activity [167]	Nuclear expression of phospho-Y1365 correlates with overall survival [170]
RON	Yes	N/A	Bladder cancer cells [175]	Transcriptional activity [175]	
ROR1/2	N/A	Cytoplasmic region 55 kDa	Transfected plasmids in HeLa cells		
EPHA4	N/A		Osteoblastic cell line [182]		
Ryk	N/A	ICD	Neural progenitor cells [185]	Neural progenitor, cell differentiation [185]	
				Transcriptional regulation of target genes: Dlx2 and Olig2 [187]	
TrkA	Yes	N/A	Pheochromocytoma 12 cells [190], glioma cells [191], parenchymal cells, hepatocytes, and activated stellate cells [192]		
Mer	Yes	N/A	Acute lymphoblastic leukemia cells [194]		
PDGFR	N/A	N/A	Early lung development or early embryonic development [196, 197]		

that underwent regeneration and in primary adrenocortical carcinomas by using immunohistochemical staining [32]. Later, nuclear EGFR was also found in other tissues such as placentas, normal mouth mucosa, and uterus from the pregnant mouse [26, 33–35] as well as tumors of the skin, breast, bladder, lung, pancreas, cervix, adrenal cortex, thyroid gland, and oral cavity [18, 26, 32, 33, 36–38]. Moreover, EGFR was also reported to localize to the nuclear membrane [34, 39, 40]. In addition to these early observations of nuclear EGFR by immunohistochemical staining, subsequent investigations further provided clear evidence showing that the full-length EGFR and its phosphorylated form exist in the nucleus [18, 26, 34, 37, 41, 42]. Aside from the full-length EGFR, a constitutively active variant of EGFR, EGFRvIII, has also been reported in the nucleus [43, 44]. Recently, a novel splice variant of EGFR called mini-LEEK (mLEEK), which lacks the extracytoplasmic, transmembrane, and tyrosine kinase domains, has been identified to function as a transcription factor in the nucleus [45].

Other EGFR family members are also found to exist in the nucleus. Nuclear rat neu/p185, the homologue of ErbB-2, was first detected in 1994. It was the first RTK shown to associate with transcriptional activity in the nucleus [46]. The human ErbB-2 was detected in the nucleus either as a full-length protein [27] or a truncated form [47]. Both the full-length and truncated forms of ErbB-3 have been observed in the nucleus [48, 49]. Finally, ErbB-4 was originally reported in the nucleus in 2000 [50], and a year later, the truncated intracellular domain of ErbB-4 (ICD) was discovered to have transcriptional function in the nucleus [30]. Collectively, these reports provide evidence to support that ErbB family receptors localize in the nucleus in addition to traditional cellular locations.

5.3 Functions of EGFR Family in the Nucleus

5.3.1 EGFR/ErbB-1

The first functional study of nuclear EGFR was reported in 2001 [26]. In this study, we demonstrated that nuclear-localized EGFR functions as a transcriptional regulator, mediating the expression of target genes such as cyclin D1 [26], and showed that nuclear EGFR binds to an AT-rich response sequence (ATRS) of the cyclin D1 promoter and stimulates its gene expression. This study is related to earlier in vitro studies which showed that EGFR interacts with supercoiled DNA and nicks double-strand DNA [51]. Although another study reported that EGFR-associated DNA-nicking activity is mediated by a M_r 100,000 protein [52], both of them still suggested that EGFR may associate with the DNA components in vitro. Other target genes of nuclear EGFR, such as inducible nitric oxide synthase (iNos) and B-Myb, were subsequently identified [18, 53]. More recently, some genes that are involved in tumorigenesis, chromosome instability, and chemoresistance, including Aurora-A [54], cyclooxygenase-2 (COX-2) [44], c-Myc [55], thymidylate synthase [56], and breast cancer resistance protein (BCRP/ABCG2) [57], were found to be regulated by nuclear EGFR.

Interestingly, EGFR does not contain a DNA-binding domain like most of the transcriptional factors do. Therefore, it was critical to identify the co-activators of nuclear EGFR to further understand its role as a transcriptional factor. Both RNA helicase A (RHA) [58] and MUC1 [59] were identified as co-regulators of the transcriptional function of nuclear EGFR. In the RHA study, EGF stimulates the interaction between EGFR and RHA in the nucleus of cancer cells, and the EGFR/RHA complex then interacts with ATRS of the target gene promoter to activate its transcription. Another co-regulator MUC1 was shown to increase the binding of EGFR with chromatin, especially to the promoter region of cyclin D1. The loss of both RHA and MUC1 significantly reduced the expression of cyclin D1. Together with earlier functional studies demonstrating that E2F1, STAT3, and STAT5 cooperate with nuclear EGFR for transcriptional regulation, EGFR is proposed to transport into the nucleus to serve as a co-activator to regulate the expression of target genes which affects several physiological and pathological phenotypes of human cancers, including proliferation, inflammation, and tumor drug resistance. It is worth mentioning that the role of nuclear EGFR family in transcriptional regulation is further supported by a recent finding that EGFR functions as a DNA-associated protein through non-biased approaches [60].

Nuclear EGFR also functions as a tyrosine kinase. In one study, EGFR was shown to bind to the chromatin to regulate DNA replication and DNA repair by phosphorylating and stabilizing proliferative cell nuclear antigen (PCNA) [61]. In another study, nuclear EGFR expression correlated with the phosphorylation of DNA-dependent protein kinase (DNA-PK), a key molecule in response to the double-strand DNA damage and DNA repair initiation in response to radiation treatment [42, 62, 63]. However, it was not yet clear whether DNA-PK serves a substrate of nuclear EGFR. A recent study suggested that nuclear EGFR expression correlated with DNA repair, which further supports the notion that nuclear EGFR may be important for DNA damage response and DNA damage repair [64].

In addition, nuclear EGFR has also been implicated in drug and radiation resistance during cancer therapy, suggesting that nuclear EGFR could be a new target for cancer treatment. David Wheeler and Chaitanya Nirodi's group reported that nuclear EGFR correlated with drug and radiation resistance, respectively [28, 65]. Non-small cell lung cancer (NSCLC) cells with mutated EGFR sensitive to the radiation treatment were shown to be defective in radiation-induced nuclear translocation of EGFR [65]. In the NSCLC cells, the expression of nuclear EGFR correlated with cetuximab-acquired resistance [28]. It was recently shown that nuclear translocation of EGFR results in TKI resistance in breast cancer cell lines through the upregulation of BCRP/ABCG2 [57].

The EGFR variant type III (EGFRvIII), which contains deletion of exons 2–7, has been reported to localize in the nucleus in primary invasive breast cancer tissues [66] and is associated with decreased overall survival in prostate cancer patients [67]. This constitutively active form of EGFR exists in the nucleus of glioblastoma and forms a complex with STAT3 to promote malignant cell transformation [43]. More recently, studies have also shown that both the full-length EGFR and the truncated EGFRvIII bind to the COX-2 promoter to regulate gene expression [44]. Another EGFR in-frame splice variant, called mLEEK, was identified in 2011 [45]. Wong and coworkers showed that mLEEK, which functions as a transcriptional regulator of target genes involved in endoplasmic reticulum (ER) stress, such as molecular chaperone GRP78/Bip, in the unfolded protein response (UPR) pathway, localizes to the nucleus and is commonly expressed in normal tissues but overexpressed in several cancer types. They further demonstrated that mLEEK controls the expression of GRP78 through direct binding with the promoter.

5.3.2 ErbB-2/HER-2

ErbB-2/HER-2 is the only receptor of EGFR family that does not have well-defined ligands. However, the ErbB-2 forms a ligands. However, with other EGFR members to conduct signal transduction [15]. ErbB-2 was the first RTK identified to have transcriptional activity in the nucleus in 1994 [46]. Its first target gene, COX-2, was later reported to be regulated by nuclear ErbB-2 through binding to the promoter of COX-2 [27]. Higher expression of nuclear ErbB-2 has been linked to anti-apoptosis, proangiogenesis, and metastasis in cancer cells. Similar to nuclear EGFR, nuclear ErbB-2 also regulates the gene expression of cyclin D1 [68]. Evidence showed that nuclear ErbB-2 forms a complex with STAT3 that plays an important role in the proliferation of breast tumors with functional progesterone receptor. A recent study further revealed that nuclear ErbB-2 interacts with micro-H2A1.2 to regulate itself gene expression [69]. In addition, nuclear ErbB-2, detected in the nucleolus, also associates with nuclear activity to function as a translational regulator of rRNA synthesis which may lead to tumor development, protein synthesis, and cell growth [70]. This is the first time that an RTK is shown to regulate RNA polymerase I-mediated in addition to polymerase II-regulated gene expression.

Another study reported the colocalization of ErbB-2 and p34(Cdc2) in the nucleus, suggesting that ErbB2 may interact with Cdc2 in the nucleus to inhibit Cdc2's activity, resulting in Taxol resistance in breast cancer [71]. Like the full-length ErbB-2, the truncated form of ErbB-2 (p95) has also been observed in the nucleus [72] and implicated in trastuzumab resistance in breast cancer patients. A recent study further indicated that TKI-induced nuclear translocation of p95 ErbB-2 contributes to TKI resistance in breast cancer cell lines [47].

Clinically, nuclear ErbB-2 has been reported to play an important role in tumorigenesis. The expression of COX-2 and nuclear ErbB-2 correlates with disease-free survival in patients who received adjuvant tamoxifen [73]. Another study showed that the overall survival in breast cancer patients is correlated with nuclear expression of ErbB-2 [74].

5.3.3 *ErbB-3*

ErbB-3 is a unique EGFR member that has low levels of kinase activity due to its functionally defective tyrosine kinase domain. However, it can still be activated by its ligand neuregulin (NRG) through heterodimerization with other ErbB members [24, 75]. In human mammary epithelial and breast cancer cell lines, the full-length ErbB-3 was reported in the nucleus [48]. Nuclear ErbB-3 was predicted to function during differentiation and polarization of cells in which ErbB-3 localized in the nucleoli. When cells were starved with serum-free media, ErbB-3 localized in the nucleus and was enriched in the nucleoli. Once cells were exposed to NRG after serum starvation, nucleoplasm and cytoplasm localized ErbB-3 were enhanced. It appeared that ErbB-3 localized to the nucleolus under serum starvation rather than on the cell surface plasma membrane and that NRG treatment changed the localization of ErbB-3 from the nucleolus to the nucleoplasm and cytoplasm. These results suggest that nucleolus-localized ErbB-3 may regulate ribosomal RNA synthesis during growth arrest. A similar phenomenon was reported in mouse lung type II epithelial cells in which ErbB-3 was detected in the nucleolar regions in serum-starved cells [76]. ErbB-3 has also been detected in the nuclei of prostate cancer cells, glial cells, and brain tissues [49, 77–81]. Moreover, clinical study of nuclear ErbB-3 suggests that nuclear staining of ErbB-3 is associated with the progression of prostate cancer [77]. In the metastatic prostate cancer specimens, ErbB-3 was also detected in the nucleus. Further study indicated that nuclear ErbB-3 may aid prostate cancer cells in survival during androgen ablation and progression of prostate cancer to the bone [78]. A truncated form of ErbB-3 was observed in the nuclei of Schwann cells, which regulates Schwann cell development through transcriptional regulation of target genes such as ezrin and high-mobility group box 1 (HMGB1) [82]. More recently, one study provided evidence to support that an 80 kDa variant of ErbB-3 promotes cyclin D1 transcription in the nucleus and activates cell proliferation [83].

5.3.4 *ErbB-4*

ErbB-4 was first detected in the nucleus of invasive breast cancers [50] in which positive immunocytochemical staining of ErbB-4 in the nucleus was detected using an antibody against the intracytoplasmic domain of ErbB-4. In the following year in 2001, Graham Carpenter and coworkers showed that the intracellular domain (ICD) of ErbB-4, released by γ-secretase cleavage, translocates to the nucleus [30]. Functional studies using GAL4 transactivation system further showed that the ErbB-4 ICD has weak but detectable transcriptional activity which is similar to the earlier findings on EGFR and rat neu [26, 46]. In addition, the nuclear-localized ErbB-4 ICD fragment also associates with STAT5A to activate the expression of β-casein in response to ligand stimulation [84]. In another study, nuclear ErbB-4 ICD was shown to bind to Eto-2, a transcriptional corepressor, and blocks Eto-2-dependent transcriptional repression through ICD kinase-independent way [85]. Nuclear ErbB-4 ICD has also been shown to play a role in astrogenesis and the development of brain through presenilin-dependent manner [86]. Immunohistochemistry staining also indicated the nuclear expression of p80 fragment of ErbB-4 in primary osteosarcoma [87]. Based on the proteinase cleavage site (called JM-a or JM-b) and the PI3K binding motif (called CYT-1 or CYT-2), four different isoforms of ErbB-4 have been identified [88]. Among them, the JM-a/CYT-2 isoform of ErbB-4 has been found in the nucleus of Madin-Darby canine kidney (MDCK) II cells in response to HB-EGF stimulation [89]. While some functions pertaining to the nuclear localization of soluble ICD fragment of ErbB-4 have been delineated, recent studies also detected the full-length ErbB-4 in the nucleus of brain samples [49] and ventricular myocytes [90]. In cardiac myocytes, nuclear-localized ErbB-4 was shown to play a role in doxorubicin-induced DNA damage response. In addition, the colocalization of ErbB-4 ICD and the HIF-1 in the nucleus was recently reported [91]. Results suggest that nuclear ICD fragment of ErbB-4 stabilizes the HIF-1 to promote its signaling.

In summary, as an important family of receptor tyrosine kinases, all of the EGFR family members have been detected in the nucleus as full-length, truncated, or mutated variants. In contrast to the traditional signaling pathway mediated by the EGFR family on the cell surface, nuclear-localized EGFR family may be a new machinery to regulate physiological and pathological function of cells.

5.4 FGFR Family

Fibroblast growth factor receptors (FGFRs) and their ligands FGFs regulate a wide range of biological functions, including cellular proliferation, survival, migration, and differentiation [92]. The FGFR family has five subfamily members: FGFR1, FGFR2, FGFR3, FGFR4, and FGFR5. FGFR5 (also known as FGFRL1) binds to FGFs but does not have tyrosine kinase domain. Intracellular trafficking of FGFRs is mainly related to the downregulation of receptors through lysosomal degradation from endosomal trafficking, which is similar to the EGFR family. So far, with the exception of FGFR4, most of the other FGFR subfamily members have been detected in the nucleus of various types of normal and cancer cells.

5.4.1 FGFR1

FGFR1 has been detected in the nuclei of subpial astrocytes in rat brain and chromaffin cells from bovine adrenal medullae (BAMC) [93, 94]. Further studies revealed that nuclear FGFR1 exists in a variety of cell types including nontransformed cells, cancer cell lines [95–101], and cochleovestibular ganglion cells [102]. FGFR1 nuclear accumulation has been observed in vivo in both developing neurons and glial cells [99–101, 103, 104] and has been shown to play important roles in glial growth, dendritic outgrowth in rat neurons, differentiation of human and rat neuronal progenitor cells (NPCs), and functional adaptations in adrenal medullary cells [100, 105]. Among these investigations of nuclear FGFRs, Michael Stachowiak and coworkers revealed the mechanism and functions of nuclear FGFR1, showing that nuclear FGFR1 releases CREB-binding protein (CBP), a general transcriptional co-activator, from ribosomal S6 kinase 1 (RSK1) and activates the transcription of target genes and chromatin remodeling [100]. The direct interaction between nuclear FGFR1 and CBP stimulates FGF-2 [97] and tyrosine hydroxylase (TH) gene expression [106]. Meanwhile, another report also demonstrated the transcriptional function of nuclear FGFR1 in which nuclear FGFR1 can upregulate cell proliferation through gene expression of cyclin D1 and c-Jun [96]. More recently, studies further revealed the mechanism by which nuclear FGFR1 regulates the transcriptional activity of nuclear FGFR1 [107, 108]. In these studies, the nuclear receptor-related 1 protein (Nurr1) was shown to be involved in the transcriptional regulation of TH gene expression mediated by nuclear FGFR1 during ventral mesencephalic dopaminergic (mDA) neuron development. The nerve growth factor IB (NGFIB/Nur77) has also been shown to play a role in nuclear FGFR1- and Nurr1-mediated regulation of developmental genes through RXR and RAR.

Although the previous studies have indicated that nuclear FGFR1 is a full-length receptor, the truncated form of FGFR1 that translocates to the nucleus was recently reported [109]. Specifically, it was shown that granzyme B (GrB) cleaves the full-length FGFR1, and the truncated 55 kDa intracellular domain then translocates into the nucleus in response to FGF-10 stimulation. This 55 kDa fragment of FGFR1 is involved in the transcriptional regulation that controls the behaviors of cancer cells.

5.4.2 FGFR2

FGFR2 was first found to localize to the nucleus of differentiating Sertoli cells during male sex determination [110]. Nuclear FGFR2 initiates the expression of Sry gene, which determines the sex of the gonad and the differentiation of cells into Sertoli cells. Later studies showed that SOX9 is required for the nuclear localization of FGFR2, which is important for the maintenance of SOX9 expression and sex determination [111]. Nuclear FGFR2 was identified in the terminal end buds during mammary gland development [112] and has been shown to play a role in hormone-independent mammary tumor growth [113]. Moreover, increased nuclear expression of FGFR2 also correlated with poor prognosis or patient genotype in breast cancer patients [114, 115]. Nuclear FGFR2 also forms a complex with STAT5 and the progesterone receptor which binds to the DNA elements in response to the progesterone to regulate mammary gland development [116]. Although specific target genes of nuclear FGFR2 have not yet been identified, these results suggest that nuclear FGFR2 has an important role in breast cancer through the transcriptional regulation.

Nuclear FGFR2 has also been detected in other cell types. For example, the accumulation of nuclear FGFR2 is associated with prostaglandin-induced osteoblast development [117]. Subsequent studies from the same laboratory showed that importin is involved in increased nuclear trafficking of FGF-2 and FGFR2, which is stimulated by prostaglandin F2α in rat osteoblasts [118]. In endometrial carcinoma, although nuclear FGFR2 was detected, no statistical significance between superficial and myoinvasive endometrial carcinoma samples was found [119]. However, in lung cancer, nuclear FGFR2 expression was significantly higher in never smoker of lung cancer patients, suggesting nuclear FGFR2 is an attractive novel target of lung cancer chemopreventive and therapeutic strategies [120].

5.4.3 FGFR3 and FGFR4

While FGFR3 was detected in the nucleus of breast epithelial cells [121], its functions have not been well investigated. The first reported nuclear FGFR3 is a 110 kDa soluble protein (FGFRIIIb) lacking the transmembrane domain and is predominantly localized to the nucleus [121]. A subsequent study further showed that FGFR3 is predominantly localized to the nucleus of human breast tissues by immunostaining. Although the authors did not clearly conclude which FGFR3 isoform localizes to the nucleus, PCR results indicated that both FGFRIIIb and full-length FGFR3 are found in these breast tissues and cell lines [122].

The truncated form of FGFR3 was reported to transport into the nucleus in an FGF-1-induced manner in transfected COS7 monkey fibroblast cells. The final soluble intracellular domain is generated by a two-step cleavage at the ectodomain and intracellular domain, yielding a soluble ICD which then translocates into the nucleus. This mechanism suggests that nuclear translocation of truncated FGFR3 may occur in FGF-induced biological events [123]. The expression pattern of FGFR3 has also been studied in normal urothelium, non-metastasized transitional cell carcinomas, and primary metastasized transitional cell carcinomas [124]. Higher nuclear staining of FGFR3 is only in the non-metastasized carcinomas, but not in the normal urothelium and metastasized carcinomas. These results may provide a possibility of nuclear FGFR3 for the diagnostic implication. Interestingly, FGFR3 was also detected on the nuclear membrane of bladder cancer cells when cells were treated with immunotoxin, suggesting that nuclear membrane-localized FGFR3 may be involved in the immunotoxin response [125].

To date, there has only been one study that documented nuclear staining of FGFR4 in the transformed glial progenitor cells of Sprague Dawley rats [126], but the function of nuclear FGFR4 is not known.

The FGFR family is another group of RTKs that has been reported to localize to the nucleus. Similar to nuclear EGFR family, nuclear-localized FGFR family is also involved in transcriptional regulation. However, among all members of the FGFR family, only FGFR1 has been well investigated for its functions in the nucleus and the mechanism underlying its nuclear trafficking. Other FGFR family members await further investigations to reveal their potential nuclear functions and clinical relevance.

5.5 VEGFR Family

Vascular endothelial growth factor receptor (VEGFR) family belongs to the RTK family and plays a role in angiogenesis. These receptors bind to vascular endothelial growth factor (VEGF) ligands and initiate signaling transduction through dimerization and autophosphorylation [127, 128]. There are three VEGFR family members, VEGFR-1 (Flt-1), VEGFR-2 (KDR/Flk-1), and VEGFR-3 (Flt-4), with all three having different functions based on their binding partners [129].

VEGF-A binds to both VEGFR-1 and VEGFR-2, but VEGFR-2 appears to mediate almost all of the known cellular processes in response to VEGF stimulation. VEGFR-1 signaling is activated in many non-endothelial cell types, including bone marrow and progenitor cells; mediates a broad spectrum of biological functions such as angiogenesis, tumor growth, and metastasis [130]; and plays a key role in vascular development and regulation of vascular permeability. Moreover, VEGFR-1 is thought to modulate VEGFR-2 signaling in endothelial cells by binding to VEGF to downregulate VEGFR-2 function [131–133]. Unlike the other VEGFR family members, VEGFR-3 does not bind to VEGF-A. Instead, it mediates lymphangiogenesis through VEGF-C and VEGF-D stimulation.

5.5.1 VEGFR-1

Nuclear VEGFR-1 was first detected in mouse hemangioma-derived cells and is associated with the nuclear expression of β-catenin [134]. Later, nuclear VEGFR-1 was found in a wide range of cell types. In MDA-MB-231 and MCF-7 breast cancer cells, VEGFR-1 was reported to colocalize with lamin A/C, which are localized in the inner nuclear membrane and expressed in the nucleoplasm [135]. Nuclear VEGFR-1 was also observed in myxoid liposarcomas (MLS), ovarian and pancreatic cancer, and normal tissues (adipose smooth muscle and umbilical cord) [136]. Another study presented more evidence of VEGFR-1 nuclear localization in gastrointestinal tissues from different species, including mouse, rat, dog, swine, and monkey [137]. The soluble fragment of VEGFR-1 cleaved by γ-secretase was identified and demonstrated to translocate into the nucleus via a presenilin-mediated manner for its role in the angiogenesis [138, 139]. Although the function of nuclear VEGFR-1 is not yet clearly understood, the discovery of nuclear VEGFR-1 in different cancer cells and normal tissues opens a new perspective for further investigation.

5.5.2 VEGFR-2

As the major mediator in response to the VEGF, VEGFR-2 is the first VEGFR family member detected in the nucleus of endothelial cells under VEGF stimulation [140]. In response to VEGF treatment, VEGFR-2 translocates to the nucleus and forms complex with endothelial nitric oxide synthase (eNOS). Immunofluorescence study showed that caveolin is involved in VEGFR-2 nuclear trafficking. These results suggest that nuclear localization of VEGFR-2/eNOS may be related to NO production in the nuclear compartment and could affect transcriptional activation. The tissue transglutaminase (tTG) is another component of VEGF/VEGFR-2 complex that is involved in the nuclear translocation of VEGFR-2 [141]. Both tyrosine kinase activity of VEGFR-2 and transglutaminase activity of tTG are required for the formation of tTG/VEGF/VEGFR-2 complex. In addition, hemodynamic forces were also found to induce nuclear translocation of VEGFR-2, which requires complex formation between VEGFR-2 and junction molecules such as VE-cadherin and β-catenin. These findings suggest that nuclear translocation of VEGFR-2 plays a role in transducing shear-stress signal into the vascular endothelial cells [142].

In 2007, one potential function of nuclear VEGFR-2 was reported in primary human umbilical vein endothelial cells (HUVEC) [143] in which nuclear VEGFR-2 appeared to be involved in endothelial cell recovery after wounding. Recently, nuclear VEGFR-2 was found to have transcriptional function through its interaction with transcriptional factor SP1 to regulate the expression of target genes including VEGFR-2 itself [144]. The identification of the transcriptional function of nuclear VEGFR-2 further supports the transcriptional activity of nuclear receptor tyrosine kinase.

Nuclear VEGFR-2 also plays important roles in neoplastic cells, including breast cancer cell line, leukemia, cancer cell line, cervical cancer cell line [145], and tumor angiogenesis [146]. In breast cancer, cervical cancer, and leukemia cell lines, VEGFR-2 was found to localize in the nucleus in response to VEGF stimulation. Moreover, immunofluorescence staining of lung tumor tissue samples also indicated that VEGFR localized to the nucleus. The observation of nuclear VEGFR-2 in cancer cells and tumor tissue samples suggests that nuclear localization of VEGFR-2 may play an important function in tumorigenesis.

5.6 Other Receptor Tyrosine Kinases

5.6.1 Insulin Receptor Family

5.6.1.1 Insulin Receptor

The insulin receptor (InsR) and the insulin-like growth factor I receptor (IGF-1R) belong to the same receptor tyrosine kinase family. Activation of InsR and IGF-1R is initiated by ligand-triggered autophosphorylation of the receptor cytoplasmic tyrosine residues and turns on a cascade of signaling transductions which control biological events such as metabolism, cell proliferation, cell differentiation, DNA repair, anti-apoptosis, survival, and growth. These events result in a range of diseases and tumorigenesis [147–149].

The earliest detection of nuclear InsR in isolated hepatocytes was documented more than 20 years ago [150]. In this study, nuclear translocation of insulin receptor was explained as a long-term effect of insulin signaling. A recent study revealed that InsR and its signal transducers are recruited to insulin-inducible gene loci to maintain mRNA synthesis of target genes, providing a possible mechanism of insulin resistance [151].

5.6.1.2 Insulin-Like Growth Factor I Receptor

IGF-1R, another member of insulin receptor family, was detected in the nucleus of hamster kidney cells [152]. This phenomenon suggested that IGFs exert their biological effects in the nucleus with IGF-1R. Later, it was shown that the levels of nuclear InsR were higher in hepatocyte cells of obese mice [153]. Although there was no mechanistic or functional study for relationship between nuclear InsR and obesity, these results suggest that nuclear translocation of InsR may be responsible for target gene expressions of insulin signaling pathway. Moreover, nuclear InsR was also reported in osteosarcoma cells [154] which supports the role of nuclear insulin signaling in transcription control, differentiation, and growth of cancer cells.

Nuclear IGF-1R has been detected in a wide spectrum of human cancer samples including kidney cancer, prostate cancer, ductal carcinoma of the breast, ductal carcinoma in situ (DCIS), NSCLC, pancreatic adenocarcinoma, colon cancer, lymphoma, uterine (malignant mixed mesodermal tumor) MMMT, and ovarian serous adenocarcinoma [155]. Importantly, clinical data analysis showed that the nuclear IGF-1R is correlated with patient survival of renal cell carcinoma. In addition, the mechanism of IGF-1R translocation into the nucleus through clathrin-mediated endocytosis was also identified in their study. The chromatin binding ability of nuclear IGF-1R suggests a potential role of IGF-1R as a transcriptional regulator. Although the target genes regulated by nuclear IGF-1R were not mentioned in their study, inhibition of IGF-1R phosphorylation and nuclear translocation by specific small-molecule inhibitors may have clinical application for cancer therapy.

Nuclear IGF-1R was also continually detected in the hepatocellular carcinoma (HCC) [156]. When cells were exposed to gefitinib (tyrosine kinase inhibitor against EGFR), IGF-1R translocated into the nucleus. Nuclear IGF-1R was shown to be associated with the upregulation of the stem cell marker CD133. These findings suggest a link between cancer drug resistance and IGF-1R nuclear translocation and implicate IGF-1R in targeted cancer therapy.

Posttranscriptional modifications play an important role in modulating cellular signaling pathways to regulate various biological responses through changes in protein activity, localization, and binding ability [157]. In addition to phosphorylation, SUMOylation of IGF-1R was recently identified [158], and SUMOylated IGF-1R is required for its nuclear accumulation induced by IGF. Mutation of several SUMO1 modification sites blocked IGF-1R from translocating into the nucleus to regulate transcription. Additional studies not only further supported the significance of SUMOylation and nuclear localization of IGF-1R in cancer cells [159] in which the level of Ubc9, a SUMO-conjugating enzyme required for the SUMOylation of IGF-1R, is higher in cancer cells than normal cells but also uncovered the mechanism of transcriptional regulation of nuclear IGF-1R through LEF1/TCF [160]. This observation suggests that overaccumulation of nuclear IGF-1R may contribute to abnormal gene expression in cancer cells and therefore play a role in tumorigenesis.

While most studies have focused on the function of nuclear IGF-1R in cancer cells, researchers working on Graves' disease recently discovered that nuclear IGF-1R is also induced by IGF in the orbital fibroblasts [161]. They proposed that glycoprotein metalloproteinase ADAM17 might be involved in the intracellular translocation of IGF-1R. Even though the mechanism of how IGF-1R translocates into the nucleus in Graves' fibroblasts is unclear and requires further investigation, these observations strongly imply that nuclear IGF-1R may play a role in the pathogenesis of immune disease.

5.6.2 HGF Receptor Family

The HGFR family includes Met, also known as the hepatocyte growth factor (HGF) receptor, and recepteur d'origine nantais (RON). The HGFR family regulates cell growth, cell motility, and morphogenesis and correlates with poor prognosis of many cancers [162, 163].

Met was first observed in the nucleus in 1996 [164]. In this study, immunohistochemistry staining of nuclear HGFR was detected in mouse liver when mice were treated with phenobarbitone, which is used to induce non-genotoxic rodent hepatocarcinogenesis. This observation suggests that nuclear-localized HGFR may be correlated with transient replication of hepatocytes caused by phenobarbitone. Ten years later, Met was reported to localize in the nucleus of normal and cancerous tissues as well as in the nucleus of cancer cell lines [165]. Positive staining of Met seemed to have come from its 60 kDa C-terminal fragment. Together, these results suggest that nuclear Met may play a role in extending signals of the Met receptor in the nucleus or may indicate a new signal transduction. In immortal liver cells, nuclear translocation of Met can be induced by HGF stimulation to enhance $Ca2^+$ signals in the nucleus mediated by Grb1 and importin β, indicating that HGF may have a specific role in bypassing signaling pathways in the cytoplasm and turning on activation directly in the nucleus [166].

Furthermore, HGFR was detected in the nucleus of an aggressive breast cancer cell line in an HGF-independent manner [167]. Consistent with a previous study [165], the nuclear HGFR observed was also a 60 kDa C-terminal fragment rather than a full-length receptor. Functional studies indicated that nuclear HGFR fragment has potential transcriptional activity. In line with the function of nuclear HGFR fragment in breast cancer cells, a bone metastatic clone derived from MDA-MB-231 showed higher levels of the 60 kDa fragment of HGFR in the nucleus with constitutive activation of β-catenin/TCF complex [168]. However, the activation of β-catenin and TCF regulated by HGF was not observed in parental MDA-MB-231 cells. Thus, targeting the nuclear Met/β-catenin/TCF signaling pathway could have potential therapeutic implication in the progression of breast cancer to bone metastases. In addition, recent studies have reported positive staining of nuclear HGFR in biliary carcinoma [169], nuclear expression of HGFR in lung cancer [170], and physiological relevance of nuclear-localized phosphorylated c-MET [171]. These findings further support the important physiological and pathological roles of nuclear HGFR and open a new avenue for further investigations.

RON is another HGF receptor family member whose expression is highly restricted in epithelial cells [172]. The binding of RON with its ligand, the HGF-like macrophage-stimulating protein (MSP), stimulates homodimerization of RON to activate signal transduction. RON has been shown to be aberrantly expressed or mutated in several cancer types, such as the bladder, breast, colon, lung, ovary, pancreas, and prostate, and is particularly associated with poor patient survival in aggressive tumors [172]. Activated RON can promote c-Src activities that mediate cell-cycle progression, angiogenesis, and survival of tumor cells [173, 174]. In bladder cancer cells, serum starvation can induce the translocation of RON to the nucleus [175], where RON forms a complex with EGFR and then passes through the nuclear pore complex mediated by importin α. The RON/EGFR complex functions as a transcriptional regulator in the nucleus. This study also identified target genes known to be involved in stress response and demonstrated the cooperation between different receptor tyrosine kinases for nuclear transport under stress response. While it is not clear whether the transcriptional regulation of nuclear RON comes from itself or nuclear EGFR, the direct nuclear transport of RON bypassing the ligand-induced signal transduction under stresses still confers a survival advantage.

5.6.3 ROR

The receptor tyrosine kinase-like orphan receptor (ROR) family plays crucial roles in the development of various organs and tissues. The human ROR family includes two members, ROR1 and ROR2. Based on the domain they contain, ROR are related to tropomyosin receptor kinase (Trk) and are also called neurotrophic tyrosine kinase receptor (NTKR) family and muscle-specific kinase (MuSK). The ROR family was named as orphan receptors because there are no identified ligands for their activation. However, recent studies have shown that the regulation of ROR2 is mediated by the frizzle domain of the Wnt5a protein and plays roles in normal development and tumorigenesis [176, 177].

The ROR1 was recently detected in the nucleus [178]. Using the bioinformatics analysis (PSORT II program), the authors predicted that ROR1 and ROR2 can be translocated into the nucleus. This prediction was supported by the detection of the cytoplasmic domain of ROR1 and ROR2 in the nucleus. The putative nuclear localization signal and the potential functions of nuclear ROR1 were identified in this study [179]. Although systematic investigation is needed to reveal the function of nuclear ROR family, the detection of nuclear ROR family provides additional evidence to support the presence of nuclear receptor tyrosine kinases.

5.6.4 Eph Receptors

Erythropoietin-producing hepatoma (Eph) receptors are activated in response to the interaction with its ligand ephrin. Both Eph receptors and ephrin are membrane-bound proteins, and thus Eph/ephrin signaling transduction requires direct cell-cell interaction. Eph receptors and ephrin ligands have been implicated in the regulation of embryonic development such as axon guidance, formation of tissue boundaries, cell migration, and segmentation. Additionally, Eph/ephrin signaling has recently been demonstrated to play critical roles in several physiological and pathological processes including long-term potentiation, angiogenesis, stem cell differentiation, and cancer [180, 181].

To date, only one member of the Eph family, EphA4, has been detected in the nucleus [182]. EphA4 was found to localize to the cell membrane, cytoplasm, and nucleus in a human osteoblastic cell line. Currently, the specific functions of nuclear EphA4 are not known, and it is not clear whether EphA4 exists in the nucleus as full-length protein or truncated fragment.

5.6.5 Ryk

The receptor-like tyrosine kinase (Ryk) is an atypical member of the family of receptor tyrosine kinases. Different from other members, the C-terminal domain of Ryk does not have detectable kinase activity [183]. Studies have indicated that Ryk functions as a co-receptor with Wnt receptor proteins and in neuron differentiation and neurite outgrowth [184]. In neural progenitor cells, upon binding to Wnt3a, Ryk is cleaved at the C-terminus to release its intracellular domain, ICD, which then translocates to the nucleus [185]. Recently, it was shown that Cdc37, a subunit of the molecular chaperone Hsp90, binds to the ICD of Ryk to stabilize it [186]. Similar to the functions of other nuclear receptor tyrosine kinases, nuclear ICD of Ryk functions as a transcriptional regulator to regulate the expression of target genes, such as Dlx2 and Olig2, which are key cell-fate determinants, promote GABAergic neuronal differentiation, and inhibit oligodendrocyte differentiation [187].

5.6.6 *Trk*

Tropomyosin receptor kinases (Trk) are receptors for nerve growth factors (NGF) such as neurotrophin. Trk regulates and controls the development and functions of the nervous system and a wide range of other biological processes [188, 189]. The Trk family comprises of TrkA (NTRK1), TrkB (NTRK2), and TrkC (NTRK3). Binding of neurotrophins to Trk receptors activates signal transduction and promotes both neuronal cell survival and death.

TrkA was first reported to accumulate in the nucleus of PC12 cells with LPA1 receptor in response to NGF stimulation [190]. Results from this report suggest that TrkA and LPA1 receptors might participate in nuclear signaling such as regulation of gene expression. In another study, phosphorylation of TrkA has been found in the nucleus of the human glioma cell line, U251, after treatment of neurotrophin [191]. In this report, the nuclear transport of phosphorylated TrkA is carried by vesicles. The phosphorylated state of nuclear TrkA suggests TrkA may potentially function as a tyrosine kinase in the nucleus. A study later confirmed the previous observation of nuclear-localized Trk by using different antibodies against either the extracellular or intracellular domains to rule out the nonspecific binding effects of antibody. In different cell types, including parenchymal cells, cells of the inflammatory infiltrate, hepatocytes, and activated stellate cells (HSC), nuclear localization of TrkA can be detected [192]. While these investigations all have observed nuclear localization of TrkA, it is not yet clear what its functions are in the nucleus.

5.6.7 *Mer*

Mer, Axl, and Tyro3 have similar structure and functional domains and form another receptor tyrosine kinase family. Mer receptor tyrosine kinases regulate a broad spectrum of physiological functions, including platelet aggregation, oncogenesis in lymphocytes, phagocytosis of apoptotic cells, and modulation of cytokine production [193]. Recent evidence indicates that Mer is a new member of receptor tyrosine kinase detected in the nucleus [194]. Long-term exposure of acute lymphoblastic leukemia cells to Gas6, a ligand of Mer receptor, results in reduced glycosylation of Mer, which increases soluble Mer and decreases chromatin-bound Mer in the nucleus, and changed the subcellular localization and even the nuclear expression pattern of Mer. These results suggest that Mer may also be functionally important in the nucleus.

5.6.8 PDGFR-Alpha

Alpha-type platelet-derived growth factor (PDGF) receptor (PDGFR-α) belongs to the PDGFR family and is a receptor for PDGF-A, PDGF-B, and PDGF-C. PDGFR-α has been shown to play an important role in the regulation of embryonic development, cell proliferation, and survival [195]. A small number of studies have reported nuclear-localized PDGFR-α in alveolar fibroblasts during early lung development or early embryonic development of mouse model [196, 197], but the functions of nuclear PDGFR-α during these developmental stages are still awaiting further investigations.

5.7 Summary

In addition to the ErbB, FGFR, and VEGFR families, eight other RTK families have also been reported to localize to the nucleus. These include the insulin receptor, HGF receptor, ROR, Trk, Eph receptor, Ryk, Mer, and PDGFR-α. Collectively, these findings open a whole new field of research in which RTKs that are originally thought to exist only on the cell surface are also found in the nucleus with specific or unknown functions. Although there has been increasing number of laboratories that are taking on this new direction of nuclear RTKs, a number of questions remain unanswered and would require continuous investigation to clarify their pathological and physiological relevance.

5.8 Conclusion

Traditionally, all receptor tyrosine kinases are thought to localize on the plasma membrane where receptors bind to their ligands to initiate the signal transduction from outside to inside of cells. Accumulating observations have provided another perspective on receptor tyrosine kinases as more than just cell surface receptors. The functions of RTKs in specific organelles and cellular components were reported by multiple studies [198–203], and 11 out of 20 receptor tyrosine kinase families were found to localize in the nucleus. Compared with functions of RTKs on the cell surface as the signaling initiator in response to ligand stimulation, nuclear-localized RTKs seem to bypass traditional signal transductions to move into the nucleus and respond to stimuli through transcriptional and translational regulation of target genes. This notion is supported by the observation of nuclear RTKs in higher proliferating normal tissues or in cancer tissues requiring higher levels of signaling to simulate cell growth, survival, and differentiation [26, 33, 105]. Stresses such as radiation treatment or chemo-treatment have also been reported to stimulate the nuclear localization of RTKs [28, 42, 204]. In addition, nuclear localization of

RTKs by immunohistochemical or immunofluorescence staining was reported in a series of normal tissues, cancer cell lines, and tumor tissues, providing stronger evidence to demonstrate that nuclear-localized RTKs have important implications in physiological conditions or tumorigenesis. It has been almost 30 years since the first receptor was detected in the nucleus [31, 150]. Efforts from researchers around the world have focused on why cell surface receptor tyrosine kinases translocate to the nucleus, how they translocate into the nucleus, and what we can learn from the non-canonical signal transduction of RTKs as understanding of the nuclear translocation events can potentially be used in clinical applications. What we have discovered so far represent just the tip of the iceberg. There is much more waiting to be further investigated and explored.

Acknowledgments This study was funded in part by grants from the National Institutes of Health (grants R01 CA109311, PO1 CA099031, and CCSG CA16672), National Breast Cancer Foundation, Inc., Breast Cancer Research Foundation, Patel Memorial Breast Cancer Endowment Fund, The University of Texas MD Anderson-China Medical University and Hospital Sister Institution Fund, Ministry of Health and Welfare, China Medical University Hospital Cancer Research Center of Excellence (MOHW103- TD-B-111-03; Taiwan), Program for Stem Cell and Regenerative Medicine Frontier Research (NSC102-2321-B-039-001; Taiwan), International Research-Intensive Centers of Excellence in Taiwan (NSC103-2911-I-002-303), and Center for Biological Pathways.

References

1. Lemmon MA, Schlessinger J. Cell signaling by receptor tyrosine kinases. Cell. 2010;141(7):1117–34. PubMed PMID: 20602996.
2. Yarden Y, Shilo BZ. SnapShot: EGFR signaling pathway. Cell. 2007;131(5):1018. PubMed PMID: 18045542.
3. Hynes NE, MacDonald G. ErbB receptors and signaling pathways in cancer. Curr Opin Cell Biol. 2009;21(2):177–84. PubMed PMID: 19208461.
4. Yarden Y. The EGFR family and its ligands in human cancer. signalling mechanisms and therapeutic opportunities. Eur J Cancer. 2001;37 Suppl 4:S3–8.
5. von Zastrow M, Sorkin A. Signaling on the endocytic pathway. Curr Opin Cell Biol. 2007;19(4):436–45. PubMed PMID: 17662591.
6. Waterman H, Yarden Y. Molecular mechanisms underlying endocytosis and sorting of ErbB receptor tyrosine kinases. FEBS Lett. 2001;490(3):142–52. PubMed PMID: 11223029.
7. Du Y, Shen J, Hsu JL, Han Z, Hsu MC, Yang CC, et al. Syntaxin 6-mediated Golgi translocation plays an important role in nuclear functions of EGFR through microtubule-dependent trafficking. Oncogene. 2014;33:756. PubMed PMID: 23376851.
8. Wang YN, Wang H, Yamaguchi H, Lee HJ, Lee HH, Hung MC. COPI-mediated retrograde trafficking from the Golgi to the ER regulates EGFR nuclear transport. Biochem Biophys Res Commun. 2010;399(4):498–504. PubMed PMID: 20674546, Pubmed Central PMCID: 2935258.
9. Wang YN, Yamaguchi H, Hsu JM, Hung MC. Nuclear trafficking of the epidermal growth factor receptor family membrane proteins. Oncogene. 2010;29(28):3997–4006. PubMed PMID: 20473332, Pubmed Central PMCID: 2904849.
10. Wang YN, Yamaguchi H, Huo L, Du Y, Lee HJ, Lee HH, et al. The translocon Sec61beta localized in the inner nuclear membrane transports membrane-embedded EGF receptor to the

nucleus. J Biol Chem. 2010;285(49):38720–9. PubMed PMID: 20937808, Pubmed Central PMCID: 2992305.
11. Wang YN, Hung MC. Nuclear functions and subcellular trafficking mechanisms of the epidermal growth factor receptor family. Cell Biosci. 2012;2(1):13. PubMed PMID: 22520625, Pubmed Central PMCID: 3418567.
12. Wang YN, Lee HH, Lee HJ, Du Y, Yamaguchi H, Hung MC. Membrane-bound trafficking regulates nuclear transport of integral epidermal growth factor receptor (EGFR) and ErbB-2. J Biol Chem. 2012;287(20):16869–79. PubMed PMID: 22451678, Pubmed Central PMCID: 3351284.
13. Yarden Y, Sliwkowski MX. Untangling the ErbB signalling network. Nat Rev Mol Cell Biol. 2001;2(2):127–37. PubMed PMID: 11252954.
14. Avraham R, Yarden Y. Feedback regulation of EGFR signalling: decision making by early and delayed loops. Nat Rev Mol Cell Biol. 2011;12(2):104–17. PubMed PMID: 21252999.
15. Citri A, Yarden Y. EGF-ERBB signalling: towards the systems level. Nat Rev Mol Cell Biol. 2006;7(7):505–16. PubMed PMID: 16829981.
16. Lo HW, Hsu SC, Hung MC. EGFR signaling pathway in breast cancers: from traditional signal transduction to direct nuclear translocalization. Breast Cancer Res Treat. 2006;95(3):211–8. PubMed PMID: 16261406.
17. Huang PH, Xu AM, White FM. Oncogenic EGFR signaling networks in glioma. Sci Signal. 2009;2(87):re6. PubMed PMID: 19738203.
18. Lo HW, Hsu SC, Ali-Seyed M, Gunduz M, Xia W, Wei Y, et al. Nuclear interaction of EGFR and STAT3 in the activation of the iNOS/NO pathway. Cancer Cell. 2005;7(6):575–89. PubMed PMID: 15950906.
19. Irmer D, Funk JO, Blaukat A. EGFR kinase domain mutations - functional impact and relevance for lung cancer therapy. Oncogene. 2007;26(39):5693–701. PubMed PMID: 17353898.
20. Gazdar AF. Activating and resistance mutations of EGFR in non-small-cell lung cancer: role in clinical response to EGFR tyrosine kinase inhibitors. Oncogene. 2009;28 Suppl 1:S24–31. PubMed PMID: 19680293.
21. Linardou H, Dahabreh IJ, Bafaloukos D, Kosmidis P, Murray S. Somatic EGFR mutations and efficacy of tyrosine kinase inhibitors in NSCLC. Nat Rev. 2009;6(6):352–66. PubMed PMID: 19483740.
22. Alimandi M, Romano A, Curia MC, Muraro R, Fedi P, Aaronson SA, et al. Cooperative signaling of ErbB3 and ErbB2 in neoplastic transformation and human mammary carcinomas. Oncogene. 1995;10(9):1813–21. PubMed PMID: 7538656.
23. Riese 2nd DJ, Stern DF. Specificity within the EGF family/ErbB receptor family signaling network. Bioessays. 1998;20(1):41–8. PubMed PMID: 9504046.
24. Baselga J, Swain SM. Novel anticancer targets: revisiting ERBB2 and discovering ERBB3. Nat Rev Cancer. 2009;9(7):463–75. PubMed PMID: 19536107.
25. Du Y, Yang H, Xu Y, Cang X, Luo C, Mao Y, et al. Conformational transition and energy landscape of ErbB4 activated by neuregulin1beta: one microsecond molecular dynamics simulations. J Am Chem Soc. 2012;134(15):6720–31. PubMed PMID: 22316159.
26. Lin SY, Makino K, Xia W, Matin A, Wen Y, Kwong KY, et al. Nuclear localization of EGF receptor and its potential new role as a transcription factor. Nat Cell Biol. 2001;3(9):802–8. PubMed PMID: 11533659.
27. Wang SC, Lien HC, Xia W, Chen IF, Lo HW, Wang Z, et al. Binding at and transactivation of the COX-2 promoter by nuclear tyrosine kinase receptor ErbB-2. Cancer Cell. 2004;6(3):251–61. PubMed PMID: 15380516.
28. Li C, Iida M, Dunn EF, Ghia AJ, Wheeler DL. Nuclear EGFR contributes to acquired resistance to cetuximab. Oncogene. 2009;28(43):3801–13. PubMed PMID: 19684613.
29. Dittmann K, Mayer C, Rodemann HP. Inhibition of radiation-induced EGFR nuclear import by C225 (Cetuximab) suppresses DNA-PK activity. Radiother Oncol. 2005;76(2):157–61. PubMed PMID: 16024112.

30. Ni CY, Murphy MP, Golde TE, Carpenter G. Gamma-Secretase cleavage and nuclear localization of ErbB-4 receptor tyrosine kinase. Science. 2001;294(5549):2179–81. PubMed PMID: 11679632.
31. Gusterson B, Cowley G, Smith JA, Ozanne B. Cellular localisation of human epidermal growth factor receptor. Cell Biol Int Rep. 1984;8(8):649–58. PubMed PMID: 6090028.
32. Kamio T, Shigematsu K, Sou H, Kawai K, Tsuchiyama H. Immunohistochemical expression of epidermal growth factor receptors in human adrenocortical carcinoma. Hum Pathol. 1990;21(3):277–82. PubMed PMID: 2312105.
33. Marti U, Burwen SJ, Wells A, Barker ME, Huling S, Feren AM, et al. Localization of epidermal growth factor receptor in hepatocyte nuclei. Hepatology. 1991;13(1):15–20. PubMed PMID: 1988335.
34. Cao H, Lei ZM, Bian L, Rao CV. Functional nuclear epidermal growth factor receptors in human choriocarcinoma JEG-3 cells and normal human placenta. Endocrinology. 1995;136(7):3163–72. PubMed PMID: 7540549.
35. Marti U, Ruchti C, Kampf J, Thomas GA, Williams ED, Peter HJ, et al. Nuclear localization of epidermal growth factor and epidermal growth factor receptors in human thyroid tissues. Thyroid. 2001;11(2):137–45. PubMed PMID: 11288982.
36. Lipponen P, Eskelinen M. Expression of epidermal growth factor receptor in bladder cancer as related to established prognostic factors, oncoprotein (c-erbB-2, p53) expression and long-term prognosis. Br J Cancer. 1994;69(6):1120–5. PubMed PMID: 7911031, Pubmed Central PMCID: 1969432.
37. Lo HW, Xia W, Wei Y, Ali-Seyed M, Huang SF, Hung MC. Novel prognostic value of nuclear epidermal growth factor receptor in breast cancer. Cancer Res. 2005;65(1):338–48. PubMed PMID: 15665312.
38. Psyrri A, Yu Z, Weinberger PM, Sasaki C, Haffty B, Camp R, et al. Quantitative determination of nuclear and cytoplasmic epidermal growth factor receptor expression in oropharyngeal squamous cell cancer by using automated quantitative analysis. Clin Cancer Res. 2005;11(16):5856–62. PubMed PMID: 16115926.
39. Wang YN, Yamaguchi H, Huo L, Du Y, Lee HJ, Lee HH, et al. The translocon SEC61{beta} localized in the inner nuclear membrane transports membrane-embedded EGF receptor to the nucleus. J Biol Chem. 2010;285:38720. PubMed PMID: 20937808.
40. Klein C, Gensburger C, Freyermuth S, Nair BC, Labourdette G, Malviya AN. A 120 kDa nuclear phospholipase Cgamma1 protein fragment is stimulated in vivo by EGF signal phosphorylating nuclear membrane EGFR. Biochemistry. 2004;43(50):15873–83. PubMed PMID: 15595842.
41. Cordero JB, Cozzolino M, Lu Y, Vidal M, Slatopolsky E, Stahl PD, et al. 1,25-Dihydroxyvitamin D down-regulates cell membrane growth- and nuclear growth-promoting signals by the epidermal growth factor receptor. J Biol Chem. 2002;277(41):38965–71. PubMed PMID: 12181310.
42. Dittmann K, Mayer C, Fehrenbacher B, Schaller M, Raju U, Milas L, et al. Radiation-induced epidermal growth factor receptor nuclear import is linked to activation of DNA-dependent protein kinase. J Biol Chem. 2005;280(35):31182–9. PubMed PMID: 16000298.
43. de la Iglesia N, Konopka G, Puram SV, Chan JA, Bachoo RM, You MJ, et al. Identification of a PTEN-regulated STAT3 brain tumor suppressor pathway. Genes Dev. 2008;22(4):449–62. PubMed PMID: 18258752, Pubmed Central PMCID: 2238667.
44. Lo HW, Cao X, Zhu H, Ali-Osman F. Cyclooxygenase-2 is a novel transcriptional target of the nuclear EGFR-STAT3 and EGFRvIII-STAT3 signaling axes. Mol Cancer Res. 2010;8(2):232–45. PubMed PMID: 20145033, Pubmed Central PMCID: 2824777.
45. Piccione EC, Lieu TJ, Gentile CF, Williams TR, Connolly AJ, Godwin AK, et al. A novel epidermal growth factor receptor variant lacking multiple domains directly activates transcription and is overexpressed in tumors. Oncogene. 2011;31:2953. PubMed PMID: 21986942, Pubmed Central PMCID: 3368095.

46. Xie Y, Hung MC. Nuclear localization of p185neu tyrosine kinase and its association with transcriptional transactivation. Biochem Biophys Res Commun. 1994;203(3):1589–98. PubMed PMID: 7945309.
47. Xia W, Liu Z, Zong R, Liu L, Zhao S, Bacus SS, et al. Truncated ErbB2 expressed in tumor cell nuclei contributes to acquired therapeutic resistance to ErbB2 kinase inhibitors. Mol Cancer Ther. 2011;10(8):1367–74. PubMed PMID: 21673090.
48. Offterdinger M, Schofer C, Weipoltshammer K, Grunt TW. c-erbB-3: a nuclear protein in mammary epithelial cells. J Cell Biol. 2002;157(6):929–39. PubMed PMID: 12045181.
49. Thompson M, Lauderdale S, Webster MJ, Chong VZ, McClintock B, Saunders R, et al. Widespread expression of ErbB2, ErbB3 and ErbB4 in non-human primate brain. Brain Res. 2007;1139:95–109. PubMed PMID: 17280647.
50. Srinivasan R, Gillett CE, Barnes DM, Gullick WJ. Nuclear expression of the c-erbB-4/HER-4 growth factor receptor in invasive breast cancers. Cancer Res. 2000;60(6):1483–7. PubMed PMID: 10749108.
51. Mroczkowski B, Mosig G, Cohen S. ATP-stimulated interaction between epidermal growth factor receptor and supercoiled DNA. Nature. 1984;309(5965):270–3. PubMed PMID: 6325948.
52. Basu M, Frick K, Sen-Majumdar A, Scher CD, Das M. EGF receptor-associated DNA-nicking activity is due to a Mr-100,000 dissociable protein. Nature. 1985;316(6029):640–1. PubMed PMID: 2993901.
53. Hanada N, Lo HW, Day CP, Pan Y, Nakajima Y, Hung MC. Co-regulation of B-Myb expression by E2F1 and EGF receptor. Mol Carcinog. 2006;45(1):10–7. PubMed PMID: 16299810.
54. Hung LY, Tseng JT, Lee YC, Xia W, Wang YN, Wu ML, et al. Nuclear epidermal growth factor receptor (EGFR) interacts with signal transducer and activator of transcription 5 (STAT5) in activating Aurora-A gene expression. Nucleic Acids Res. 2008;36(13):4337–51. PubMed PMID: 18586824, Pubmed Central PMCID: 2490761.
55. Jaganathan S, Yue P, Paladino DC, Bogdanovic J, Huo Q, Turkson J. A functional nuclear epidermal growth factor receptor, SRC and Stat3 heteromeric complex in pancreatic cancer cells. PLoS One. 2011;6(5):e19605. PubMed PMID: 21573184, Pubmed Central PMCID: 3088706.
56. Kim HP, Yoon YK, Kim JW, Han SW, Hur HS, Park J, et al. Lapatinib, a dual EGFR and HER2 tyrosine kinase inhibitor, downregulates thymidylate synthase by inhibiting the nuclear translocation of EGFR and HER2. PLoS One. 2009;4(6):e5933. PubMed PMID: 19529774, Pubmed Central PMCID: 2691960.
57. Huang WC, Chen YJ, Li LY, Wei YL, Hsu SC, Tsai SL, et al. Nuclear translocation of epidermal growth factor receptor by Akt-dependent phosphorylation enhances breast cancer-resistant protein expression in gefitinib-resistant cells. J Biol Chem. 2011;286(23):20558–68. PubMed PMID: 21487020, Pubmed Central PMCID: 3121497.
58. Huo L, Wang YN, Xia W, Hsu SC, Lai CC, Li LY, et al. RNA helicase A is a DNA-binding partner for EGFR-mediated transcriptional activation in the nucleus. Proc Natl Acad Sci U S A. 2010;107(37):16125–30. PubMed PMID: 20802156.
59. Bitler BG, Goverdhan A, Schroeder JA. MUC1 regulates nuclear localization and function of the epidermal growth factor receptor. J Cell Sci. 2010;123:1716–23. PubMed PMID: 20406885.
60. Hu S, Xie Z, Onishi A, Yu X, Jiang L, Lin J, et al. Profiling the human protein-DNA interactome reveals ERK2 as a transcriptional repressor of interferon signaling. Cell. 2009;139(3):610–22. PubMed PMID: 19879846, Pubmed Central PMCID: 2774939.
61. Wang SC, Nakajima Y, Yu YL, Xia W, Chen CT, Yang CC, et al. Tyrosine phosphorylation controls PCNA function through protein stability. Nat Cell Biol. 2006;8(12):1359–68. PubMed PMID: 17115032.
62. Dittmann K, Mayer C, Kehlbach R, Rothmund MC, Peter RH. Radiation-induced lipid peroxidation activates src kinase and triggers nuclear EGFR transport. Radiother Oncol. 2009;92(3):379–82. PubMed PMID: 19560222.

63. Dittmann K, Mayer C, Kehlbach R, Rodemann HP. Radiation-induced caveolin-1 associated EGFR internalization is linked with nuclear EGFR transport and activation of DNA-PK. Mol Cancer. 2008;7:69. PubMed PMID: 18789131, Pubmed Central PMCID: 2546440.
64. Lee PC, Lee HJ, Kakadiya R, Sanjiv K, Su TL, Lee TC. Multidrug-resistant cells overexpressing P-glycoprotein are susceptible to DNA crosslinking agents due to attenuated Src/nuclear EGFR cascade-activated DNA repair activity. Oncogene. 2013;32:1144. PubMed PMID: 22525278.
65. Das AK, Chen BP, Story MD, Sato M, Minna JD, Chen DJ, et al. Somatic mutations in the tyrosine kinase domain of epidermal growth factor receptor (EGFR) abrogate EGFR-mediated radioprotection in non-small cell lung carcinoma. Cancer Res. 2007;67(11):5267–74. PubMed PMID: 17545606.
66. Ge H, Gong X, Tang CK. Evidence of high incidence of EGFRvIII expression and coexpression with EGFR in human invasive breast cancer by laser capture microdissection and immunohistochemical analysis. Int J Cancer. 2002;98(3):357–61. PubMed PMID: 11920586.
67. Edwards J, Traynor P, Munro AF, Pirret CF, Dunne B, Bartlett JM. The role of HER1-HER4 and EGFRvIII in hormone-refractory prostate cancer. Clin Cancer Res. 2006;12(1):123–30. PubMed PMID: 16397033.
68. Beguelin W, Diaz Flaque MC, Proietti CJ, Cayrol F, Rivas MA, Tkach M, et al. Progesterone receptor induces ErbB-2 nuclear translocation to promote breast cancer growth via a novel transcriptional effect: ErbB-2 function as a coactivator of Stat3. Mol Cell Biol. 2010;30(23):5456–72. PubMed PMID: 20876300, Pubmed Central PMCID: 2976427.
69. Li X, Kuang J, Shen Y, Majer MM, Nelson CC, Parsawar K, et al. The atypical histone macroH2A1.2 interacts with HER-2 protein in cancer cells. J Biol Chem. 2012;287(27):23171–83.
70. Li LY, Chen H, Hsieh YH, Wang YN, Chu HJ, Chen YH, et al. Nuclear ErbB2 enhances translation and cell growth by activating transcription of ribosomal RNA genes. Cancer Res. 2011;71(12):4269–79. PubMed PMID: 21555369, Pubmed Central PMCID: 3117049.
71. Tan M, Jing T, Lan KH, Neal CL, Li P, Lee S, et al. Phosphorylation on tyrosine-15 of p34(Cdc2) by ErbB2 inhibits p34(Cdc2) activation and is involved in resistance to taxol-induced apoptosis. Mol Cell. 2002;9(5):993–1004. PubMed PMID: 12049736.
72. Scaltriti M, Rojo F, Ocana A, Anido J, Guzman M, Cortes J, et al. Expression of p95HER2, a truncated form of the HER2 receptor, and response to anti-HER2 therapies in breast cancer. J Natl Cancer Inst. 2007;99(8):628–38. PubMed PMID: 17440164.
73. Dillon MF, Stafford AT, Kelly G, Redmond AM, McIlroy M, Crotty TB, et al. Cyclooxygenase-2 predicts adverse effects of tamoxifen: a possible mechanism of role for nuclear HER2 in breast cancer patients. Endocr Relat Cancer. 2008;15(3):745–53. PubMed PMID: 18469157.
74. Schillaci R, Guzman P, Cayrol F, Beguelin W, Diaz Flaque MC, Proietti CJ, et al. Clinical relevance of ErbB-2/HER2 nuclear expression in breast cancer. BMC Cancer. 2012;12:74. PubMed PMID: 22356700, Pubmed Central PMCID: 3342900.
75. Sithanandam G, Anderson LM. The ERBB3 receptor in cancer and cancer gene therapy. Cancer Gene Ther. 2008;15(7):413–48. PubMed PMID: 18404164, Pubmed Central PMCID: 2761714.
76. Zscheppang K, Korenbaum E, Bueter W, Ramadurai SM, Nielsen HC, Dammann CE. ErbB receptor dimerization, localization, and co-localization in mouse lung type II epithelial cells. Pediatr Pulmonol. 2006;41(12):1205–12. PubMed PMID: 17063476.
77. Koumakpayi IH, Diallo JS, Le Page C, Lessard L, Gleave M, Begin LR, et al. Expression and nuclear localization of ErbB3 in prostate cancer. Clin Cancer Res. 2006;12(9):2730–7. PubMed PMID: 16675564.
78. Cheng CJ, Ye XC, Vakar-Lopez F, Kim J, Tu SM, Chen DT, et al. Bone microenvironment and androgen status modulate subcellular localization of ErbB3 in prostate cancer cells. Mol Cancer Res. 2007;5(7):675–84. PubMed PMID: 17634423, Pubmed Central PMCID: 2000833.

79. Koumakpayi IH, Le Page C, Delvoye N, Saad F, Mes-Masson AM. Macropinocytosis inhibitors and Arf6 regulate ErbB3 nuclear localization in prostate cancer cells. Mol Carcinog. 2011;50(11):901–12. PubMed PMID: 21438025.
80. Koumakpayi IH, Diallo JS, Le Page C, Lessard L, Filali-Mouhim A, Begin LR, et al. Low nuclear ErbB3 predicts biochemical recurrence in patients with prostate cancer. BJU Int. 2007;100(2):303–9. PubMed PMID: 17532856.
81. Raabe TD, Deadwyler G, Varga JW, Devries GH. Localization of neuregulin isoforms and erbB receptors in myelinating glial cells. Glia. 2004;45(2):197–207. PubMed PMID: 14730713.
82. Adilakshmi T, Ness-Myers J, Madrid-Aliste C, Fiser A, Tapinos N. A nuclear variant of ErbB3 receptor tyrosine kinase regulates ezrin distribution and Schwann cell myelination. J Neurosci. 2011;31(13):5106–19. PubMed PMID: 21451047, Pubmed Central PMCID: 3086203.
83. Andrique L, Fauvin D, El Maassarani M, Colasson H, Vannier B, Seite P. ErbB3(80 kDa), a nuclear variant of the ErbB3 receptor, binds to the Cyclin D1 promoter to activate cell proliferation but is negatively controlled by p14ARF. Cell Signal. 2012;24(5):1074–85. PubMed PMID: 22261253.
84. Williams CC, Allison JG, Vidal GA, Burow ME, Beckman BS, Marrero L, et al. The ERBB4/HER4 receptor tyrosine kinase regulates gene expression by functioning as a STAT5A nuclear chaperone. J Cell Biol. 2004;167(3):469–78. PubMed PMID: 15534001, Pubmed Central PMCID: 2172499.
85. Linggi B, Carpenter G. ErbB-4 s80 intracellular domain abrogates ETO2-dependent transcriptional repression. J Biol Chem. 2006;281(35):25373–80. PubMed PMID: 16815842.
86. Sardi SP, Murtie J, Koirala S, Patten BA, Corfas G. Presenilin-dependent ErbB4 nuclear signaling regulates the timing of astrogenesis in the developing brain. Cell. 2006;127(1): 185–97. PubMed PMID: 17018285.
87. Hughes DP, Thomas DG, Giordano TJ, Baker LH, McDonagh KT. Cell surface expression of epidermal growth factor receptor and Her-2 with nuclear expression of Her-4 in primary osteosarcoma. Cancer Res. 2004;64(6):2047–53. PubMed PMID: 15026342.
88. Carpenter G. ErbB-4: mechanism of action and biology. Exp Cell Res. 2003;284(1):66–77. PubMed PMID: 12648466.
89. Zeng F, Zhang MZ, Singh AB, Zent R, Harris RC. ErbB4 isoforms selectively regulate growth factor induced Madin-Darby canine kidney cell tubulogenesis. Mol Biol Cell. 2007;18(11):4446–56. PubMed PMID: 17761534, Pubmed Central PMCID: 2043549.
90. Icli B, Bharti A, Pentassuglia L, Peng X, Sawyer DB. ErbB4 localization to cardiac myocyte nuclei, and its role in myocyte DNA damage response. Biochem Biophys Res Commun. 2012;418(1):116–21. PubMed PMID: 22244893, Pubmed Central PMCID: 3273580.
91. Paatero I, Jokilammi A, Heikkinen PT, Iljin K, Kallioniemi OP, Jones FE, et al. Interaction with ErbB4 promotes hypoxia-inducible factor-1alpha signaling. J Biol Chem. 2012;287(13):9659–71. PubMed PMID: 22308027, Pubmed Central PMCID: 3322979.
92. Turner N, Grose R. Fibroblast growth factor signalling: from development to cancer. Nat Rev Cancer. 2010;10(2):116–29. PubMed PMID: 20094046.
93. Gonzalez AM, Berry M, Maher PA, Logan A, Baird A. A comprehensive analysis of the distribution of FGF-2 and FGFR1 in the rat brain. Brain Res. 1995;701(1–2):201–26. PubMed PMID: 8925285.
94. Stachowiak MK, Moffett J, Joy A, Puchacz E, Florkiewicz R, Stachowiak EK. Regulation of bFGF gene expression and subcellular distribution of bFGF protein in adrenal medullary cells. J Cell Biol. 1994;127(1):203–23. PubMed PMID: 7929563, Pubmed Central PMCID: 2120178.
95. Stachowiak MK, Maher PA, Joy A, Mordechai E, Stachowiak EK. Nuclear accumulation of fibroblast growth factor receptors is regulated by multiple signals in adrenal medullary cells. Mol Biol Cell. 1996;7(8):1299–317. PubMed PMID: 8856671.

96. Reilly JF, Maher PA. Importin beta-mediated nuclear import of fibroblast growth factor receptor: role in cell proliferation. J Cell Biol. 2001;152(6):1307–12. PubMed PMID: 11257130.
97. Peng H, Moffett J, Myers J, Fang X, Stachowiak EK, Maher P, et al. Novel nuclear signaling pathway mediates activation of fibroblast growth factor-2 gene by type 1 and type 2 angiotensin II receptors. Mol Biol Cell. 2001;12(2):449–62. PubMed PMID: 11179427, Pubmed Central PMCID: 30955.
98. Bryant DM, Wylie FG, Stow JL. Regulation of endocytosis, nuclear translocation, and signaling of fibroblast growth factor receptor 1 by E-cadherin. Mol Biol Cell. 2005;16(1):14–23. PubMed PMID: 15509650.
99. Clarke WE, Berry M, Smith C, Kent A, Logan A. Coordination of fibroblast growth factor receptor 1 (FGFR1) and fibroblast growth factor-2 (FGF-2) trafficking to nuclei of reactive astrocytes around cerebral lesions in adult rats. Mol Cell Neurosci. 2001;17(1):17–30. PubMed PMID: 11161466.
100. Fang X, Stachowiak EK, Dunham-Ems SM, Klejbor I, Stachowiak MK. Control of CREB-binding protein signaling by nuclear fibroblast growth factor receptor-1: a novel mechanism of gene regulation. J Biol Chem. 2005;280(31):28451–62. PubMed PMID: 15929978.
101. Leadbeater WE, Gonzalez AM, Logaras N, Berry M, Turnbull JE, Logan A. Intracellular trafficking in neurones and glia of fibroblast growth factor-2, fibroblast growth factor receptor 1 and heparan sulphate proteoglycans in the injured adult rat cerebral cortex. J Neurochem. 2006;96(4):1189–200. PubMed PMID: 16417571.
102. Bilak MM, Hossain WA, Morest DK. Intracellular fibroblast growth factor produces effects different from those of extracellular application on development of avian cochleovestibular ganglion cells in vitro. J Neurosci Res. 2003;71(5):629–47. PubMed PMID: 12584722.
103. Klimaschewski L, Meisinger C, Grothe C. Localization and regulation of basic fibroblast growth factor (FGF-2) and FGF receptor-1 in rat superior cervical ganglion after axotomy. J Neurobiol. 1999;38(4):499–506. PubMed PMID: 10084685.
104. Horbinski C, Stachowiak EK, Chandrasekaran V, Miuzukoshi E, Higgins D, Stachowiak MK. Bone morphogenetic protein-7 stimulates initial dendritic growth in sympathetic neurons through an intracellular fibroblast growth factor signaling pathway. J Neurochem. 2002;80(1):54–63. PubMed PMID: 11796743.
105. Stachowiak MK, Fang X, Myers JM, Dunham SM, Berezney R, Maher PA, et al. Integrative nuclear FGFR1 signaling (INFS) as a part of a universal "feed-forward-and-gate" signaling module that controls cell growth and differentiation. J Cell Biochem. 2003;90(4):662–91. PubMed PMID: 14587025.
106. Peng H, Myers J, Fang X, Stachowiak EK, Maher PA, Martins GG, et al. Integrative nuclear FGFR1 signaling (INFS) pathway mediates activation of the tyrosine hydroxylase gene by angiotensin II, depolarization and protein kinase C. J Neurochem. 2002;81(3):506–24. PubMed PMID: 12065659.
107. Baron O, Forthmann B, Lee YW, Terranova C, Ratzka A, Stachowiak EK, et al. Cooperation of nuclear fibroblast growth factor receptor 1 and nurr1 offers new interactive mechanism in postmitotic development of mesencephalic dopaminergic neurons. J Biol Chem. 2012;287(24):19827–40. PubMed PMID: 22514272.
108. Lee YW, Terranova C, Birkaya B, Narla S, Kehoe D, Parikh A, et al. A novel nuclear FGF Receptor-1 partnership with retinoid and Nur receptors during developmental gene programming of embryonic stem cells. J Cell Biochem. 2012;113:2920. PubMed PMID: 22539306.
109. Chioni AM, Grose R. FGFR1 cleavage and nuclear translocation regulates breast cancer cell behavior. J Cell Biol. 2012;197(6):801–17. PubMed PMID: 22665522, Pubmed Central PMCID: 3373409.
110. Schmahl J, Kim Y, Colvin JS, Ornitz DM, Capel B. Fgf9 induces proliferation and nuclear localization of FGFR2 in Sertoli precursors during male sex determination. Development. 2004;131(15):3627–36. PubMed PMID: 15229180.

111. Bagheri-Fam S, Sim H, Bernard P, Jayakody I, Taketo MM, Scherer G, et al. Loss of Fgfr2 leads to partial XY sex reversal. Dev Biol. 2008;314(1):71–83. PubMed PMID: 18155190.
112. Lu P, Ewald AJ, Martin GR, Werb Z. Genetic mosaic analysis reveals FGF receptor 2 function in terminal end buds during mammary gland branching morphogenesis. Dev Biol. 2008;321(1):77–87. PubMed PMID: 18585375, Pubmed Central PMCID: 2582391.
113. Giulianelli S, Cerliani JP, Lamb CA, Fabris VT, Bottino MC, Gorostiaga MA, et al. Carcinoma-associated fibroblasts activate progesterone receptors and induce hormone independent mammary tumor growth: A role for the FGF-2/FGFR-2 axis. Int J Cancer. 2008;123(11):2518–31. PubMed PMID: 18767044.
114. Sun S, Jiang Y, Zhang G, Song H, Zhang X, Zhang Y, et al. Increased expression of fibroblastic growth factor receptor 2 is correlated with poor prognosis in patients with breast cancer. J Surg Oncol. 2012;105(8):773–9. PubMed PMID: 22006548.
115. Martin AJ, Grant A, Ashfield AM, Palmer CN, Baker L, Quinlan PR, et al. FGFR2 protein expression in breast cancer: nuclear localisation and correlation with patient genotype. BMC Res Notes. 2011;4:72. PubMed PMID: 21418638, Pubmed Central PMCID: 3073906.
116. Cerliani JP, Guillardoy T, Giulianelli S, Vaque JP, Gutkind JS, Vanzulli SI, et al. Interaction between FGFR-2, STAT5, and progesterone receptors in breast cancer. Cancer Res. 2011;71(10):3720–31. PubMed PMID: 21464042.
117. Sabbieti MG, Marchetti L, Gabrielli MG, Menghi M, Materazzi S, Menghi G, et al. Prostaglandins differently regulate FGF-2 and FGF receptor expression and induce nuclear translocation in osteoblasts via MAPK kinase. Cell Tissue Res. 2005;319(2):267–78. PubMed PMID: 15654655.
118. Marchetti L, Sabbieti MG, Agas D, Menghi M, Materazzi G, Menghi G, et al. PGF2alpha increases FGF-2 and FGFR2 trafficking in Py1a rat osteoblasts via clathrin independent and importin beta dependent pathway. J Cell Biochem. 2006;97(6):1379–92. PubMed PMID: 16365892.
119. Gatius S, Velasco A, Azueta A, Santacana M, Pallares J, Valls J, et al. FGFR2 alterations in endometrial carcinoma. Mod Pathol. 2011;24(11):1500–10. PubMed PMID: 21725289.
120. Behrens C, Lin HY, Lee JJ, Raso MG, Hong WK, Wistuba II, et al. Immunohistochemical expression of basic fibroblast growth factor and fibroblast growth factor receptors 1 and 2 in the pathogenesis of lung cancer. Clin Cancer Res. 2008;14(19):6014–22. PubMed PMID: 18829480.
121. Johnston CL, Cox HC, Gomm JJ, Coombes RC. Fibroblast growth factor receptors (FGFRs) localize in different cellular compartments. A splice variant of FGFR-3 localizes to the nucleus. J Biol Chem. 1995;270(51):30643–50.
122. Zammit C, Barnard R, Gomm J, Coope R, Shousha S, Coombes C, et al. Altered intracellular localization of fibroblast growth factor receptor 3 in human breast cancer. J Pathol. 2001;194(1):27–34. PubMed PMID: 11329138.
123. Degnin CR, Laederich MB, Horton WA. Ligand activation leads to regulated intramembrane proteolysis of fibroblast growth factor receptor 3. Mol Biol Cell. 2011;22(20):3861–73. PubMed PMID: 21865593, Pubmed Central PMCID: 3192865.
124. Rotterud R, Fossa SD, Nesland JM. Protein networking in bladder cancer: immunoreactivity for FGFR3, EGFR, ERBB2, KAI1, PTEN, and RAS in normal and malignant urothelium. Histol Histopathol. 2007;22(4):349–63. PubMed PMID: 17290345.
125. Martinez-Torrecuadrada JL, Cheung LH, Lopez-Serra P, Barderas R, Canamero M, Ferreiro S, et al. Antitumor activity of fibroblast growth factor receptor 3-specific immunotoxins in a xenograft mouse model of bladder carcinoma is mediated by apoptosis. Mol Cancer Ther. 2008;7(4):862–73. PubMed PMID: 18413799.
126. Kokkinakis DM, Rushing EJ, Shareef MM, Ahmed MM, Yang S, Singha UK, et al. Physiology and gene expression characteristics of carcinogen-initiated and tumor-transformed glial progenitor cells derived from the CNS of methylnitrosourea (MNU)-treated Sprague-Dawley rats. J Neuropathol Exp Neurol. 2004;63(11):1182–99. PubMed PMID: 15581186.
127. Ferrara N, Gerber HP, LeCouter J. The biology of VEGF and its receptors. Nat Med. 2003;9(6):669–76. PubMed PMID: 12778165.

128. Olsson AK, Dimberg A, Kreuger J, Claesson-Welsh L. VEGF receptor signalling - in control of vascular function. Nat Rev Mol Cell Biol. 2006;7(5):359–71. PubMed PMID: 16633338.
129. Neufeld G, Cohen T, Gengrinovitch S, Poltorak Z. Vascular endothelial growth factor (VEGF) and its receptors. FASEB J. 1999;13(1):9–22. PubMed PMID: 9872925.
130. Holmes K, Roberts OL, Thomas AM, Cross MJ. Vascular endothelial growth factor receptor-2: structure, function, intracellular signalling and therapeutic inhibition. Cell Signal. 2007;19(10):2003–12. PubMed PMID: 17658244.
131. Rahimi N. VEGFR-1 and VEGFR-2: two non-identical twins with a unique physiognomy. Front Biosci. 2006;11:818–29. PubMed PMID: 16146773, Pubmed Central PMCID: 1360224.
132. Sato Y, Kanno S, Oda N, Abe M, Ito M, Shitara K, et al. Properties of two VEGF receptors, Flt-1 and KDR, in signal transduction. Ann N Y Acad Sci. 2000;902:201–5. discussion 5–7, PubMed PMID: 10865839.
133. Luttun A, Tjwa M, Carmeliet P. Placental growth factor (PlGF) and its receptor Flt-1 (VEGFR-1): novel therapeutic targets for angiogenic disorders. Ann N Y Acad Sci. 2002;979:80–93. PubMed PMID: 12543719.
134. Ilan N, Tucker A, Madri JA. Vascular endothelial growth factor expression, beta-catenin tyrosine phosphorylation, and endothelial proliferative behavior: a pathway for transformation? Lab Invest. 2003;83(8):1105–15. PubMed PMID: 12920240.
135. Lee TH, Seng S, Sekine M, Hinton C, Fu Y, Avraham HK, et al. Vascular endothelial growth factor mediates intracrine survival in human breast carcinoma cells through internally expressed VEGFR1/FLT1. PLoS Med. 2007;4(6):e186. PubMed PMID: 17550303, Pubmed Central PMCID: 1885450.
136. Andersson MK, Goransson M, Olofsson A, Andersson C, Aman P. Nuclear expression of FLT1 and its ligand PGF in FUS-DDIT3 carrying myxoid liposarcomas suggests the existence of an intracrine signaling loop. BMC Cancer. 2010;10:249. PubMed PMID: 20515481, Pubmed Central PMCID: 2889895.
137. Hagedorn A, Germann PG, Junker-Walker U, Tomovic A, Seewald W, Polkinghorne A, et al. Immunohistochemical study about the Flt-1/VEGFR1 expression in the gastrointestinal tract of mouse, rat, dog, swine and monkey. Exp Toxicol Pathol. 2005;57(2):149–59. PubMed PMID: 16325525.
138. Cai J, Chen Z, Ruan Q, Han S, Liu L, Qi X, et al. Gamma-Secretase and presenilin mediate cleavage and phosphorylation of vascular endothelial growth factor receptor-1. J Biol Chem. 2011;286(49):42514–23. PubMed PMID: 22016384, Pubmed Central PMCID: 3234916.
139. Cai J, Jiang WG, Grant MB, Boulton M. Pigment epithelium-derived factor inhibits angiogenesis via regulated intracellular proteolysis of vascular endothelial growth factor receptor 1. J Biol Chem. 2006;281(6):3604–13. PubMed PMID: 16339148.
140. Feng Y, Venema VJ, Venema RC, Tsai N, Caldwell RB. VEGF induces nuclear translocation of Flk-1/KDR, endothelial nitric oxide synthase, and caveolin-1 in vascular endothelial cells. Biochem Biophys Res Commun. 1999;256(1):192–7. PubMed PMID: 10066445.
141. Dardik R, Inbal A. Complex formation between tissue transglutaminase II (tTG) and vascular endothelial growth factor receptor 2 (VEGFR-2): proposed mechanism for modulation of endothelial cell response to VEGF. Exp Cell Res. 2006;312(16):2973–82. PubMed PMID: 16914140.
142. Shay-Salit A, Shushy M, Wolfovitz E, Yahav H, Breviario F, Dejana E, et al. VEGF receptor 2 and the adherens junction as a mechanical transducer in vascular endothelial cells. Proc Natl Acad Sci U S A. 2002;99(14):9462–7. PubMed PMID: 12080144, Pubmed Central PMCID: 123163.
143. Santos SC, Miguel C, Domingues I, Calado A, Zhu Z, Wu Y, et al. VEGF and VEGFR-2 (KDR) internalization is required for endothelial recovery during wound healing. Exp Cell Res. 2007;313(8):1561–74. PubMed PMID: 17382929.
144. Domingues I, Rino J, Demmers JA, de Lanerolle P, Santos SC. VEGFR2 translocates to the nucleus to regulate its own transcription. PLoS One. 2011;6(9):e25668. PubMed PMID: 21980525, Pubmed Central PMCID: 3182252.

145. Blazquez C, Cook N, Micklem K, Harris AL, Gatter KC, Pezzella F. Phosphorylated KDR can be located in the nucleus of neoplastic cells. Cell Res. 2006;16(1):93–8. PubMed PMID: 16467880.
146. Zhang Z, Neiva KG, Lingen MW, Ellis LM, Nor JE. VEGF-dependent tumor angiogenesis requires inverse and reciprocal regulation of VEGFR1 and VEGFR2. Cell Death Differ. 2010;17(3):499–512. PubMed PMID: 19834490, Pubmed Central PMCID: 2822115.
147. Belfiore A, Frasca F, Pandini G, Sciacca L, Vigneri R. Insulin receptor isoforms and insulin receptor/insulin-like growth factor receptor hybrids in physiology and disease. Endocr Rev. 2009;30(6):586–623. PubMed PMID: 19752219.
148. Gualco E, Wang JY, Del Valle L, Urbanska K, Peruzzi F, Khalili K, et al. IGF-IR in neuroprotection and brain tumors. Front Biosci. 2009;14:352–75. PubMed PMID: 19273072, Pubmed Central PMCID: 2679154.
149. Pollak M. The insulin receptor/insulin-like growth factor receptor family as a therapeutic target in oncology. Clin Cancer Res. 2012;18(1):40–50. PubMed PMID: 22215905.
150. Podlecki DA, Smith RM, Kao M, Tsai P, Huecksteadt T, Brandenburg D, et al. Nuclear translocation of the insulin receptor. A possible mediator of insulin's long term effects. J Biol Chem. 1987;262(7):3362–8. PubMed PMID: 3546306.
151. Nelson JD, LeBoeuf RC, Bomsztyk K. Direct recruitment of insulin receptor and ERK signaling cascade to insulin-inducible gene loci. Diabetes. 2011;60(1):127–37. PubMed PMID: 20929976, Pubmed Central PMCID: 3012164.
152. Chen CW, Roy D. Up-regulation of nuclear IGF-I receptor by short term exposure of stilbene estrogen, diethylstilbestrol. Mol Cell Endocrinol. 1996;118(1–2):1–8. PubMed PMID: 8735585.
153. Gletsu NA, Field CJ, Clandinin MT. Obese mice have higher insulin receptor levels in the hepatocyte cell nucleus following insulin stimulation in vivo with an oral glucose meal. Biochim Biophys Acta. 1999;1454(3):251–60. PubMed PMID: 10452959.
154. Seol KC, Kim SJ. Nuclear matrix association of insulin receptor and IRS-1 by insulin in osteoblast-like UMR-106 cells. Biochem Biophys Res Commun. 2003;306(4):898–904. PubMed PMID: 12821126.
155. Aleksic T, Chitnis MM, Perestenko OV, Gao S, Thomas PH, Turner GD, et al. Type 1 insulin-like growth factor receptor translocates to the nucleus of human tumor cells. Cancer Res. 2010;70(16):6412–9. PubMed PMID: 20710042, Pubmed Central PMCID: 2981028.
156. Bodzin AS, Wei Z, Hurtt R, Gu T, Doria C. Gefitinib resistance in HCC mahlavu cells: upregulation of CD133 expression, activation of IGF-1R signaling pathway, and enhancement of IGF-1R nuclear translocation. J Cell Physiol. 2012;227(7):2947–52. PubMed PMID: 21959795.
157. Deribe YL, Pawson T, Dikic I. Post-translational modifications in signal integration. Nat Struct Mol Biol. 2010;17(6):666–72. PubMed PMID: 20495563.
158. Sehat B, Tofigh A, Lin Y, Trocme E, Liljedahl U, Lagergren J, et al. SUMOylation mediates the nuclear translocation and signaling of the IGF-1 receptor. Sci Signal. 2010;3(108):ra10. PubMed PMID: 20145208.
159. Deng H, Lin Y, Badin M, Vasilcanu D, Stromberg T, Jernberg-Wiklund H, et al. Over-accumulation of nuclear IGF-1 receptor in tumor cells requires elevated expression of the receptor and the SUMO-conjugating enzyme Ubc9. Biochem Biophys Res Commun. 2011;404(2):667–71. PubMed PMID: 21147068.
160. Warsito D, Sjostrom S, Andersson S, Larsson O, Sehat B. Nuclear IGF1R is a transcriptional co-activator of LEF1/TCF. EMBO Rep. 2012;13(3):244–50. PubMed PMID: 22261717, Pubmed Central PMCID: 3323138.
161. Hoa N, Tsui S, Afifiyan NF, Sinha Hikim A, Li B, Douglas RS, et al. Nuclear targeting of IGF-1 receptor in orbital fibroblasts from Graves' disease: apparent role of ADAM17. PLoS One. 2012;7(4):e34173. PubMed PMID: 22506015, Pubmed Central PMCID: 3323600.
162. Comoglio PM, Boccaccio C. The HGF receptor family: unconventional signal transducers for invasive cell growth. Genes Cells. 1996;1(4):347–54. PubMed PMID: 9135079.
163. Birchmeier C, Birchmeier W, Gherardi E, Vande Woude GF. Met, metastasis, motility and more. Nat Rev Mol Cell Biol. 2003;4(12):915–25. PubMed PMID: 14685170.

164. Orton TC, Doughty SE, Kalinowski AE, Lord PG, Wadsworth PF. Expression of growth factors and growth factor receptors in the liver of C57BL/10 J mice following administration of phenobarbitone. Carcinogenesis. 1996;17(5):973–81. PubMed PMID: 8640946.
165. Pozner-Moulis S, Pappas DJ, Rimm DL. Met, the hepatocyte growth factor receptor, localizes to the nucleus in cells at low density. Cancer Res. 2006;66(16):7976–82. PubMed PMID: 16912172.
166. Gomes DA, Rodrigues MA, Leite MF, Gomez MV, Varnai P, Balla T, et al. c-Met must translocate to the nucleus to initiate calcium signals. J Biol Chem. 2008;283(7):4344–51. PubMed PMID: 18073207, Pubmed Central PMCID: 2825875.
167. Matteucci E, Bendinelli P, Desiderio MA. Nuclear localization of active HGF receptor Met in aggressive MDA-MB231 breast carcinoma cells. Carcinogenesis. 2009;30(6):937–45. PubMed PMID: 19357348.
168. Previdi S, Maroni P, Matteucci E, Broggini M, Bendinelli P, Desiderio MA. Interaction between human-breast cancer metastasis and bone microenvironment through activated hepatocyte growth factor/Met and beta-catenin/Wnt pathways. Eur J Cancer. 2010;46(9):1679–91. PubMed PMID: 20350802.
169. Sanada Y, Osada S, Tokuyama Y, Tanaka Y, Takahashi T, Yamaguchi K, et al. Critical role of c-Met and Ki67 in progress of biliary carcinoma. Am Surg. 2010;76(4):372–9. PubMed PMID: 20420246.
170. Tretiakova M, Salama AK, Karrison T, Ferguson MK, Husain AN, Vokes EE, et al. MET and phosphorylated MET as potential biomarkers in lung cancer. J Environ Pathol Toxicol Oncol. 2011;30(4):341–54. PubMed PMID: 22181983.
171. Levallet G, Vaisse-Lesteven M, Le Stang N, Ilg AG, Brochard P, Astoul P, et al. Plasma cell membrane localization of c-MET predicts longer survival in patients with malignant mesothelioma: a series of 157 cases from the MESOPATH Group. J Thorac Oncol. 2012;7(3):599–606. PubMed PMID: 22246193.
172. Wang MH, Padhye SS, Guin S, Ma Q, Zhou YQ. Potential therapeutics specific to c-MET/RON receptor tyrosine kinases for molecular targeting in cancer therapy. Acta Pharmacol Sin. 2010;31(9):1181–8. PubMed PMID: 20694025.
173. Danilkovitch-Miagkova A, Angeloni D, Skeel A, Donley S, Lerman M, Leonard EJ. Integrin-mediated RON growth factor receptor phosphorylation requires tyrosine kinase activity of both the receptor and c-Src. J Biol Chem. 2000;275(20):14783–6. PubMed PMID: 10747844.
174. Feres KJ, Ischenko I, Hayman MJ. The RON receptor tyrosine kinase promotes MSP-independent cell spreading and survival in breast epithelial cells. Oncogene. 2009;28(2): 279–88. PubMed PMID: 18836480.
175. Liu HS, Hsu PY, Lai MD, Chang HY, Ho CL, Cheng HL, et al. An unusual function of RON receptor tyrosine kinase as a transcriptional regulator in cooperation with EGFR in human cancer cells. Carcinogenesis. 2010;31(8):1456–64. PubMed PMID: 20498137, Pubmed Central PMCID: 2915629.
176. Rebagay G, Yan S, Liu C, Cheung NK. ROR1 and ROR2 in human malignancies: potentials for targeted therapy. Front Oncol. 2012;2:34. PubMed PMID: 22655270.
177. Green JL, Kuntz SG, Sternberg PW. Ror receptor tyrosine kinases: orphans no more. Trends Cell Biol. 2008;18(11):536–44. PubMed PMID: 18848778.
178. Tseng HC, Lyu PC, Lin WC. Nuclear localization of orphan receptor protein kinase (Ror1) is mediated through the juxtamembrane domain. BMC Cell Biol. 2010;11:48. PubMed PMID: 20587074.
179. Tseng HC, Kao HW, Ho MR, Chen YR, Lin TW, Lyu PC, et al. Cytoskeleton network and cellular migration modulated by nuclear-localized receptor tyrosine kinase ROR1. Anticancer Res. 2011;31(12):4239–49. PubMed PMID: 22199287.
180. Pasquale EB. Eph receptors and ephrins in cancer: bidirectional signalling and beyond. Nat Rev Cancer. 2010;10(3):165–80. PubMed PMID: 20179713.
181. Kullander K, Klein R. Mechanisms and functions of Eph and ephrin signalling. Nat Rev Mol Cell Biol. 2002;3(7):475–86. PubMed PMID: 12094214.

182. Kuroda C, Kubota S, Kawata K, Aoyama E, Sumiyoshi K, Oka M, et al. Distribution, gene expression, and functional role of EphA4 during ossification. Biochem Biophys Res Commun. 2008;374(1):22–7. PubMed PMID: 18601903.
183. Halford MM, Stacker SA. Revelations of the RYK receptor. Bioessays. 2001;23(1):34–45. PubMed PMID: 11135307.
184. Cadigan KM, Liu YI. Wnt signaling: complexity at the surface. J Cell Sci. 2006;119(Pt 3):395–402. PubMed PMID: 16443747.
185. Lyu J, Yamamoto V, Lu W. Cleavage of the Wnt receptor Ryk regulates neuronal differentiation during cortical neurogenesis. Dev Cell. 2008;15(5):773–80. PubMed PMID: 19000841.
186. Lyu J, Wesselschmidt RL, Lu W. Cdc37 regulates Ryk signaling by stabilizing the cleaved Ryk intracellular domain. J Biol Chem. 2009;284(19):12940–8. PubMed PMID: 19269974.
187. Zhong J, Kim HT, Lyu J, Yoshikawa K, Nakafuku M, Lu W. The Wnt receptor Ryk controls specification of GABAergic neurons versus oligodendrocytes during telencephalon development. Development. 2011;138(3):409–19. PubMed PMID: 21205786, Pubmed Central PMCID: 3014630.
188. Patapoutian A, Reichardt LF. Trk receptors: mediators of neurotrophin action. Curr Opin Neurobiol. 2001;11(3):272–80. PubMed PMID: 11399424.
189. Klesse LJ, Parada LF. Trks: signal transduction and intracellular pathways. Microsc Res Tech. 1999;45(4–5):210–6. PubMed PMID: 10383113.
190. Moughal NA, Waters C, Sambi B, Pyne S, Pyne NJ. Nerve growth factor signaling involves interaction between the Trk A receptor and lysophosphatidate receptor 1 systems: nuclear translocation of the lysophosphatidate receptor 1 and Trk A receptors in pheochromocytoma 12 cells. Cell Signal. 2004;16(1):127–36. PubMed PMID: 14607283.
191. Gong A, Zhang Z, Xiao D, Yang Y, Wang Y, Chen Y. Localization of phosphorylated TrkA in carrier vesicles involved in its nuclear translocation in U251 cell line. Sci China C Life Sci. 2007;50(2):141–6. PubMed PMID: 17447019.
192. Bonacchi A, Taddei ML, Petrai I, Efsen E, Defranco R, Nosi D, et al. Nuclear localization of TRK-A in liver cells. Histol Histopathol. 2008;23(3):327–40. PubMed PMID: 18072090.
193. Linger RM, DeRyckere D, Brandao L, Sawczyn KK, Jacobsen KM, Liang X, et al. Mer receptor tyrosine kinase is a novel therapeutic target in pediatric B-cell acute lymphoblastic leukemia. Blood. 2009;114(13):2678–87. PubMed PMID: 19643988.
194. Migdall-Wilson J, Bates C, Schlegel J, Brandao L, Linger RM, DeRyckere D, et al. Prolonged exposure to a Mer ligand in leukemia: Gas6 favors expression of a partial Mer glycoform and reveals a novel role for Mer in the nucleus. PLoS One. 2012;7(2):e31635. PubMed PMID: 22363695, Pubmed Central PMCID: 3282750.
195. Andrae J, Gallini R, Betsholtz C. Role of platelet-derived growth factors in physiology and medicine. Genes Dev. 2008;22(10):1276–312. PubMed PMID: 18483217, Pubmed Central PMCID: 2732412.
196. Hamilton TG, Klinghoffer RA, Corrin PD, Soriano P. Evolutionary divergence of platelet-derived growth factor alpha receptor signaling mechanisms. Mol Cell Biol. 2003;23(11):4013–25. PubMed PMID: 12748302, Pubmed Central PMCID: 155222.
197. Chen L, Acciani T, Le Cras T, Lutzko C, Perl AK. Dynamic regulation of platelet-derived growth factor receptor alpha expression in alveolar fibroblasts during realveolarization. Am J Respir Cell Mol Biol. 2012;47(4):517–27. PubMed PMID: 22652199, Pubmed Central PMCID: 3488620.
198. Mittar S, Ulyatt C, Howell GJ, Bruns AF, Zachary I, Walker JH, et al. VEGFR1 receptor tyrosine kinase localization to the Golgi apparatus is calcium-dependent. Exp Cell Res. 2009;315(5):877–89. PubMed PMID: 19162007.
199. Kermorgant S, Parker PJ. Receptor trafficking controls weak signal delivery: a strategy used by c-Met for STAT3 nuclear accumulation. J Cell Biol. 2008;182(5):855–63. PubMed PMID: 18779368.

200. Akgoz M, Kalyanaraman V, Gautam N. Receptor-mediated reversible translocation of the G protein betagamma complex from the plasma membrane to the Golgi complex. J Biol Chem. 2004;279(49):51541–4. PubMed PMID: 15448129.
201. Robertson BJ, Park RD, Snider MD. Role of vesicular traffic in the transport of surface transferrin receptor to the Golgi complex in cultured human cells. Arch Biochem Biophys. 1992;292(1):190–8. PubMed PMID: 1727635.
202. Demory ML, Boerner JL, Davidson R, Faust W, Miyake T, Lee I, et al. Epidermal growth factor receptor translocation to the mitochondria: regulation and effect. J Biol Chem. 2009;284(52):36592–604. PubMed PMID: 19840943.
203. Cao X, Zhu H, Ali-Osman F, Lo HW. EGFR and EGFRvIII undergo stress- and EGFR kinase inhibitor-induced mitochondrial translocalization: a potential mechanism of EGFR-driven antagonism of apoptosis. Mol Cancer. 2011;10:26. PubMed PMID: 21388543.
204. Liccardi G, Hartley JA, Hochhauser D. EGFR nuclear translocation modulates DNA repair following cisplatin and ionizing radiation treatment. Cancer Res. 2011;71(3):1103–14. PubMed PMID: 21266349, Pubmed Central PMCID: 3033323.
205. Hadzisejdic I, et al. Nuclear EGFR in ductal invasive breast cancer: correlation with cyclin-D1 and prognosis. Mod Pathol. 2010;23(3):393–403.
206. Li CF, et al. EGFR Nuclear Import in Gallbladder Carcinoma: Nuclear Phosphorylated EGFR Upregulates iNOS Expression and Confers Independent Prognostic Impact. Ann Surg Oncol. 2011.
207. Nowsheen S, Bonner JA, Yang ES. The poly(ADP-Ribose) polymerase inhibitor ABT-888 reduces radiation-induced nuclear EGFR and augments head and neck tumor response to radiotherapy. Radiother Oncol. 2011;99(3):331–8.
208. Dekanic A, et al. Strong nuclear EGFR expression in colorectal carcinomas is associated with cyclin-D1 but not with gene EGFR amplification. Diagn Pathol. 2011;6:108.
209. Xia W, et al. Nuclear expression of epidermal growth factor receptor is a novel prognostic value in patients with ovarian cancer. Mol Carcinog. 2009;48(7):610–7.
210. Hoshino M, et al. Nuclear expression of phosphorylated EGFR is associated with poor prognosis of patients with esophageal squamous cell carcinoma. Pathobiology. 2007;74(1):15–21.
211. Kim JH, et al. Expression of HER-2 and nuclear localization of HER-3 protein in canine mammary tumors: histopathological and immunohistochemical study. Vet J. 2011;189(3): 318–22.
212. Feng S, et al. Nuclear localization of a complex of fibroblast growth factor(FGF)-1 and an NH2-terminal fragment of FGF receptor isoforms R4 and R1alpha in human liver cells. Biochim Biophys Acta. 1996;1310(1):67–73.
213. Stachowiak EK, et al. Nuclear accumulation of fibroblast growth factor receptors in human glial cells--association with cell proliferation. Oncogene. 1997;14(18):2201–11.

Chapter 6
Computational and Modeling Aspects of RTK Networks

Yehoshua Enuka, Morris E. Feldman, and Yosef Yarden

6.1 Introduction

One major challenge in studying signaling pathways is to decipher the exact mechanisms that underlie the translation of an extracellular signaling cue to a specific cellular outcome. This goal is particularly challenging when dealing with receptor tyrosine kinase (RTK) signaling pathways, because they are highly complex and interconnected [1]. In 2005, a survey of the ERBB signaling network, a prototypical RTK network, assigned 122 proteins to the network with 211 interactions among these proteins [2]. Although biochemistry provides tools to study local and direct (short-range) interactions within networks, and cell biology allows the elucidation of spatial input–output relationships, experimental techniques are poorly suited to drawing long-range and indirect connections, as well as dynamic aspects of complex networks [3, 4]. Computational modeling aims to bridge this gap in understanding by revealing the hidden functional and topological properties of networks.

The bulk of this review will focus on the ERBB network, as an example of other RTK signaling networks. We will describe each of the three main layers of the ERBB network, as well as delve into the insights that have been gained into that layers using computational modeling. As we examine the layers of the ERBB network, we will mainly highlight results from mass-action ordinary differential equation models (see Box 6.1). Because mass-action models require extensive information on all of the components of a network, including their kinetic parameters, and this level of granularity is not available for the entire ERBB network, most mass-action modeling studies are limited in scope to modeling only a single layer of the ERBB network. To study the entire ERBB network as a whole, where the experimental data coverage is less complete, requires

Y. Enuka • M.E. Feldman • Y. Yarden, Ph.D. (✉)
Department of Biological Regulation, Weizmann Institute of Science, Rehovot 76100, Israel
e-mail: yosef.yarden@weizmann.ac.il

D.L. Wheeler, Y. Yarden (eds.), *Receptor Tyrosine Kinases: Structure, Functions and Role in Human Disease*, DOI 10.1007/978-1-4939-2053-2_6

abstract modeling methods. Hence, following our treatment of the individual layers of the network, we will turn to the entire ERBB network as a whole and highlight the findings of several abstract modeling techniques. In the final section of this chapter, we will describe how modeling strategies can help in finding potential targets for pharmaceutical manipulation.

Box 6.1 Mass-Action Modeling Using Ordinary Differential Equations

Mass-action modeling is based on ordinary differential equations (ODEs, also called rate equations), which express the overall rate of change in the concentrations of cellular components. The dynamics of the signaling cascade is modeled by an ordered set of rate equations. The rate equations are coupled, meaning that a change in the concentration of one component causes a change in another component. Rate equations generally have the form:

$$\frac{\mathrm{d}C}{\mathrm{d}t} = \text{generation} - \text{consumption}$$

The above equation states that the change in concentration of a certain component over time is dependent upon the rate of generation minus the rate of consumption of the component. The generation and consumption terms can be either a constant or a function of the concentration of cellular components, such as mRNAs, proteins, and small molecules. Usually, the set of ODEs is too complicated for an analytical solution. Therefore, ODEs are solved using numerical methods based on the initial concentrations of the components involved in the signaling cascade and their kinetic parameters.

It is important noting that ODEs have only a single independent variable, usually time. When another independent variable, such as space, should be included, partial differential equations become necessary.

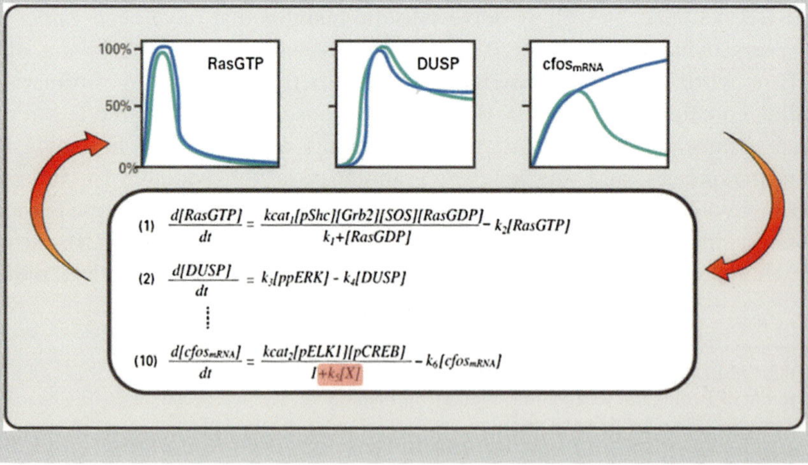

(continued)

Box 6.1 (continued)

This mass-action model is derived from experimental data (green line) and then is used to predict the dynamics of the signaling components (blue line). When model predictions do not fit experimental data, the model is refined. For example, the model by Nakakuki et al. predicted that knockdown of DUSP would lead to increased duration of Fos expression [6]. However, experimental data did not fit this prediction. Hence, the model was refined to include a yet unknown repressor of Fos transcription (see equation 10).

6.2 Topic 1: The Multilayered ERBB Signaling Network

Although the ERBB network is relatively complex and involves many interacting proteins, its basic topology is well understood and serves as a basis on which to build computational models (see Fig. 6.1). The ERBB network has a "bow-tie" topology consisting of three layers [1]: (1) the input layer, where the ligand activates one or more receptors. The activated receptors then signal to (2) the signal processing layer, also called giant strong component network, or the core process [5], which integrates and decodes the signals that are further propagated to (3) the output layer, where transcriptional, biochemical, and metabolic routes culminate and determine the cell fate.

In the following sections, we highlight the *emergent behavior* of each of the layers that comprise the bow-tie structure of the ERBB network, as revealed by mass-action computational models. Mass-action modeling, also known as mechanistic computational modeling, is the main technique used to model the layers of the ERBB network. The mass-action method uses ordinary differential equations (ODEs), which are also known as chemical-kinetics equations, to decipher the regulatory mechanisms underlying ERBB signaling. Mechanistic models are based upon two parameters that must be either measured or estimated for every *species* in the model, namely, initial concentrations and rate constants. When experimental data for these parameters is missing, it is necessary to estimate their values. When parameters must be estimated, simulation techniques are used to ensure that the final output is robust against variations in these parameters (see, e.g., [6]). The mass-action modeling approach has been employed in many models of the ERBB network, some of which will be discussed in detail below [7–14].

The process of modeling is a recursive process; first a model is built based on current knowledge, and then new hypotheses are derived. Later, these hypotheses are examined using available tools, and finally the model is validated or else rejected. If the model is validated, it substantiates the regulatory mechanisms that appear in the model. However, if the model is rejected, it calls for model refinement by changing the regulatory mechanisms that are included in the model. For example, Box 6.1 relates to a model-driven hypothesis that has been experimentally rejected and eventually led to model refinement.

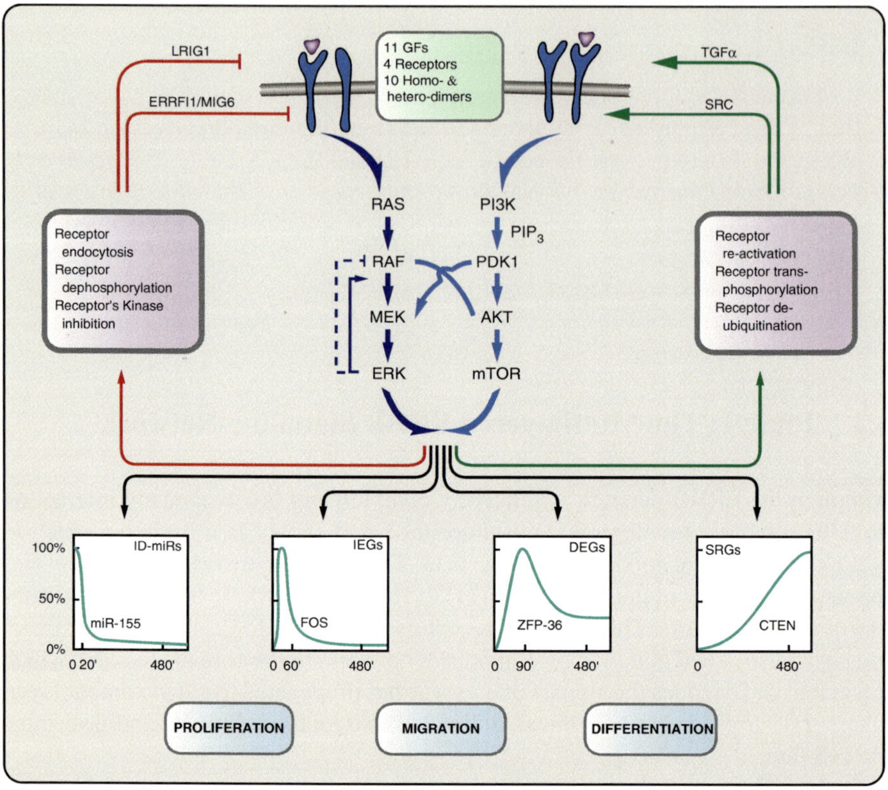

Fig. 6.1 A schematic representation of the ERBB network. The input layer comprises 11 EGF-like growth factors, each binds specific homo- and heterodimers of the four ERBB proteins. ERBB2/HER2 serves as a coreceptor, which is unable to bind a soluble ligand, but its heterodimers are characterized by enhanced signaling due to weaker negative feedback loops. ERBB signals are funneled into a web of intracellular signaling cascades, such as the one leading to ERK activation and another cascade culminating in mTOR stimulation. In combination, these multiple biochemical pathways regulate transcription factors and repressors, in a way that instigates a ligand-specific gene expression program. The program initiates by immediate downregulation of 24 micro-RNAs (denoted immediately downregulated micro-RNAs, ID-miRs). Downregulation of the ID-miRs permits transcription of several immediate early genes (IEGs, e.g., EGR1). The wave of IEG's is later repressed by the delayed early genes (DEGs), a group encoding phosphatases and transcription repressors, and this translates to regulation of a group of secondary response genes (SRGs). Importantly, ERBB activation is coupled to positive feedback loops, such as the secretion of several EGF-like ligands, which restimulate the network. Reciprocally, ERBB activation initiates several negative feedback loops, such as receptor ubiquitination and degradation, as well as upregulation of a kinase-inhibitory protein called ERRFI1/Mig6

6.2.1 The Input Layer

6.2.1.1 Combinatorial Complexity of the Input Layer

The input layer of the ERBB signaling network consists of 11 ligands, 4 receptors, of which three can directly bind a ligand, as well as eight or more intracellular binding partners [15]. Different receptor-ligand complexes often drive distinct biological outcomes [16]. The four receptors, especially upon ligand binding, assemble into several types of homo- and heterodimers. Furthermore, each receptor has at least nine tyrosine residues that may be dynamically phosphorylated [17]. Additionally, each receptor-ligand complex may interact with different combinations of intracellular binding partners, thus generating a very large number of potential multicomponent protein complexes [18]. Complicating this system even further, each protein complex has distinct functional properties in terms of ligand binding affinity, endocytic routing, and receptor activation. Hence, the complexity of the regulatory landscape of the network increases in a combinatorial manner as additional components are considered.

To tackle the complexity of a system, which like the ERBB network has so many combinatorial interaction possibilities, one of the main principles that must be addressed is the *granularity* of the model, which is the depth and density of species that can be included in the model [15]. For example, the utilization of mass spectrometry has enabled researchers to investigate the spatiotemporal dynamics of protein isoforms, covalent modifications (e.g., phosphorylations and acetylations), and protein complex formation [19] and has thus greatly enhanced the granularity of signaling pathway maps [20]. However, in reality most models must make compromises regarding the granularity of the system, meaning that some species and processes are lumped together, due to limited computing power and limitations of the computational methods [8]. To illustrate the issue of granularity, we will contrast two models of the ERBB network. One model was proposed by the group of Kholodenko, who lumped together several species in the input layer by representing all autophosphorylation sites as a single "virtual phosphorylation site" [7, 10]. This decreased the complexity of the model by limiting the number of species to 25 and the number of reactions to 47. In contrast, the group led by Hlavacek considered all possible species and reactions [21], leading them to a much higher resolution and allowing them to predict different dynamics of phosphorylation of individual tyrosines in response to EGF. Surprisingly, however, both models are quite consistent with the experimental data. Nevertheless, it is possible that challenging the models with distinct initial inputs would reveal differences due to the granularity of each model.

Because of the debatable contribution of considering all possible species, models of the input layer, as well as downstream layers, often use simplified network formats that merely resemble the actual biological system [22]. Notably, not only are species combined together but also processes; for example, Resat et al. lumped together the process of recycling of ERBB receptors to the plasma membrane, meaning that the passage of the receptors through the endosomal compartment is ignored [23]. The justification for this omission is that recycling receptors that pass through the endosomal compartment are no longer active, and therefore, they no longer have an effect on signaling. In conclusion, lumping together species and processes is often a legitimate and necessary tool to decrease the complexity of the network without compromising the predictive power of the model.

The potency of ERBB homo- and heterodimers varies greatly, but the basis for this variation is often not clearly understood from the basic properties of the RTKs themselves. For instance, ERBB2-ERBB3 heterodimers are considered to be the most potent signaling pair of the ERBB network [24], and ERBB2 overexpression has been shown to elevate signaling to the extracellular-regulated kinase (ERK), an important output signal of the processing unit. However, the exact mechanism responsible for signal amplification by ERBB2 was not clear based on our knowledge of the ERBB2 molecule itself. Two complementary computational models have therefore addressed the question, "Is ERBB2 a more potent activator of the downstream network or are there simply more active ERBB2 dimers on the cell surface?" Hendriks et al. showed that ERBB1 and ERBB2 homodimers have essentially the same potency per molecule to activate ERK, thus excluding the possibility of greater potency of ERBB2 containing dimers [25]. A later model by Birtwistle et al. further clarified the role of ERBB2 on EGF signaling [7]. Their model simulated ERBB2 overexpression and showed that upon overexpression, the balance shifted towards more ERBB1-ERBB2 heterodimers. These heterodimers did not undergo ligand-induced internalization, thus causing a more sustained signal on the plasma membrane and subsequently more prolonged ERK activation, which has a critical effect on the overall network output (see the section on the processing unit).

6.2.1.2 Positive Feedback Circuit Acting at the Input Layer

Autocrine and *paracrine* loops are important positive feedback circuits that enhance the input layer. Activation of the ERBB network causes transcription of multiple ERBB ligands such as TGF-α and HB-EGF [26], which then re-feed the network. Sometimes, other signaling pathways participate in this autocrine activation, as in the case of the cleavage of HB-EGF to the mature form upon activation of G-protein-coupled receptors [27].

6.2.1.3 Endocytosis: A Negative Feedback Circuit Acting at the Input Layer

The major negative feedback loop acting at the input layer is *endocytosis* of active ligand-receptor complexes through clathrin-coated pits and other routes, followed by either recycling or degradation in lysosomes. Early studies of the EGFR receptor utilized computational methods combined with biochemical tools, such as site-directed mutagenesis, to decipher the internalization process. One study found that the internalization of EGFR depends on ligand binding to the receptor and specific carboxyl-terminal sequences responsible for bringing the receptor-ligand complex to coated pits [28]. Another early computationally aided study was able to resolve a disagreement between two models that tried to explain accelerated receptor degradation upon ligand stimulation [29]. One model claimed that accelerated receptor degradation is the result of induced internalization [30], while the other model proposed that it is the result of the activation of the kinase domain of the receptor, followed by a reduced recycling rate to the plasma membrane [31]. The study by Opresko et al. found that accelerated receptor degradation results from both models acting at the same time. Receptors are induced to internalize upon activation, and once internalized, activation of the kinase domain exposes a buried sequence that reduces the recycling rate by targeting the receptor for lysosomal degradation.

The early studies that modeled endocytosis separately from signaling were sufficient to decipher the basic mechanisms of endocytosis as an attenuator of EGFR signaling. However, it has been shown that as internalized receptors transit through several different compartments, these receptors might remain active. Therefore, internalized receptors cannot be merely considered as silenced receptors. The internalized receptors' kinase activity is determined by the specific ligand-receptor complex; the receptor microenvironment, such as intravesicular pH [32]; as well as substrate accessibility in different cellular compartments. Distinct endocytic compartments, including clathrin-, caveolae- and Rab-enriched compartments, can deliver differential sets of proteins to diverse cellular targets, generating specific signaling outputs [33].

6.2.1.3.1 Specific Models of EGFR Endocytosis

Clearly, more sophisticated models would be necessary to integrate the complex interplay between EGFR trafficking and signal transduction. An important model by Resat et al. showed that at low EGF concentrations, constitutive endocytosis dominates receptor trafficking and sorting properties, while at high EGF concentrations, ligand-induced endocytosis dominates these processes [23]. Moreover, this model showed that ligand availability determines the distribution of phosphorylated receptors between the plasma membrane and endosomes. Finally, they compared the signaling activity induced by EGF and TGF-α and found that TGF-α-induced signals are strongly biased towards the plasma membrane compared to EGF-induced signals, because TGF-α dissociates from internalized receptors, and the receptors are then preferentially recycled. EGF on the other hand remains bound to ERBB1 even after it has been internalized, and therefore, the EGF-induced signal remains active throughout the endocytic cycle.

Another ODE model of the input layer has been introduced by Hendriks et al., who studied the effect of HER2 overexpression on the level of activated EGFR in the plasma membrane [34]. They concluded that HER2 overexpression increases the rate of EGFR recycling and decreases the rate of EGFR internalization, thereby increasing overall levels of activated EGFR at the plasma membrane. Later, a deterministic model developed by Shankaran et al. showed that co-expression of ERBB1, ERBB2, and ERBB3 reduces signal attenuation and shifts signaling towards the plasma membrane [35], which resembles the effect of TGF-α binding to EGFR mentioned earlier. In addition, they demonstrated that when these receptors are co-expressed, ERBB2-ERBB3 dimers are favored. Moreover, their model reaffirmed the conclusion of a previous study showing that ERBB2-ERBB3 heterodimers are more potent, in terms of cell response, than ERBB2-ERBB1 heterodimers.

ERBB1 signaling is negatively regulated through ubiquitylation by recruitment of the E3 ubiquitin ligase CBL to ERBB1 phosphorylated at tyrosine 1045 or by indirect recruitment of CBL through GRB2 [36, 37]. The EGFR-CBL complex then recruits several proteins, among which are endophilin and clathrin-interacting proteins that together are able to drive clathrin assembly and budding that leads to EGFR endocytosis. The small GTPase RhoA might enhance EGFR signaling by downregulating endophilin-mediated endocytosis of EGFR, and possibly by other mechanisms, such as activating the MEKK1 kinase [38]. To decipher the contribution of RhoA to EGFR endocytosis, Ung et al. extended a previously published ODE model by including RhoA's cross talk with endophilin and MEKK1 [39]. Their simulation predictions fit experimental results of the cross talk between CBL, internalization of EGFR, ERK activation, and the effect of PP2A and MKP3 on ERK activation. Next, they simulated the effect of overexpressing RhoA and its effector kinase ROCK, on EGFR endocytosis. Overexpression of RhoA and ROCK prolonged ERK activation, as expected, but surprisingly reduced the maximal ERK activation. This reduction was explained by negative feedback from active ERK to active SOS. Overall, their model suggests that RhoA plays a role in determining the duration of ERK activation by two mechanisms: inhibition of endophilin-mediated EGFR endocytosis and MEKK1-mediated ERK activation. It should be noted, however, that these predictions are still awaiting experimental validation.

Phosphotyrosine phosphatases (PTPs) act at the input layer by reversing tyrosine phosphorylation on EGFR and other RTKs and therefore might form another negative feedback mechanism [40]. Currently, however, the exact regulation of phosphatases acting at the input layer is still incompletely understood. A model by Kholodenko et al. suggested that experimentally observed pEGFR transients are not the result of receptor-mediated activation of specific PTPs but rather the result of EGFR phosphotyrosine residues being masked by phosphotyrosine binding proteins [10]. This masking might prevent dephosphorylation of the corresponding phosphotyrosine residues by PTPs, hence lead to the accumulation of pEGFR.

6.2.2 The Signal Processing Core of the ERBB Network

The ERBB signal processing unit is composed of several three-layer kinase cascades. Each layer consists of a kinase that is able to phosphorylate, and thereby activate, the kinase of the next tier. Interestingly, although the processing unit is composed of components that are shared among several signaling pathways, it is still able to signal to the correct output. For example, EGF activates ERBB receptors while the nerve growth factor, NGF, activates the TrkA receptor, but both signals are funneled to the *mitogen-activated protein kinase* (MAPK) module. However, while EGF transiently activates ERK, which is the last step in the enzymatic cascade of the MAPK module, NGF causes sustained activation of ERK. This difference in the dynamics of ERK activation seems to be the crucial determinant of cell fate, because transient activation by EGF leads to proliferation while sustained activation by NGF gives rise to the differentiation of PC12 pheochromocytoma cells [41]. The same effect of the duration of ERK activation can also be seen in other cell lines, in which EGF causes proliferation of cells through transient activation of downstream signaling, while neuregulin-1 (NRG1), a ligand of ERBB3 and ERBB4, causes differentiation of cells through sustained activation of the network [42, 43].

The different durations of ERK activation under EGF vs. NGF stimulus might be explained by different topological arrangements of the network [44]. Modular response analysis [45] was used to examine the associations between pairs of modules in the MAPK cascade in order to reveal different wirings of the network under EGF and NGF stimuli. Under EGF stimulation, the MAPK network exhibits a negative feedback between ERK and RAF, while with NGF stimulation, ERK positively signals backwards to RAF through protein kinase C, thereby refeeding the enzymatic cascade. The authors further showed that the positive feedback of ERK on RAF under NGF treatment results in a switch-like response of ERK activation.

6.2.2.1 Ultrasensitivity and Bistability of the Signal Processing Core

Cell fate decisions require conversion of a graded input into discrete stable states. Within the signal processing layer of the ERBB network, the MAPK cascade may behave in a *bistable* manner, which means that it is able to switch between two distinct stable states. To make the states stable, the network must also possess *hysteresis*, meaning that a stimulus must change beyond a certain threshold in order to cause a change in the output state of the network. In addition to hysteresis, another important aspect of bistable systems is *ultrasensitivity* [46]. Due to signal amplification, as the signal descends down the MAPK cascade, stimulus/response curve becomes steeper. The signal at the lower layer of the cascade has the properties of a cooperative enzyme and requires only a 2.5-fold increase in stimulus in order to drive a ninefold increase, from 10 to 90 % of the MAPK response. Thus, the relation between EGFR activation and ERK phosphorylation is highly nonlinear. Importantly, due to the combination of this nonlinearity and saturation, a wide range of EGF concentrations give rise to similar levels of ERK activation [12]. The nonlinear dependence of downstream signaling on receptor activity was further supported by a computational model from the Lauffenburger group [25].

It is well known that positive feedback loops are involved in the generation of the ultrasensitive response and bistability of the processing layer [47, 48]. However, are feedback loops required for the emergence of this important attribute of biological networks? Recent computational models shed new light on this issue. MAPK activation requires phosphorylation on both a threonine and a tyrosine in a TxY motif, and phosphorylation and dephosphorylation of these two amino acids must occur by a two-collision mechanism. This means that once the first phosphosite is phosphorylated, the monophosphorylated product is released from the enzyme, and the second phosphosite may be phosphorylated only after the enzyme and the monophosphorylated species re-collide again. It was shown by Markevich et al. that such a double phosphorylation mechanism is sufficient by itself to generate a bistable system [49], suggesting that the emergent properties of the signal processing core, such as bistability and ultrasensitivity, might be inherent to the two-collision MAPK activation mechanism.

6.2.2.2 Phosphatases: Constant and Induced Attenuation of the Signal Processing Core

A major attenuator of signals within the processing unit is dephosphorylation by specific phosphatases. NRG-stimulated signaling provides an example of the complex role of constitutively active phosphatases in ERBB signaling. NRG binds to ERBB4 and subsequently stimulates signals that involve activation of the ERK cascade, as well as the PI3K-Akt pathway. These two pathways are interconnected as it was shown that activated Akt decreases the level of ERK activation [50, 51]. This was also found to be true in the case of NRG-stimulated ERBB4 signaling [9]. A constitutively active phosphatase that participates in the Raf-Akt cross talk, protein phosphatase 2A (PP2A), has contradicting effects on ERK signaling. On the one hand, it is negatively regulating the MAPK cascade by dephosphorylation of MEK, while on the other hand, it has an enhancing effect on the MAPK cascade by dephosphorylating Akt. Hatakeyama et al. modeled the cross talk between the Akt and the ERK cascades to further define the significance of PP2A in ERK signaling. Their simulations agreed with experimental results of ERK signaling only when PP2A activity, along with the Raf-Akt cross talk, was included in the model [9].

The core process may also be attenuated by MAPK phosphatases (MKPs, also called DUSPs) that are transcriptionally induced by EGF. Upon simulation, the MAPK cascade in the core process is subject to attenuation by newly transcribed genes; among these are the dual specificity phosphatases (DUSP) [52]. Subsets of these phosphatases are involved in different pathways (e.g., DUSP3, DUSP4, DUSP6, and DUSP7 are ERK-specific while DUSP1 and DUSP10 are p38- and JNK-specific). DUSPs facilitate strict regulation of signal duration. The negative feedback of DUSP on ERK activation is manifested in two ways: (1) dephosphorylation ERK in the cytoplasm and (2) sequestering of ERK in the nucleus [53]. The effect of DUSPs on ERK signaling has previously been modeled [6]. It was established earlier that active ERK phosphorylates and stabilizes the DUSP protein product, and subsequently DUSP translocates to the nucleus and dephosphorylates active ERK molecules [54]. Nakakuki et al. used a simple model that included only the late steps of ERBB signaling, thus excluding earlier steps in the signaling cascade. They found that EGF and NRG cause transient and sustained activation of ERK signaling, respectively. Their model reaffirmed that c-Fos is transiently expressed in response to both stimuli [6]. What might be the element that ensures that c-Fos expression remains transient in response to either stimulus, although in the case of NRG the upstream signaling is sustained? Their model predicted that although NRG causes sustained activation of ERK, *nuclear* activity of ERK is transient due to a larger DUSP upregulation caused by the NRG stimulation. Indeed, inhibition of DUSP by using RNA interference sustained nuclear ERK activity, but the output of ERK activity, c-Fos mRNA expression, remained essentially the same. According to the authors, this could be explained by the activity of an additional and as yet unknown negative regulator of c-Fos.

6.2.3 The Transcriptional Output Layer

The information from the core process flows to the output layer and dictates a specific transcriptional response that determines cell fate. In the ERBB network, the kinetics of the core process downstream of the receptor is a major determinant of cell fate [16]. Different kinetics of ERBB activation, and subsequent ERK activation, cause different responses in a group of genes called immediate early genes (IEGs), one of which is c-Fos. c-Fos responds much more strongly to sustained ERK activation upon NGF or NRG stimulation than to transient ERK activation resulting from EGF [43]. Distinct network responses to different stimuli can often be attributed to network motifs, like positive or negative feedback loops. In the case of c-Fos, sustained ERK activation is required to generate active c-Fos because c-Fos is embedded in a network motif called a coherent feedforward loop (CFL) [55]. ERK activation leads to transcription and translation of c-Fos, but without phosphorylation, the c-Fos protein is unstable and it undergoes rapid degradation. If ERK is still active once the c-Fos protein is produced, then c-Fos will be phosphorylated and stabilized, leading to differentiation. If, however, ERK activation is transient, then only little c-Fos will accumulate [56]. This network motif is called a coherent feedforward loop, because the ERK signal first activates transcription and then it feeds forward to activate the newly synthesized c-Fos protein. This kind of network topology is responsible for a “sign-sensitive delay.” Reduction of the signal is immediately reflected as loss of output, while induction of the signal causes activation of the output in a delayed manner, since the activators must accumulate in order to act coordinately. The ERK signal that eventually drives production of active, phosphorylated c-Fos, is only slightly longer in response to NRG compared to EGF stimulation, but thanks to the CFL, this is translated to the all-or-none phosphorylated c-Fos response and finally to distinct cell fates (differentiation versus proliferation).

As described above, while much progress has been gained in understanding the ERBB network specifically at the input and core process layers, understanding of the output layer is still lagging behind due to some methodological difficulties, which are described elsewhere [57]. Quantitative mechanistic modeling of the output layer would require more detailed data than is currently available in terms of metabolic, biochemical, and transcriptional processes. So far, only the c-Fos expression response has been modeled in a mechanistic way. Naturally, there is a need to include transcriptional, as well as biochemical, cytoskeletal, and metabolic responses in predictive models, to gain deeper insights into ERBB signaling. For example, at the posttranscriptional level, a group of 24 micro-RNAs was found to be immediately downregulated following EGF stimulation, while the mRNAs that they target are simultaneously upregulated [58]. This dynamic response suggests that the induction of immediate early genes is enhanced by the repression of specific micro-RNAs.

6.2.4 The ERBB Network Is Robust, yet Fragile

Complex signaling networks, such as the ERBB network, are robust in their behavior, meaning that their response to certain stimuli is reproducible and quite resistant to internal and external perturbations. The identification and definition of network motifs and topology may help uncover mechanisms that confer robustness to the ERBB network. Robustness is a key feature of networks sharing bow-tie structure [5, 59], and computational models might help to shed light on the exact topological organization that contributes to robustness. For example, a few models have shown that negative feedback loops confer robustness in face of parameter perturbations [6, 60, 61]. In addition, it was postulated that ERK is robust to changes because of its involvement in various negative feedback loops [7]. Other more complex network motifs might also confer robustness: one CFL nested in yet another CFL appears to confer robustness of phosphorylated c-Fos, as an output, to otherwise noisy pERK signaling [6]. Indeed an important virtue of robust complex systems is noise filtering, which may also be achieved by negative feedback loops, while switch-like behavior may be achieved by positive feedback loops [62].

There are inherent trade-offs between robustness, fragility, performance, and resource demands [63, 64]. In other words, increasing network robustness towards certain perturbations might expose fragility to other perturbations. Biological systems have thus evolved into a state termed "robust, yet fragile" [65]. The robustness-fragility trade-off is reflected in the inhomogeneous scale-free topology of signaling networks, where signaling networks usually rely on a few highly connected hubs. Eliminating one of these network hubs will most likely have disastrous effects on network functionality [66]. Disease states likely take advantage of the fragile aspects of otherwise robust networks.

6.3 Topic 2: Abstract Computational Modeling Techniques

Thus far, we have presented the contribution of mass-action ordinary differential equation modeling to understanding the ERBB network. Such models usually require extensive prior knowledge of the system of interest, including topology, as well as accurately determined kinetic parameters. However, when such detailed data is not available for modeling, alternative approaches are needed (Table 6.1).

Table 6.1 A list of computational models and the methods used for modeling

Model (reference)	Computational method	Unique properties
[28]	Sequence alignment and secondary structure prediction	The internalization of EGFR depends on ligand binding and specific carboxyl-terminal sequences responsible for binding to coated pits
[29]	Linear regression and nonlinear parameter estimation	Accelerated receptor degradation is the result of two concomitant processes mediated by specific carboxyl-terminal sequences
[46, 67]	Deterministic ODEs	Description of the switch-like behavior of the MAPK cascade. The last enzyme in the cascade exhibits an ultrasensitive behavior
[4]	Deterministic ODEs	The bistable behavior of the interconnected phospholipase C to protein kinase C, as well as the Ras-Raf-MAPK pathways, is robust over a range of kinetic parameters
[10]	Deterministic ODEs	Phosphorylated tyrosine residues bound by their ligand proteins are protected from dephosphorylation. This gives rise to a transient pattern of EGFR phosphorylation. In addition, the kinetic model predicted that EGFR-mediated phosphorylation of Shc would decrease the binding affinity of Shc to EGFR and induce the dissociation of Shc from EGFR. This prediction was corroborated by experimental evidence [68]
[69]	Deterministic ODEs	A kinase that is activated by EGF downstream of MEK forms a negative feedback loop and determines whether ERK signaling is transient or sustained. Consequently, ERK signaling duration determines if the cell proliferates or differentiates
[12]	Deterministic ODEs	The ERK signal is maximal over a broad range of EGF concentrations. ERK's activation amplitude is a nonlinear function of EGFR-ligand complex levels
[70]	Deterministic ODEs	Positive feedback formed by autocrine signaling loops determines the amplitude and duration of the Ras-MAPK cascade. The signaling in the cell is context dependent
[9]	Deterministic ODEs	Description of the ERK and PI3K pathways stimulated by NRG and inhibition of Raf-1 and downstream ERK by Akt. NRG-stimulated signaling is controlled by protein phosphatase 2A
[23]	Stochastic	A model that integrates endocytosis and signaling of EGFR
[71]	Decision tree	ERK signaling and substrate fibronectin levels are important determinants of migration speed
[72]	ODEs	An autocrine ligand accumulation that takes into account cell density can predict the concentration of ligand as a function of time

(continued)

Table 6.1 (continued)

Model (reference)	Computational method	Unique properties
[20]	Unsupervised learning (clustering)	Clustering of phosphorylation dynamics in response to EGF reveals new proteins taking part in EGFR signaling
[73]	Fuzzy c-means clustering	An improved mass spectrometry method reveals many new EGF-stimulated phosphorylation sites on proteins belonging to various functional clusters such as transcription factors and actin-binding proteins
[35]	Deterministic ODEs	Dimerization and endocytic properties of ERBB1–ERBB3 heterodimers
[7]	Deterministic ODEs	EGF that binds to ERBB1 and NRG that binds to ERBB3 and ERBB4 lead to the activation of ERK and Akt. HER2 overexpression leads to sustained EGF signaling
[43]	Generalization of the multiplicative decomposition model	The difference between early transcriptional response of EGF and NRG is not qualitative but quantitative. These rather graded quantitative differences dictate discrete cell fates
[74]	Deterministic ODEs	Synergistic activation of the ERK pathway by EGF and the insulin receptor, due to an increase in the levels of PIP3 as a result of EGF and insulin co-activation of PI3K. This facilitates recruitment of GAB1 and increases activation of the Ras pathway. Experimentally validated by the addition of wortmannin (a PI3K inhibitor) that abrogated this synergy
[75]	Deterministic ODE	AKT phosphorylation is more sensitive to ERBB3 than to other ERBB receptors. This prediction was validated by using a monoclonal antibody against ERBB3
[76]	Bayesian networks	A microwestern array (MWA) method that enables investigation of large-scale protein modifications or expression levels
[77]	Unsupervised learning (clustering)	A web interface for exploring correlations between functional annotations and phosphorylation dynamics enabled function prediction of previously uncharacterized proteins
[6]	Deterministic ODEs	Comparison of ERK and c-Fos expression upon EGF and NRG stimulation. Predicted new inhibitory mechanisms that control the transient expression of c-Fos
[78]	Exhaustive clustering	A program that finds the parameter sets that yield the best cluster sets in terms of enrichment of biological features

The field of exploring different modeling approaches is constantly advancing, and new modeling techniques have been developed to cope with low information data that is often coupled to data obtained using high-throughput techniques, such as microarrays and mass spectrometry.

6.3.1 Data-Driven Modeling, Principle Component Analysis and Clustering

As mentioned earlier, the connection between upstream signaling cascades and transcriptional responses is incompletely understood. Using high-throughput methods that produce large amounts of data, such as microarrays, one can measure the transcriptional behavior of thousands of genes in response to ERBB signaling. However, we still do not know all of the interacting transcription factors that participate in the transcriptional response to ERBB signaling, and we also lack kinetic details of the transcriptional response even for the transcription factors that we do know. This situation calls for a different modeling approach than mechanistic computational modeling. For example, Nagashima et al. compared the transcriptional response of MCF-7 cells to EGF or NRG stimulation [43]. They used a statistical method, based on a multiplicative decomposition model, to estimate expression profiles of thousands of genes. Then, they selected genes whose transcripts significantly changed in response to EGF or NRG stimulation. Their analysis revealed that the two stimuli (EGF and NRG) activate essentially the same groups of genes, but to different extents. Specifically, transcription of immediate early genes, including regulators of ERBB signaling, is higher in response to NRG than to EGF. Stronger immediate early gene transcription may prolong intracellular signaling through more potent and stable positive feedback.

Another approach that was used to decipher the connections between upstream signaling and final cellular outcome is *unsupervised learning*, also called clustering or data partitioning. The main principle of this approach is to choose a subset of the data that is enriched for a particular feature. This approach has been successfully implemented in a few studies that performed time-series measurements of phosphorylation levels using mass spectrometry. The study by Zhang et al. used *self-organizing maps* to reveal new protein kinase substrates, as well as new phosphorylation sites [20]. Olsen et al. improved the method, and thereby significantly increased the number of identified phosphopeptides that dynamically change in response to EGF [73]. They found that *fuzzy c-means* clustering yielded better clustering than other clustering approaches. Using another clustering approach, the authors were able to globally characterize the transcriptional regulators that are activated by EGF-dependent phosphorylation. The role of specific phosphorylation sites was then inferred from their time-resolved phosphorylation kinetics. Based on this approach, the authors suggested that phosphorylation sites on A-RAF and MEK2 might impose negative feedback on the MAPK cascade.

In order to correctly assemble subsets of input data that will show significant enrichment of a common feature, the parameters used for data clustering must be optimized. It is essentially impossible to manually select parameters (i.e., number of clusters and choice of distance metric) that will yield the best clustering. The recently developed MCAM framework [78] addresses this challenge by exhaustively applying combinations of the different parameters and evaluating the outcome of all clustering solutions. When applied to dynamical phosphorylation measurements, it was found that different combinations of parameters were needed in order to maximize the enrichment score of different predefined biological features. Therefore, the best parameter combination is dependent upon the desired set of features to be enriched. In terms of biological inference, this method is able to raise new hypotheses regarding the function of specific phosphorylation events. For example, they found that ENO1, an enolase considered to be cytoplasmic, might be also localized to lamellipodia and participate in the cytoskeletal response to EGF. In addition, the authors compared phosphorylation dynamics of wild-type cells and HER2-overexpressing cells, stimulated with either EGF or NRG. Their analysis revealed numerous significantly changing associations between phosphopeptides, possibly indicating network rewiring in HER2-overexpressing cells.

6.3.2 Decision Tree Modeling

Decision trees are tree-like graphs that form an extension of *Boolean models* and impose directionality to models. They are especially useful when constructing signal-response models, particularly when the data is sparse and mainly qualitative, as is often the case when attempting to model large-scale networks. A typical decision tree contains nodes that resemble the components of the network, whereas the branches resemble different activation states or decisions. The tree often starts from a single node that hierarchically expands through branches to other nodes, until it reaches the response nodes. These might be either the cellular output or activation states of certain proteins. Progress through the tree is made by deciding which branch to take out of each node.

Decision-tree models proved to be successful in deciphering the regulatory mechanisms of migrating cells under treatment with two or more simultaneous stimuli. Two or more such stimuli can act either synergistically, antagonistically, or additively, leading to specific outcomes. For example, in the presence of EGF, fibroblast migration is highly dependent on fibronectin concentration [79]. Further analysis of this phenomenon, using a decision-tree model [71], was able to identify ERK signals as important determinants of migration speed. In a subsequent study, the authors constructed an improved decision-tree model that was based on additional experimental measurements [80]. Using the improved model, they were able to predict, and later validate, that reducing activation of myosin light chains increased the motility of human breast carcinoma cells.

Bayesian inference is another method that can model large-scale data and reveal causal physical or functional relationships among components that comprise a network. The main advantages of Bayesian models are their ability to handle noisy data and their robustness to the omission of variables from the model [81]. Dynamic Bayesian models can reveal complex interactions in the network, including feedback loops [76, 82, 83].

6.4 Concluding Remarks: How Relevant Are Computational Models to Understanding Mechanisms of Disease?

Disease states may exploit network robustness to maintain the disease despite intervention by the immune system or by pharmacological agents. Such robustness of disease states is called co-opted robustness. As mentioned above, there is an inherent trade-off between robustness and fragility. Hence, it is beneficial to identify the fragile components in pathological systems and to target these components [59]. For example, adaptation of bacterial chemotaxis is robust to changes in the concentrations of individual proteins that comprise the chemotaxis network, but is fragile when subject to simultaneous elimination of two proteins [84]. In RTK signaling, some cancer cells simultaneously co-activate ERBB1 and c-Met, and thus, the ERBB network in these cells is robust to EGFR kinase inhibitors when applied as monotherapy [85]. Hence, targeting more than one component might be needed in order to oppose the co-opted robustness of these cancer cells. Modeling may assist finding such fragile points in otherwise robust disease networks. Because the ERBB network is frequently involved in epithelial and in brain tumors, and drugs targeting the network have emerged as effective anticancer drugs [86], sensitivity or resistance to ERBB interceptors provides many potential applications of computational models to pathological states.

One useful example is ERBB2, which is overexpressed in about 25 % of breast and gastric tumors [87]. Previous ERBB2 modeling found that the expression level of ERBB2 in normal cells is just slightly below the concentrations that would allow for efficient heterodimerization with EGFR/ERBB1 [34]. This might help explain why ERBB2 overexpression exerts such a profound effect on cell physiology. Furthermore, as mentioned earlier, the model by [7] suggests that ERBB1:ERBB2 heterodimers escape negative regulation leading to enhanced EGF signaling and subsequent sustained activation of ERK [7]. Therefore, HER2 is a natural target for therapeutic intervention using either monoclonal antibodies, either single monoclonals or a combination of two antibodies, or kinase inhibitors specific to this ligandless coreceptor of ERBB1, ERBB3, and ERBB4 [88, 89].

Glossary

Autocrine Secretion of a ligand that activates a receptor on the secreting cell itself.

Boolean model A discrete modeling method that produces a qualitative dynamic description of system behavior.

Emergent behavior The complex behavior of a system that arises from functional interactions of simple components.

Endocytosis The process of inward folding of the plasma membrane, thereby creating intracellular vesicles that uptake materials from outside a cell.

Fuzzy c-means A clustering method where data points are assigned a score for belonging to each cluster. Thus, it enables to identify which data points fit better to particular clusters.

Granularity The detailed level of the data included in a model. Higher model's granularity means that the system is broken up into many smaller parts.

Mitogen-activated protein kinase (MAPK) A group of four kinases that are activated in a cascade. The dynamical activation pattern of these kinases leads to diverse cell fates.

Paracrine Secretion of a ligand that stimulates adjacent cells.

Self-organizing map A computational method based on unsupervised learning that clusters data into groups. Additionally, it is able to graphically present the relationships among different groups.

Species Forms of macromolecules, such as protein isoforms originating from RNA splicing, posttransnationally modified proteins and protein complexes.

Ultrasensitivity The output response of an enzymatic reaction where, above a certain threshold, a small increase in input causes a large increase in output response. The ultrasensitive stimulus/response curve is steeper than the hyperbolic Michaelis-Menten curve. It enables bistable behavior of signaling networks.

Unsupervised learning A computational approach that makes no prior assumptions on data organization. This method might reveal unanticipated biological insights.

References

1. Citri A, Yarden Y. EGF-ERBB signalling: towards the systems level. Nat Rev Mol Cell Biol. 2006;7:505–16.
2. Oda K, Matsuoka Y, Funahashi A, Kitano H. A comprehensive pathway map of epidermal growth factor receptor signaling. Mol Syst Biol. 2005;1:2005.0010.
3. Alon U. An introduction to systems biology: design principles of biological circuits. Boca Raton, FL: CRC Press, Taylor and Francis; 2006.
4. Bhalla US, Iyengar R. Functional modules in biological signalling networks. Novartis Found Symp. 2001;239:4–13. discussion 13–15, 45–51.
5. Kitano H. Biological robustness. Nat Rev Genet. 2004;5:826–37.
6. Nakakuki T, Birtwistle MR, Saeki Y, Yumoto N, Ide K, Nagashima T, Brusch L, Ogunnaike BA, Okada-Hatakeyama M, Kholodenko BN. Ligand-specific c-Fos expression emerges from the spatiotemporal control of ErbB network dynamics. Cell. 2010;141:884–96.

7. Birtwistle MR, Hatakeyama M, Yumoto N, Ogunnaike BA, Hoek JB, Kholodenko BN. Ligand-dependent responses of the ErbB signaling network: experimental and modeling analyses. Mol Syst Biol. 2007;3:144.
8. Chen WW, Schoeberl B, Jasper PJ, Niepel M, Nielsen UB, Lauffenburger DA, Sorger PK. Input-output behavior of ErbB signaling pathways as revealed by a mass action model trained against dynamic data. Mol Syst Biol. 2009;5:239.
9. Hatakeyama M, Kimura S, Naka T, Kawasaki T, Yumoto N, Ichikawa M, Kim JH, Saito K, Saeki M, Shirouzu M, et al. A computational model on the modulation of mitogen-activated protein kinase (MAPK) and Akt pathways in heregulin-induced ErbB signalling. Biochem J. 2003;373:451–63.
10. Kholodenko BN, Demin OV, Moehren G, Hoek JB. Quantification of short term signaling by the epidermal growth factor receptor. J Biol Chem. 1999;274:30169–81.
11. Purvis J, Ilango V, Radhakrishnan R. Role of network branching in eliciting differential short-term signaling responses in the hypersensitive epidermal growth factor receptor mutants implicated in lung cancer. Biotechnol Prog. 2008;24:540–53.
12. Schoeberl B, Eichler-Jonsson C, Gilles ED, Muller G. Computational modeling of the dynamics of the MAP kinase cascade activated by surface and internalized EGF receptors. Nat Biotechnol. 2002;20:370–5.
13. Shin SY, Rath O, Choo SM, Fee F, McFerran B, Kolch W, Cho KH. Positive- and negative-feedback regulations coordinate the dynamic behavior of the Ras-Raf-MEK-ERK signal transduction pathway. J Cell Sci. 2009;122:425–35.
14. Tyson JJ, Chen K, Novak B. Network dynamics and cell physiology. Nat Rev Mol Cell Biol. 2001;2:908–16.
15. Aldridge BB, Burke JM, Lauffenburger DA, Sorger PK. Physicochemical modelling of cell signalling pathways. Nat Cell Biol. 2006;8:1195–203.
16. Avraham R, Yarden Y. Feedback regulation of EGFR signalling: decision making by early and delayed loops. Nat Rev Mol Cell Biol. 2011;12:104–17.
17. Jorissen RN, Walker F, Pouliot N, Garrett TP, Ward CW, Burgess AW. Epidermal growth factor receptor: mechanisms of activation and signalling. Exp Cell Res. 2003;284:31–53.
18. Hlavacek WS, Faeder JR, Blinov ML, Perelson AS, Goldstein B. The complexity of complexes in signal transduction. Biotechnol Bioeng. 2003;84:783–94.
19. Lewitzky M, Simister PC, Feller SM. Beyond 'furballs' and 'dumpling soups' - towards a molecular architecture of signaling complexes and networks. FEBS Lett. 2012;586:2740–50.
20. Zhang Y, Wolf-Yadlin A, Ross PL, Pappin DJ, Rush J, Lauffenburger DA, White FM. Time-resolved mass spectrometry of tyrosine phosphorylation sites in the epidermal growth factor receptor signaling network reveals dynamic modules. Mol Cell Proteomics. 2005;4:1240–50.
21. Blinov ML, Faeder JR, Goldstein B, Hlavacek WS. A network model of early events in epidermal growth factor receptor signaling that accounts for combinatorial complexity. Biosystems. 2006;83:136–51.
22. Snoep JL, Bruggeman F, Olivier BG, Westerhoff HV. Towards building the silicon cell: a modular approach. Biosystems. 2006;83:207–16.
23. Resat H, Ewald JA, Dixon DA, Wiley HS. An integrated model of epidermal growth factor receptor trafficking and signal transduction. Biophys J. 2003;85:730–43.
24. Pinkas-Kramarski R, Soussan L, Waterman H, Levkowitz G, Alroy I, Klapper L, Lavi S, Seger R, Ratzkin BJ, Sela M, et al. Diversification of Neu differentiation factor and epidermal growth factor signaling by combinatorial receptor interactions. EMBO J. 1996;15:2452–67.
25. Hendriks BS, Orr G, Wells A, Wiley HS, Lauffenburger DA. Parsing ERK activation reveals quantitatively equivalent contributions from epidermal growth factor receptor and HER2 in human mammary epithelial cells. J Biol Chem. 2005;280:6157–69.
26. Schulze A, Lehmann K, Jefferies HB, McMahon M, Downward J. Analysis of the transcriptional program induced by Raf in epithelial cells. Genes Dev. 2001;15:981–94.
27. Schafer B, Gschwind A, Ullrich A. Multiple G-protein-coupled receptor signals converge on the epidermal growth factor receptor to promote migration and invasion. Oncogene. 2004;23:991–9.

28. Chang CP, Lazar CS, Walsh BJ, Komuro M, Collawn JF, Kuhn LA, Tainer JA, Trowbridge IS, Farquhar MG, Rosenfeld MG, et al. Ligand-induced internalization of the epidermal growth factor receptor is mediated by multiple endocytic codes analogous to the tyrosine motif found in constitutively internalized receptors. J Biol Chem. 1993;268:19312–20.
29. Opresko LK, Chang CP, Will BH, Burke PM, Gill GN, Wiley HS. Endocytosis and lysosomal targeting of epidermal growth factor receptors are mediated by distinct sequences independent of the tyrosine kinase domain. J Biol Chem. 1995;270:4325–33.
30. Wiley HS, Herbst JJ, Walsh BJ, Lauffenburger DA, Rosenfeld MG, Gill GN. The role of tyrosine kinase activity in endocytosis, compartmentation, and down-regulation of the epidermal growth factor receptor. J Biol Chem. 1991;266:11083–94.
31. Felder S, Miller K, Moehren G, Ullrich A, Schlessinger J, Hopkins CR. Kinase activity controls the sorting of the epidermal growth factor receptor within the multivesicular body. Cell. 1990;61:623–34.
32. French AR, Tadaki DK, Niyogi SK, Lauffenburger DA. Intracellular trafficking of epidermal growth factor family ligands is directly influenced by the pH sensitivity of the receptor/ligand interaction. J Biol Chem. 1995;270:4334–40.
33. Howe CL, Mobley WC. Signaling endosome hypothesis: a cellular mechanism for long distance communication. J Neurobiol. 2004;58:207–16.
34. Hendriks BS, Opresko LK, Wiley HS, Lauffenburger D. Quantitative analysis of HER2-mediated effects on HER2 and epidermal growth factor receptor endocytosis: distribution of homo- and heterodimers depends on relative HER2 levels. J Biol Chem. 2003;278:23343–51.
35. Shankaran H, Wiley HS, Resat H. Modeling the effects of HER/ErbB1-3 coexpression on receptor dimerization and biological response. Biophys J. 2006;90:3993–4009.
36. Levkowitz G, Waterman H, Ettenberg SA, Katz M, Tsygankov AY, Alroy I, Lavi S, Iwai K, Reiss Y, Ciechanover A, et al. Ubiquitin ligase activity and tyrosine phosphorylation underlie suppression of growth factor signaling by c-Cbl/Sli-1. Mol Cell. 1999;4:1029–40.
37. Levkowitz G, Waterman H, Zamir E, Kam Z, Oved S, Langdon WY, Beguinot L, Geiger B, Yarden Y. c-Cbl/Sli-1 regulates endocytic sorting and ubiquitination of the epidermal growth factor receptor. Genes Dev. 1998;12:3663–74.
38. Gallagher ED, Gutowski S, Sternweis PC, Cobb MH. RhoA binds to the amino terminus of MEKK1 and regulates its kinase activity. J Biol Chem. 2004;279:1872–7.
39. Ung CY, Li H, Ma XH, Jia J, Li BW, Low BC, Chen YZ. Simulation of the regulation of EGFR endocytosis and EGFR-ERK signaling by endophilin-mediated RhoA-EGFR crosstalk. FEBS Lett. 2008;582:2283–90.
40. Tarcic G, Boguslavsky SK, Wakim J, Kiuchi T, Liu A, Reinitz F, Nathanson D, Takahashi T, Mischel PS, Ng T, et al. An unbiased screen identifies DEP-1 tumor suppressor as a phosphatase controlling EGFR endocytosis. Curr Biol. 2009;19:1788–98.
41. Marshall CJ. Specificity of receptor tyrosine kinase signaling: transient versus sustained extracellular signal-regulated kinase activation. Cell. 1995;80:179–85.
42. Lessor T, Yoo JY, Davis M, Hamburger AW. Regulation of heregulin beta1-induced differentiation in a human breast carcinoma cell line by the extracellular-regulated kinase (ERK) pathway. J Cell Biochem. 1998;70:587–95.
43. Nagashima T, Shimodaira H, Ide K, Nakakuki T, Tani Y, Takahashi K, Yumoto N, Hatakeyama M. Quantitative transcriptional control of ErbB receptor signaling undergoes graded to biphasic response for cell differentiation. J Biol Chem. 2007;282:4045–56.
44. Santos SD, Verveer PJ, Bastiaens PI. Growth factor-induced MAPK network topology shapes Erk response determining PC-12 cell fate. Nat Cell Biol. 2007;9:324–30.
45. Kholodenko BN, Kiyatkin A, Bruggeman FJ, Sontag E, Westerhoff HV, Hoek JB. Untangling the wires: a strategy to trace functional interactions in signaling and gene networks. Proc Natl Acad Sci U S A. 2002;99:12841–6.
46. Huang CY, Ferrell Jr JE. Ultrasensitivity in the mitogen-activated protein kinase cascade. Proc Natl Acad Sci U S A. 1996;93:10078–83.
47. Bhalla US, Ram PT, Iyengar R. MAP kinase phosphatase as a locus of flexibility in a mitogen-activated protein kinase signaling network. Science. 2002;297:1018–23.

48. Ferrell Jr JE. Self-perpetuating states in signal transduction: positive feedback, double-negative feedback and bistability. Curr Opin Cell Biol. 2002;14:140–8.
49. Markevich NI, Hoek JB, Kholodenko BN. Signaling switches and bistability arising from multisite phosphorylation in protein kinase cascades. J Cell Biol. 2004;164:353–9.
50. Moelling K, Schad K, Bosse M, Zimmermann S, Schweneker M. Regulation of Raf-Akt Cross-talk. J Biol Chem. 2002;277:31099–106.
51. Zimmermann S, Moelling K. Phosphorylation and regulation of Raf by Akt (protein kinase B). Science. 1999;286:1741–4.
52. Amit I, Citri A, Shay T, Lu Y, Katz M, Zhang F, Tarcic G, Siwak D, Lahad J, Jacob-Hirsch J, et al. A module of negative feedback regulators defines growth factor signaling. Nat Genet. 2007;39:503–12.
53. Pouyssegur J, Lenormand P. Fidelity and spatio-temporal control in MAP kinase (ERKs) signalling. Eur J Biochem. 2003;270:3291–9.
54. Brondello JM, McKenzie FR, Sun H, Tonks NK, Pouyssegur J. Constitutive MAP kinase phosphatase (MKP-1) expression blocks G1 specific gene transcription and S-phase entry in fibroblasts. Oncogene. 1995;10:1895–904.
55. Alon U. Network motifs: theory and experimental approaches. Nat Rev Genet. 2007; 8:450–61.
56. Murphy LO, Blenis J. MAPK signal specificity: the right place at the right time. Trends Biochem Sci. 2006;31:268–75.
57. Lazzara MJ, Lauffenburger DA. Quantitative modeling perspectives on the ErbB system of cell regulatory processes. Exp Cell Res. 2009;315:717–25.
58. Avraham R, Sas-Chen A, Manor O, Steinfeld I, Shalgi R, Tarcic G, Bossel N, Zeisel A, Amit I, Zwang Y, et al. EGF decreases the abundance of microRNAs that restrain oncogenic transcription factors. Sci Signal. 2010;3:ra43.
59. Amit I, Wides R, Yarden Y. Evolvable signaling networks of receptor tyrosine kinases: relevance of robustness to malignancy and to cancer therapy. Mol Syst Biol. 2007;3:151.
60. Shankaran H, Ippolito DL, Chrisler WB, Resat H, Bollinger N, Opresko LK, Wiley HS. Rapid and sustained nuclear-cytoplasmic ERK oscillations induced by epidermal growth factor. Mol Syst Biol. 2009;5:332.
61. Sturm OE, Orton R, Grindlay J, Birtwistle M, Vyshemirsky V, Gilbert D, Calder M, Pitt A, Kholodenko B, Kolch W. The mammalian MAPK/ERK pathway exhibits properties of a negative feedback amplifier. Sci Signal. 2010;3:ra90.
62. Xiong W, Ferrell Jr JE. A positive-feedback-based bistable 'memory module' that governs a cell fate decision. Nature. 2003;426:460–5.
63. Csete ME, Doyle JC. Reverse engineering of biological complexity. Science. 2002; 295:1664–9.
64. Kitano H. The theory of biological robustness and its implication in cancer. Ernst Schering Res Found Workshop. 2007; (61):69.
65. Carlson JM, Doyle J. Complexity and robustness. Proc Natl Acad Sci U S A. 2002;99 Suppl 1: 2538–45.
66. Jeong H, Mason SP, Barabasi AL, Oltvai ZN. Lethality and centrality in protein networks. Nature. 2001;411:41–2.
67. Mahalingam D, Swords R, Carew JS, Nawrocki ST, Bhalla K, Giles FJ. Targeting HSP90 for cancer therapy. Br J Cancer. 2009;100:1523–9.
68. Suenaga A, Kiyatkin AB, Hatakeyama M, Futatsugi N, Okimoto N, Hirano Y, Narumi T, Kawai A, Susukita R, Koishi T, et al. Tyr-317 phosphorylation increases Shc structural rigidity and reduces coupling of domain motions remote from the phosphorylation site as revealed by molecular dynamics simulations. J Biol Chem. 2004;279:4657–62.
69. Brightman FA, Fell DA. Differential feedback regulation of the MAPK cascade underlies the quantitative differences in EGF and NGF signalling in PC12 cells. FEBS Lett. 2000; 482:169–74.
70. Shvartsman SY, Muratov CB, Lauffenburger DA. Modeling and computational analysis of EGF receptor-mediated cell communication in Drosophila oogenesis. Development. 2002;129: 2577–89.

71. Hautaniemi S, Kharait S, Iwabu A, Wells A, Lauffenburger DA. Modeling of signal-response cascades using decision tree analysis. Bioinformatics. 2005;21:2027–35.
72. Monine MI, Berezhkovskii AM, Joslin EJ, Wiley HS, Lauffenburger DA, Shvartsman SY. Ligand accumulation in autocrine cell cultures. Biophys J. 2005;88:2384–90.
73. Olsen JV, Blagoev B, Gnad F, Macek B, Kumar C, Mortensen P, Mann M. Global, in vivo, and site-specific phosphorylation dynamics in signaling networks. Cell. 2006;127:635–48.
74. Borisov N, Aksamitiene E, Kiyatkin A, Legewie S, Berkhout J, Maiwald T, Kaimachnikov NP, Timmer J, Hoek JB, Kholodenko BN. Systems-level interactions between insulin-EGF networks amplify mitogenic signaling. Mol Syst Biol. 2009;5:256.
75. Schoeberl B, Pace EA, Fitzgerald JB, Harms BD, Xu L, Nie L, Linggi B, Kalra A, Paragas V, Bukhalid R, et al. Therapeutically targeting ErbB3: a key node in ligand-induced activation of the ErbB receptor-PI3K axis. Sci Signal. 2009;2:ra31.
76. Ciaccio MF, Wagner JP, Chuu CP, Lauffenburger DA, Jones RB. Systems analysis of EGF receptor signaling dynamics with microwestern arrays. Nat Methods. 2010;7:148–55.
77. Naegle KM, White FM, Lauffenburger DA, Yaffe MB. Robust co-regulation of tyrosine phosphorylation sites on proteins reveals novel protein interactions. Mol Biosyst. 2012;8: 2771–82.
78. Naegle KM, Welsch RE, Yaffe MB, White FM, Lauffenburger DA. MCAM: multiple clustering analysis methodology for deriving hypotheses and insights from high-throughput proteomic datasets. PLoS Comput Biol. 2011;7:e1002119.
79. Maheshwari G, Wells A, Griffith LG, Lauffenburger DA. Biophysical integration of effects of epidermal growth factor and fibronectin on fibroblast migration. Biophys J. 1999;76: 2814–23.
80. Kharait S, Hautaniemi S, Wu S, Iwabu A, Lauffenburger DA, Wells A. Decision tree modeling predicts effects of inhibiting contractility signaling on cell motility. BMC Syst Biol. 2007;1:9.
81. Pe'er D. Bayesian network analysis of signaling networks: a primer. Sci STKE. 2005;2005:pl4.
82. Sachs K, Gifford D, Jaakkola T, Sorger P, Lauffenburger DA. Bayesian network approach to cell signaling pathway modeling. Sci STKE. 2002;2002:pe38.
83. Vinayagam A, Stelzl U, Foulle R, Plassmann S, Zenkner M, Timm J, Assmus HE, Andrade-Navarro MA, Wanker EE. A directed protein interaction network for investigating intracellular signal transduction. Sci Signal. 2011;4:rs8.
84. Alon U, Surette MG, Barkai N, Leibler S. Robustness in bacterial chemotaxis. Nature. 1999;397:168–71.
85. Xu L, Kikuchi E, Xu C, Ebi H, Ercan D, Cheng KA, Padera R, Engelman JA, Janne PA, Shapiro GI, et al. Combined EGFR/MET or EGFR/HSP90 inhibition is effective in the treatment of lung cancers codriven by mutant EGFR containing T790M and MET. Cancer Res. 2012;72:3302–11.
86. Yarden Y, Pines G. The ERBB network: at last, cancer therapy meets systems biology. Nat Rev Cancer. 2012;12:553–63.
87. Slamon DJ, Clark GM, Wong SG, Levin WJ, Ullrich A, McGuire WL. Human breast cancer: correlation of relapse and survival with amplification of the HER-2/neu oncogene. Science. 1987;235:177–82.
88. Arteaga CL, Sliwkowski MX, Osborne CK, Perez EA, Puglisi F, Gianni L. Treatment of HER2-positive breast cancer: current status and future perspectives. Nat Rev Clin Oncol. 2012;9:16–32.
89. Baselga J. Targeting tyrosine kinases in cancer: the second wave. Science. 2006;312:1175–8.

Chapter 7
Endocytosis and Endosomal Sorting of Receptor Tyrosine Kinases

Alexander Sorkin and Arola Fortian

7.1 Introduction

Receptor tyrosine kinases (RTKs), as other integral membrane proteins, are co-translationally inserted into the membrane of the endoplasmic reticulum, transported to the Golgi complex where they are N-glycosylated, and then delivered to the plasma membrane. The function of RTKs is controlled at multiple steps of their intracellular trafficking. At the cell surface, RTKs are activated by binding of their ligands (growth factors) and the subsequent behavior of an activated RTK is regulated by endocytosis. First, growth factor binding increases the rate of RTK endocytosis (internalization). Second, after internalization, activated RTKs are efficiently sorted from endosomes to lysosomes for proteolytic degradation (Fig. 7.1). Examples from all RTK subfamilies studied to date demonstrate accelerated lysosomal degradation of these receptors upon their activation. Acceleration of both sorting processes, at the cell surface and in endosomes, upon RTK activation results in the RTK protein downregulation, which is the distinguishable characteristic of the endocytic trafficking of this receptor family.

Endocytosis is the major regulator of intensity, duration, and specificity of RTK signaling. Dysregulation of the endocytic trafficking of RTKs leading to their aberrant signaling has been observed in a variety of cancer cells [1, 2]. Various modes of reciprocal interactions between endocytosis and signaling systems have been extensively reviewed [3–5] and will not be discussed in this chapter. Hence we shall focus our discussion on the molecular mechanisms implicated in RTK internalization and post-endocytic sorting. In recent years significant progress

A. Sorkin, Ph.D. (✉) • A. Fortian, Ph.D.
Department of Cell Biology, University of Pittsburgh School of Medicine,
Pittsburgh, PA 15261, USA
e-mail: sorkin@pitt.edu

D.L. Wheeler, Y. Yarden (eds.), *Receptor Tyrosine Kinases: Structure, Functions and Role in Human Disease*, DOI 10.1007/978-1-4939-2053-2_7

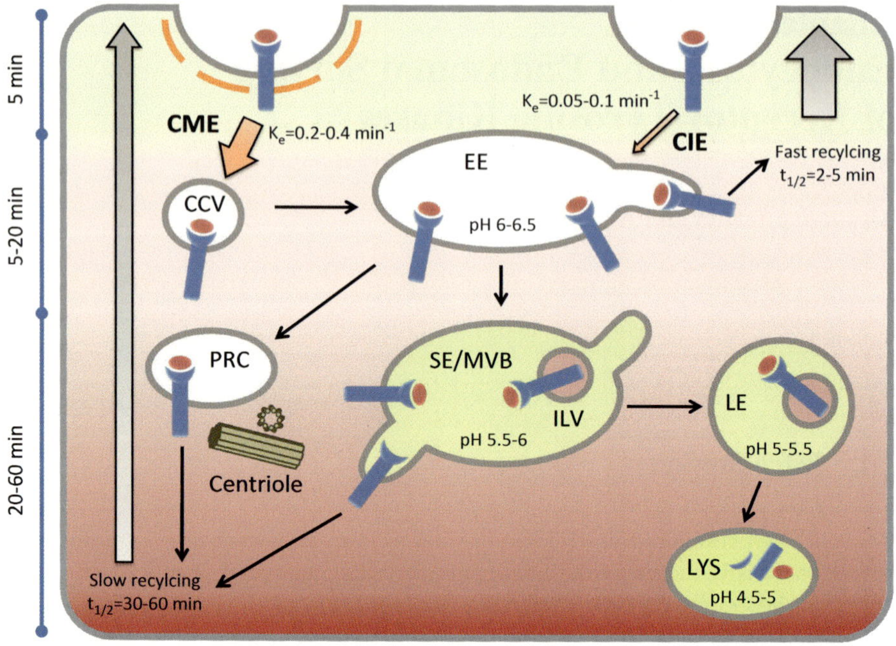

Fig. 7.1 Pathways of RTK endocytosis. Ligand-activated receptors are internalized by clathrin-mediated endocytosis (CME) via clathrin-coated pits (CCP) and clathrin-coated vesicles (CCVs) or by clathrin-independent endocytosis (CIE). Endocytic vesicles formed by both mechanisms are fused to early endosomes (EE). Early endosomes mature into sorting endosomes (SE) or multivesicular bodies (MVB) where RTKs are incorporated into intraluminal vesicles (ILV) by membrane invagination and sorted to late endosomes and lysosomes (Lys). RTK is recycled from peripheral endosomes, from tubular extensions of sorting endosomes and MVBs, and from the pericentriolar recycling compartment (PRC). Recycling rates from PRC and MVB are lower than the recycling rate from early endosomes. Typical ranges of the internalization rate constant k_e and intraendosomal pH values are shown. Typical half-life times of RTK recycling are shown in minutes

in understanding the endocytic mechanisms of RTK and other proteins has been made, and regrettably, it is impossible to cite all excellent studies of endocytosis in one review article. However, mechanisms and molecular components that are *specific* to endocytosis and intracellular sorting of RTKs remain to be poorly defined. Many original observations related to RTK trafficking have been made in studies of epidermal growth factor (EGF) receptor (EGFR). Stemming from EGFR as the prototypic model system, we will discuss below how the activation of various RTKs at the cell surface results in their endocytosis, recycling, and targeting to lysosomes.

7.2 Pathways and Kinetics of RTK Endocytic Trafficking

7.2.1 *Internalization*

RTKs are constitutively internalized and recycled back to the cell surface as part of basic endocytosis process and turnover of integral membrane proteins. Constitutive recycling of RTKs is on average 4–5-fold faster than their constitutive internalization. Correspondingly, a pool of RTKs at the cell surface is 4–5-fold larger than the internalized pool. Binding of a ligand to an RTK increases the rate of receptor internalization (Fig. 7.1). Most of RTKs freely diffuse in the plane of the plasma membrane, and each receptor is capable of reaching all endocytic sites at the cell surface within seconds. In some cases, a substantial pool of RTK is immobile and, therefore, incapable of internalization. Such surface retention can be due to association with specialized plasma membrane structures such as membrane protrusions (microvilli) or caveolae [6–9]. Interaction with Na^+-K^+ exchanger has been implicated in the retention of EGFR in the plasma membrane [10]. ErbB2, the closest homologue of EGFR, binds to the plasma membrane protein erbin [11]. Binding to erbin and accumulation in microvilli result in an inefficient endocytosis of ErbB2, which lacks its own ligand but is activated by homo- or hetero-dimerization with other ErbBs [12]. Reversely, in some experimental systems, RTKs have been shown to constitutively internalize with the high rate that has not appreciably increased the ligand binding [13–15]. Thus, activation-induced acceleration of RTK internalization is determined by the contribution of both the surface retention and endocytosis-accelerating mechanisms. Therefore, the extent by which endocytosis is accelerated by the ligand may vary for different RTKs.

Clathrin-mediated endocytosis (CME) is the major pathway of internalization of many plasma membrane proteins and the best studied mechanistically and structurally [16]. Accumulation of ligand-activated EGFR, neurotrophic tyrosine kinase receptor type 1 (TrkA), insulin receptor (IR), and other RTKs in clathrin-coated pits (CCPs) has been observed [17–20]. Internalization of EGFR and several other RTKs was blocked by small interfering RNAs (siRNAs) to clathrin heavy chain and chemical inhibitors of clathrin [21–25]. Together, various morphological and biochemical studies demonstrate that CME is the main portal of cell entry of activated RTKs, although coated pit localization of many RTKs is yet to be demonstrated. However, a number of endocytic pathways that do not involve clathrin have also been described [26].

Kinetics analysis of EGFR endocytosis suggested that CME is saturated when a large number of surface EGFRs are activated by EGF. CME is the fastest internalization pathway (rate constant k_e ~ up to 0.2–0.6 min^{-1}). The contribution of a slower clathrin-independent endocytosis (CIE) increases with the increase in the number of active (ligand-occupied) EGFRs at the cell surface due to high EGF concentration and high EGFR expression levels, which leads to overall reduction in the apparent internalization rate to 0.05–0.10 min^{-1} [27, 28]. This hypothesis was supported experimentally by the demonstration that siRNA depletion of clathrin or expression of a dominant-negative dynamin mutant significantly inhibits EGFR internalization only when low concentrations of EGF were used [29, 30]. Similar saturation of the rapid internalization pathway and reduced rates of IR and insulin-like growth factor 1 receptor (IGF-1R) internalization with increasing the ligand concentration were observed [31, 32]. Whether saturation of rapid endocytosis is a general phenomenon for the RTK family is unclear mostly because measurement of internalization rates of RTKs without the interference of recycling is technically difficult, and therefore, changes in the apparent rates of RTK internalization observed in many studies could be due to changes in either internalization or recycling.

7.2.2 Post-endocytic Sorting of RTKs: Recycling Versus Degradation

Endocytic vesicles carrying activated RTKs formed by CME and CIE fuse with early endosomes (EEs). EE is the heterogenous-shaped compartments with vesicular and tubular subcompartments [33–36]. EEs undergo homotypical fusion and move along microtubules from the cell periphery toward the centriole/Golgi area (Fig. 7.1). Fusion results in enlargement of the vesicular part (most of the endosomal volume) and decreasing the amount of tubular extensions (most of the membrane area of the endosome) of EE, as part of the process called endosomal maturation. Maturation further involves involution (inward invagination) of the endosomal membrane leading to the appearance of intraluminal vesicles (ILVs) that are characteristic of multivesicular endosomes or bodies (MVBs). These morphological changes are accompanied by changes in biochemical composition of endosomes and increasing the luminal acidification of endosomes. Sorting of RTKs takes place in EE and MVB (Fig. 7.1). RTKs that are located in the limiting membrane of EEs and MVBs are capable of recycling back to the cell surface through tubular extensions of these endosomes. RTKs that are packaged into ILVs are incapable of recycling and remain in MVB that mature into late endosomes. Maturation into late endosomes involves removal of EE components and enrichment in proteases and other lysosomal enzymes that are delivered from the Golgi complex. Degradation of RTKs begins in late endosomes and continues after fusion of late endosomes with "primary" lysosomes carrying proteolytic enzymes, which results in the formation of mature lysosomes.

In neuronal cells, sorting of neurotrophin-bound receptors, such as Trk receptors, must take place after their signaling activity is completed and, therefore, must be spatially separated from the site of endocytosis in the axonal termini. Thus, endosomes containing active Trk are transported from the axonal termini to the soma by the microtubule-dependent, dynein-motor-mediated transport. This mechanism allows rapid movement of Trk-containing "signaling" endosomes for up to several meter distance. The mechanisms and function of the retrograde transport of Trk receptors were recently reviewed [37, 38].

Because many activated and internalized RTKs are efficiently degraded due to targeting to late endosome and lysosomes, it is generally assumed that these receptors are not recycled. However, significant amount of recycling of ligand-bound IR [39], EGFR [40], and TrkA [41] has been observed. EGF-EGFR complexes are shown to be rapidly recycled back to the cell surface from peripheral EE and with a slower kinetics from pericentriolar endosomes (Fig. 7.1) [40]. The fast recycling pathway or "retroendocytosis" is mediated by constitutive formation of recycling carriers from peripheral EE, whereas a slower pathway presumably involves a temperature-dependent sorting step in MVBs and specialized pericentriolar recycling compartments. In all the abovementioned experiments, recycling was monitored by appearance of internalized ligand-receptor complexes at the cell surface and, therefore, corresponds to the recycling of ligand-RTK complexes that are not dissociated in endosomes. In addition, unoccupied EGFRs that release EGF in endosomes were shown to recycle back to the plasma membrane [42]. It is estimated that ligand-occupied receptors recycle with a slower rate than unoccupied EGFR [43]. Furthermore, a pool of unoccupied, and therefore, recycled EGFR in endosomes is significantly larger when EGFR is activated by ligands like transforming growth factor-α that displays a higher pH sensitivity of binding and dissociates from the receptor in endosomes, as compared with EGF that remains mostly receptor bound [44, 45].

In general, recycling pathways and kinetics of RTKs are not well studied due to, as mentioned above, technical difficulties in measuring the rates of internalization and recycling without mutual interference of these two processes. While recycling is often considered to be the default pathway of endosomal sorting, EGF degradation has been demonstrated to be a saturable process, much like the internalization step [46]. Assuming that the majority of internalized EGF-EGFR complexes remain intact, EGF degradation is proportional to lysosomal targeting and degradation of EGFR. We hypothesize that because of the saturability of the lysosomal sorting process, the size of the pool of recycling internalized RTKs is likely to be proportional to the expression levels of RTKs. In other words, high overexpression, such as the case of EGFR in some tumor cells, would result in the highest extent of RTK recycling. For instance, in cells expressing high levels of EGFR and stimulated with high EGF concentrations, up to 80 % of internalized EGF-EGFR complexes were recycled [40].

7.2.3 Downregulation as the Result of Increased Endocytosis and Degradation

The rates of internalization, recycling, and degradation determine the turnover rates or half-life of a RTK protein. Degradation rate counterbalances the rate of biosynthesis, thus maintaining a constant level of RTK expression in the cell. Acceleration of internalization and degradation, without significant changes in the biosynthetic rate, results in the dramatic decrease in the surface pools and total amounts of RTKs in the cell and, consequently, in the downregulation of these RTKs. The rates of both constitutive and ligand-induced turnover vary among RTKs and depend on the cell type. However, importantly, due to the saturability of both internalization and degradation steps, half-life of RTKs positively correlates with the expression level of the RTK. For example, the constitutive $t_{1/2}$ of the colony-stimulating factor 1 receptor (CSF-1R) in macrophages is about 1 h and, after activation, decreases to 15 min [47]. The same parameters for EGFR overexpressed in carcinoma cells are 24 h or longer in the absence and 8 h in the presence of EGF [48, 49].

7.3 Mechanisms of Clathrin-Mediated Endocytosis

7.3.1 Role of Kinase Activity and Ubiquitination

Binding of growth factors to RTK leads to activation of the tyrosine kinase, which results in phosphorylation of cytoplasmic substrates and tyrosine residues in the cytoplasmic domain of RTKs. Kinase activity has been shown to be necessary for the recruitment to CCPs and rapid internalization of EGFR, IR, fibroblast growth factor receptor (FGFR), platelet-derived growth factor receptor (PDGFR)-β, vascular endothelial growth factor receptor (VEGFR), and other RTKs [13, 50–56]. The EGFR kinase inhibitor was found to inhibit internalization of the endogenous EGFR in ex vivo assays in human airway tissues [57]. Under certain experimental conditions in cell culture, e.g., high occupancy of surface EGFR by EGF leading to saturation of the CME pathway, significant endocytosis of kinase-inactive EGFR can be observed [58, 59].

Because mechanisms of RTK internalization are not fully understood, the role of kinase activity is not well defined. One of the predominant models of RTK endocytosis suggests that RTK is recruited into CCP through the interaction of ubiquitin moieties conjugated to RTK with the ubiquitin-binding proteins located in CCP (Fig. 7.2a). Ubiquitination is the posttranslational modification by covalent attachment of a 76-amino acid ubiquitin polypeptide, predominantly, to lysine residues in the substrate (Fig. 7.2). Ubiquitin moieties act as sorting signals recognized by proteins containing ubiquitin-binding domains (UBD) [60]. In this model, kinase activity serves to create binding sites (phosphotyrosine-containing sequence motifs) in the RTK cytoplasmic domains for adaptor proteins containing Src homology 2 (SH2) and phosphotyrosine-binding (PTB) domains that mediate RTK ubiquitination (Figs. 7.2 and 7.3).

The first RTKs found to be ubiquitinated were PDGFR-β and EGFR [61, 62]. In subsequent studies ubiquitin conjugation was detected in many other RTKs and, therefore, can be classified as a highly characteristic posttranslational modification of this receptor family. Several exceptions have been reported. For example, activation of ErbB2 at the cell surface by homo- and hetero-dimerization does not lead to significant ubiquitination [12]. Likewise, anaplastic lymphoma kinase, ALK, does not appear to be ubiquitinated when activated, and the slow turnover of ALK is related to its tumorigenic activity [63].

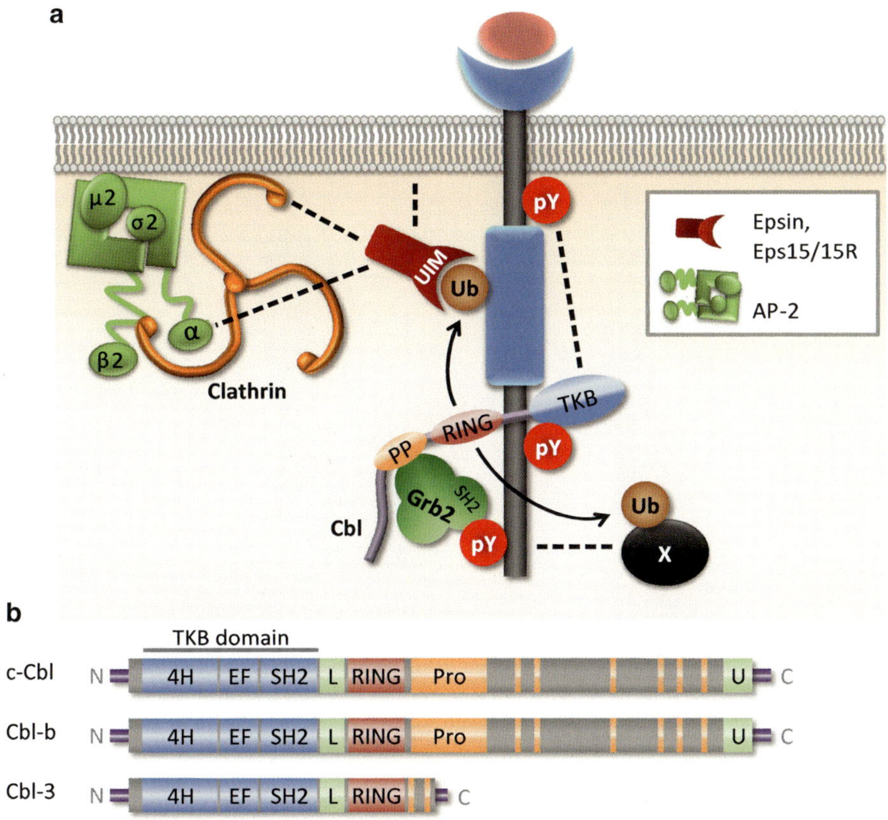

Fig. 7.2 RTKs interactions with Grb2 and Cbl leading to ubiquitination. (**a**) Interaction of an activated RTK (example of EGFR and c-Met) with c-Cbl/Cbl-b and Grb2. TKB domain of Cbl binds to phosphotyrosine (pY)-containing motif in the juxtamembrane or C-terminal domain of the receptor. Proline-rich motifs (PP) in the C-terminus of Cbl interact with SH3 domains of Grb2. SH2 domain of Grb2 binds to specific pY-containing motifs in the receptor. Cbl facilitates transfer of ubiquitin (Ub) from E2 ubiquitin-conjugating enzyme (UbcH5) bound to the RING domain (not shown) to lysines in the kinase domain of the receptor. Proteins with ubiquitin-interacting motifs (UIM, a type of UBDs), such as epsin, Eps15, and Eps15R, interact with ubiquitinated receptors and also bind to the appendage domains of AP-2 and the terminal domain of clathrin heavy chains (assembled in triskelions). The hinge domain of AP-2 (between the trunk and appendages) binds to the terminal domain of clathrin heavy chain. This network of interactions provides a platform for RTK recruitment into CCP. (**b**) Schematic representation of three mammalian isoforms of Cbl: c-Cbl, Cbl-b, and Cbl-3 (also called Cbl-c). The amino-terminal tyrosine-kinase-binding domain (TKB) is composed of three different subdomains: a four-helix bundle (4H), a calcium-binding EF hand, and a divergent SH2 domain. The RING finger domain (RING) is important for recruitment of E2 enzymes. The region between the TKB and the RING finger is referred to as the linker region (L). Proline-rich motifs (Pro) mediate interactions with SH3 domains. UBA and LZ domains (U) are responsible for ubiquitin binding and Cbl dimerization, respectively

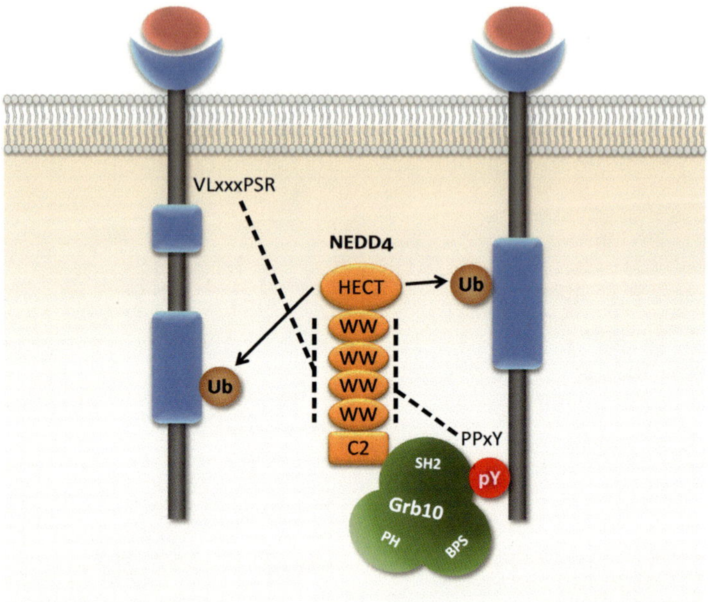

Fig. 7.3 RTK interactions with NEDD4 ligases leading to ubiquitination. NEDD4 E3 ligases conjugate ubiquitin (Ub) to RTKs with a single-domain kinase (EGFR, IR/IGF-1R, Met, Trk subtypes) (**a**) and RTKs in which the kinase domain contains an insert (FGFR, c-kit, PDGFR, CSFR, VEGFR) (**b**). NEDD4 family E3 ligases contain C2 (calcium-dependent lipid-binding domain), two to four WW domains, and catalytic HECT domain. WW domains bind PPxY motifs or unconventional sequences (example of FGFR is shown) in RTKs. NEDD4 interacts with IR and IGF-1R via the Grb10 adaptor. Grb10 contains an SH2 domain that simultaneously binds both the phosphotyrosine motif in RTKs and the C2 domain in NEDD4

Proteins can be modified by attachment of a single ubiquitin to a single lysine residue (monoubiquitination) or several lysines (multi-monoubiquitination). Lysines in the ubiquitin itself can further serve as substrates for ubiquitin conjugation resulting in the formation of polyubiquitin chains. Mass spectrometry studies revealed that EGFR is monoubiquitinated and also ubiquitinated by Lys63-linked ubiquitin chains [64]. Similarly, TrkA was proposed to be either monoubiquitinated or Lys63-chain polyubiquitinated [65, 66]. In contrast, IGF-1R was found to be Lys48- and Lys29-polyubiquitinated during antibody-driven downregulation [67].

7.3.2 Cbl E3 Ubiquitin Ligase

Ubiquitination is catalyzed sequentially by the E1 ubiquitin-activating enzyme, E2 ubiquitin-conjugating enzyme, and E3 ubiquitin ligase [68]. E3 binds to the substrate and facilitates conjugation of ubiquitin to the substrate. Two largest families of E3s are the ligases containing RING finger and HECT domains. Cbl proteins are RING E3s that are most relevant to RTKs. Cbl proteins bind and ubiquitinate EGFR, FGFR, PDGFR, VEGFR, CSF-1R, hepatocyte growth factor (c-Met), macrophage-stimulating protein (Ron), and ephrin (EphR) receptors [69–75]. There are three mammalian species of Cbl: c-Cbl, Cbl-b, and Cbl-3. All forms contain the tyrosine-kinase-binding domain (TKB) allowing direct interaction of Cbl with phosphotyrosine-containing motifs in RTKs (Fig. 7.2b). TKB is connected to the RING domain by the linker sequence, and both the RING and linker are involved in the interaction with E2 enzymes. Long forms, c-Cbl and Cbl-b, also contain C-terminal proline-rich domain interacting with Src Homology 3 (SH3) domains (Fig. 7.2b). Thus, c-Cbl and Cbl-b bind EGFR, c-Met, PDGFR-β, stem cell factor receptor (c-kit), and other RTKs indirectly through the interaction of their proline-rich sequences with the SH3 domain of Grb2 adaptor protein, which binds to these receptors through its SH2 domain (Fig. 7.2a) [76–81].

Functional and morphological studies demonstrated that Cbl and Grb2 are the key mediators of RTK ubiquitination and endocytosis. Grb2 and Cbl are co-localized with and remained bound to activated EGFR in CCPs [80, 82, 83]. c-Cbl knockout, siRNA interference depletion of c-Cbl and Cbl-b, and overexpression of ligase-inactive c-Cbl mutants inhibit ubiquitination and internalization of CSF-1R, EGFR, c-kit and, other RTKs [47, 64, 73, 84]. Likewise, multiple experimental approaches in mammalian cells have demonstrated the importance of Grb2 in ubiquitination and internalization of RTKs via Cbl [79–82, 85]. In *Drosophila* and *C. elegans*, D-Cbl L binds to Grb2/Drk and Sli-1/Cbl binds to SEM-5/Grb2 to promote EGFR endocytosis [86, 87].

Ubiquitination of RTKs can be indirectly regulated by proteins that modulate activity of Cbl. For example, Cbl turnover is regulated by its ubiquitination mediated by HECT domain E3 ligases, and the ligase activity of Cbl is negatively regulated by interactions with Sprouty2 and p85Cool-1 [69, 88–91]. Tyrosine phosphorylation of Cbl promotes its ligase activity [92]. This phosphorylation can be a partial explanation for the requirement of RTK kinase activity for RTK endocytosis. Collectively, these studies suggest that Grb2-Cbl interaction has a key role in ubiquitination, recruitment to CCPs, and internalization of various types of RTKs.

With all the data demonstrating RTK ubiquitination, the model of ubiquitin-dependent RTK internalization is not supported by the direct evidence. Ubiquitin-conjugation sites were mapped and mutated in EGFR and IGF-1R [64, 67]. Comprehensive mutagenesis of lysines in FGFR2 was also performed [93]. Surprisingly, these studies revealed that ubiquitination of EGFR and FGFR2 is not essential for their internalization. In contrast, mutation of lysines in the IGF-1R significantly reduced antibody-induced internalization of this receptor [67].

While several lines of evidences demonstrate the role of Grb2 and Cbl in internalization of EGFR and several other RTKs, this evidence contradicts to the demonstration of normal internalization of non-ubiquitinated receptor mutants. One possibility to reconcile these data is to assume that an unknown receptor-associated protein is ubiquitinated by Cbl and mediates receptor internalization in the absence of ubiquitination of the receptor (Fig. 7.2a). Another possibility is that Grb2 and Cbl may mediate internalization via an ubiquitination-independent mechanism. For instance, leucine repeat kinase Lrrk1 binds to EGFR through Grb2 and promotes EGFR endocytosis [94]. It has also been shown that the SH2 domain of Grb2 binds to Tom1L protein that interacts with clathrin, thus linking EGFR to CCPs [95]. However, in this model it is unclear how the single Grb2 SH2 domain binds simultaneously to EGFR and Tom1L. Upon PDGF stimulation, Grb2 binds to PDGFR-β, and the SH3 domain of Grb2 interacts with dynamin, a GTPase essential for fission of coated vesicles, to facilitate internalization of this receptor [96]. Several studies of Cbl-interacting adaptor protein CIN85 showed that its SH3 domains interact with endophilin, a CCP protein, and a number of other proteins. CIN85 has been implicated in internalization of various RTKs [97–102]. However, the effects of knockdown or knockout of CIN85 on RTK internalization, as well as CIN85 localization in CCPs, have not been demonstrated. In conclusion, whereas the importance of Cbl and Grb2 in RTK endocytosis is well established, the role of RTK ubiquitination in this process remains unclear. At the same time, ubiquitination-independent internalization mechanisms involving Grb2 and Cbl are not well supported by the functional and localization data.

Another line of evidence supporting the role of ubiquitination in RTK endocytosis is based on the effects of mutations and depletion/knockout of the components of CCPs that recognize ubiquitinated endocytic cargo. Three proteins containing UBDs residing in CCPs (epsin-1/2, Eps15, and Eps15R, a protein highly homologous to Eps15) have been proposed to be involved in RTK internalization (Fig. 7.2a). Binding of these proteins to ubiquitin moieties in vitro and their co-immunoprecipitation with EGFR have been demonstrated [29, 103]. Moreover, it was reported that siRNA knockdown of epsin-1 completely abolished EGF internalization [104]. Other studies, however, reported that individual or simultaneous depletion of epsin-1, Eps15, and Eps15R did not specifically block the CME of EGFR [23, 29, 105]. Other UBD-containing adaptors called ankyrin repeat domain (Ankrd) 13A, B, and D proteins have recently been shown to bind K63-ubiquitin chains and implicated in EGFR internalization [106]. However, localization of Ancrd13s in coated pits was not demonstrated. Epsin and Eps15/Eps15R contain multiple protein-protein interaction domains, which complicates the functional analysis of their role in internalization due to the pleiotropic effects of overexpressed mutants of these proteins on the interaction network of endocytic proteins [107]. Interestingly, epsin was found distributed throughout the entire clathrin coat, whereas Eps15 is located at the edge of the coat, suggesting that epsin and Eps15/Eps15R may have distinct functions during cargo recruitment and vesicle formation [104, 108]. Clearly, better understanding of the function of UBD-containing proteins in the redundant or sequential recognition of ubiquitinated RTKs would require high-resolution live-cell imaging analysis during the process of RTK recruitment into the CCP.

7.3.3 NEDD4 Family and Other E3 Ligases

The second major class of E3 ubiquitin ligases involved in RTK endocytosis are the HECT domain E3s. HECT domain recruits E2 enzymes, which results in the transfer of ubiquitin from E2 to a conserved cysteine residue in the HECT domain and then to lysine residues of the substrate. NEDD4 family of E3 ligases are the most "endocytosis-relevant" HECT E3s. These ligases bear WW domains capable of binding to the PPxY (x, any amino acid) sequence motifs of the substrate (Fig. 7.3) [109]. NEDD4-2 was shown to bind to the carboxyl-terminal PPxY motif of TrkA and ubiquitinate this receptor [66]. AIP4/Itch, another NEDD4 family ligase, binds to the PPxY motif in ErbB4 [110]. NEDD4 E3 also ubiquitinate proteins that do not have PPxY motifs but these interaction mechanisms are mainly unknown [111]. For example, NEDD4-1 binds to an unconventional sequence of FGFR1 and ubiquitinates this receptor (Fig. 7.3) [112]. Overexpression of dominant-negative mutants or siRNA depletion of NEDD4 family E3s inhibited endocytosis of TrkA, ErbB4, and FGFR1 [66, 110, 112]. NEDD4-1 binds to IGF-1R and IR indirectly through the SH2 adaptor Grb10 that interacts with phosphotyrosine-containing motifs of these receptors [113, 114] (Fig. 7.2). The SH2 domain of Grb10 can simultaneously bind the C2 domain of NEDD4-1 and a phosphotyrosine motif in IGF-1R [115]. The role of the NEDD4-1/Grb10 complex was demonstrated in experiments in which expression of dominant-interfering mutants of Grb10 and NEDD4-1 inhibited IGF-1R endocytosis [113, 114]. More recently, however, genetic knockout of NEDD4-1 had an opposite effect—reduction of IGF-1R on the cell surface of mouse fibroblasts, suggesting that the role of NEDD4-1 in this system is more complex [116]. It was proposed that NEDD4-1 ubiquitinates Grb10 and targets Grb10 for degradation [116]. In NEDD4-1 knockout Grb10 is upregulated which results in increased endocytosis of IGF-1R. Such regulation would imply that in the absence of NEDD4-1, another E3 binds to Grb10 and ubiquitinates IGFR-1. IGF-1R can also be ubiquitinated by the RING domain E3 ligase, Mdm2, which binds to IGF-1R through the interaction with β-arrestins, and by c-Cbl [117, 118]. Indeed, siRNA knockdown of individual E3s implicated in IGF-1R ubiquitination did not affect IGF-1R endocytosis [67].

The IGF-1R example of the redundant role of multiple E3s in ubiquitination of a single RTK can be extended to EGFR (Cbl, Smurf2) [73, 119], ErbB4 (Itch, NEDD4, and WW1) [110, 120, 121], and TrkA (Cbl, NEDD4-2, TRAF-6/UbcH7) [65, 66, 122]. Ubiquitination of a single RTK by different E3 ligases may differentially effect trafficking [67, 119]. It is conceivable that the relative contribution of different E3s is cell-type specific and depends on experimental conditions, and, in some cases, the effects on substrate ubiquitination can be indirect. An interesting example is the "multilevel" regulation of an EGFR homologue, ErbB3, ubiquitinated by the RING family E3 Nrdp1. Nrdp1 ubiquitinates ErbB3 in a ligand (neuregulin)-independent manner. Neuregulin stimulation augments this ubiquitination by increasing the stability of Nrdp1 that is itself controlled by ubiquitination [123].

7.3.4 Tyrosine-Based and Di-leucine Motifs in RTK Endocytosis

Many endocytic cargo proteins possess specific amino acid motifs in the cytoplasmic domains that are recognized by adaptor proteins that in turn interact with clathrin heavy chain, the main structural component of CCP. The principal adaptor in CCP is AP-2. AP-2 is a heterotetrameric complex consisting of α, β2, μ2, and σ2 subunits assembled in the core or "trunk" domain and two appendage domains connected to the trunk by flexible linkers (hinge domains) (Figs. 7.2 and 7.4). Several RTKs have been proposed to utilize AP-2 for entering to the CCP (Fig. 7.4). IR has the di-leucine D(E)xxxLL(I) (x, any amino acid residue) ("LL") internalization motif [124–126], which directly binds to the trunk domain of AP-2 [127] (Fig. 7.4). The NPxY motif in its unphosphorylated state was shown to mediate internalization of IR, IGF-1R, and PDGFR β [32, 128–130]. These motifs interact with PTB domains

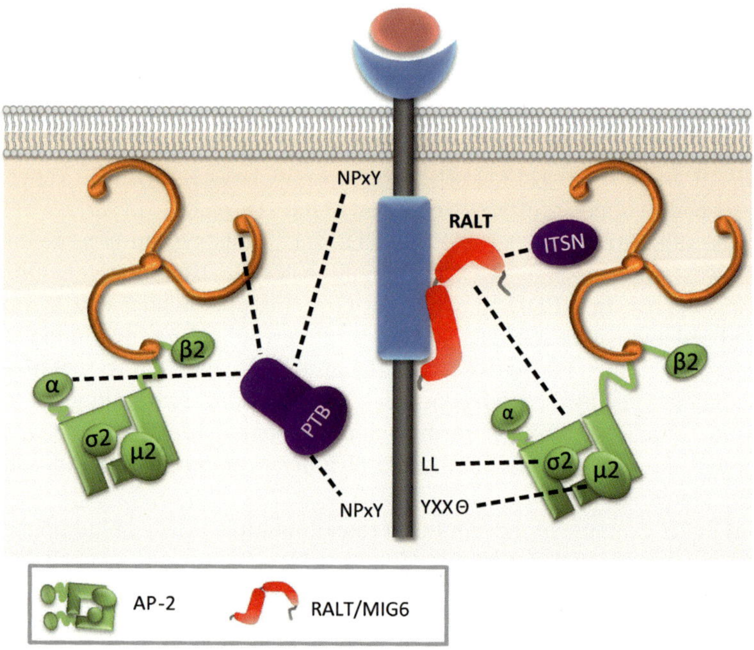

Fig. 7.4 Interactions of RTKs with AP-2 implicated in internalization. Direct interactions of YxxΘ and di-leucine (LL) motifs of RTKs with, respectively, μ2 and σ2 subunits of the AP-2 complex are shown. The hinge domain of β2 subunit of AP-2 binds to the terminal domain of clathrin heavy chain. The NPxY motifs of RTKs are recognized by monomeric adaptor proteins containing PTB domains. These monomeric adaptors can bind to α and β2 (not shown) appendage domains of AP-2 and/or directly bind to the terminal domain of clathrin. EGFR can interact with AP-2 indirectly through receptor-associated late transducer (RALT/MIG-6) that simultaneously binds to the EGFR kinase domain and AP-2 or intersectin, a protein recruited to clathrin coats through interactions of its SH3 and EH (Eps15 homology) domains

present in several monomeric adaptor proteins, such as Disabled2, ARH, and NUMB, which in turn bind to the appendage domains of AP-2 and/or the terminal domain of clathrin heavy chain (Fig. 7.4) [131]. The YxxΘ (YRAL) (Θ, bulky hydrophobic residue) motif of EGFR is responsible for binding to the μ2 subunit of AP-2 [132] (Fig. 7.4).

Although the importance of LL and NPxY motifs for internalization of IR, IGF-1R, and PDGFR has been demonstrated by site-directed mutagenesis, the interaction of these receptors with AP-2 has not been demonstrated. EGFR is the only RTK for which interaction with AP-2 can be detected by co-immunoprecipitation and in vitro pull-down assays [133–135]. EGFR:AP-2 interaction was dependent on the YRAL motif of the receptor. However, mutation of these EGFR residues or residues located in the tyrosine-motif-binding interface of μ2 did not affect EGFR internalization [132, 134]. Additionally, it was shown that EGFR can phosphorylate Tyr6 in the β2 subunit of AP-2, and this can be considered as another evidence of EGFR:AP-2 interaction in the intact cell. Interestingly, tyrosine phosphorylation of β2 Tyr6 was predicted to facilitate conformational changes in the trunk domain of AP-2 necessary for the interaction with the LL motif [127]. The atypical LL motif in the carboxyl-terminus of EGFR was shown to be important for AP-2 phosphorylation, suggesting the role of this LL motif in EGFR:AP-2 interaction [136]. However, mutation of this motif did not affect clathrin-dependent internalization of EGFR. Overall, understanding of the role of tyrosine- and di-leucine based internalization motifs and AP-2 in RTK internalization is incomplete, and the contribution of RTK:AP-2 interaction in ligand-induced RTK endocytosis remains to be defined.

Recently a mechanism of internalization involving EGFR:AP-2 interaction but not requiring receptor tyrosine phosphorylation was proposed [137]. A tumor suppressor protein, RALT/MIG-6, binds ligand-activated EGFR and inhibits its kinase activity. At the same time, RALT/MIG-6 mediates EGFR internalization by binding to AP-2 and intersectin, another protein located in CCPs [137] (Fig. 7.4). AP-2 interaction with EGFR has also been shown to mediate ligand-independent EGFR endocytosis induced by activation of p38/MAPK [138]. Potential involvement of AP-2 in EGFR endocytosis was demonstrated in experiments in which mutations in the LL motif of EGFR inhibited kinase-independent EGFR endocytosis induced by high EGF concentrations [139]. In summary, with the exception of β2/AP-2 phosphorylation by the EGFR, the nature of the dependence of RTK:AP-2 interactions on ligand-binding and RTK activity is unknown. At the same time, accumulating evidence implicates RTK:AP-2 interactions in kinase-independent mechanisms of internalization.

7.4 Mechanisms of Clathrin-Independent Endocytosis

As discussed in the first section of this chapter, it is well established that EGFR can be internalized by clathrin-independent mechanisms. The CIE of FGFR, IR, IGFR-1, c-Met, and VEGFR has also been demonstrated [55, 118, 140–143]. Whereas molecular mechanisms of various types of CIE are not well defined [26], the mechanisms and pathways of RTK CIE can be roughly divided into two groups. One group can be classified as the pinocytosis-like endocytosis. This type of endocytosis typically involves actin cytoskeleton rearrangement and membrane ruffling but does not involve specific targeting of receptors into endocytic vesicles. Ruffling-associated endocytosis of EGFR was observed in early studies in epidermoid carcinoma A-431 cells that express very high levels of EGFR [35]. Formation of large endocytic structures of a macropinocytic nature in clathrin-independent but EGFR-activity-dependent manner was later demonstrated in COS cells [144]. CIE of EGFR mediated by dorsal plasma membrane ruffles is most well described [145]. In this case, extensive membrane ruffling in response to EGFR activation leads to formation of morphologically heterologous, tubular-vesicular endosomal compartments that move toward the nucleus. Similar morphological appearance of heterologously shaped endosomes containing NGF and TrkA receptors was observed in PC12 cells and neurons [146]. This endocytosis does not involve CCP but requires a protein called Pincher. Furthermore, Barbieri and co-workers described the clathrin-independent mechanism of EGFR endocytosis that, similar to the dorsal-ruffle-mediated CIE, requires the activity of phosphoinositol-3-kinase (PI3K) [147]. The latter study also implicated RIN1 (Ras binding and Rab5 GEF) and EGF-induced Rab5 activity in this type of CIE. Likewise, RIN1 has been implicated in ephrin RTK (EphA4) endocytosis [148]. An Eph receptor system involves activation of Eph endocytosis by a membrane-anchored ligand (ephrin) [149]. In neurons, endocytosis of ephrin-EphA4 receptor complex is mediated by the Rho family guanine exchange factor (GEF), Vav2, and is actin-dependent [150]. Whether EphA4 endocytosis mediated by Vav2 or RIN1 occurs via CCP is, however, unknown.

Another mechanism implicated in the CIE of RTKs can be defined by its sensitivity to inhibitors of caveolae and cholesterol-disrupting agents [29, 118, 142]. Originally, caveolae was proposed to be the site of EGFR endocytosis [29] but in later studies the same group of investigators classified this pathway more loosely as a lipid raft/cholesterol-dependent endocytosis [151]. In general, the lack of experimental approaches to inhibit specific pathways of CIE hinders the analysis of the contribution and functional role of CIE in RTK trafficking. Observations of CIE in RTK studies tend to be made under experimental conditions favoring receptor overexpression and by using high concentrations of growth factors. Therefore, we propose that CME is the major pathway of RTK endocytosis in cells with normal levels of RTK proteins that are stimulated with concentrations of growth factors equivalent to those present in vivo.

7.5 Molecular Mechanisms of RTK Endosomal Sorting by ESCRTs

Sorting of internalized RTK begins in early endosomes where ubiquitinated RTK is recognized by ESCRT (endosomal sorting complex required for transport)-0. The main component of ESCRT-0 is a heterodimeric complex of hepatocyte growth factor regulated tyrosine kinase substrate (Hrs) and signal-transducing adaptor

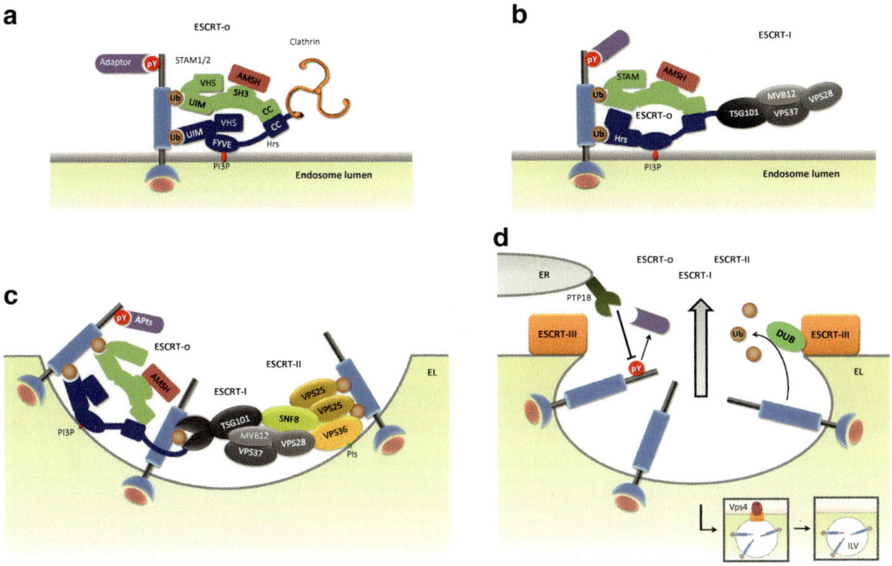

Fig. 7.5 Hypothetic mechanism of ESCRT-mediated RTK sorting in MVB. (**a**) Ubiquitin moieties of RTK bind to UIMs of Hrs and STAM1/2 (ESCRT-0 complex). The FYVE domain of Hrs binds phosphoinositide-3-phosphate (PI3P). Hrs and STAM interact via their coiled-coil (CC) domains. In addition, Hrs and STAM contain VHS and SH3 domains, respectively. AMSH, a Lys63-polyubiquitin-specific DUB, interacts with the SH3 domain of STAM. Leucine motif ("clathrin-binding box") of Hrs binds to the terminal domains of clathrin heavy chains leading to the assembly of clathrin triskelions into flat lattices. Signaling adaptors containing SH2 and PTB domains (*Adaptor*) and E3 ubiquitin ligases (E3) remain associated with phosphotyrosine-containing motifs (pY) of RTKs. EL, endosomal lumen. (**b**) A component of ESCRT-I, TSG101, is recruited onto the endosomal membrane from cytosol by means of the interaction with the carboxyl-terminal domain of Hrs, followed by translocation of other ESCRT-I components to the endosome. (**c**) ESCRT-I and ESCRT-II mediate invagination of the limiting membrane of the endosome. The vacuolar protein sorting 28 (VPS28) interacts with VPS36 protein, and this interaction mediates the association of ESCRT-I with the ESCRT-II complex. VPS36 binds to phosphoinositol lipids (in *green*). Only main components and interactions of ESCRT-0 and ESCRT-I are shown. (**d**) ESCRT-III is recruited by binding to ESCRT-II. Cargo is deubiquitinated by ESCRT-III-associated DUBs. ESCRT-III assembles around the neck of the forming vesicle to promote its scission from the limiting membrane and formation of an intraluminal vesicle (ILV) (see *bottom-right* panels). During vesicle formation, signaling adaptors and ubiquitin ligases are removed, presumably, through dephosphorylation of RTKs by PTBs. Phosphotyrosine phosphatase non-receptor type 1 (PTPN1; also known as PTP1B) associated with endoplasmic reticulum (ER) is proposed to dephosphorylate EGFR during ILV formation. After vesicle scission, ESCRT-III is released from the membrane by AAA-ATPase Vps4

molecules 1 and 2 (STAM1/2) (Fig. 7.5a). Hrs is associated with the endosomal membrane by means of interaction of its FYVE domain with phosphoinositol-3-phosphate, a lipid that is enriched in endosomes. Two UBDs in Hrs and another UBD in STAM are thought to bind ubiquitinated RTKs. Hrs also possesses the clathrin-binding motif. Recruitment of clathrin to endosomes by Hrs is proposed to result in assembly of double-layered flat clathrin lattices [152]. It is also proposed that this ESCRT-0/clathrin microdomain limits free diffusion of RTKs in the endosomal membrane, thus retaining RTKs in the vesicular compartment of endosomes and preventing RTK distribution within the entire endosomal membrane area, which is mostly constituted by tubular extensions of endosomes. Because tubular extensions mediate recycling of membrane components from endosomes to plasma membrane, RTKs that are associated with Hrs microdomains are incapable of recycling.

In support to the model that recycling is the default pathway of cargo that is not trapped by ESCRT-0 in endosomes, RTK are often found in the same tubular membrane compartments with nutrient carrier receptors, such as the transferrin receptor, suggesting common recycling pathways. Recycling from peripheral early and pericentriolar endosomes is regulated by Rab4 and Rab11 (and homologous Rabs, such as Rab25), respectively [153]. Specific "recycling" sequence motifs have been described for other classes of receptors but have not been found in RTKs [154]. Recently, however, GGA3 adaptor protein was shown to bind c-Met via the Crk adaptor and promote c-Met recycling, suggesting that specific "recycling-targeting" mechanisms may exist [155]. In contrast, it has been proposed that ubiquitinated EGFR interacts with the UBD of GGA3 in early endosomes prior to the EGFR interaction with ESCRT-0, and this interaction is important for EGFR degradation, not recycling [156].

ESCRT-0 facilitates recruitment of multiprotein complex ESCRT-I onto the endosomal membrane, which in turn results in recruitment of ESCRT-II and ESCRT-III (Fig. 7.5b, c). ESCRT-I and ESCRT-II cooperate to promote endosomal membrane invagination and ultimately facilitate packaging of the cargo into intraluminal vesicles (ILV), which bud toward the endosome interior, in a process mediated by ESCRT-III (Fig. 7.5d) [157–159]. Remarkable advances have been made in elucidating the atomic structures of various ESCRT components; however, the mechanics of ILV formation and the mechanisms by which ubiquitinated cargo is sequentially passed from ESCRT-0 to late ESCRTs are not fully understood.

The model of ESCRT-dependent sorting of RTKs is supported by several lines of experimental evidence. First, degradation of EGFR, FGFR2, and IGF-1R is highly sensitive to the level of their ubiquitination. Mutations of ubiquitin-conjugation or Cbl-binding sites dramatically inhibited degradation of RTKs [64, 67, 73, 93]. Interaction with Hrs and incorporation into ILVs of non-ubiquitinated EGFR mutant were highly inefficient [160]. In general, mutations, knockdowns, and knockouts of E3 ligases, E3 adaptors, such as Grb2 and Grb10, and proteins that modulate activity of E3s have strong effects on the ligand-induced degradation of RTKs. For example, Cbl-binding protein, such as Sprouty2, and low-density lipoprotein receptor-like proteins, Lrig1 and CIN85, regulate the rate of lysosomal degradation of EGFR [88, 161–164]. The effects of overexpression of dominant-negative mutants of Hrs and other ESCRT components, their downregulation by knockouts and knockdowns, and a microRNA on the lysosomal degradation of RTKs have been well documented [165–170]. Correspondingly, inhibition of the RTK kinase activity and ubiquitination by threonine phosphorylation in the juxtamembrane domain of EGFR by protein kinase C (PKC) or mutations of ubiquitination sites elevate recycling of EGFR [160, 171]. Similarly, activation of PKC promoted the recycling of PDGFR-β [172].

It should be noted that even non-ubiquitinated mutants of RTKs can be incorporated into ILVs and slowly degrade by a lysosomal mechanism, implying that other ubiquitination-independent mechanisms of MVB sorting exist. For instance, the LL motif in the carboxyl-terminus of EGFR was shown to be important for EGFR degradation [136]. The binding of RALT/MIG-6 associated with EGFR to the endosomal SNARE syntaxin 8 also promotes EGFR degradation [173]. The mechanisms underlying ubiquitin-independent sorting are not understood, although it is proposed that non-ubiquitinated cargo can bind Alix, a component of ESCRT-III, thus bypassing the dependence on the cargo interaction with ESCRT-0.

Because ubiquitin-UBD interactions are typically of low affinity, binding of ubiquitinated RTK to ESCRT-0 is likely to involve multivalent interactions. Therefore, multi-monoubiquitination and polyubiquitination of RTKs are proposed to be necessary for the efficient interaction of RTK with ESCRT-0 and RTK sorting in MVB [174]. Existence of multiple UBDs in a single Hrs-STAM complex and additionally in other proteins associated with this complex, such as the splicing form of Eps15 [175], supports this hypothesis. EGFR and TrkA were shown to be ubiquitinated by Lys63-polyubiquitin chains [64, 65]. The critical role of Lys63-ubiquitin chains in MVB sorting and lysosomal degradation of other ubiquitinated cargo has been proposed [176, 177]. It is well accepted that in contrast to Lys48-linked chains that target proteins to proteasome, Lys63-linked polyubiquitin chains have non-proteasomal functions [178]. Importantly, Lys63-linked di-ubiquitin but not Lys48 chains has conformation similar to that of monoubiquitin and is recognized by UBDs that also recognize monoubiquitin [179]. We propose that Lys63-linked polyubiquitination is the major molecular signal involved in the endocytic sorting of mammalian RTKs, although this hypothesis requires direct experimental prove.

While a role of ESCRT-0 in the recruitment of ubiquitinated cargo is widely accepted, it is less clear how this cargo is transferred from ESCRT-0 to ESCRT-I and ESCRT-II (Fig. 7.5c). It is hypothesized that ubiquitinated cargo, like RTK, dissociates from ESCRT-0 and becomes associated with ESCRT-I, whose component TSG101 has an UBD. However, interactions of RTKs with TSG101 and ESCRT-II have not been experimentally demonstrated. Both ESCRT-I and ESCRT-II are predicted to interact with ubiquitin with much weaker affinity than in case of ubiquitin interaction with ESCRT-0. ESCRT-III components do not have UBDs. It is further proposed that the assembled multimeric ESCRT-III modules surround the cargo-containing membrane and keep this cargo within the invaginating membrane that will give rise to an ILV. This model implies that (1) RTK must be ligand bound in order to maintain the kinase activity, phosphorylation, Cbl association, and ubiquitination, at least during the passage through ESCRTs-0–II; (2) ubiquitin must be removed from the cargo prior to its packaging into ILVs to avoid the loss of ubiquitin, perhaps, at the stage of assembled ESCRT-III, when ubiquitin interactions are not anymore necessary; and (3) RTKs must be dephosphorylated before their incorporation into ILVs to avoid degradation of phosphotyrosine-bound adaptor proteins associated with RTKs. The mechanisms of such tightly tuned RTK deubiquitination and dephosphorylation during MVB sorting are not defined.

Electron microscopy demonstrated that EGF remains bound to receptors in ILVs [180, 181]. Since adaptors that are associated with internalized EGFR, such as Grb2 and Shc, have not been detected in ILVs, EGFR dephosphorylation must be very efficient to counteract the kinase activity of the receptor. PTP1B, a phosphotyrosine phosphatase present in endoplasmic reticulum, has been implicated in dephosphorylation of EGFR in MVBs [182]. The activity of PTP1B involves formation of the membrane contact sites between endoplasmic reticulum and MVB membranes [182]. At what stage of the ESCRT-mediated sorting process EGFR dephosphorylation takes place and the status of EGFR phosphorylation in ILVs remain unknown.

Several deubiquitination enzymes (DUBs) have been proposed to remove ubiquitin from RTKs. DUBs have at least three functions before and during the ESCRT-mediated sorting process: (1) deubiquitination of RTKs prior to and during RTK-ESCRT interactions, thus preventing these interactions; (2) removal of ubiquitin from RTKs "primed" to be incorporated into ILVs; and (3) deubiquitination of the individual components of ESCRTs. AMSH (*a*ssociated *m*olecule of *SH*3 domain of STAM) was shown to associate with ESCRT-0 and also with the ESCRT-III component CHMP3 [183, 184]. It was proposed that AMSH directly deubiquitinates EGFR and PDGFR and therefore decreases rates of their degradation [167, 185, 186]. POH1, a DUB associated with the proteasome, has been implicated in deubiquitination of ErbB2 [187]. Recently, Cezanne 1 and 2 were shown to directly control EGFR ubiquitination [188]. Another DUB involved in cargo sorting is Usp8, which affects EGFR degradation by controlling the stability of STAM [189] and Eps15 [190]. However, Usp8 siRNA knockdown has had opposite effects on EGFR and c-Met degradation in cell culture in different studies [191, 192]. A conditional mouse knockout of Usp8 in the liver resulted in low levels of EGFR, c-Met, and ErbB3, and this effect was proposed to be due to the downregulation of ESCRT components in the absence of Usp8 [193]. The precise role of various DUBs in the regulation of RTK sorting remains to be determined.

7.6 Concluding Remarks

A variety of molecular mechanisms of endocytosis and post-endocytic sorting of ligand-activated RTKs have been proposed, studied, and elucidated in recent years. However, our understanding of these mechanisms is still not sufficient to generate experimental tools and develop approaches that would allow effective inhibition of endocytosis of a specific RTK without the interference with endocytosis of other proteins. Lack of such tools and approaches hinders the ability to define the role of RTK endocytosis in the regulation of signaling networks. By contrast, many experimental approaches are established to test whether and how signaling by RTKs regulates an RTK-specific and general endocytic machineries, and whether there are compartments that are specializing in handling RTK endocytic trafficking, such as RTK-specific endocytic sites (coats) and endosomes. A number of examples demonstrating an active role of RTK signaling during endocytosis have been reported.

The existence of CCPs that are specialized for EGFR endocytosis has been proposed in several studies [25, 194, 195]. Also, EGF-induced increase in the number of coated pits has been demonstrated in PC12 cells [196]. Activation of EphB2 receptor resulted in phosphorylation of synaptojanin, a phosphatidylinositol 5′-phosphatase that is involved in CME, which influenced the dynamics of transferrin endocytosis [197]. During endosomal sorting, EGF was proposed to promote the formation of ILVs in MVB, and this effect required phosphorylation of annexin 1 and EGFR ubiquitination [160, 198]. This type of studies will be expanded in the future by using careful and quantitative analyses of the effects of RTK activity on the endocytosis system under physiological experimental conditions.

The incomplete understanding and a relatively slow progress in defining specific mechanism of RTK endocytosis is largely due to the substantial mechanistic redundancies in the RTK endocytic system. The internalization step of RTK trafficking, in particular, is mediated and regulated by a multitude of mechanisms that are often redundant. One explanation behind the robustness of the endocytic process is the existence of clathrin-dependent and clathrin-independent mechanisms. Furthermore, CME can itself be regulated by several redundant mechanisms. In early studies EGFR was considered to be the prototypic and "straightforward" model of stimulus-induced endocytosis. Studies for the past 25 years demonstrated the frustrating complexity of this receptor endocytosis. The CME of EGFR involves ubiquitination, binding to AP-2, acetylation, and additional undefined mechanisms, and simultaneous inhibition of all these mechanisms is necessary to substantially suppress EGFR internalization [199]. In a similar manner, AP-2-binding motifs and ubiquitination have been implicated in internalization of other RTKs. Furthermore, there is significant redundancy in the contribution of E3 ligases to RTK ubiquitination. The redundancy of endocytosis mechanisms observed in mammalian cultured cells may be attributed to a high adaptability of cells growing in vitro. It is possible that endocytosis mechanisms in vivo display less redundancy and the contribution of individual mechanisms depends on the developmental and differentiation status of cells in the tissue. Therefore, mechanistic analysis of endocytic pathways in vivo in various cells and tissues may help to better define individual molecular mechanisms of RTK endocytosis and determine the contribution of these mechanisms in the regulation of physiologically relevant and specific cellular systems.

Acknowledgments This work was supported by NCI grants CA089151 and CA112219.

References

1. Mosesson Y, Mills GB, Yarden Y. Derailed endocytosis: an emerging feature of cancer. Nat Rev. 2008;8(11):835–50.
2. Joffre C, Barrow R, Menard L, Calleja V, Hart IR, Kermorgant S. A direct role for Met endocytosis in tumorigenesis. Nat Cell Biol. 2011;13(7):827–37.
3. Sorkin A, von Zastrow M. Endocytosis and signalling: intertwining molecular networks. Nat Rev Mol Cell Biol. 2009;10(9):609–22.

4. Platta HW, Stenmark H. Endocytosis and signaling. Curr Opin Cell Biol. 2011;23(4):393–403.
5. Scita G, Di Fiore PP. The endocytic matrix. Nature. 2010;463(7280):464–73.
6. Hommelgaard AM, Lerdrup M, van Deurs B. Association with membrane protrusions makes ErbB2 an internalization-resistant receptor. Mol Biol Cell. 2004;15(4):1557–67.
7. Foti M, Moukil MA, Dudognon P, Carpentier JL. Insulin and IGF-1 receptor trafficking and signalling. Novartis Found Symp. 2004;262:125–41. discussion 41–7, 265–8.
8. Mineo C, Gill GN, Anderson RG. Regulated migration of epidermal growth factor receptor from caveolae. J Biol Chem. 1999;274(43):30636–43.
9. Liu P, Ying Y, Ko YG, Anderson RG. Localization of platelet-derived growth factor-stimulated phosphorylation cascade to caveolae. J Biol Chem. 1996;271(17):10299–303.
10. Lazar CS, Cresson CM, Lauffenburger DA, Gill GN. The Na+/H+ exchanger regulatory factor stabilizes epidermal growth factor receptors at the cell surface. Mol Biol Cell. 2004;15(12):5470–80.
11. Borg JP, Marchetto S, Le Bivic A, Ollendorff V, Jaulin-Bastard F, Saito H, et al. ERBIN: a basolateral PDZ protein that interacts with the mammalian ERBB2/HER2 receptor. Nat Cell Biol. 2000;2(7):407–14.
12. Yarden Y, Sliwkowski MX. Untangling the ErbB signalling network. Nat Rev Mol Cell Biol. 2001;2(2):127–37.
13. McClain DA. Mechanism and role of insulin receptor endocytosis. Am J Med Sci. 1992;304(3):192–201.
14. Jopling HM, Howell GJ, Gamper N, Ponnambalam S. The VEGFR2 receptor tyrosine kinase undergoes constitutive endosome-to-plasma membrane recycling. Biochem Biophys Res Commun. 2011;410(2):170–6.
15. Burke PM, Wiley HS. Human mammary epithelial cells rapidly exchange empty EGFR between surface and intracellular pools. J Cell Physiol. 1999;180(3):448–60.
16. McMahon HT, Boucrot E. Molecular mechanism and physiological functions of clathrin-mediated endocytosis. Nat Rev Mol Cell Biol. 2011;12(8):517–33.
17. Gorden P, Carpentier JL, Cohen S, Orci L. Epidermal growth factor: morphological demonstration of binding, internalization, and lysosomal association in human fibroblasts. Proc Natl Acad Sci U S A. 1978;75(10):5025–9.
18. Carpentier JL, Gorden P, Anderson RG, Goldstein JL, Brown MS, Cohen S, et al. Co-localization of 125I-epidermal growth factor and ferritin-low density lipoprotein in coated pits: a quantitative electron microscopic study in normal and mutant human fibroblasts. J Cell Biol. 1982;95(1):73–7.
19. Bogdanovic E, Coombs N, Dumont DJ. Oligomerized Tie2 localizes to clathrin-coated pits in response to angiopoietin-1. Histochem Cell Biol. 2009;132(2):225–37.
20. Beattie EC, Howe CL, Wilde A, Brodsky FM, Mobley WC. NGF signals through TrkA to increase clathrin at the plasma membrane and enhance clathrin-mediated membrane trafficking. J Neurosci. 2000;20(19):7325–33.
21. Lampugnani MG, Orsenigo F, Gagliani MC, Tacchetti C, Dejana E. Vascular endothelial cadherin controls VEGFR-2 internalization and signaling from intracellular compartments. J Cell Biol. 2006;174(4):593–604.
22. Zheng J, Shen WH, Lu TJ, Zhou Y, Chen Q, Wang Z, et al. Clathrin-dependent endocytosis is required for TrkB-dependent Akt-mediated neuronal protection and dendritic growth. J Biol Chem. 2008;283(19):13280–8.
23. Huang F, Khvorova A, Marshall W, Sorkin A. Analysis of clathrin-mediated endocytosis of epidermal growth factor receptor by RNA interference. J Biol Chem. 2004;279(16):16657–61.
24. von Kleist L, Stahlschmidt W, Bulut H, Gromova K, Puchkov D, Robertson MJ, et al. Role of the clathrin terminal domain in regulating coated pit dynamics revealed by small molecule inhibition. Cell. 2011;146(3):471–84.
25. Motley A, Bright NA, Seaman MN, Robinson MS. Clathrin-mediated endocytosis in AP-2-depleted cells. J Cell Biol. 2003;162(5):909–18.
26. Sandvig K, Pust S, Skotland T, van Deurs B. Clathrin-independent endocytosis: mechanisms and function. Curr Opin Cell Biol. 2011;23(4):413–20.

27. Wiley HS. Anomalous binding of epidermal growth factor to A431 cells is due to the effect of high receptor densities and a saturable endocytic system. J Cell Biol. 1988;107:801–10.
28. Lund KA, Opresko LK, Strarbuck C, Walsh BJ, Wiley HS. Quantitative analysis of the endocytic system involved in hormone-induced receptor internalization. J Biol Chem. 1990;265:15713–23.
29. Sigismund S, Woelk T, Puri C, Maspero E, Tacchetti C, Transidico P, et al. Clathrin-independent endocytosis of ubiquitinated cargos. Proc Natl Acad Sci U S A. 2005;102(8): 2760–5.
30. Jiang X, Sorkin A. Epidermal growth factor receptor internalization through clathrin-coated pits requires Cbl RING finger and proline-rich domains but not receptor polyubiquitylation. Traffic. 2003;4(8):529–43.
31. Backer JM, Shoelson SE, Haring E, White MF. Insulin receptors internalize by a rapid, saturable pathway requiring receptor autophosphorylation and an intact juxtamembrane region. J Cell Biol. 1991;115(6):1535–45.
32. Prager D, Li HL, Yamasaki H, Melmed S. Human insulin-like growth factor I receptor internalization. Role of the juxtamembrane domain. J Biol Chem. 1994;269(16):11934–7.
33. Miller K, Beardmore J, Kanety H, Schlessinger J, Hopkins CR. Localization of the epidermal growth factor (EGF) receptor within the endosome of EGF-stimulated epidermoid carcinoma (A431) cells. J Cell Biol. 1986;102(2):500–9.
34. Beguinot L, Lyall RM, Willingham MC, Pastan I. Down-regulation of the epidermal growth factor receptor in KB cells is due to receptor internalization and subsequent degradation in lysosomes. Proc Natl Acad Sci U S A. 1984;81(8):2384–8.
35. Haigler HT, McKanna JA, Cohen S. Rapid stimulation of pinocytosis in human carcinoma cells A-431 by epidermal growth factor. J Cell Biol. 1979;83(1):82–90.
36. Hopkins CR, Gibson A, Shipman M, Miller K. Movement of internalized ligand-receptor complexes along a continuous endosomal reticulum. Nature. 1990;346(6282):335–9.
37. Wu C, Cui B, He L, Chen L, Mobley WC. The coming of age of axonal neurotrophin signaling endosomes. J Proteomics. 2009;72(1):46–55.
38. Chowdary PD, Che DL, Cui B. Neurotrophin signaling via long-distance axonal transport. Annu Rev Phys Chem. 2012;63:571.
39. Huecksteadt T, Olefsky JM, Brandenberg D, Heidenreich KA. Recycling of photoaffinity-labeled insulin receptors in rat adipocytes. Dissociation of insulin-receptor complexes is not required for receptor recycling. J Biol Chem. 1986;261(19):8655–9.
40. Sorkin A, Krolenko S, Kudrjavtceva N, Lazebnik J, Teslenko L, Soderquist AM, et al. Recycling of epidermal growth factor-receptor complexes in A431 cells: identification of dual pathways. J Cell Biol. 1991;112(1):55–63.
41. Zapf-Colby A, Olefsky JM. Nerve growth factor processing and trafficking events following TrkA-mediated endocytosis. Endocrinology. 1998;139(7):3232–40.
42. Masui H, Castro L, Mendelsohn J. Consumption of EGF by A431 cells: evidence for receptor recycling. J Cell Biol. 1993;120(1):85–93.
43. Resat H, Ewald JA, Dixon DA, Wiley HS. An integrated model of epidermal growth factor receptor trafficking and signal transduction. Biophys J. 2003;85(2):730–43.
44. French AR, Tadaki DK, Niyogi SK, Lauffenburger DA. Intracellular trafficking of epidermal growth factor family ligands is directly influenced by the pH sensitivity of the receptor/ligand interaction. J Biol Chem. 1995;270(9):4334–40.
45. Roepstorff K, Grandal MV, Henriksen L, Knudsen SL, Lerdrup M, Grovdal L, et al. Differential effects of EGFR ligands on endocytic sorting of the receptor. Traffic. 2009;10(8): 1115–27.
46. French AR, Sudlow GP, Wiley HS, Lauffenburger DA. Postendocytic trafficking of epidermal growth factor-receptor complexes is mediated through saturable and specific endosomal interactions. J Biol Chem. 1994;269(22):15749–55.
47. Lee PS, Wang Y, Dominguez MG, Yeung YG, Murphy MA, Bowtell DD, et al. The Cbl protooncoprotein stimulates CSF-1 receptor multiubiquitination and endocytosis, and attenuates macrophage proliferation. EMBO J. 1999;18(13):3616–28.
48. Stoscheck CM, Carpenter G. Characterization of the metabolic turnover of epidermal growth factor receptor protein in A-431 cells. J Cell Physiol. 1984;120(3):296–302.

49. Duex JE, Sorkin A. RNA interference screen identifies Usp18 as a regulator of epidermal growth factor receptor synthesis. Mol Biol Cell. 2009;20(6):1833–44.
50. Sorkina T, Huang F, Beguinot L, Sorkin A. Effect of tyrosine kinase inhibitors on clathrin-coated pit recruitment and internalization of epidermal growth factor receptor. J Biol Chem. 2002;277(30):27433–41.
51. Glenney Jr JR, Chen WS, Lazar CS, Walton GM, Zokas LM, Rosenfeld MG, et al. Ligand-induced endocytosis of the EGF receptor is blocked by mutational inactivation and by micro-injection of anti-phosphotyrosine antibodies. Cell. 1988;52:675–84.
52. Sorokin A, Mohammadi M, Huang J, Schlessinger J. Internalization of fibroblast growth factor receptor is inhibited by a point mutation at tyrosine 766. J Biol Chem. 1994;269(25):17056–61.
53. Lamaze C, Schmid SL. Recruitment of epidermal growth factor receptors into coated pits requires their activated tyrosine kinase. J Cell Biol. 1995;129:47–54.
54. Sorkin A, Westermark B, Heldin CH, Claesson-Welsh L. Effect of receptor kinase inactivation on the rate of internalization and degradation of PDGF and the PDGF beta-receptor. J Cell Biol. 1991;112(3):469–78.
55. Wiedlocha A, Sorensen V. Signaling, internalization, and intracellular activity of fibroblast growth factor. Curr Top Microbiol Immunol. 2004;286:45–79.
56. Dougher M, Terman BI. Autophosphorylation of KDR in the kinase domain is required for maximal VEGF-stimulated kinase activity and receptor internalization. Oncogene. 1999;18(8):1619–27.
57. Wolff M, Tetzlaff K, Nivens MC, Schneider FJ, Jung B, Hohlfeld J, et al. In vivo inhibition of epidermal growth factor receptor autophosphorylation prevents receptor internalization. Exp Cell Res. 2011;317(1):42–50.
58. Honegger AM, Dull TJ, Felder S, Obberghen EV, Bellot F, Szapary D, et al. Point mutation at the ATP binding site of EGF receptor abolishes protein-tyrosine kinase activity and alters cellular routing. Cell. 1987;51:199–209.
59. Wang Q, Villeneuve G, Wang Z. Control of epidermal growth factor receptor endocytosis by receptor dimerization, rather than receptor kinase activation. EMBO Rep. 2005;6(10):942–8.
60. Acconcia F, Sigismund S, Polo S. Ubiquitin in trafficking: the network at work. Exp Cell Res. 2009;315(9):1610–8.
61. Galcheva-Gargova Z, Theroux SJ, Davis RJ. The epidermal growth factor receptor is covalently linked to ubiquitin. Oncogene. 1995;11(12):2649–55.
62. Mori S, Heldin CH, Claesson-Welsh L. Ligand-induced polyubiquitination of the platelet-derived growth factor beta-receptor. J Biol Chem. 1992;267(9):6429–34.
63. Serresi M, Piccinini G, Pierpaoli E, Fazioli F. A ligand-inducible anaplastic lymphoma kinase chimera is endocytosis impaired. Oncogene. 2004;23(5):1098–108.
64. Huang F, Kirkpatrick D, Jiang X, Gygi S, Sorkin A. Differential regulation of EGF receptor internalization and degradation by multiubiquitination within the kinase domain. Mol Cell. 2006;21(6):737–48.
65. Geetha T, Jiang J, Wooten MW. Lysine 63 polyubiquitination of the nerve growth factor receptor TrkA directs internalization and signaling. Mol Cell. 2005;20(2):301–12.
66. Arevalo JC, Waite J, Rajagopal R, Beyna M, Chen ZY, Lee FS, et al. Cell survival through Trk neurotrophin receptors is differentially regulated by ubiquitination. Neuron. 2006;50(4):549–59.
67. Mao Y, Shang Y, Pham VC, Ernst JA, Lill JR, Scales SJ, et al. Poly-ubiquitination of the insulin-like growth factor I receptor (IGF-IR) activation loop promotes antibody-induced receptor internalization and down-regulation. J Biol Chem. 2011;286:41852.
68. Pickart CM. Back to the future with ubiquitin. Cell. 2004;116(2):181–90.
69. Marmor MD, Yarden Y. Role of protein ubiquitylation in regulating endocytosis of receptor tyrosine kinases. Oncogene. 2004;23(11):2057–70.
70. Duval M, Bedard-Goulet S, Delisle C, Gratton JP. Vascular endothelial growth factor-dependent down-regulation of Flk-1/KDR involves Cbl-mediated ubiquitination.

Consequences on nitric oxide production from endothelial cells. J Biol Chem. 2003;278(22):20091–7.
71. Miyake S, Mullane-Robinson KP, Lill NL, Douillard P, Band H. Cbl-mediated negative regulation of platelet-derived growth factor receptor-dependent cell proliferation. A critical role for Cbl tyrosine kinase-binding domain. J Biol Chem. 1999;274(23):16619–28.
72. Fasen K, Cerretti DP, Huynh-Do U. Ligand binding induces Cbl-dependent EphB1 receptor degradation through the lysosomal pathway. Traffic. 2008;9(2):251–66.
73. Levkowitz G, Waterman H, Ettenberg SA, Katz M, Tsygankov AY, Alroy I, et al. Ubiquitin ligase activity and tyrosine phosphorylation underlie suppression of growth factor signaling by c-Cbl/Sli-1. Mol Cell. 1999;4(6):1029–40.
74. Penengo L, Rubin C, Yarden Y, Gaudino G. c-Cbl is a critical modulator of the Ron tyrosine kinase receptor. Oncogene. 2003;22(24):3669–79.
75. Wilhelmsen K, Burkhalter S, van der Geer P. C-Cbl binds the CSF-1 receptor at tyrosine 973, a novel phosphorylation site in the receptor's carboxy-terminus. Oncogene. 2002;21(7):1079–89.
76. Batzer AG, Rotin D, Urena JM, Skolnik EY, Schlessinger J. Hierarchy of binding sites for Grb2 and Shc on the epidermal growth factor receptor. Mol Cell Biol. 1994;14(8):5192–201.
77. Waterman H, Katz M, Rubin C, Shtiegman K, Lavi S, Elson A, et al. A mutant EGF-receptor defective in ubiquitylation and endocytosis unveils a role for Grb2 in negative signaling. EMBO J. 2002;21(3):303–13.
78. Sun J, Pedersen M, Bengtsson S, Ronnstrand L. Grb2 mediates negative regulation of stem cell factor receptor/c-Kit signaling by recruitment of Cbl. Exp Cell Res. 2007;313(18):3935–42.
79. Peschard P, Fournier TM, Lamorte L, Naujokas MA, Band H, Langdon WY, et al. Mutation of the c-Cbl TKB domain binding site on the Met receptor tyrosine kinase converts it into a transforming protein. Mol Cell. 2001;8(5):995–1004.
80. Jiang X, Huang F, Marusyk A, Sorkin A. Grb2 regulates internalization of EGF receptors through clathrin-coated pits. Mol Biol Cell. 2003;14(3):858–70.
81. Li N, Lorinczi M, Ireton K, Elferink LA. Specific Grb2-mediated interactions regulate clathrin-dependent endocytosis of the cMet-tyrosine kinase. J Biol Chem. 2007;282(23):16764–75.
82. Johannessen LE, Pedersen NM, Pedersen KW, Madshus IH, Stang E. Activation of the epidermal growth factor (EGF) receptor induces formation of EGF receptor- and Grb2-containing clathrin-coated pits. Mol Cell Biol. 2006;26(2):389–401.
83. de Melker AA, van der Horst G, Calafat J, Jansen H, Borst J. c-Cbl ubiquitinates the EGF receptor at the plasma membrane and remains receptor associated throughout the endocytic route. J Cell Sci. 2001;114(Pt 11):2167–78.
84. Zeng S, Xu Z, Lipkowitz S, Longley JB. Regulation of stem cell factor receptor signaling by Cbl family proteins (Cbl-b/c-Cbl). Blood. 2005;105(1):226–32.
85. Huang F, Sorkin A. Growth factor receptor binding protein 2-mediated recruitment of the RING domain of Cbl to the epidermal growth factor receptor is essential and sufficient to support receptor endocytosis. Mol Biol Cell. 2005;16(3):1268–81.
86. Wang PY, Pai LM. D-Cbl binding to Drk leads to dose-dependent down-regulation of EGFR signaling and increases receptor-ligand endocytosis. PLoS One. 2011;6(2):e17097.
87. Yoon CH, Chang C, Hopper NA, Lesa GM, Sternberg PW. Requirements of multiple domains of SLI-1, a Caenorhabditis elegans homologue of c-Cbl, and an inhibitory tyrosine in LET-23 in regulating vulval differentiation. Mol Biol Cell. 2000;11(11):4019–31.
88. Wong ES, Fong CW, Lim J, Yusoff P, Low BC, Langdon WY, et al. Sprouty2 attenuates epidermal growth factor receptor ubiquitylation and endocytosis, and consequently enhances Ras/ERK signalling. EMBO J. 2002;21(18):4796–808.
89. Thien CB, Langdon WY. Negative regulation of PTK signalling by Cbl proteins. Growth Factors. 2005;23(2):161–7.
90. Bao J, Gur G, Yarden Y. Src promotes destruction of c-Cbl: implications for oncogenic synergy between Src and growth factor receptors. Proc Natl Acad Sci U S A. 2003;100(5):2438–43.
91. Wu WJ, Tu S, Cerione RA. Activated Cdc42 sequesters c-Cbl and prevents EGF receptor degradation. Cell. 2003;114(6):715–25.

92. Kobashigawa Y, Tomitaka A, Kumeta H, Noda NN, Yamaguchi M, Inagaki F. Autoinhibition and phosphorylation-induced activation mechanisms of human cancer and autoimmune disease-related E3 protein Cbl-b. Proc Natl Acad Sci U S A. 2011;108(51):20579–84.
93. Haugsten EM, Malecki J, Bjorklund SM, Olsnes S, Wesche J. Ubiquitination of fibroblast growth factor receptor 1 is required for its intracellular sorting but not for its endocytosis. Mol Biol Cell. 2008;19(8):3390–403.
94. Hanafusa H, Ishikawa K, Kedashiro S, Saigo T, Iemura S, Natsume T, et al. Leucine-rich repeat kinase LRRK1 regulates endosomal trafficking of the EGF receptor. Nat Commun. 2011;2:158.
95. Liu NS, Loo LS, Loh E, Seet LF, Hong W. Participation of Tom1L1 in EGF-stimulated endocytosis of EGF receptor. EMBO J. 2009;28(22):3485–99.
96. Kawada K, Upadhyay G, Ferandon S, Janarthanan S, Hall M, Vilardaga JP, et al. Cell migration is regulated by platelet-derived growth factor receptor endocytosis. Mol Cell Biol. 2009;29(16):4508–18.
97. Szymkiewicz I, Kowanetz K, Soubeyran P, Dinarina A, Lipkowitz S, Dikic I. CIN85 participates in Cbl-b-mediated down-regulation of receptor tyrosine kinases. J Biol Chem. 2002;277(42):39666–72.
98. Buchse T, Horras N, Lenfert E, Krystal G, Korbel S, Schumann M, et al. CIN85 interacting proteins in B cells - specific role for SHIP-1. Mol Cell Proteomics. 2011;10:M110.006239.
99. Petrelli A, Gilestro GF, Lanzardo S, Comoglio PM, Migone N, Giordano S. The endophilin-CIN85-Cbl complex mediates ligand-dependent downregulation of c-Met. Nature. 2002;416(6877):187–90.
100. Soubeyran P, Kowanetz K, Szymkiewicz I, Langdon WY, Dikic I. Cbl-CIN85-endophilin complex mediates ligand-induced downregulation of EGF receptors. Nature. 2002;416(6877): 183–7.
101. Kowanetz K, Szymkiewicz I, Haglund K, Kowanetz M, Husnjak K, Taylor JD, et al. Identification of a novel proline-arginine motif involved in CIN85-dependent clustering of Cbl and down-regulation of epidermal growth factor receptors. J Biol Chem. 2003;278(41):39735–46.
102. Schmidt MH, Hoeller D, Yu J, Furnari FB, Cavenee WK, Dikic I, et al. Alix/AIP1 antagonizes epidermal growth factor receptor downregulation by the Cbl-SETA/CIN85 complex. Mol Cell Biol. 2004;24(20):8981–93.
103. Hawryluk MJ, Keyel PA, Mishra SK, Watkins SC, Heuser JE, Traub LM. Epsin 1 is a polyubiquitin-selective clathrin-associated sorting protein. Traffic. 2006;7(3):262–81.
104. Kazazic M, Bertelsen V, Pedersen KW, Vuong TT, Grandal MV, Rodland MS, et al. Epsin 1 is involved in recruitment of ubiquitinated EGF receptors into clathrin-coated pits. Traffic. 2009;10(2):235–45.
105. Chen H, Ko G, Zatti A, Di Giacomo G, Liu L, Raiteri E, et al. Embryonic arrest at midgestation and disruption of Notch signaling produced by the absence of both epsin 1 and epsin 2 in mice. Proc Natl Acad Sci U S A. 2009;106(33):13838–43.
106. Tanno H, Yamaguchi T, Goto E, Ishido S, Komada M. The Ankrd 13 family of UIM-bearing proteins regulates EGF receptor endocytosis from the plasma membrane. Mol Biol Cell. 2012;23:1343.
107. Sugiyama S, Kishida S, Chayama K, Koyama S, Kikuchi A. Ubiquitin-interacting motifs of Epsin are involved in the regulation of insulin-dependent endocytosis. J Biochem. 2005;137(3):355–64.
108. Tebar F, Sorkina T, Sorkin A, Ericsson M, Kirchhausen T. Eps15 is a component of clathrin-coated pits and vesicles and is located at the rim of coated pits. J Biol Chem. 1996;271(46):28727–30.
109. Rotin D, Kumar S. Physiological functions of the HECT family of ubiquitin ligases. Nat Rev Mol Cell Biol. 2009;10(6):398–409.
110. Sundvall M, Korhonen A, Paatero I, Gaudio E, Melino G, Croce CM, et al. Isoform-specific monoubiquitination, endocytosis, and degradation of alternatively spliced ErbB4 isoforms. Proc Natl Acad Sci U S A. 2008;105(11):4162–7.

111. Persaud A, Alberts P, Amsen EM, Xiong X, Wasmuth J, Saadon Z, et al. Comparison of substrate specificity of the ubiquitin ligases Nedd4 and Nedd4-2 using proteome arrays. Mol Syst Biol. 2009;5:333.
112. Persaud A, Alberts P, Hayes M, Guettler S, Clarke I, Sicheri F, et al. Nedd4-1 binds and ubiquitylates activated FGFR1 to control its endocytosis and function. EMBO J. 2011;30(16):3259–73.
113. Vecchione A, Marchese A, Henry P, Rotin D, Morrione A. The Grb10/Nedd4 complex regulates ligand-induced ubiquitination and stability of the insulin-like growth factor I receptor. Mol Cell Biol. 2003;23(9):3363–72.
114. Monami G, Emiliozzi V, Morrione A. Grb10/Nedd4-mediated multiubiquitination of the insulin-like growth factor receptor regulates receptor internalization. J Cell Physiol. 2008;216(2):426–37.
115. Huang Q, Szebenyi DM. Structural basis for the interaction between the growth factor-binding protein GRB10 and the E3 ubiquitin ligase NEDD4. J Biol Chem. 2010;285(53):42130–9.
116. Cao XR, Lill NL, Boase N, Shi PP, Croucher DR, Shan H, et al. Nedd4 controls animal growth by regulating IGF-1 signaling. Sci Signal. 2008;1(38):ra5.
117. Girnita L, Shenoy SK, Sehat B, Vasilcanu R, Girnita A, Lefkowitz RJ, et al. {beta}-Arrestin is crucial for ubiquitination and down-regulation of the insulin-like growth factor-1 receptor by acting as adaptor for the MDM2 E3 ligase. J Biol Chem. 2005;280(26):24412–9.
118. Sehat B, Andersson S, Girnita L, Larsson O. Identification of c-Cbl as a new ligase for insulin-like growth factor-I receptor with distinct roles from Mdm2 in receptor ubiquitination and endocytosis. Cancer Res. 2008;68(14):5669–77.
119. Ray D, Ahsan A, Helman A, Chen G, Hegde A, Gurjar SR, et al. Regulation of EGFR protein stability by the HECT-type ubiquitin ligase SMURF2. Neoplasia. 2011;13(7):570–8.
120. Li Y, Zhou Z, Alimandi M, Chen C. WW domain containing E3 ubiquitin protein ligase 1 targets the full-length ErbB4 for ubiquitin-mediated degradation in breast cancer. Oncogene. 2009;28(33):2948–58.
121. Zeng F, Xu J, Harris RC. Nedd4 mediates ErbB4 JM-a/CYT-1 ICD ubiquitination and degradation in MDCK II cells. FASEB J. 2009;23(6):1935–45.
122. Takahashi Y, Shimokawa N, Esmaeili-Mahani S, Morita A, Masuda H, Iwasaki T, et al. Ligand-induced downregulation of TrkA is partly regulated through ubiquitination by Cbl. FEBS Lett. 2011;585(12):1741–7.
123. Cao Z, Wu X, Yen L, Sweeney C, Carraway 3rd KL. Neuregulin-induced ErbB3 downregulation is mediated by a protein stability cascade involving the E3 ubiquitin ligase Nrdp1. Mol Cell Biol. 2007;27(6):2180–8.
124. Haft CR, Klausner RD, Taylor SI. Involvement of dileucine motifs in the internalization and degradation of the insulin receptor. J Biol Chem. 1994;269(42):26286–94.
125. Hamer I, Haft CR, Paccaud JP, Maeder C, Taylor S, Carpentier JL. Dual role of a dileucine motif in insulin receptor endocytosis. J Biol Chem. 1997;272(35):21685–91.
126. Morrison P, Chung KC, Rosner MR. Mutation of Di-leucine residues in the juxtamembrane region alters EGF receptor expression. Biochemistry. 1996;35(46):14618–24.
127. Kelly BT, McCoy AJ, Spate K, Miller SE, Evans PR, Honing S, et al. A structural explanation for the binding of endocytic dileucine motifs by the AP2 complex. Nature. 2008;456(7224):976–9.
128. Backer JM, Shoelson SE, Weiss MA, Hua QX, Cheatham RB, Haring E, et al. The insulin receptor juxtamembrane region contains two independent tyrosine/beta-turn internalization signals. J Cell Biol. 1992;118(4):831–9.
129. Chen WJ, Goldstein JL, Brown MS. NPXY, a sequence often found in cytoplasmic tails, is required for coated pit-mediated internalization of the low density lipoprotein receptor. J Biol Chem. 1990;265(6):3116–23.
130. Wu H, Windmiller DA, Wang L, Backer JM. YXXM motifs in the PDGF-beta receptor serve dual roles as phosphoinositide 3-kinase binding motifs and tyrosine-based endocytic sorting signals. J Biol Chem. 2003;278(42):40425–8.

131. Traub LM. Tickets to ride: selecting cargo for clathrin-regulated internalization. Nat Rev Mol Cell Biol. 2009;10(9):583–96.
132. Sorkin A, Mazzotti M, Sorkina T, Scotto L, Beguinot L. Epidermal growth factor receptor interaction with clathrin adaptors is mediated by the Tyr974-containing internalization motif. J Biol Chem. 1996;271(23):13377–84.
133. Sorkin A, Carpenter G. Interaction of activated EGF receptors with coated pit adaptins. Science. 1993;261(5121):612–5.
134. Nesterov A, Carter RE, Sorkina T, Gill GN, Sorkin A. Inhibition of the receptor-binding function of clathrin adaptor protein AP-2 by dominant-negative mutant mu2 subunit and its effects on endocytosis. EMBO J. 1999;18(9):2489–99.
135. Boll W, Gallusser A, Kirchhausen T. Role of the regulatory domain of the EGF-receptor cytoplasmic tail in selective binding of the clathrin-associated complex AP-2. Curr Biol. 1995;5(10):1168–78.
136. Huang F, Jiang X, Sorkin A. Tyrosine phosphorylation of the beta2 subunit of clathrin adaptor complex AP-2 reveals the role of a di-leucine motif in the epidermal growth factor receptor trafficking. J Biol Chem. 2003;278(44):43411–7.
137. Frosi Y, Anastasi S, Ballaro C, Varsano G, Castellani L, Maspero E, et al. A two-tiered mechanism of EGFR inhibition by RALT/MIG6 via kinase suppression and receptor degradation. J Cell Biol. 2010;189(3):557–71.
138. Grandal MV, Grovdal LM, Henriksen L, Andersen MH, Holst MR, Madshus IH, et al. Differential roles of Grb2 and AP-2 in p38 MAPK- and EGF-induced EGFR internalization. Traffic. 2012;13:576.
139. Wang Q, Zhu F, Wang Z. Identification of EGF receptor C-terminal sequences 1005-1017 and di-leucine motif 1010LL1011 as essential in EGF receptor endocytosis. Exp Cell Res. 2007;313(15):3349–63.
140. Haugsten EM, Zakrzewska M, Brech A, Pust S, Olsnes S, Sandvig K, et al. Clathrin- and dynamin-independent endocytosis of FGFR3 – implications for signalling. PLoS One. 2011;6(7):e21708.
141. Lanahan AA, Hermans K, Claes F, Kerley-Hamilton JS, Zhuang ZW, Giordano FJ, et al. VEGF receptor 2 endocytic trafficking regulates arterial morphogenesis. Dev Cell. 2010;18(5):713–24.
142. Salani B, Passalacqua M, Maffioli S, Briatore L, Hamoudane M, Contini P, et al. IGF-IR internalizes with Caveolin-1 and PTRF/Cavin in HaCat cells. PLoS One. 2010;5(11):e14157.
143. Abella JV, Parachoniak CA, Sangwan V, Park M. Dorsal ruffle microdomains potentiate Met receptor tyrosine kinase signaling and down-regulation. J Biol Chem. 2010;285(32): 24956–67.
144. Yamazaki T, Zaal K, Hailey D, Presley J, Lippincott-Schwartz J, Samelson LE. Role of Grb2 in EGF-stimulated EGFR internalization. J Cell Sci. 2002;115(Pt 9):1791–802.
145. Orth JD, Krueger EW, Weller SG, McNiven MA. A novel endocytic mechanism of epidermal growth factor receptor sequestration and internalization. Cancer Res. 2006; 66(7):3603–10.
146. Valdez G, Philippidou P, Rosenbaum J, Akmentin W, Shao Y, Halegoua S. Trk-signaling endosomes are generated by Rac-dependent macroendocytosis. Proc Natl Acad Sci U S A. 2007;104(30):12270–5.
147. Barbieri MA, Roberts RL, Gumusboga A, Highfield H, Alvarez-Dominguez C, Wells A, et al. Epidermal growth factor and membrane trafficking. EGF receptor activation of endocytosis requires Rab5a. J Cell Biol. 2000;151(3):539–50.
148. Deininger K, Eder M, Kramer ER, Zieglgansberger W, Dodt HU, Dornmair K, et al. The Rab5 guanylate exchange factor Rin1 regulates endocytosis of the EphA4 receptor in mature excitatory neurons. Proc Natl Acad Sci U S A. 2008;105(34):12539–44.
149. Pitulescu ME, Adams RH. Eph/ephrin molecules – a hub for signaling and endocytosis. Genes Dev. 2010;24(22):2480–92.
150. Cowan CW, Shao YR, Sahin M, Shamah SM, Lin MZ, Greer PL, et al. Vav family GEFs link activated Ephs to endocytosis and axon guidance. Neuron. 2005;46(2):205–17.

151. Sigismund S, Argenzio E, Tosoni D, Cavallaro E, Polo S, Di Fiore PP. Clathrin-mediated internalization is essential for sustained EGFR signaling but dispensable for degradation. Dev Cell. 2008;15(2):209–19.
152. Sachse M, Urbe S, Oorschot V, Strous GJ, Klumperman J. Bilayered clathrin coats on endosomal vacuoles are involved in protein sorting toward lysosomes. Mol Biol Cell. 2002;13(4):1313–28.
153. Grant BD, Donaldson JG. Pathways and mechanisms of endocytic recycling. Nat Rev Mol Cell Biol. 2009;10(9):597–608.
154. Hanyaloglu AC, von Zastrow M. A novel sorting sequence in the beta2-adrenergic receptor switches recycling from default to the Hrs-dependent mechanism. J Biol Chem. 2007;282(5):3095–104.
155. Parachoniak CA, Luo Y, Abella JV, Keen JH, Park M. GGA3 functions as a switch to promote Met receptor recycling, essential for sustained ERK and cell migration. Dev Cell. 2011;20(6):751–63.
156. Puertollano R, Bonifacino JS. Interactions of GGA3 with the ubiquitin sorting machinery. Nat Cell Biol. 2004;6(3):244–51.
157. Teis D, Saksena S, Emr SD. SnapShot: the ESCRT machinery. Cell. 2009;137(1):182–e1.
158. Wollert T, Hurley JH. Molecular mechanism of multivesicular body biogenesis by ESCRT complexes. Nature. 2010;464(7290):864–9.
159. Henne WM, Buchkovich NJ, Emr SD. The ESCRT pathway. Dev Cell. 2011;21(1):77–91.
160. Eden ER, Huang F, Sorkin A, Futter CE. The role of EGF receptor ubiquitination in regulating its intracellular traffic. Traffic. 2012;13:329.
161. Ronning SB, Pedersen NM, Madshus IH, Stang E. CIN85 regulates ubiquitination and degradative endosomal sorting of the EGF receptor. Exp Cell Res. 2011;317(13):1804–16.
162. Dikic I, Schmidt MH. Malfunctions within the Cbl interactome uncouple receptor tyrosine kinases from destructive transport. Eur J Cell Biol. 2007;86(9):505–12.
163. Takayama Y, May P, Anderson RG, Herz J. Low density lipoprotein receptor-related protein 1 (LRP1) controls endocytosis and c-CBL-mediated ubiquitination of the platelet-derived growth factor receptor beta (PDGFR beta). J Biol Chem. 2005;280(18):18504–10.
164. Gur G, Rubin C, Katz M, Amit I, Citri A, Nilsson J, et al. LRIG1 restricts growth factor signaling by enhancing receptor ubiquitylation and degradation. EMBO J. 2004;23(16):3270–81.
165. Bache KG, Raiborg C, Mehlum A, Stenmark H. STAM and Hrs are subunits of a multivalent ubiquitin-binding complex on early endosomes. J Biol Chem. 2003;278(14):12513–21.
166. Jekely G, Rorth P. Hrs mediates downregulation of multiple signalling receptors in Drosophila. EMBO Rep. 2003;4(12):1163–8.
167. Bowers K, Piper SC, Edeling MA, Gray SR, Owen DJ, Lehner PJ, et al. Degradation of endocytosed epidermal growth factor and virally ubiquitinated major histocompatibility complex class I is independent of mammalian ESCRTII. J Biol Chem. 2006;281(8):5094–105.
168. Ewan LC, Jopling HM, Jia H, Mittar S, Bagherzadeh A, Howell GJ, et al. Intrinsic tyrosine kinase activity is required for vascular endothelial growth factor receptor 2 ubiquitination, sorting and degradation in endothelial cells. Traffic. 2006;7(9):1270–82.
169. Belleudi F, Leone L, Maggio M, Torrisi MR. Hrs regulates the endocytic sorting of the fibroblast growth factor receptor 2b. Exp Cell Res. 2009;315(13):2181–91.
170. Wurdinger T, Tannous BA, Saydam O, Skog J, Grau S, Soutschek J, et al. miR-296 regulates growth factor receptor overexpression in angiogenic endothelial cells. Cancer Cell. 2008; 14(5):382–93.
171. Bao J, Alroy I, Waterman H, Schejter ED, Brodie C, Gruenberg J, et al. Threonine phosphorylation diverts internalized epidermal growth factor receptors from a degradative pathway to the recycling endosome. J Biol Chem. 2000;275(34):26178–86.
172. Hellberg C, Schmees C, Karlsson S, Ahgren A, Heldin CH. Activation of protein kinase C alpha is necessary for sorting the PDGF beta-receptor to Rab4a-dependent recycling. Mol Biol Cell. 2009;20(12):2856–63.
173. Ying H, Zheng H, Scott K, Wiedemeyer R, Yan H, Lim C, et al. MIG-6 controls EGFR trafficking and suppresses gliomagenesis. Proc Natl Acad Sci U S A. 2010;107(15):6912–7.

174. Ren X, Hurley JH. VHS domains of ESCRT-0 cooperate in high-avidity binding to polyubiquitinated cargo. EMBO J. 2010;29(6):1045–54.
175. Roxrud I, Raiborg C, Pedersen NM, Stang E, Stenmark H. An endosomally localized isoform of Eps15 interacts with Hrs to mediate degradation of epidermal growth factor receptor. J Cell Biol. 2008;180(6):1205–18.
176. Duncan LM, Piper S, Dodd RB, Saville MK, Sanderson CM, Luzio JP, et al. Lysine-63-linked ubiquitination is required for endolysosomal degradation of class I molecules. EMBO J. 2006;25(8):1635–45.
177. Lauwers E, Jacob C, Andre B. K63-linked ubiquitin chains as a specific signal for protein sorting into the multivesicular body pathway. J Cell Biol. 2009;185(3):493–502.
178. Adhikari A, Chen ZJ. Diversity of polyubiquitin chains. Dev Cell. 2009;16(4):485–6.
179. Varadan R, Assfalg M, Haririnia A, Raasi S, Pickart C, Fushman D. Solution conformation of Lys63-linked di-ubiquitin chain provides clues to functional diversity of polyubiquitin signaling. J Biol Chem. 2004;279(8):7055–63.
180. McKanna JA, Haigler HT, Cohen S. Hormone receptor topology and dynamics: morphological analysis using ferritin-labeled epidermal growth factor. Proc Natl Acad Sci U S A. 1979;76(11):5689–93.
181. Carpentier JL, White MF, Orci L, Kahn CR. Direct visualization of the phosphorylated epidermal growth factor receptor during its internalization in A-431 cells. J Cell Biol. 1987;105:2751–62.
182. Eden ER, White IJ, Tsapara A, Futter CE. Membrane contacts between endosomes and ER provide sites for PTP1B-epidermal growth factor receptor interaction. Nat Cell Biol. 2010;12(3):267–72.
183. Ma YM, Boucrot E, Villen J, el Affar B, Gygi SP, Gottlinger HG, et al. Targeting of AMSH to endosomes is required for epidermal growth factor receptor degradation. J Biol Chem. 2007;282(13):9805–12.
184. Tanaka N, Kaneko K, Asao H, Kasai H, Endo Y, Fujita T, et al. Possible involvement of a novel STAM-associated molecule “AMSH” in intracellular signal transduction mediated by cytokines. J Biol Chem. 1999;274(27):19129–35.
185. McCullough J, Clague MJ, Urbe S. AMSH is an endosome-associated ubiquitin isopeptidase. J Cell Biol. 2004;166(4):487–92.
186. McCullough J, Row PE, Lorenzo O, Doherty M, Beynon R, Clague MJ, et al. Activation of the endosome-associated ubiquitin isopeptidase AMSH by STAM, a component of the multivesicular body-sorting machinery. Curr Biol. 2006;16(2):160–5.
187. Liu H, Buus R, Clague MJ, Urbe S. Regulation of ErbB2 receptor status by the proteasomal DUB POH1. PLoS One. 2009;4(5):e5544.
188. Pareja F, Ferraro DA, Rubin C, Cohen-Dvashi H, Zhang F, Aulmann S, et al. Deubiquitination of EGFR by Cezanne-1 contributes to cancer progression. Oncogene. 2012;31:4599.
189. Urbe S, McCullough J, Row P, Prior IA, Welchman R, Clague MJ. Control of growth factor receptor dynamics by reversible ubiquitination. Biochem Soc Trans. 2006;34(Pt 5):754–6.
190. Mizuno E, Kobayashi K, Yamamoto A, Kitamura N, Komada M. A deubiquitinating enzyme UBPY regulates the level of protein ubiquitination on endosomes. Traffic. 2006;7(8):1017–31.
191. Mizuno E, Iura T, Mukai A, Yoshimori T, Kitamura N, Komada M. Regulation of epidermal growth factor receptor down-regulation by UBPY-mediated deubiquitination at endosomes. Mol Biol Cell. 2005;16(11):5163–74.
192. Row PE, Prior IA, McCullough J, Clague MJ, Urbe S. The ubiquitin isopeptidase UBPY regulates endosomal ubiquitin dynamics and is essential for receptor down-regulation. J Biol Chem. 2006;281(18):12618–24.
193. Niendorf S, Oksche A, Kisser A, Lohler J, Prinz M, Schorle H, et al. Essential role of ubiquitin-specific protease 8 for receptor tyrosine kinase stability and endocytic trafficking in vivo. Mol Cell Biol. 2007;27(13):5029–39.
194. Cao TT, Mays RW, von Zastrow M. Regulated endocytosis of G-protein-coupled receptors by a biochemically and functionally distinct subpopulation of clathrin-coated pits. J Biol Chem. 1998;273:24592–602.

195. Confalonieri S, Salcini AE, Puri C, Tacchetti C, Di Fiore PP. Tyrosine phosphorylation of Eps15 is required for ligand-regulated, but not constitutive, endocytosis. J Cell Biol. 2000;150(4):905–12.
196. Connolly JL, Green SA, Greene LA. Comparison of rapid changes in surface morphology and coated pit formation of PC12 cells in response to nerve growth factor, epidermal growth factor, and dibutyryl cyclic AMP. J Cell Biol. 1984;98(2):457–65.
197. Irie F, Okuno M, Pasquale EB, Yamaguchi Y. EphrinB-EphB signalling regulates clathrin-mediated endocytosis through tyrosine phosphorylation of synaptojanin 1. Nat Cell Biol. 2005;7(5):501–9.
198. White IJ, Bailey LM, Aghakhani MR, Moss SE, Futter CE. EGF stimulates annexin 1-dependent inward vesiculation in a multivesicular endosome subpopulation. EMBO J. 2006;25(1):1–12.
199. Goh LK, Huang F, Kim W, Gygi S, Sorkin A. Multiple mechanisms collectively regulate clathrin-mediated endocytosis of the epidermal growth factor receptor. J Cell Biol. 2010;189(5):871–83.

Chapter 8
Structural Features of the Receptor Tyrosine Kinase Ectodomains

Michael C. Lawrence and Colin W. Ward

8.1 The Modular Structure of the RTK Ectodomains

Fifty-eight receptor tyrosine kinases (RTKs, EC 2.7.10.1) occur in the human genome, defined by their sequence containing an extracellular region ("ectodomain"), a transmembrane segment and an intracellular tyrosine kinase domain [1]. These fifty-eight RTKs are divided into twenty phylogenic families, distinguished from each other largely on the basis of the primary structure of their ectodomains [2–4]. The salient feature of the RTK ectodomain is its modular domain architecture (Fig. 8.1). The variation in type, number and order of the constituent ectodomain domains most likely arose through exon shuffling, gene duplication and/or gene loss events in the early stages of vertebrate evolution [5–7] and, indeed, phylogenic classification of RTKs corresponds closely to their gene intron/exon structure [1]. The expansion of the RTK superfamily was relatively rapid, with the ancestral RTK occurring earlier than Metazoa. The first RTK to be detected outside of Metazoa was that of choanoflagellate *Monosiga brevicollis* [8]. This protein displays a similar modular architecture to that found in the modern vertebrate RTKs and appears closely related to the fibroblast growth factor receptor family (FGFR) and to the angiopoietin receptor family (Tie).

All RTKs are monomeric, single-chain polypeptides, with the exception of those of the insulin receptor (InsR) family and of Met. Members of the InsR family are

M.C. Lawrence, Ph.D. (✉)
Walter+Eliza Hall Institute of Medical Research, 1G Royal Parade,
Parkville, VIC 3052, Australia

Department of Medical Biology, University of Melbourne, Royal Parade,
Parkville, VIC, Australia
e-mail: lawrence@wehi.edu.au

C.W. Ward, Ph.D.
Walter+Eliza Hall Institute of Medical Research, 1G Royal Parade,
Parkville, VIC 3052, Australia

D.L. Wheeler, Y. Yarden (eds.), *Receptor Tyrosine Kinases: Structure, Functions and Role in Human Disease*, DOI 10.1007/978-1-4939-2053-2_8

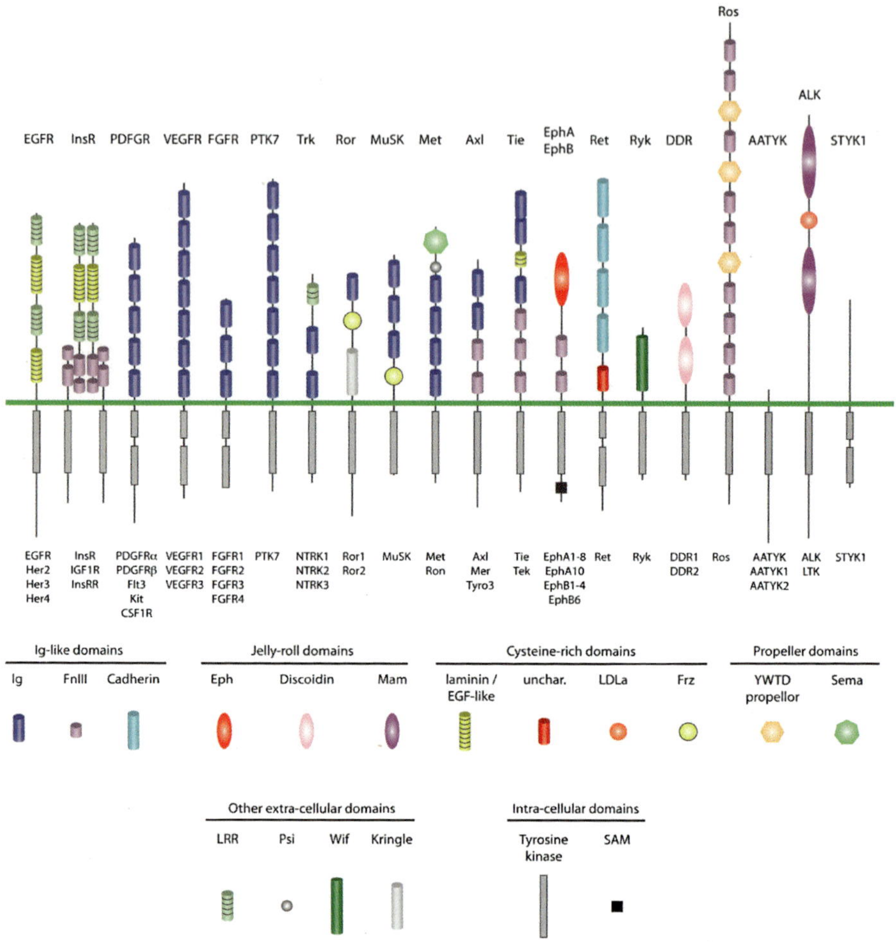

Fig. 8.1 Modular domain architecture of the RTK family ectodomains. Adapted from Figure 1 of [2], used by permission

disulphide-linked $(\alpha/\beta)_2$ homodimers [9] and Met is a disulphide-linked α/β monomer [10]. The size of RTK ectodomains vary considerably across the superfamily, ranging from fewer than 40 residues in the human apoptosis-associated tyrosine kinase family (AATYK—also referred to as lemur tyrosine kinase (LMR1 or LMTK) family) to more than 1,830 residues in the human proto-oncogene tyrosine-protein kinase (Ros).

The individual modules that comprise the RTK ectodomains are “domains,” in the sense of being spatially distinct structural units capable of folding independently [11]. Domain types are traditionally categorized by fold topology, disulphide bond arrangement and/or core motifs of invariant (or highly conserved) residues that reflect a common evolutionary relationship. Sixteen different domain types occur within the ectodomains of the human RTK superfamily (Fig. 8.1). The most common of these are of the immunoglobulin-like (Ig-like) fold type, including Ig domains themselves [12], fibronectin Type III (FnIII) domains [13] and cadherin domains [14]. The second most common domain type is of the jelly-roll or Swiss-roll fold topology [15]; domains of this type include the ephrin-binding domain [16], the discoidin domain [17, 18] and the meprin/A-5 protein receptor/protein tyrosine phosphatase μ ("Mam") domain [19]. The remainder of domain types occur in more restricted subsets of the RTKs [2]—these include leucine-rich repeat domains [20], laminin/EGF-like cysteine-rich domains [11], and other cysteine-rich domain types (the low-density lipoprotein receptor class A domains [21], frizzled domains [22] and the as-yet-uncharacterized cysteine-rich domains of the Ret family [23]), kringle domains [24], Wnt-inhibitory-factor domains [25], YWTD propeller domains [26, 27], plexin-semaphorin-integrin ("Psi") domains [28] and semaphorin domains [29].

The overall conformation adopted by the RTK ectodomain assembly varies considerably across the superfamily, with the linear domain arrangement shown in Fig. 8.1 being misleading in some cases. Three-dimensional structures that have been determined to date (see Tables 8.1 and 8.2) have shown that ectodomains are typically more extended than compact, facilitated in part by the fact that the majority of the constituent domain topologies are such that their respective N- and C-termini are positioned at opposite ends of the module. Examples of extended ectodomain structures include the ephrin receptor EphA2, the stem cell factor receptor Kit and the nerve growth factor receptor NTrk1 [30–32]. However, more compact conformations also exist. InsR and members of the epidermal growth factor receptor (EGFR) family exhibit a folded-over conformation in their *apo* form, with members of the latter family adopting an extended structure on binding ligand [33]. The six N-terminal domains (denoted Ig1, Ig2, EGF1, EGF2, EGF3 and Ig3, respectively) of the ectodomain of the angiopoietin receptor Tek (also termed Tie2) have a compact structure, with the Ig1 domain folded back over the Ig2, Ig3, EGF1 and EGF2 domains [34]. How these domains are positioned with respect to the remaining three (fibronectin) modules of the Tek ectodomain is not yet known. The linker polypeptide segments between ectodomain modules are in general short, often comprising no more than one or two residues and resulting in the consecutive domains abutting each other. This arrangement reduces (or eliminates) conformational flexibility at the domain-domain junction. An extended ectodomain structure offers the propensity for large-scale conformational change—allowing, for example, the relevant modules of the ectodomain or its bound ligand to reposition to interact with membrane-anchored co-receptors.

8.2 The Domain Architectures

This section describes in more detail the individual domain architectures that occur within the RTK ectodomains and serves as nonspecialist structural biological background for *Volume II*. The section provides for each domain type both an illustration of the fold topology and a three-dimensional cartoon of a representative example. A common representation format and colour scheme has been adopted for easy cross-comparison (see Box 8.1). For further detail, the reader is encouraged to consult the many electronic databases of domain and protein structure—in particular, the PDB [35], DALI [36], PFAM [37], SMART [38] and SCOP [39].

Box 8.1

In Panels (a) of Figs. 8.2, 8.3, 8.4, 8.5, 8.6, 8.7, 8.8, 8.9, 8.10, 8.11, 8.12, 8.13, 8.14, 8.15, 8.16, and 8.17, β-strands are represented as solid arrows, with the arrowhead at the C-terminal end of each strand. The strands within each β-sheet are shown in the same colour. Helical secondary structure elements, where they are judged to be regularly occurring elements of the fold within the RTK superfamily, are shown as blue rectangles. Dashed red lines indicate disulphide bonds. Polypeptide N- and C-termini are represented by a solid blue dot and a solid red dot, respectively. Polypeptide segments interconnecting secondary structure elements are shown as black lines. Where appropriate, the conventional numbering or lettering of individual sheets, strands or helices is also given. Given the large variation across the RTK superfamily, no attempt has been made to place the various components of each fold on a common scale.

Panels (b) of Figs. 8.2, 8.3, 8.4, 8.5, 8.6, 8.7, 8.8, 8.9, 8.10, 8.11, 8.12, 8.13, 8.14, 8.15, 8.16, and 8.17 are cartoons of the three-dimensional structure of representative examples of the respective domain types found within the RTK superfamily. These are shown as ribbon diagrams, in the style original developed Jane Richardson [93]. The colour scheme of these elements follows that of the associated Panel (a). The side chains of cysteine residues in involved in disulphide bonds are shown in stick representation (red), while polypeptide loop segments that are disordered in the experimentally derived structures are indicated by dashed black lines. These Panels were generated using Dino3D [94] and rendered using POVRAY [95].

8.2.1 The Immunoglobulin (Ig) Domain

The Ig domain is the most common domain type found in the RTK ectodomains (Fig. 8.1), occurring in the platelet-derived growth factor receptor (PDGFR) family, vascular epithelial growth factor receptor (VEGFR) family, protein tyrosine kinase 7 (PTK7), nerve growth factor receptor (Trk) family, receptor tyrosine kinase-like orphan receptor (Ror) family, muscle-specific kinase (MuSK) family, Met family, Axl family and Tie family. As their name implies, Ig domains are also found in the proteins of the immune system, which exhibit a similar modular architecture to that seen in the RTKs [40]. Ig domains are two β-sheet sandwich structures with an anti-parallel, Greek-key topology, with the two sheets often connected by a disulphide bond. Ig domains can be further subcategorized on the basis of the location and length of their edge-of-sheet strands [41–44]; however, the core structure of all Ig domains is recognised to consist of strands B and E in one sheet and strands F and C in the other [12]. If a D strand exists, then it is associated with the sheet B and E. Ig domains within the RTK superfamily most commonly belong to the so-called

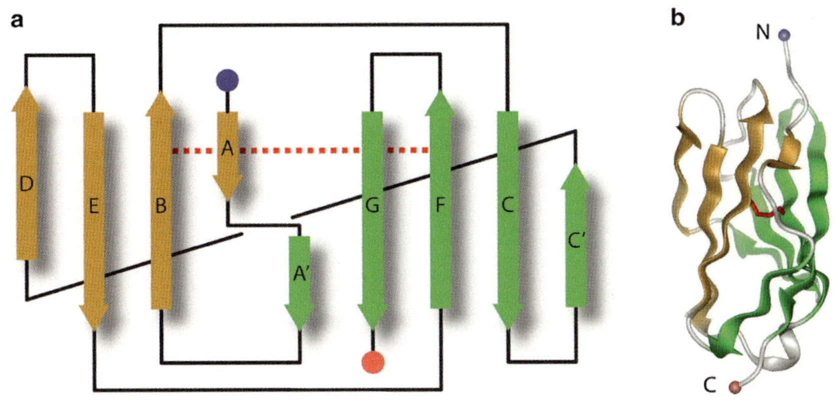

Fig. 8.2 Architecture of immunoglobulin (Ig) domains. (**a**) Topology diagram of an (I-set) Ig domain. (**b**) Three-dimensional cartoon representation of the (I-set) C-terminal Ig domain of Tek [34], PDB entry 2GY5. The depiction style is described in Box 8.1

I-set [41] of the Ig domain superfamily, which have one β-sheet consisting of strands A, B, E and D and the other of strands A′, G, F, C and C′. The canonical topology of an I-set Ig domain is shown in Fig. 8.2a and the three-dimensional fold of a representative Ig domain, namely, that of the third Ig domain of Tek—an I-set type [34]—is shown in Fig. 8.2b.

8.2.2 The Fibronectin Type III (FnIII) Domain

The FnIII domain [11, 13] is related to the Ig domain in that it is also a two β-sheet sandwich structure based on an antiparallel, Greek-key topology. However, while both Ig and FnIII domains have a common core consisting of strands B and E in one of the sheets and strands C, F and G in the other, the D strand in an FnIII domain is relocated to the opposite sheet to that which is located in an Ig domain. In the FnIII domain, the D strand is denoted as strand C′ (compare Figs. 8.2a and 8.3a). Members of the InsR family each contain three FnIII domains, members of the Axl family two, members of the Tie family three and Ros nine (Fig. 8.1). An example of the three-dimensional fold of an RTK FnIII domain—namely, that of the first FnIII domain of InsR [45]—is shown in Fig. 8.3b.

8.2.3 The Cadherin Domain

The cadherin ("calcium-dependent adhesion") domain is a two β-sheet sandwich Ig superfamily member (Fig. 8.4a). It exhibits an antiparallel Greek-key topology with one sheet consisting of strands A, G, F, C, C′ and C′′ and the second of strands

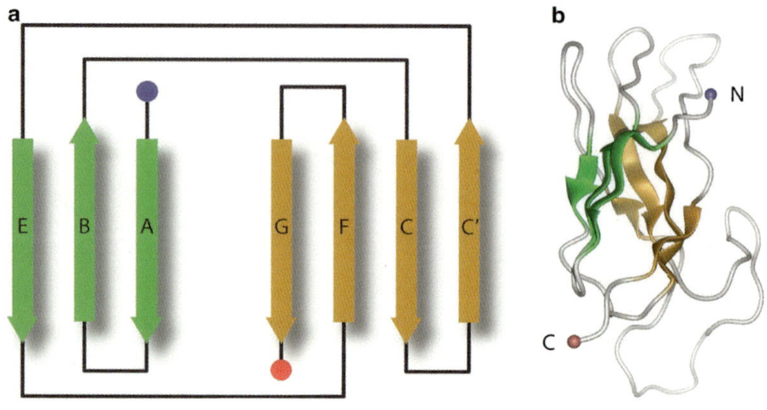

Fig. 8.3 Architecture of fibronectin Type III (FnIII) domains. (**a**) Topology diagram of an FnIII domain. (**b**) Three-dimensional cartoon representation of the first FnIII domain of InsR [45, 96]. PDB entry 3LOH. The depiction style is described in Box 8.1

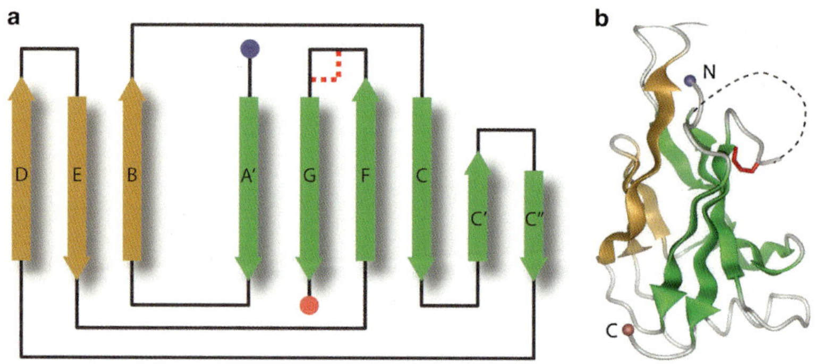

Fig. 8.4 Architecture of cadherin domains. (**a**) Topology diagram of a cadherin domain. (**b**) Three-dimensional cartoon representation of the N-terminal cadherin domain of Ret [46]. PDB entry 2X2U. The depiction style is described in Box 8.1

B, E and D [14]. As its name suggests, cadherin domains occur in a wide variety of proteins involved in cell-cell adhesion. The cadherin domains occur in only one RTK, namely, Ret ("rearranged during transfection"), which contains four cadherin modules. The three-dimensional fold of the N-terminal cadherin domain of Ret [46] is shown in Fig. 8.4b.

8.2.4 The Ephrin-Binding (Eph) Domain

The Eph domain [16] has a jelly-roll β-sandwich folding topology [15]. It is comprised of eleven strands, with the first β-sheet consisting (in order) of strands D, A/E, M, G, K and J and the second β-sheet consisting (in order) of strand C, F/F′, L,

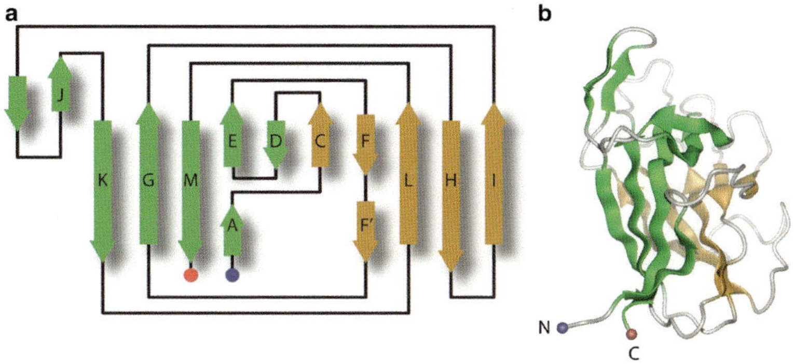

Fig. 8.5 Architecture of ephrin-binding domains. (**a**) Topology diagram of an ephrin-binding domain. (**b**) Three-dimensional cartoon representation of the N-terminal ephrin-binding domain of EphB4 [97]. PDB entry 2HLE. The depiction style is described in Box 8.1

H and I, where the "/" indicates a split strand (Fig. 8.5a). A single Eph domain occurs in each member of the ephrin-interacting Type A and Type B receptors (namely, EphA1-8, EphA10, EphB1-4 and EphB6) [47]. An example of the three-dimensional fold of an ephrin-binding domain—namely, that of EphB4 [48]—is shown in Fig. 8.5b.

8.2.5 *The Discoidin Domain*

The discoidin domain, like the ephrin-binding domain, has a jelly-roll folding topology but is smaller in size than the latter. The first β-sheet of the discoidin domain consists (in order) of strands 1, 2, 7, 4 and 5 and the second β-sheet consists (in order) of strands 8, 3 and 6. A disulphide bond links the N- and C-terminal segments of the domain and a further disulphide bond links the loop between strands 3 and 4 to the loop between strands 7 and 8 (Fig. 8.6a) [17, 18]. Two discoidin domains occur in both discoidin receptors 1 and 2 (DDR1 and DDR2); they are not found elsewhere within the RTK superfamily. The three-dimensional fold of the N-terminal discoidin domain of DDR2 [18] is shown in Fig. 8.6b.

8.2.6 *The Mam Domain*

The Mam domain is the third domain type within the RTK superfamily that is based on the jelly-roll folding topology. It occurs in one RTK family, namely, the anaplastic lymphoma kinase (ALK) family, which comprises ALK and the leukocyte receptor tyrosine kinase (LTK). These receptors both have two Mam domains. No

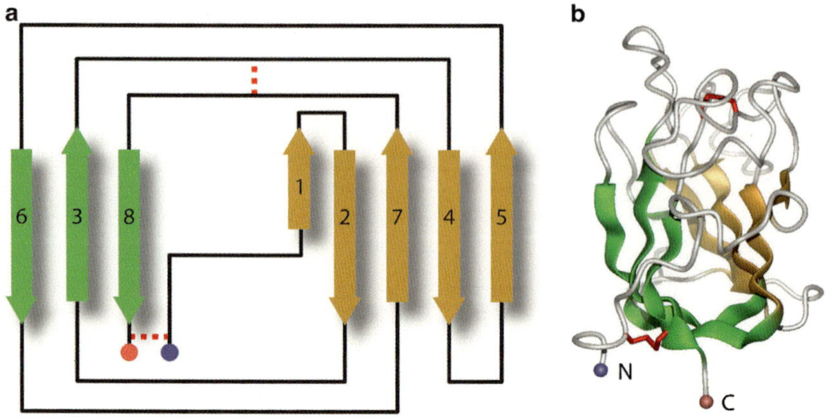

Fig. 8.6 Architecture of discoidin domains. (**a**) Topology diagram of a discoidin domain. (**b**) Three-dimensional cartoon representation of the N-terminal discoidin domain of DDR2 [17]. PDB entry 2WUH. The depiction style is described in Box 8.1

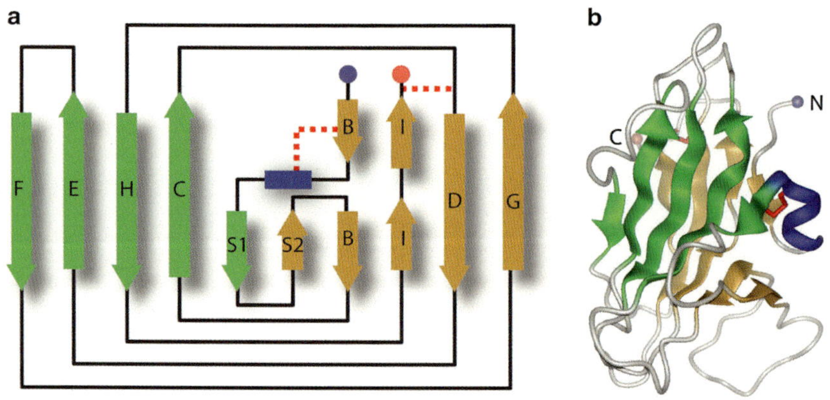

Fig. 8.7 Architecture of Mam domains. (**a**) Topology diagram of a Mam domain. (**b**) Three-dimensional cartoon representation of the Mam domain from receptor protein tyrosine phosphatase μ [19]. PDB entry 2V5Y. The depiction style is described in Box 8.1

three-dimensional structures of ectodomain fragments of either of these receptors are currently available. The topology of the Mam domain from the receptor protein tyrosine phosphatase μ [19] is thus illustrated here (Fig. 8.7a). It consists of ten strands arranged into two sheets, the first of which consists (in order) of strands S2, B, I, D and G and the second of which consists (in order) of strands S1, C, H, E and F. Disulphide bonds link the N-terminal portion of the split strand B to a small helical segment immediately upstream of the short strand S1 and a further disulphide bond links the loop between strands C and D to the C-terminus of the domain. The three-dimensional fold is shown in Fig. 8.7b.

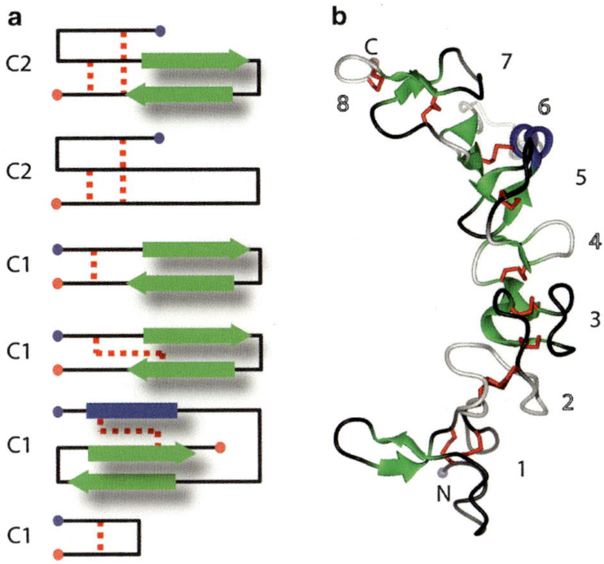

Fig. 8.8 Architecture of laminin-/epidermal growth factor (EGF)-type cysteine-rich domains. (**a**) Examples of the secondary structure and disulphide topology of C1 and C2 laminin-/EGF-type cysteine-rich domains, based on those found within the cysteine-rich domain of IGF-1R [50]. (**b**) Three-dimensional cartoon representation of the intact cysteine-rich domain of IGF-1R, showing its assembly from eight laminin-/EGF-type cysteine-rich domains [50]. Random coil regions of modules 1, 3, 5 and 7 are shown in a darker shade for clarity. PDB entry 1IGR. The depiction style is described in Box 8.1

8.2.7 Laminin/Epidermal Growth Factor-Like (EGF-Like) Cysteine-Rich Domains

There are at least four types of cysteine-rich domains: the laminin/EGF-like type, the LDLa (low-density lipoprotein class A repeat) domain, the frizzled type and an as-yet-uncharacterized type found in Ret. The laminin/EGF-like cysteine-rich domains occur in the InsR, EGFR and Tie receptor families of RTKs. They exhibit an irregular architecture, made up of sub-modules C1 and C2 that contain one or two disulphide bonds, respectively. The disulphide connectivity in the C2 sub-module is 1-3, 2-4, similar to that seen in laminin, EGF, EGFR and the tumour necrosis factor (TNF) receptor [23, 49]. The cysteine-rich domain of laminin has a C2-C1-C1 sub-module arrangement. InsR and EGFR family members each contain a single cysteine-rich domain (between their respective L1 and L2 leucine-rich repeat domains) that consists of eight such sub-modules in the order C2-C2-C2-C1-C1-C1-C1-C1′. The EGFR family members contain also a second cysteine-rich domain that consists of seven such sub-modules in a different order, namely, C2-C1-C1-C2-C1-C1-C2 [50]. Examples of the topologies of the sub-modules of a cysteine-rich domain, namely, those that occur within IGF-1R [50], are shown in Fig. 8.8a. Their assembly into an intact cysteine-rich domain is shown in Fig. 8.8b.

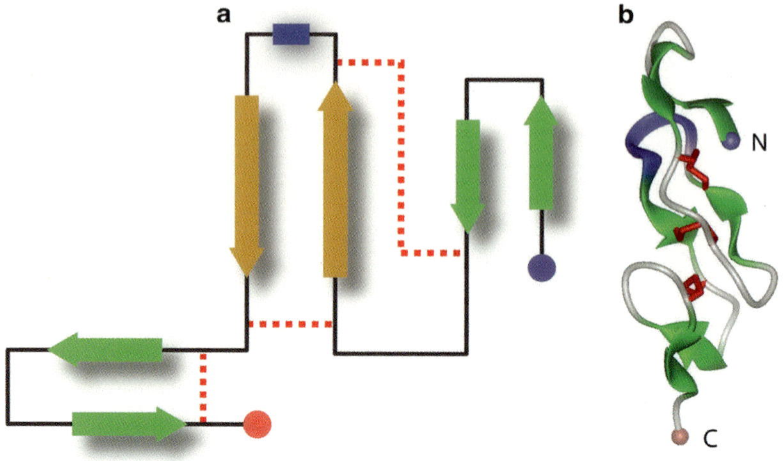

Fig. 8.9 Architecture of C2-C1 epidermal growth factor (EGF)-type modules. (**a**) Topology diagram of a C2-C1 EGF-type domain. (**b**) Three-dimensional cartoon representation of Tek [34]. PDB entry 2GY5. The depiction style is described in Box 8.1

The ectodomains of Tie1 and Tek each contain three C2-C1 EGF-like modules, which consist of a major two-strand, antiparallel β-sheet stabilized by three disulphide bonds in a 1-3, 2-4 (C2), 5-6 (C1) pattern [11]. In the case of Tek, the core of the EGF domains is elaborated by an additional smaller, two-strand antiparallel β-sheet at its N-terminal end and an α-helix [34]. The fold topology of the C2-C1 module pair of the N-terminal EGF-like domain of Tek and its three-dimensional structure are shown in Fig. 8.9a, b, respectively.

8.2.8 The LDLa Domain

The LDLa (low-density lipoprotein class A repeat) domain is of the cysteine-rich type. It contains three-disulphide bridges with a 1-3, 2-5, 4-6 pairing (Fig. 8.10a) and a highly conserved cluster of negatively charged amino acids, many of which are clustered on one face of the module domain [21]. LDLa domains occur only within the ALK family of the RTKs, with a single copy of the domain within both ALK and LTK. No three-dimensional structures are available for these domains from either of these receptors. An example of the three-dimensional structure of an LDLa domain, namely, that from the LDL receptor [21], is shown in Fig. 8.10b.

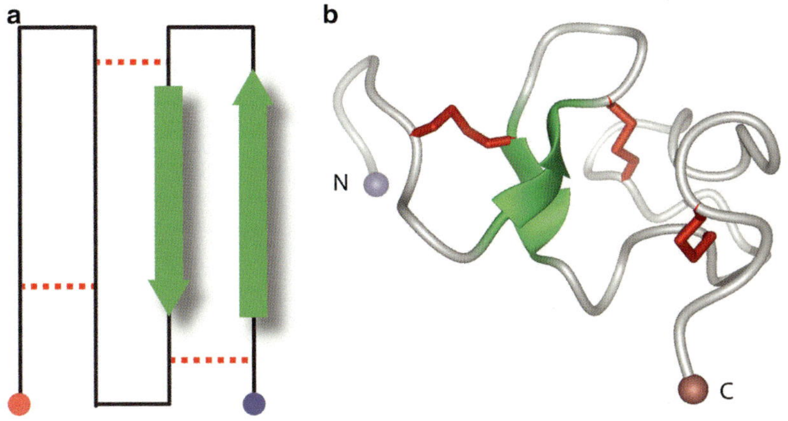

Fig. 8.10 Architecture of low-density lipoprotein receptor A (LDLa) cysteine-rich domains. (**a**) Topology diagram of an LDLa domain. (**b**) Three-dimensional cartoon representation of the N-terminal LDLa cysteine-rich domain of low-density lipoprotein receptor [21]. PDB entry 1LDL. The depiction style is described in Box 8.1

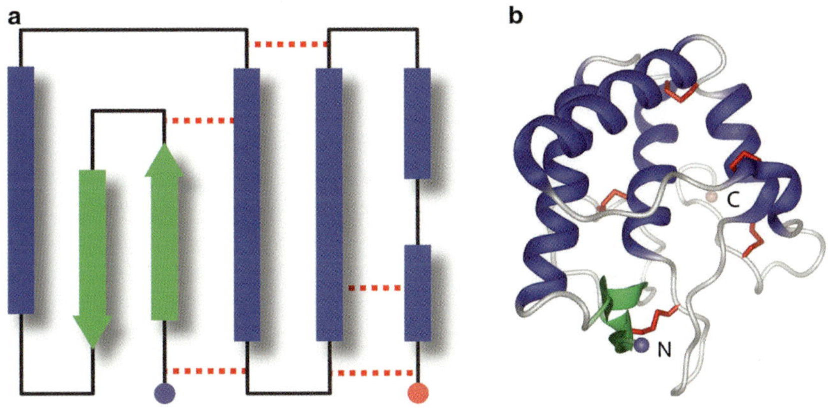

Fig. 8.11 Architecture of frizzled cysteine-rich domains. (**a**) Topology diagram of a frizzled domain. (**b**) Three-dimensional cartoon representation of the frizzled domain of MuSK [52]. PDB entry 3HKL. The depiction style is described in Box 8.1

8.2.9 The Frizzled Domain

The frizzled domain is a cysteine-rich domain [22, 51], the cysteines of which form five disulphide bridges with a defined spacing and pattern, interspersed with predominantly α-helical secondary structure elements (Fig. 8.11a). A single frizzled domain occurs in each member of the MuSK and ROR families (Fig. 8.1). An example of the three-dimensional structure of a frizzled domain, namely, that of MuSK [52], is shown in Fig. 8.11b.

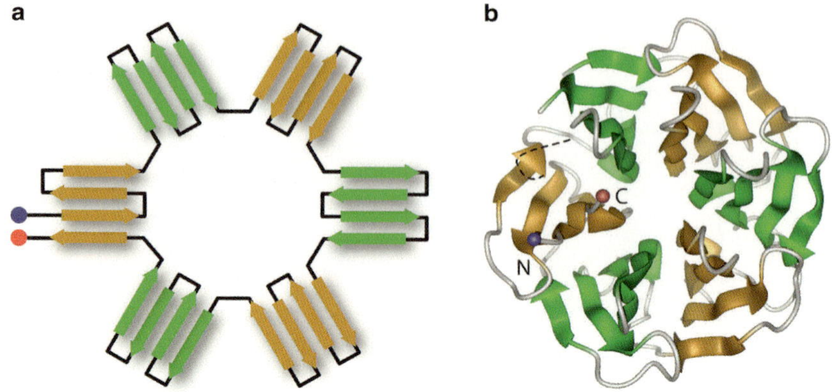

Fig. 8.12 Architecture of YWTD propeller domains. (**a**) Topology diagram of a YWTD propeller domain. (**b**) Three-dimensional cartoon representation of the YWTD propeller domain of LDLR-A [27]. PDB entry 1IJQ. The depiction style is described in Box 8.1

8.2.10 The YWTD Propeller Domain

The first structure of a YWTD (Tyr-Trp-Thr-Asp) propeller domain to be determined was that of the low-density lipoprotein receptor class A (LDLR-A) [26, 27], wherein it was shown to have six blades, each of which is a four-stranded, antiparallel β-sheet consisting of about 43 residues. Close to the N-terminus of each sheet is the four-residue YWTD consensus motif. YWTD propellers are abundant across the broader class of extracellular receptors, but occur in only a single RTK, namely, the proto-oncogene receptor tyrosine kinase Ros, which contains three YWTD propeller domains. No three-dimensional structure exists for any of these domains of Ros. The topology of the YWTD propeller domain as found in LDLR-A is shown in Fig. 8.12a and its three-dimensional structure in Fig. 8.12b [27].

8.2.11 The Semaphorin (Sema) Domain

The Sema domain is a seven-stranded β-propeller, with each blade consisting of a four-stranded, antiparallel β-sheet (Fig. 8.13a). Its name reflects its occurrence in semaphorins [29]. Within the RTK superfamily, a single Sema domain occurs in both members of the Met family—Met and the *recepteur d'origine nantais* (Ron). In both receptors, the Sema domain contains seven disulphide bonds as well as a furin processing site between the fourth and fifth blade of the propeller [53, 54]. The three-dimensional structure of the Sema domain of Met [53] is illustrated in Fig. 8.13b.

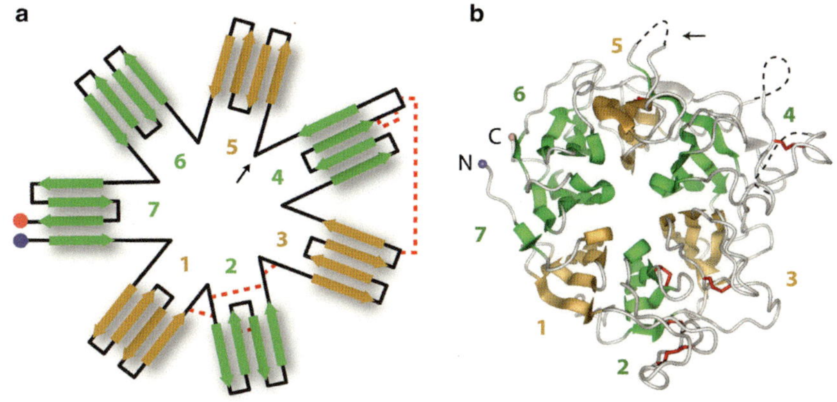

Fig. 8.13 Architecture of semaphorin (Sema) domains. (**a**) Topology diagram of a Sema domain. (**b**) Three-dimensional cartoon representation of the Sema domain of the Met [53]. The location of the proteolytic cleavage site between the fourth and fifth β-sheet is *arrowed*. PDB entry 1SHY. The depiction style is described in Box 8.1

8.2.12 The Leucine-Rich Repeat (LRR) Domain

The LRR domain is a single-stranded, right-handed β-helix (Fig. 8.14a). InsR and EGFR family members each contain two LRR domains (termed L1 and L2, respectively) separated by a cysteine-rich domain. The L1 and L2 LRR domains of InsR and EGFR each contain five leucine-rich repeat motifs plus one partial leucine-rich repeat motif [20], with each leucine-rich repeat motif corresponding to one turn of the β-helix. The N- and C-terminal turns of the β-helix form capping elements to the β-helix, with a disulphide bond linking the respective terminal turn of the β-helix to that immediately adjacent. These capping elements also contain either one or two α-helical segments. The β-helix is roughly rectangular in cross section, with canonical β-sheet assembly occurring within three of the four surfaces. LRR domains also occur within the Trk family. In these latter receptors, there is single LRR domain, which consists of only three leucine-rich repeat motifs, with the assembled barrel having approximately two-and-a-half turns [32]. The capping elements of the Trk LRR domains also contain disulphide bonds, but, unlike those in the InsR and EGFR families, they do not connect adjacent turns of the β-helix, providing instead connections to residues immediately adjacent to the LRR domain [32]. In all three families (InsR, EGFR and Trk), the leucine-rich character of the LRR domains is associated with the close packing of leucine and other hydrophobic residue side chains into the core of the β-helix. An example of the three-dimensional fold of an LRR domain, namely, that of the N-terminal LRR domain of InsR [55], is shown in Fig. 8.14b.

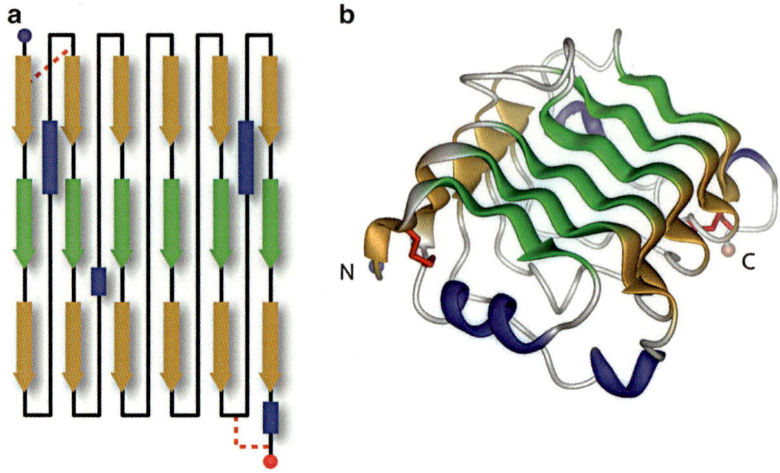

Fig. 8.14 Architecture of leucine-rich-repeat domains. (**a**) Topology diagram of a leucine-rich-repeat domain. (**b**) Three-dimensional cartoon representation of the N-terminal leucine-rich-repeat domain of InsR [55]. PDB entry 2HR7. The depiction style is described in Box 8.1

8.2.13 The Psi Domain

The Psi domain is found in many signalling proteins—its name reflects its occurrence in *p*lexins, *s*emaphorins and *i*ntegrins [28]. Like the Sema domain, it occurs within the Met family of RTKs, with each member containing a single copy (Fig. 8.1). Psi domains consist of a three-stranded, antiparallel β-sheet and two short α-helices, cross-linked by up to four disulphide bonds as shown in Fig. 8.15a. The three-dimensional solution structure of the Psi domain of Met [56] is illustrated in Fig. 8.15b.

8.2.14 The Wnt-Inhibitory-Factor (Wif) Domain

The Wif domain occurs in a single RTK, namely, related-to-tyrosine-kinase (RYK), and only in single copy [25]. It is related in structure to Wnt-inhibitory-factor 1 (Wif-1), the structure of which has been determined [57] and shown to exhibit an immunoglobulin-like fold with nine β-strands. In the case of Wif-1, two of the strands (6 and 9) are interrupted by elements of irregular secondary structure, while a single α-helix occurs in the loops strands 2 and 3—whether these elaborations are present in the Wif domain of RYK is unknown. The topology of the Wif domain as it exists in Wif-1 is shown in Fig. 8.16a and its three-dimensional structure in Fig. 8.16b [57].

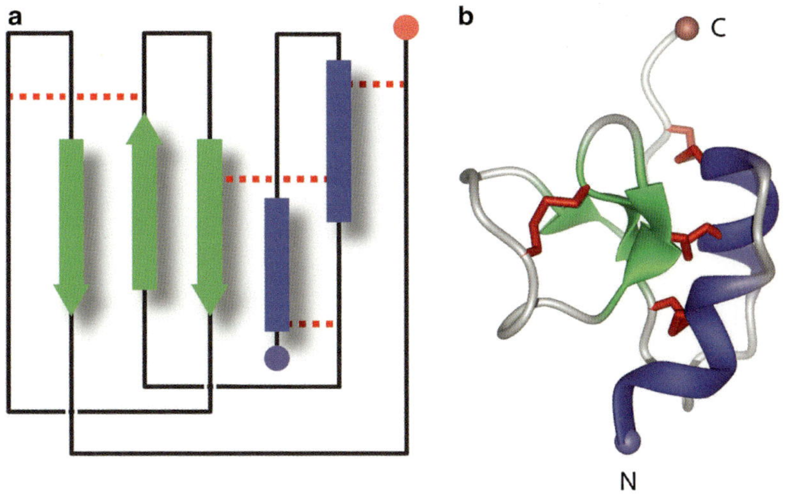

Fig. 8.15 Architecture of Psi domains. (**a**) Topology diagram of a Psi domain. (**b**) Three-dimensional cartoon representation of the Psi domain of Met [56]. PDB entry 1SHY. The depiction style is described in Box 8.1

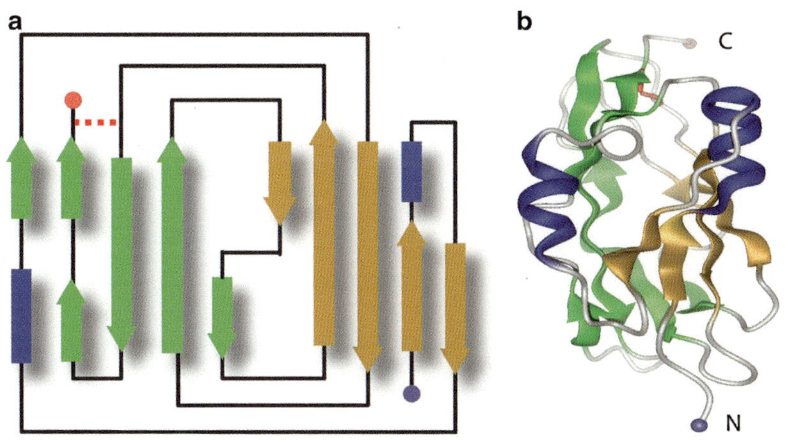

Fig. 8.16 Architecture of Wnt-inhibitory-factor (Wif) domains. (**a**) Topology diagram of a Wif domain. (**b**) Three-dimensional cartoon representation of Wnt-inhibitory-factor 1 [57]. PDB entry 2D3J. The depiction style is described in Box 8.1

8.2.15 The Kringle Domain

The kringle domain occurs in the ROR family of RTKs and in each member only in single copy. Kringle domains are common in many proteins related to blood clotting [24]. The defining motif has three disulphide bonds and three loops, with one loop containing a two-stranded, antiparallel β-sheet (Fig. 8.17a). Two of the

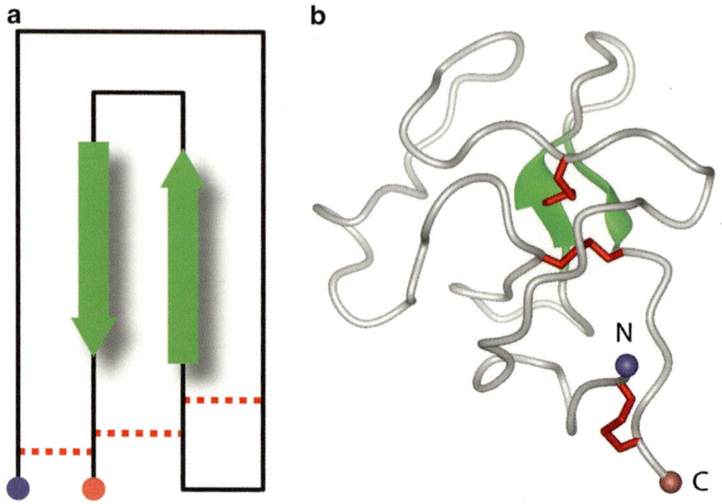

Fig. 8.17 Architecture of kringle domains. (**a**) Topology diagram of a kringle domain. (**b**) Three-dimensional cartoon representation of the kringle domain from *apo*-lipoprotein A [58]. PDB entry 1I71. The depiction style is described in Box 8.1

disulphide bonds are located in close proximity at the core of the fold. No structure exists for the kringle domains of either ROR1 or ROR2; an example of the three-dimensional structure of a kringle domain, namely, that of apolipoprotein A [58], is shown in Fig. 8.17b.

8.3 Mechanisms of Receptor Activation

8.3.1 Multiple Modes of Ligand Binding and Receptor Activation

The varying, multi-domain architecture of the RTKs reflects not only the wide variety of ligands that signal into the superfamily but also the wide variety of modes of ligand binding and mechanisms of receptor activation [2]. A unifying theme for most RTKs (where sufficiently characterized) is that ligand binding leads to receptor dimerization, and that this process initiates signal transduction, which results in activation of the intracellular tyrosine kinase domains. The InsR family is an exception, as these receptors are preformed, disulphide-linked dimers. A further theme is that ligand binding is often accompanied by conformational change within the RTK ectodomain, with such change being integral to the process of dimerization and ultimately to signal transduction. Other modes of ligand binding and receptor activation also occur. For example, Tek is understood to function through the formation of ligand-induced clusters [34]. An interesting variation on both these mechanisms is that of the ephrin receptors (Eph), with the ephrin ligand itself being a membrane-anchored protein. While the formation of an $(\mathrm{Eph/ephrin})_2$ homodimer results in forward (RTK) signalling as well as in reverse signalling via phosphorylation of the ephrin cytoplasmic tail [59], the formation of higher-order clusters of Eph/ephrin complexes is considered essential for physiological signalling [60].

The large repertoire of ectodomain structures coupled with an equally-large range of ligands means that there are very few, if any, unifying themes in the mode of ligand binding across the twenty RTK families. In many cases, ligand binding occurs at sites on the receptor that tolerate structural diversity. For example, in the case of Ig domains, ligand-binding sites often occur in the loop regions between β-strands. These sites occur near the inter-domain boundaries and ligand binding may thus effect change in relative disposition of the juxtaposing domains. Peculiarities also occur, for example, (a) a β-strand of the receptor extending the β-sheet of a ligand, as seen in the EGFR/transforming growth factor complex [61], (b) a surface-bound secondary structure element of the receptor re-orienting to engage ligand, as seen in the case of the InsR, where the α-chain C-terminal helix re-positions upon the surface of the first leucine-rich repeat domain to engage the incoming insulin molecule [62], and (c) the receptor itself inducing conformational change within the ligand, as seen in the case of insulin engaging InsR [62].

Commensurate with the diversity of ectodomain structure is variation in the way in which ligand binding induces RTK dimerization across the RTK superfamily. Four major modalities have been discerned [2], which can be described as follows.

8.3.2 Ligand-Mediated Dimerization Without Receptor Contacts

Here, a homodimeric ligand acts as a cross-bridge between two receptor monomers, with no receptor monomer-monomer contacts being formed, nor there being any change in the conformation of the receptor ectodomains monomers (Fig. 8.18a). An example of such an RTK is neurotrophic tyrosine kinase receptor type 1 (NTrk1), which binds nerve growth factor (NGF) via its membrane-proximal Ig domain [63]. The crystal structure of the intact NTrk1 ectodomain in complex with NGF [32] reveals an NGF dimer cross-linking the respective membrane proximal domains (denoted Ig-C2) of two NTrk1 monomers, with no further inter-ectodomain contacts occurring between the remaining Ig domains (denoted Ig-C1) or between the leucine-rich repeat domains (LRR) of the receptor (Fig. 8.18b). A crystal structure of the Ig-C1 domain of NTrk2 in complex with neurotrophin [64] reveals a similar arrangement to that seen in the NTrk1/NGF complex, suggesting that NTrk2 activation also falls within this category.

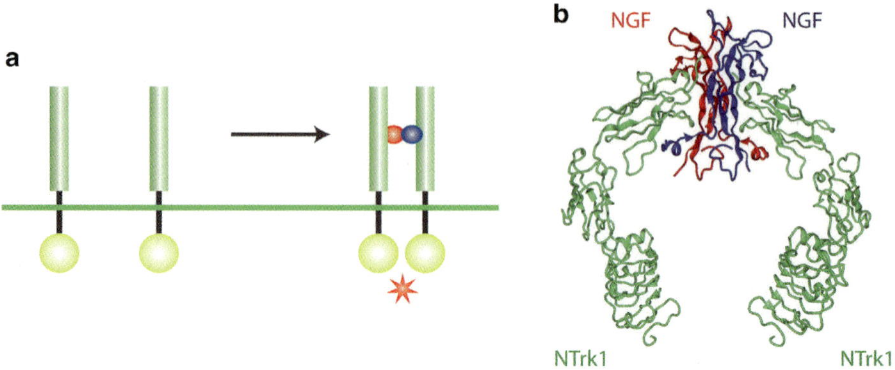

Fig. 8.18 RTK activation by ligand-mediated dimerization. (**a**) Schematic representation of a dimeric ligand cross-linking two receptor monomers. (**b**) An example of ligand-mediated dimerization: dimeric NGF (*blue/red*) cross-linking a pair of NTrk1 monomers (*green*) [32]. PDB entry 2IGF

8.3.3 Ligand-Mediated Dimerization with Receptor Contacts

Here, as above, a dimeric ligand cross-links two receptor monomers, without inducing conformational change within the ligand-binding domains of the receptor monomers. However, the dimerization of the receptor monomers brings together receptor domains that are remote from the ligand-binding site, which then re-orient to stabilize the ligand-bound receptor homodimer (Fig. 8.19a). A number of RTKs are understood to function in this fashion. Crystallographic studies have shown that the vascular endothelial growth factor C (VEGF-C), a disulphide-linked dimer [65], binds at the junction of the second and third Ig domains of vascular endothelial growth factor receptor 2 (VEGFR-2), thus forming a symmetric cross bridge between two VEGFR-2 monomers [66]. While the ectodomain of the *apo*-VEGFR-2 is highly flexible, the cross-bridged structure appears to be stabilized by symmetric interactions between pairs of VEGFR-2 Ig domains that lie membrane-proximal to the ligand-binding site [67]. Related crystal structures are available for VEGF-A bound to the second Ig domain of VEGFR-1 [68] and placenta growth factor bound to the second Ig domain of VEGFR-1 [69].

Stem cell factor (SCF) is the ligand for the RTK Kit within the PDGFR family. SCF is a dimer of four-helix bundles that interacts with the three N-terminal Ig domains of Kit without inducing conformational change in these three domains [31, 70]. However, upon SCF binding, the two membrane-proximal domains of Kit re-orient with respect to the ligand-binding Ig domains and form a homotypic interaction that allows activation of the tyrosine kinase domains (Fig. 8.19b). Colony stimulating factor (CSF) interacting with colony-stimulating factor 1 receptor (CSF1R, also termed FMS and also within the PDGFR family) likely functions in a similar way, but here dimerization via SCF is dependent upon formation of membrane proximal contacts between Ig domains 4 and between Ig domains 5 [71].

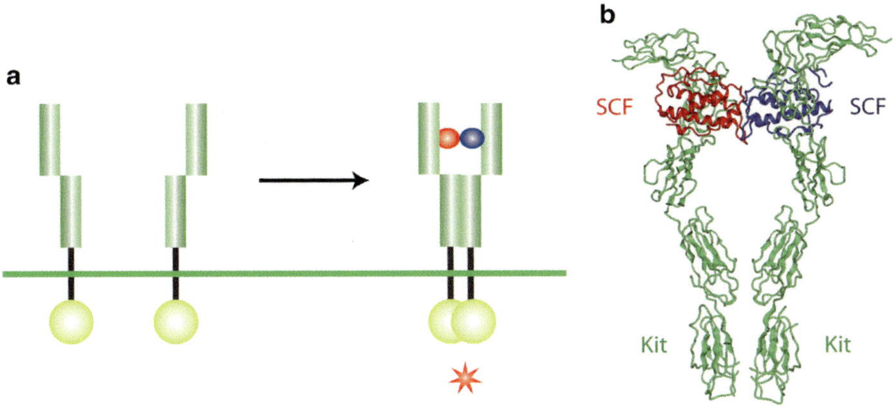

Fig. 8.19 RTK activation by ligand-mediated dimerization with receptor contacts. (**a**) Schematic representation of a dimeric ligand cross-linking two receptor monomers resulting in further receptor-receptor ectodomain contacts. (**b**) An example of receptor cross-linking with further receptor-receptor contacts: dimeric SCF (*blue/red*) binding to a pair of Kit monomers [31]. PDB entry 2E9W

8.3.4 Dimerization via Multiple Ligand-to-Receptor Contacts

Here, a monomeric ligand cross-links two receptor monomers, with its respective binding sites on the two receptor monomers being distinct. In addition, concomitant receptor monomer-monomer interactions then stabilize the complex (Fig. 8.20a). Receptors that are known to function in this fashion are the FGFRs. Fibroblast growth factors (FGFs) are monomeric ligands that interact with a major binding site on each receptor monomer [72]. These 1:1 complexes then associate into a 2:2 dimer [73], with the dimer being stabilized by further interactions between each FGF and the alternate FGFR monomer and by a heparin molecule interacting with both the FGFs and the FGFRs (Fig. 8.20b).

Activation of InsR and IGF-1R also fall into this category, though here the receptor is a disulphide-linked homodimer [74], with ligand binding resulting in conformational change within the pre-existing receptor homodimer [75, 76]. The primary ligand-binding site (Site 1) of InsR consists of the first LRR domain of one InsR monomer and the C-terminal segment of the α-chain ("αCT") of the alternate receptor monomer. The secondary ligand-binding site (Site 2′) is proposed to lie at the junction of the first and second FnIII domains of the alternate monomer (denoted above by the prime superscript) to that contributing the LRR domain. Ligand binding to InsR and IGF-1R is negatively cooperative [77], in the sense that ligand cross-linking the (Site 1/Site 2′) pair reduces the ligand affinity of the opposite (Site 1′/Site 2) pair. Negative cooperativity in this context implies the loss of twofold symmetry in the ligand/receptor complex. Interestingly, the InsR and IGF-1R receptor monomers can also form functional, disulphide-linked heterodimers [78].

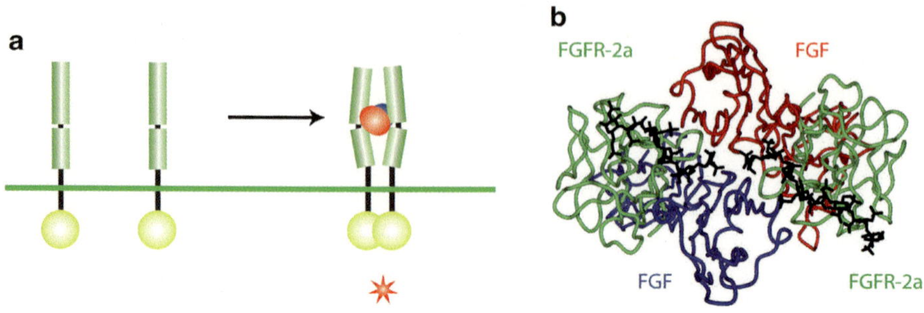

Fig. 8.20 RTK activation by monomeric ligand cross-linking receptor monomers. (**a**) Schematic representation of a monomeric ligands cross-linking receptor monomers. (**b**) An example of two copies of monomeric FGF (*blue/red*) cross-linking two FGFR-2a receptor monomers (*green*), with further cross-linking by heparin (*black*) [98]. PDB entry 1FQ9

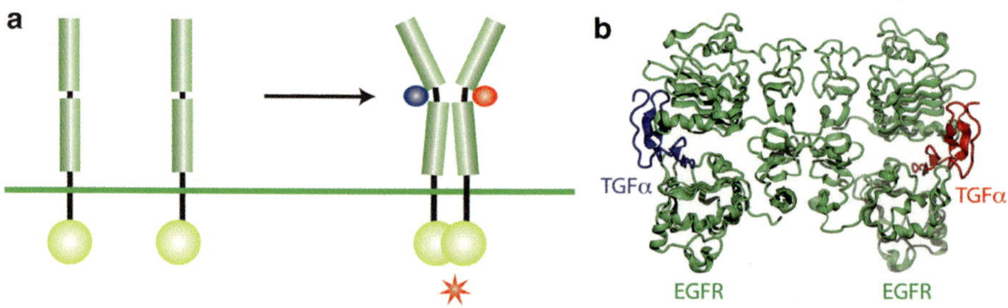

Fig. 8.21 RTK activation by receptor-mediated dimerization. (**a**) Schematic representation of receptor-mediated dimerization. (**b**) An example of receptor-mediated dimerization: Transforming growth factor α (TGFα) binding to EGFR [61]. PDB entry 1MOX

A further example is the activation of the Axl via the vitamin K-dependent growth arrest specific protein 6 (GAS6). Here, a pair of GAS6 monomers both independently cross-link a pair of Axl monomers [79].

8.3.5 Receptor-Mediated Dimerization

Here, a monomeric ligand binds to a receptor monomer, inducing conformational change in the receptor monomer. The resultant receptor monomer is now in an activated state that allows it to dimerize with a second receptor monomer to generate the signalling-active form, without there being any interaction between the ligands themselves and without the ligands cross-linking the receptor monomers. Receptors that are known to function in this fashion are the EGFR/ErbB family. EGF binding cross links the first and second leucine-rich domains of EGFR, stabilizing a conformational change from a folded-over structure to an extended structure that allows dimerization of EGFR monomers via their first cysteine-rich (CR1) domains [61] (Fig. 8.21). In the *apo*-receptor, the dimer-forming region of the CR1 domain is inaccessible, as the ectodomain adopts a folded-over conformation wherein the C-terminal cysteine-rich CR2 domain is folded over the CR1 dimerization domain [80–82]. Once activated, further EGFR monomer-monomer interactions are also understood to occur between the respective EGFR CR2 domains [83].

8.3.6 *Other Modes of Activation*

While the above four modes provide a framework for understanding RTK activation, the mode of functioning for many RTKs remains partially or completely unknown. However, it is clear that other paradigms must exist. Met is one such example. Hepatocyte growth factor (HGF) is the natural ligand of Met and, in its active form, is a two-chain α/β disulphide-linked heterodimer [84, 85]. A crystal structure of the Met Sema domain in complex with the β-chain of HGF [53] suggests a model by which 2:2 complexes of Met/HGF might be formed (consistent with Modes 1 and 2 above), but such complexes have thus far not been detected in solution [86]. Met signalling may thus depend on as-yet undiscerned interactions between the membrane-proximal Ig domains of the receptor, on membrane anchoring, or on unknown interactions with further proteins [86]. DDRs bind collagen via their discoidin domains, with the receptors being non-covalent dimers in the absence of collagen [87]. While a three-dimensional structure of a discoidin domain bound to collagen exists, the mechanism of activation is as yet unclear. Conformational change may arise from collagen cross-linking the DRR monomers with 2:1 stoichiometry, or from the formation of a 2:2 stoichiometric complex with no cross-linking of the DDR monomers [17].

8.4 Conclusion

Understanding the structural biology of RTK ligand binding and activation is a challenging task, with difficulty in obtaining recombinant RTK ectodomains or ectodomain fragments in sufficient quantities for structural analysis delaying progress in many cases. Glycosylation of the ectodomain modules as well as conformational flexibility can present major difficulties for structure determination by crystallography [88–90]. Conformational change upon ligand binding as revealed in crystal structures of RTKs may not necessarily correspond to that occurring in the intact, membrane-embedded RTK, adding yet a further degree of difficulty to understanding these events. Recent developments in high-resolution cryo-transmission electron microscopy [91] and x-ray free electron lasers [92] offer further scope for progress.

Acknowledgments This work was supported by the Australian National Health and Medical Research Council (NHMRC) Project Grant No. 1005896, NHMRC Independent Research Institutes Infrastructure Support Scheme Grant 361646, a Victorian State Government Operational Infrastructure Support Grant and the Harold and Pip Appel Fund.

Appendix

Table 8.1 Unliganded RTK ectodomain structures in the PDB[a]

Receptor family	Receptor	Receptor domains	PDB ID
Axl	Mer	Fibronectin Type III domain 2	2DBJ
	Tyro3	Immunoglobulin-like domain 1 and 2	1RHF
DDR	DDR1	Discoidin domains 1 and 2	4AG4
	DDR2	Discoidin domain 1	2Z4F
EGFR	Her2	Leucine-rich repeat domain 1, cysteine-rich domain 1 and leucine-rich repeat domain 2	2A91
	Her3	Leucine-rich repeat domain 1, cysteine-rich domain 1, leucine-rich repeat domain 2 and cysteine-rich domain 2	1M6B
	Her4	Leucine-rich repeat domain 1, cysteine-rich domain 1, leucine-rich repeat domain 2 and cysteine-rich domain 2	2AHX
	Her4	Leucine-rich repeat domain 1, cysteine-rich domain 1 and leucine-rich repeat domain 2	3U2P
EphA	EphA1	Fibronectin Type III domain 2	1X5A
	EphA2	Ephrin-binding domain, cysteine-rich domain and fibronectin Type III domains 1 and 2	2X10
	EphA2	Ephrin-binding domain	3C8X
	EphA2	Ephrin-binding domain, cysteine-rich domain and fibronectin Type III domains 1 and 2	3FL7
	EphA2	Ephrin-binding domain	3HPN
	EphA4	Ephrin-binding domain	2LW8
	EphA4	Ephrin-binding domain	2WO1
	EphA4	Ephrin-binding domain	3CKH
	EphA4	Ephrin-binding domain, cysteine-rich domain and fibronectin Type III domains 1 and 2	4BK4
	EphA5	Ephrin-binding domain	4ET7
	EphA7	Ephrin-binding domain	3NRU
	EphA8	Fibronectin Type III domain 2	1X5L
EphB	EphB1	Fibronectin Type III domain 2	2DJS
	EphB2	Ephrin-binding domain	1NUK
	EphB2	Ephrin-binding domain	3ETP
	EphB3	Ephrin-binding domain	3P1I
	EphB4	Fibronectin Type III domain 2	2E7H
FGFR	FGFR1	Immunoglobulin-like domain 1	2CKN
	FGFR1	Immunoglobulin-like domain 1	2CR3
	FGFR2	Immunoglobulin-like domain 1	4HWU
	FGFR2	Immunoglobulin-like domain 2	1WVZ
	FGFR2	Immunoglobulin-like domain 2	3CAF
	FGFR2	Immunoglobulin-like domain 2	3DAR
	FGFR2	Immunoglobulin-like domain 2	3EUU
InsR	IGF1R	Leucine-rich repeat domain 1, cysteine-rich domain and leucine-rich repeat domain 2	1IGR
	InsR	Leucine-rich repeat domain 1, cysteine-rich domain and leucine-rich repeat domain 2	2HR7

(continued)

Table 8.1 (continued)

Receptor family	Receptor	Receptor domains	PDB ID
	InsR	Leucine-rich repeat domain 1, cysteine-rich domain, leucine-rich repeat domain 2 and fibronectin Type III domains 1–3	2DTG
	InsR	Leucine-rich repeat domain 1, cysteine-rich domain, leucine-rich repeat domain 2 and fibronectin Type III domains 1–3 (plus αCT peptide)	3LOH
Met	Met	Psi domain	1SSL
	Ron	Semaphorin and psi domains	4FWW
MuSK	MuSK	Frizzled domain	3HKL
PDGFR	Kit	Immunoglobulin-like domains 1–5	2EC8
Ret	Ret	Cadherin domains 1 and 2	2X2U
Tie	Tek	Immunoglobulin-like domains 1 and 2, cysteine-rich domain, immunoglobulin-like domain 3 and fibronectin Type III domain 1	4K0V
	Tie	Immunoglobulin-like domains 1 and 2, cysteine-rich domain and immunoglobulin-like domain 3	2GY5
Trk	NTrk1	Immunoglobulin-like domain 2	1HE7
	NTrk1	Immunoglobulin-like domain 2	1WWA
	NTrk2	Immunoglobulin-like domain 2	1WWB
	NTrk3	Immunoglobulin-like domain 2	1WWC
VEGFR	VEGFR1	Immunoglobulin-like domain 2	1QSV
	VEGFR1	Immunoglobulin-like domain 2	1QSZ
	VEGFR2	Immunoglobulin-like domain 7	3KVQ
	VEGFR3	Immunoglobulin-like domain 4 and 5	4BSJ

[a]Excluded are structures formed as complexes with exogenous ligands, peptides, antibody fragments (Fabs) or biologics, except where, in the case of certain Fabs, these are integral to the structure determination of the RTK component

Table 8.2 Liganded RTK ectodomain structures in the PDB[a]

Receptor family	Receptor	Receptor domains	Ligand	PDB ID
Axl	Axl	Immunoglobulin-like domains 1 and 2	Growth arrest specific protein 6 (laminin G-like domains 1 and 2)	2C5D
DDR	DDR2	Discoidin domain 1	Collagen peptide	2WUH
EGFR	EGFR	Leucine-rich repeat domain 1 domain, cysteine-rich 1 and leucine-rich repeat domain 2	Epidermal growth factor	1IVO
	EGFR	Leucine-rich repeat domain 1, cysteine-rich domain, leucine-rich repeat domain 2 and cysteine-rich domain 2	Epidermal growth factor	1NQL

(continued)

Table 8.2 (continued)

Receptor family	Receptor	Receptor domains	Ligand	PDB ID
	EGFR	Leucine-rich repeat domain 1, cysteine-rich domain 1, leucine-rich repeat domain 2 and cysteine-rich domain 2	Epidermal growth factor	3NJP
	EGFR	Leucine-rich repeat domain 1, cysteine-rich domain 1, leucine-rich repeat domain 2 and cysteine-rich domain 2	Spitz (epidermal growth factor-like domain)	3LTF
	EGFR	Leucine-rich repeat domain 1, cysteine-rich domain 1, leucine-rich repeat domain 2 and cysteine-rich domain 2	Spitz (epidermal growth factor-like domain)	3LTG
	EGFR	Leucine-rich repeat domain 1, cysteine-rich domain and leucine-rich repeat domain 2	Transforming growth factor α	1MOX
	Her4	Leucine-rich repeat domain 1, cysteine-rich domain 1, leucine-rich repeat domain 2 and cysteine-rich domain 2	Neuregulin-1 β	3U7U
EphA	EphA2	Ephrin-binding domain	Ephrin-A1	3CZU
	EphA2	Ephrin-binding domain	Ephrin-A1	3HEI
	EphA2	Ephrin-binding domain and cysteine-rich domain	Ephrin-A1	3MBW
	EphA2	Ephrin-binding domain, cysteine-rich domain and fibronectin Type III domains 1 and 2	Ephrin-A5	2X11
	EphA2	Ephrin-binding domain, cysteine-rich domain and fibronectin Type III domain 1	Ephrin-A5	3MX0
	EphA4	Ephrin-binding domain, cysteine-rich domain and fibronectin Type III domains 1 and 2	Ephrin-A5	4BK5
	EphA4	Ephrin-binding domain, cysteine-rich domain and fibronectin Type III domains 1 and 2	Ephrin-A5	4BKA
	EphA4	Ephrin-binding domain, cysteine-rich domain and fibronectin Type III domains 1 and 2	Ephrin-B3	4BKF
	EphA4	Ephrin-binding domain	Ephrin-A2	2WO3
	EphA4	Ephrin-binding domain	Ephrin-B2	2WO2
	EphA4	Ephrin-binding domain	Ephrin-B2	3GXU
	EphB2	Ephrin-binding domain	Ephrin-B2	1KGY
	EphB2	Ephrin-binding domain	Ephrin-A5	1SHW
	EphB4	Ephrin-binding domain	Ephrin-B2	2HLE

(continued)

Table 8.2 (continued)

Receptor family	Receptor	Receptor domains	Ligand	PDB ID
FGFR	FGFR1	Immunoglobulin-like domains 2 and 3	Fibroblast growth factor-1	3OJV
	FGFR1	Immunoglobulin-like domains 2 and 3	Fibroblast growth factor-1	1EVT
	FGFR1	Immunoglobulin-like domains 2 and 3	Fibroblast growth factor-2	1CVS
	FGFR1	Immunoglobulin-like domains 2 and 3	Fibroblast growth factor-2 and heparin	1FQ9
	FGFR2	Immunoglobulin-like domains 2 and 3	Fibroblast growth factor-1	3OJM
	FGFR2	Immunoglobulin-like domains 2 and 3	Fibroblast growth factor-10	1NUN
	FGFR2	Immunoglobulin-like domains 2 and 3	Fibroblast growth factor-1	1DJS
	FGFR2	Immunoglobulin-like domains 2 and 3	Fibroblast growth factor-1 and heparin	1E0O
	FGFR2	Immunoglobulin-like domain 2	Fibroblast growth factor-1	3CU1
	FGFR2	Immunoglobulin-like domains 2 and 3	Fibroblast growth factor-1	3OJ2
	FGFR2	Immunoglobulin-like domains 2 and 3	Fibroblast growth factor-2	1EV2
	FGFR2	Immunoglobulin-like domains 2 and 3	Fibroblast growth factor-2	1II4
	FGFR2	Immunoglobulin-like domains 2 and 3	Fibroblast growth factor-2	1IIL
	FGFR2	Immunoglobulin-like domains 2 and 3	Fibroblast growth factor-8b	2FDB
	FGFR3	Immunoglobulin-like domains 2 and 3	Fibroblast growth factor-1	1RY7
InsR	InsR	Leucine-rich repeat domain 1, cysteine-rich domain and α-chain C-terminal peptide	Insulin	3W11
	InsR	Leucine-rich repeat domain 1, cysteine-rich domain and α-chain C-terminal peptide	High-affinity insulin derivative	3W12
	InsR	Leucine-rich repeat domain 1, cysteine-rich domain and long α-chain C-terminal peptide	High-affinity insulin derivative	3W13
	InsR	Leucine-rich repeat domain 1, cysteine-rich domain, leucine-rich repeat domain 2, fibronectin Type III domain 1 and α-chain C-terminal peptide	Insulin	3W14
Met	Met	Semaphorin and psi domains	Hepatocyte growth factor-β-chain	1SHY

(continued)

Table 8.2 (continued)

Receptor family	Receptor	Receptor domains	Ligand	PDB ID
PDGFR	CSF1R	Immunoglobulin-like domains 1–3	Macrophage colony stimulating factor-1	3EJJ
	CSF1R	Immunoglobulin-like domains 1–3	Interleukin-34	4DKD
	CSF1R	Immunoglobulin-like domains 1–3	Interleukin-34	4EXP
	Flt3	Immunoglobulin-like domains 1–4	SL cytokine	3QS7
	Flt3	Immunoglobulin-like domains 1–5	SL cytokine	3QS9
	Kit	Immunoglobulin-like domains 1–3	Stem cell factor	2O26
	Kit	Immunoglobulin-like domains 1–5	Stem cell factor	2E9W
	PDGFRB	Immunoglobulin-like domains 1–3	Platelet-derived growth factor subunit B	3MJG
Tie	Tek	Immunoglobulin-like domains 1 and 2, cysteine-rich domain and immunoglobulin-like domain 3	Angiopoietin-2 (receptor binding domain)	2GY7
Trk	NTRK1	Immunoglobulin-like domain 2	Nerve growth factor	1WWW
	NTRK1	Leucine-rich repeat domain and immunoglobulin-like domains 1 and 2	Nerve growth factor	2IFG
	NTRK2	Immunoglobulin-like domain 2	Neurotrophin 4/5	1HCF
VEGFR	VEGFR1	Immunoglobulin-like domain 2	Vascular epithelial growth factor-A	1FLT
	VEGFR1	Immunoglobulin-like domain 2	Vascular epithelial growth factor-A	1QTY
	VEGFR1	Immunoglobulin-like domain 2	Placental growth factor	1RV6
	VEGFR2	Immunoglobulin-like domains 2 and 3	Vascular epithelial growth factor-A	3V2A
	VEGFR2	Immunoglobulin-like domains 2 and 3	Vascular epithelial growth factor-C	2X1W
	VEGFR2	Immunoglobulin-like domains 2 and 3	Vascular epithelial growth factor-C	2X1X
	VEGFR2	Immunoglobulin-like domains 2 and 3	Vascular epithelial growth factor-E	3V6B
	VEGFR3	Immunoglobulin-like domains 1 and 2	Vascular epithelial growth factor-C	4BSK

[a]Excluded are structures formed as complexes with exogenous ligands, peptides, antibody fragments (Fabs) or biologics, except where, in the case of certain Fabs, these are integral to the structure determination of the RTK component

References

1. Robinson DR, Wu YM, Lin SF. The protein tyrosine kinase family of the human genome. Oncogene. 2000;19(49):5548–57.
2. Lemmon MA, Schlessinger J. Cell signaling by receptor tyrosine kinases. Cell. 2010;141(7):1117–34.
3. Schlessinger J. Cell signaling by receptor tyrosine kinases. Cell. 2000;103(2):211–25.
4. Hubbard SR, Till JH. Protein tyrosine kinase structure and function. Annu Rev Biochem. 2000;69:373–98.
5. D'Aniello S, Irimia M, Maeso I, Pascual-Anaya J, Jiménez-Delgado S, Bertrand S, et al. Gene expansion and retention leads to a diverse tyrosine kinase superfamily in amphioxus. Mol Biol Evol. 2008;25(9):1841–54.
6. Gu J, Gu X. Natural history and functional divergence of protein tyrosine kinases. Gene. 2003;317(1–2):49–57.
7. Grassot J, Gouy M, Perrière G, Mouchiroud G. Origin and molecular evolution of receptor tyrosine kinases with immunoglobulin-like domains. Mol Biol Evol. 2006;23(6): 1232–41.
8. King N, Carroll SB. A receptor tyrosine kinase from choanoflagellates: molecular insights into early animal evolution. Proc Natl Acad Sci. 2001;98(26):15032–7.
9. Adams TE, Epa VC, Garrett TP, Ward CW. Structure and function of the type 1 insulin-like growth factor receptor. Cell Mol Life Sci. 2000;57(7):1050–93.
10. Komada M, Hatsuzawa K, Shibamoto S, Ito F, Nakayama K, Kitamura N. Proteolytic processing of the hepatocyte growth factor/scatter factor receptor by furin. FEBS Lett. 1993;328(1–2): 25–9.
11. Bork P, Downing AK, Kieffer B, Campbell ID. Structure and distribution of modules in extracellular proteins. Q Rev Biophys. 1996;29(2):119–67.
12. Bork P, Holm L, Sander C. The immunoglobulin fold. Structural classification, sequence patterns and common core. J Mol Biol. 1994;242(4):309–20.
13. Campbell ID, Spitzfaden C. Building proteins with fibronectin type III modules. Structure. 1994;2(5):333–7.
14. Shapiro L, Kwong PD, Fannon AM, Colman DR, Hendrickson WA. Considerations on the folding topology and evolutionary origin of cadherin domains. Proc Natl Acad Sci. 1995;92(15):6793–7.
15. Richardson JS. The anatomy and taxonomy of protein structure. Adv Protein Chem. 1981;34: 167–339.
16. Himanen JP, Henkemeyer M, Nikolov DB. Crystal structure of the ligand-binding domain of the receptor tyrosine kinase EphB2. Nature. 1998;396(6710):486–91.
17. Carafoli F, Bihan D, Stathopoulos S, Konitsiotis AD, Kvansakul M, Farndale RW, et al. Crystallographic insight into collagen recognition by discoidin domain receptor 2. Structure. 2009;17(12):1573–81.
18. Ichikawa O, Osawa M, Nishida N, Goshima N, Nomura N, Shimada I. Structural basis of the collagen-binding mode of discoidin domain receptor 2. EMBO J. 2007;26(18):4168–76.
19. Aricescu AR, Hon WC, Siebold C, Lu W, van der Merwe PA, Jones EY. Molecular analysis of receptor protein tyrosine phosphatase mu-mediated cell adhesion. EMBO J. 2006;25(4): 701–12.
20. Ward CW, Garrett TP. The relationship between the L1 and L2 domains of the insulin and epidermal growth factor receptors and leucine-rich repeat modules. BMC Bioinform. 2001;2:4.
21. Daly NL, Scanlon MJ, Djordjevic JT, Kroon PA, Smith R. Three-dimensional structure of a cysteine-rich repeat from the low-density lipoprotein receptor. Proc Natl Acad Sci U S A. 1995;92(14):6334–8.
22. Rehn M, Pihlajaniemi T, Hofmann K, Bucher P. The frizzled motif: in how many different protein families does it occur? Trends Biochem Sci. 1998;23(11):415–7.

23. Ward CW, Hoyne PA, Flegg RH. Insulin and epidermal growth factor receptors contain the cysteine repeat motif found in the tumor necrosis factor receptor. Proteins. 1995;22(2): 141–53.
24. Park CH, Tulinsky A. Three-dimensional structure of the kringle sequence: structure of prothrombin fragment 1. Biochemistry. 1986;25(14):3977–82.
25. László P. The WIF, module. Trends Biochem Sci. 2000;25(1):12–3.
26. Springer TA. An extracellular beta-propeller module predicted in lipoprotein and scavenger receptors, tyrosine kinases, epidermal growth factor precursor, and extracellular matrix components. J Mol Biol. 1998;283(4):837–62.
27. Jeon H, Meng W, Takagi J, Eck MJ, Springer TA, Blacklow SC. Implications for familial hypercholesterolemia from the structure of the LDL receptor YWTD-EGF domain pair. Nat Struct Biol. 2001;8(6):499–504.
28. Bork P, Doerks T, Springer TA, Snel B. Domains in plexins: links to integrins and transcription factors. Trends Biochem Sci. 1999;24(7):261–3.
29. Winberg ML, Noordermeer JN, Tamagnone L, Comoglio PM, Spriggs MK, Tessier-Lavigne M, et al. Plexin A is a neuronal semaphorin receptor that controls axon guidance. Cell. 1998;95(7):903–16.
30. Seiradake E, Harlos K, Sutton G, Aricescu AR, Jones EY. An extracellular steric seeding mechanism for Eph-ephrin signaling platform assembly. Nat Struct Mol Biol. 2010;17(4):398–402.
31. Yuzawa S, Opatowsky Y, Zhang Z, Mandiyan V, Lax I, Schlessinger J. Structural basis for activation of the receptor tyrosine kinase KIT by stem cell factor. Cell. 2007;130(2):323–34.
32. Wehrman T, He X, Raab B, Dukipatti A, Blau H, Garcia KC. Structural and mechanistic insights into nerve growth factor interactions with the TrkA and p75 receptors. Neuron. 2007;53(1):25–38.
33. Ward CW, Lawrence MC, Streltsov VA, Adams TE, McKern NM. The insulin and EGF receptor structures: new insights into ligand-induced receptor activation. Trends Biochem Sci. 2007;32(3):129–37.
34. Barton WA, Tzvetkova-Robev D, Miranda EP, Kolev MV, Rajashankar KR, Himanen JP, et al. Crystal structures of the Tie2 receptor ectodomain and the angiopoietin-2-Tie2 complex. Nat Struct Mol Biol. 2006;13(6):524–32.
35. Berman HM, Westbrook J, Feng Z, Gilliland G, Bhat TN, Weissig H, et al. The protein data bank. Nucleic Acids Res. 2000;28(1):235–42.
36. Holm L, Sander C. Protein structure comparison by alignment of distance matrices. J Mol Biol. 1993;233(1):123–38.
37. Finn RD, Mistry J, Tate J, Coggill P, Heger A, Pollington JE, et al. The Pfam protein families database. Nucleic Acids Res. 2010;38(Database issue):D211–22.
38. Letunic I, Doerks T, Bork P. SMART 7: recent updates to the protein domain annotation resource. Nucleic Acids Res. 2012;40(Database issue):D302–5.
39. Andreeva A, Howorth D, Chandonia JM, Brenner SE, Hubbard TJ, Chothia C, et al. Data growth and its impact on the SCOP database: new developments. Nucleic Acids Res. 2008;36(Database issue):D419–25.
40. Gelfand IM, Chothia C, Kister AE. Immunoglobulin fold: structures of proteins in the immunoglobulin superfamily. Chichester: John Wiley & Sons, Ltd; 2001.
41. Harpaz Y, Chothia C. Many of the immunoglobulin superfamily domains in cell adhesion molecules and surface receptors belong to a new structural set which is close to that containing variable domains. J Mol Biol. 1994;238(4):528–39.
42. Smith DK, Xue H. Sequence profiles of immunoglobulin and immunoglobulin-like domains. J Mol Biol. 1997;274(4):530–45.
43. Mott HR, Nietlispach D, Hopkins LJ, Mirey G, Camonis JH, Owen D. Structure of the GTPase-binding domain of Sec5 and elucidation of its Ral binding site. J Biol Chem. 2003;278(19):17053–9.
44. Wiesmann C, Muller YA, de Vos AM. Ligand-binding sites in Ig-like domains of receptor tyrosine kinases. J Mol Med. 2000;78(5):247–60.

45. McKern NM, Lawrence MC, Streltsov VA, Lou MZ, Adams TE, Lovrecz GO, et al. Structure of the insulin receptor ectodomain reveals a folded-over conformation. Nature. 2006; 443(7108):218–21.
46. Kjaer S, Hanrahan S, Totty N, McDonald NQ. Mammal-restricted elements predispose human RET to folding impairment by HSCR mutations. Nat Struct Mol Biol. 2010;17(6):726–31.
47. Himanen JP, Saha N, Nikolov DB. Cell-cell signaling via Eph receptors and ephrins. Curr Opin Cell Biol. 2007;19(5):534–42.
48. Chrencik JE, Brooun A, Kraus ML, Recht MI, Kolatkar AR, Han GW, et al. Structural and biophysical characterization of the EphB4*ephrinB2 protein-protein interaction and receptor specificity. J Biol Chem. 2006;281(38):28185–92.
49. Ward CW, Garrett TP. Structural relationships between the insulin receptor and epidermal growth factor receptor families and other proteins. Curr Opin Drug Discov Devel. 2004;7(5):630–8.
50. Garrett TP, McKern NM, Lou M, Frenkel MJ, Bentley JD, Lovrecz GO, et al. Crystal structure of the first three domains of the type-1 insulin-like growth factor receptor. Nature. 1998; 394(6691):395–9.
51. Xu YK, Nusse R. The Frizzled CRD domain is conserved in diverse proteins including several receptor tyrosine kinases. Curr Biol. 1998;8(12):R405–6.
52. Stiegler AL, Burden SJ, Hubbard SR. Crystal structure of the frizzled-like cysteine-rich domain of the receptor tyrosine kinase MuSK. J Mol Biol. 2009;393(1):1–9.
53. Stamos J, Lazarus RA, Yao X, Kirchhofer D, Wiesmann C. Crystal structure of the HGF beta-chain in complex with the Sema domain of the Met receptor. EMBO J. 2004;23(12): 2325–35.
54. Chao KL, Tsai IW, Chen C, Herzberg O. Crystal structure of the Sema-PSI extracellular domain of human RON receptor tyrosine kinase. PLoS One. 2012;7(7):e41912.
55. Lou M, Garrett TP, McKern NM, Hoyne PA, Epa VC, Bentley JD, et al. The first three domains of the insulin receptor differ structurally from the insulin-like growth factor 1 receptor in the regions governing ligand specificity. Proc Natl Acad Sci U S A. 2006;103(33):12429–34.
56. Kozlov G, Perreault A, Schrag JD, Park M, Cygler M, Gehring K, et al. Insights into function of PSI domains from structure of the Met receptor PSI domain. Biochem Biophys Res Commun. 2004;321(1):234–40.
57. Liepinsh E, Banyai L, Patthy L, Otting G. NMR structure of the WIF domain of the human Wnt-inhibitory factor-1. J Mol Biol. 2006;357(3):942–50.
58. Ye Q, Rahman MN, Koschinsky ML, Jia Z. High-resolution crystal structure of apolipoprotein(a) kringle IV type 7: insights into ligand binding. Protein Sci. 2001;10(6):1124–9.
59. Himanen JP, Nikolov DB. Eph signaling: a structural view. Trends Neurosci. 2003;26(1): 46–51.
60. Stein E, Lane AA, Cerretti DP, Schoecklmann HO, Schroff AD, Van Etten RL, et al. Eph receptors discriminate specific ligand oligomers to determine alternative signaling complexes, attachment, and assembly responses. Genes Dev. 1998;12(5):667–78.
61. Garrett TP, McKern NM, Lou M, Elleman TC, Adams TE, Lovrecz GO, et al. Crystal structure of a truncated epidermal growth factor receptor extracellular domain bound to transforming growth factor alpha. Cell. 2002;110(6):763–73.
62. Menting JG, Whittaker J, Margetts MB, Whittaker LJ, Kong GK, Smith BJ, et al. How insulin engages its primary binding site on the insulin receptor. Nature. 2013;493(7431):241–5.
63. Wiesmann C, Ultsch MH, Bass SH, de Vos AM. Crystal structure of nerve growth factor in complex with the ligand-binding domain of the TrkA receptor. Nature. 1999;401(6749):184–8.
64. Banfield MJ, Naylor RL, Robertson AG, Allen SJ, Dawbarn D, Brady RL. Specificity in Trk receptor:neurotrophin interactions: the crystal structure of TrkB-d5 in complex with neurotrophin-4/5. Structure. 2001;9(12):1191–9.
65. Vitt UA, Hsu SY, Hsueh AJ. Evolution and classification of cystine knot-containing hormones and related extracellular signaling molecules. Mol Endocrinol. 2001;15(5):681–94.
66. Leppanen VM, Prota AE, Jeltsch M, Anisimov A, Kalkkinen N, Strandin T, et al. Structural determinants of growth factor binding and specificity by VEGF receptor 2. Proc Natl Acad Sci U S A. 2010;107(6):2425–30.

67. Ruch C, Skiniotis G, Steinmetz MO, Walz T, Ballmer-Hofer K. Structure of a VEGF-VEGF receptor complex determined by electron microscopy. Nat Struct Mol Biol. 2007;14(3):249–50.
68. Wiesmann C, Fuh G, Christinger HW, Eigenbrot C, Wells JA, de Vos AM. Crystal structure at 1.7 A resolution of VEGF in complex with domain 2 of the Flt-1 receptor. Cell. 1997;91(5):695–704.
69. Christinger HW, Fuh G, de Vos AM, Wiesmann C. The crystal structure of placental growth factor in complex with domain 2 of vascular endothelial growth factor receptor-1. J Biol Chem. 2004;279(11):10382–8.
70. Liu H, Chen X, Focia PJ, He X. Structural basis for stem cell factor-KIT signaling and activation of class III receptor tyrosine kinases. EMBO J. 2007;26(3):891–901.
71. Chen X, Liu H, Focia PJ, Shim AH, He X. Structure of macrophage colony stimulating factor bound to FMS: diverse signaling assemblies of class III receptor tyrosine kinases. Proc Natl Acad Sci U S A. 2008;105(47):18267–72.
72. Stauber DJ, DiGabriele AD, Hendrickson WA. Structural interactions of fibroblast growth factor receptor with its ligands. Proc Natl Acad Sci U S A. 2000;97(1):49–54.
73. Pellegrini L, Burke DF, von Delft F, Mulloy B, Blundell TL. Crystal structure of fibroblast growth factor receptor ectodomain bound to ligand and heparin. Nature. 2000;407(6807): 1029–34.
74. McKern NM, Lou M, Frenkel MJ, Verkuylen A, Bentley JD, Lovrecz GO, et al. Crystallization of the first three domains of the human insulin-like growth factor-1 receptor. Protein Sci. 1997;6(12):2663–6.
75. Lawrence MC, McKern NM, Ward CW. Insulin receptor structure and its implications for the IGF-1 receptor. Curr Opin Struct Biol. 2007;17:699–705.
76. Ward CW, Menting JG, Lawrence MC. The insulin receptor changes conformation in unforeseen ways on ligand binding: sharpening the picture of insulin receptor activation. Bioessays. 2013;35:945.
77. De Meyts P, Roth J, Neville Jr DM, Gavin 3rd JR, Lesniak MA. Insulin interactions with its receptors: experimental evidence for negative cooperativity. Biochem Biophys Res Commun. 1973;55(1):154–61.
78. Belfiore A, Frasca F, Pandini G, Sciacca L, Vigneri R. Insulin receptor isoforms and insulin receptor/insulin-like growth factor receptor hybrids in physiology and disease. Endocr Rev. 2009;30(6):586–623.
79. Sasaki T, Knyazev PG, Clout NJ, Cheburkin Y, Gohring W, Ullrich A, et al. Structural basis for Gas6-Axl signalling. EMBO J. 2006;25(1):80–7.
80. Cho HS, Mason K, Ramyar KX, Stanley AM, Gabelli SB, Denney Jr DW, et al. Structure of the extracellular region of HER2 alone and in complex with the Herceptin Fab. Nature. 2003;421(6924):756–60.
81. Ferguson KM, Berger MB, Mendrola JM, Cho HS, Leahy DJ, Lemmon MA. EGF activates its receptor by removing interactions that autoinhibit ectodomain dimerization. Mol Cell. 2003;11(2):507–17.
82. Bouyain S, Longo PA, Li S, Ferguson KM, Leahy DJ. The extracellular region of ErbB4 adopts a tethered conformation in the absence of ligand. Proc Natl Acad Sci U S A. 2005;102(42):15024–9.
83. Burgess AW, Cho HS, Eigenbrot C, Ferguson KM, Garrett TP, Leahy DJ, et al. An open-and-shut case? Recent insights into the activation of EGF/ErbB receptors. Mol Cell. 2003;12(3):541–52.
84. Nakamura T, Nishizawa T, Hagiya M, Seki T, Shimonishi M, Sugimura A, et al. Molecular cloning and expression of human hepatocyte growth factor. Nature. 1989;342(6248):440–3.
85. Hartmann G, Naldini L, Weidner KM, Sachs M, Vigna E, Comoglio PM, et al. A functional domain in the heavy chain of scatter factor/hepatocyte growth factor binds the c-Met receptor and induces cell dissociation but not mitogenesis. Proc Natl Acad Sci U S A. 1992;89(23): 11574–8.
86. Gherardi E, Youles ME, Miguel RN, Blundell TL, Iamele L, Gough J, et al. Functional map and domain structure of MET, the product of the c-met protooncogene and receptor for hepatocyte growth factor/scatter factor. Proc Natl Acad Sci U S A. 2003;100(21):12039–44.

87. Noordeen NA, Carafoli F, Hohenester E, Horton MA, Leitinger B. A transmembrane leucine zipper is required for activation of the dimeric receptor tyrosine kinase DDR1. J Biol Chem. 2006;281(32):22744–51.
88. Stura EA, Nemerow GR, Wilson IA. Strategies in the crystallization of glycoproteins and protein complexes. J Cryst Growth. 1992;122:273–85.
89. Baker HM, Day CL, Norris GE, Baker EN. Enzymatic deglycosylation as a tool for crystallization of mammalian binding proteins. Acta Crystallogr. 1994;50(Pt 4):380–4.
90. Sousa R. Use of glycerol, polyols and other protein structure stabilizing agents in protein crystallization. Acta Crystallogr. 1995;51(Pt 3):271–7.
91. Orlova EV, Saibil HR. Structural analysis of macromolecular assemblies by electron microscopy. Chem Rev. 2011;111(12):7710–48.
92. Chapman HN, Fromme P, Barty A, White TA, Kirian RA, Aquila A, et al. Femtosecond X-ray protein nanocrystallography. Nature. 2011;470(7332):73–7.
93. Richardson JS. Early ribbon drawings of proteins. Nat Struct Biol. 2000;7(8):624–5.
94. Philippsen A. DINO: visualizing structural biology 2000. Available from: http://www.bioz.unibas.ch/~xray/dino.
95. Persistence of Vision Pty Ltd. Persistence of Vision Raytracer (Version 3.7) 2004. Available from: http://www.povray.org/download/.
96. Smith BJ, Huang K, Kong G, Chan SJ, Nakagawa S, Menting JG, et al. Structural resolution of a tandem hormone-binding element in the insulin receptor and its implications for design of peptide agonists. Proc Natl Acad Sci U S A. 2010;107(15):6771–6.
97. Chrencik JE, Brooun A, Recht MI, Kraus ML, Koolpe M, Kolatkar AR, et al. Structure and thermodynamic characterization of the EphB4/Ephrin-B2 antagonist peptide complex reveals the determinants for receptor specificity. Structure. 2006;14(2):321–30.
98. Schlessinger J, Plotnikov AN, Ibrahimi OA, Eliseenkova AV, Yeh BK, Yayon A, et al. Crystal structure of a ternary FGF-FGFR-heparin complex reveals a dual role for heparin in FGFR binding and dimerization. Mol Cell. 2000;6(3):743–50.

Chapter 9
Structural Features of the Kinase Domain

Dániel Süveges and Natalia Jura

9.1 The Overall Architecture of the Kinase Domain

The structure of the kinase domain has been repetitively captured by X-ray crystallographers contributing unique crystals structures of more than 25 % of all ~500 kinases present in the human genome [1]. The analysis of these structures reveals a highly conserved fold shared by all kinases, which includes serine/threonine and tyrosine kinases, in both their soluble and membrane-spanning forms. Despite their smaller number relative to the serine/threonine kinases (less than 20 % of all kinases [2]), tyrosine kinases are particularly well represented in the pool of solved structures due to their frequent deregulation in diseases that motivates structural investigation. It was a serine/threonine kinase, however, protein kinase A (PKA), that revealed the kinase architecture for the first time [3]. This architecture can be best described by the presence of two structurally distinct lobes, the N-terminal lobe (N-lobe) and the C-terminal lobe (C-lobe), which both build up the catalytic site of the kinase where the nucleotide adenosine-5′-triphosphate (ATP), bivalent cations, and the substrate tyrosine residue meet (Fig. 9.1a). The N-lobe and C-lobe are connected by a linker region called the hinge region, which enables the relative motions of the two lobes. These motions are critical for catalysis by spatially organizing the catalytic residues into a highly conserved arrangement that enables phosphotransfer.

D. Süveges
Department of Cellular and Molecular Pharmacology, Cardiovascular Research Institute, University of California, San Francisco, San Francisco, CA, USA

N. Jura, Ph.D. (✉)
Department of Cellular and Molecular Pharmacology, Cardiovascular Research Institute, University of California, San Francisco, 555 Mission Bay BLVD South, Rm 452, San Francisco, CA 94158, USA
e-mail: Natalia.Jura@ucsf.edu

D.L. Wheeler, Y. Yarden (eds.), *Receptor Tyrosine Kinases: Structure, Functions and Role in Human Disease*, DOI 10.1007/978-1-4939-2053-2_9

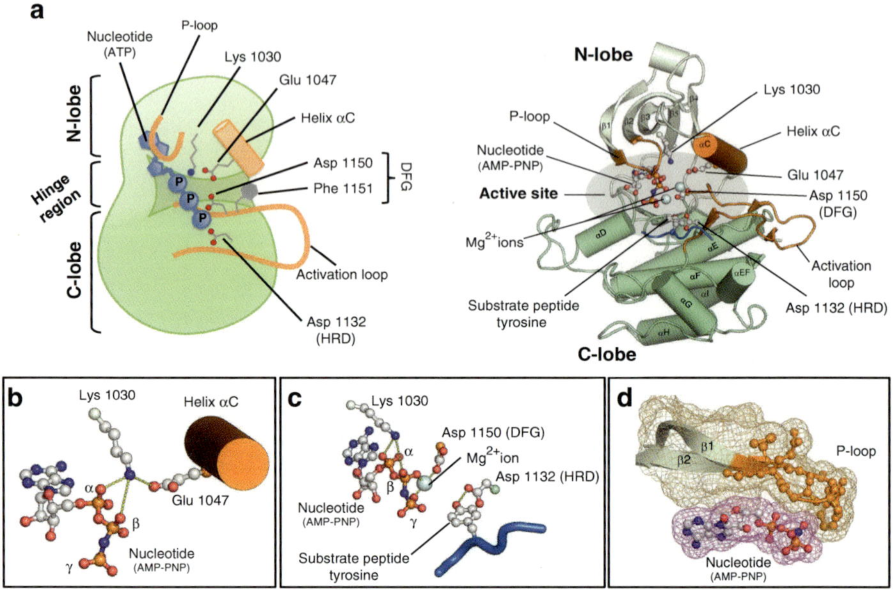

Fig. 9.1 Architecture of the kinase domain. (**a**) The cartoon and detailed structural representation of an overall kinase domain architecture based on the example of the structure of an insulin receptor kinase in the active conformation (PDB ID: 1IR3). The α-helices and β-strands are numbered as previously described [23]. (**b**) Coordination of the α- and β-phosphate groups of the nucleotide (AMP-PNP, as seen in the structure of the insulin receptor kinase, PDB ID: 1IR3) by the catalytic lysine and glutamate. (**c**) Catalytically critical interactions of the DFG and HRD aspartates with the nucleotide and the substrate. (**d**) The mesh surface representation depicts close interaction of the P-loop with the nucleotide in the active site

The N-lobe and the C-lobe are structurally distinct (Fig. 9.1a). In the majority of kinases, the N-lobe consists of five beta strands (denoted β1-β5) and one helix (αC). The helix αC plays a major role in catalysis and in the transition of the kinase between inactive and active states. The N-lobe also contains an important glycine-rich loop, connecting the β1 and β2 strands, called the phosphate-binding loop or the P-loop. The P-loop participates in coordination of β- and γ-phosphates of ATP.

In contrast to the N-lobe, the C-lobe is dominated by helices (denoted αD through αI). The C-lobe provides a channel for the substrate binding and directs the substrate towards the active site. During the catalytic cycle, the C-lobe is less dynamic than the N-lobe, with the exception of one essential structural component, called the activation loop. As in case of helix αC, the conformation of the activation loop changes during the catalytic cycle. The N-terminal portion of the activation loop contains a region spanning a highly conserved sequence Asp-Phe-Gly, called the DFG motif, that participates in coordination of metal cations. The middle part of the activation loop is more divergent between the kinases, both in length and sequence. It participates in substrate binding and in proper presentation of the substrate for phosphorylation. The activation loop also contains one or more phosphorylation sites, which play an important function in kinase activation process. As discussed in later sections, kinases can differ diametrically in their dependence on this regulatory step.

9.2 The Structural Basis for Phosphorylation

9.2.1 Architecture of the ATP-Binding Site

The tyrosine phosphorylation reaction involves the transfer of the γ-phosphate from the ATP onto the hydroxyl group of a tyrosine residue. For this to happen, ATP needs to be bound in a catalytically competent state that is compatible with phosphotransfer. ATP binding is therefore controlled by highly conserved residues located in the active site of the kinase, which occupies a deep cleft located between the N-lobe and the C-lobe of the kinase domain (Fig. 9.1a). These critical residues involve a lysine—glutamate pair that forms a salt bridge in the active state (Fig. 9.1b). The lysine residue (Lys 1030 in the insulin receptor (IR) tyrosine kinase—for reference IR numbering is used until otherwise noted) is contributed by the conserved VAIK motif located on β3 strand and the glutamate (Glu 1047) by helix αC. The orientation of helix αC changes during activation allowing for formation of the Lys 1030–Glu 1047 salt bridge, which enables positioning of the lysine side chain for interaction with both α- and β-phosphates of ATP. Another important residue for ATP positioning is an aspartate residue from the DFG motif (Asp 1150) located at the N-terminus of the activation loop (Fig. 9.1c). The DFG aspartate coordinates a metal ion, which interacts then with the β- and γ-phosphate groups of ATP. In addition to their role in orienting ATP, the bivalent metal cations, most frequently magnesium, are important for neutralization of the negative charge on ATP [4].

ATP is further stabilized in the active site through the glycine-rich P-loop containing sequence GxGxΦG, where Φ is usually a phenylalanine or a tyrosine, while x is less defined. Although the adenine base of ATP is buried in the active site cleft, the phosphates are localized towards the front of the cleft where the P-loop plays a major role in stabilizing their binding. This is largely achieved thanks to the glycine residues that provide flexibility and allow the P-loop to make close contact with the phosphates of the nucleotide through backbone interaction (Fig. 9.1d). Together with the metal cation and the catalytic lysine (Lys 1030), P-loop also contributes to the neutralization of the negative charge on ATP [5]. In addition to the P-loop, another component participating in the organization of the active site is the aspartate residue located in the HRD motif (Asp 1132). The HRD-aspartate plays a role in both orienting the metal cations and the hydroxyl group of the substrate tyrosine for catalysis. Thus, it is a coordinated effort of the catalytic lysine, aspartates in the DFG and HRD motifs, and the P-loop to position the β- and γ-phosphate groups of ATP for hydrolysis and subsequent phosphorylation of the substrate residue.

9.2.2 The Hydrophobic Spines

The simultaneous coordination of all structural elements in the active state of a kinase for efficient catalysis requires proper positioning of the residues along extended distance (~25 Å)—spanning from the catalytic lysine in the N-lobe to the activation loop in the C-lobe. What assures this communication during kinase activation? Only recently, two highly conserved intramolecular networks of connectivities between the N-lobe and the C-lobe were discovered to be essential in controlling the proper positioning of ATP and the substrate peptide for catalysis [6, 7]. These networks involve a set of hydrophobic residues spanning both lobes of the kinase, which align in the active conformation to form two independent spines: regulatory and catalytic (Fig. 9.2). Those hydrophobic spines assemble in the presence of ATP and the substrate peptide assuring proper positioning of the catalytic components. In the inactive state of the kinase, the spines are disassembled.

The regulatory spine is composed of four residues Leu 301 in strand β5 (residue numbering is provided for the Abelson (Abl) tyrosine kinase, PDB ID: 2G2I), Met 290 in helix αC, Phe 382 in the DFG motif, and His 361 in the HRD motif. This spine is anchored to helix αF, which is buried in the center of the C-lobe through a hydrogen bond between a conserved aspartate residue, Asp 421, and the hydroxyl group of a histidine residue in the HRD motif [7]. The assembly of the regulatory spine is coupled to the conformation of the activation loop and therefore critical for substrate binding [6]. The catalytic spine assures connectivity between the N-lobe and the C-lobe upon nucleotide binding [7]. It incorporates the adenine ring of ATP and the residues located in helix αF (Leu 428 and Ile 432), helix αD (Leu 323), and Cys 369 in the C-lobe and the residues in the N-lobe: Val 371 and Leu 370 in the β7 strand, Val 256 in β2 strand, and Ala 269 in β3 strand. Through helix αF, assembly of the regulatory spine is linked to the assembly of the catalytic spine, which emphasizes the essential role of this helix for kinase catalysis.

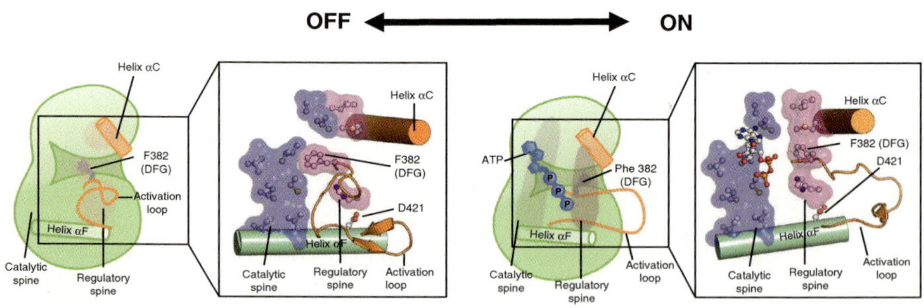

Fig. 9.2 Assembly of the hydrophobic spines. Cartoon representation of the hydrophobic spines (the regulatory and the catalytic spine) in the disassembled state in the inactive kinase conformation (*left*) and in the assembled state in the active conformation (*right*). The insets on the right of the cartoons show surface representation of the residues corresponding to the regulatory and catalytic spines in the inactive (PDB code: 1OPJ) and in the active structure (PDB code: 2G2I) of the Abl kinase

The flexibility of the spines' assembly is critical for normal function of a kinase. This is best evidenced by secondary mutations found in cancer patients treated with kinase inhibitors [8]. These mutations target the residue denoted as the gatekeeper residue (Thr 315 in Abl), so named because of its importance for inhibitor specificity [9]. Specifically, in patients treated with kinase inhibitors against an oncogenic fusion of the Abl kinase, BCR-Abl, and the epidermal growth factor (EGF) receptor, secondary mutations that replace the gatekeeper residue (Thr 315 in BCR-Abl, Thr 766 in EGF receptor) with a bulkier hydrophobic residue (Ile in BCR-Abl, Met in EGF receptor) are common. The gatekeeper residue is located in the direct vicinity of the regulatory spine (in BCR-Abl the gatekeeper residue is directly adjacent to Leu 301). The replacement of the gatekeeper by a bulkier hydrophobic residue allows for its direct interaction with the regulatory spine and consequently results in the extension and further stabilization of the spine. This likely reduces the ability of the regulatory spine to disassemble, thereby shifting the equilibrium towards the active conformation of the kinase [8].

9.3 Mechanisms for Kinase Activation

9.3.1 Inactive State of the Kinase

Generally, in the absence of an activating stimulus, kinases remain in the inactive states. These states are stabilized by a network of inhibitory interactions that are provided either intra- or intermolecularly. This inhibition, aimed to reduce basal activity of the enzyme, is an essential regulatory step in kinase signaling. Structural studies of the inactive states of kinases have yielded insights into these inhibitory mechanisms. In many cases, the inactive structures helped to elucidate how the transition between the inactive and active states occurs in kinases. In addition, these structural insights highlighted a remarkable diversity of the inactive states adopted by kinases. This is in contrast to the catalytically active kinase conformations, which tend to be relatively similar between all kinases because they have to satisfy the structural requirements of efficient catalysis.

The first inactive conformation of a kinase to be solved was of a serine/threonine kinase, cyclin-dependent kinase 2 (CDK2) [10]. It was subsequently observed in the structure of the Src tyrosine kinase and therefore denoted as CDK-/Src-like inactive conformation. In this conformation the relative orientation of the N-lobe and C-lobe changes to more compact than observed in the active conformation (Fig. 9.3a). The helix αC is displaced from the active site and moves outward from the N-lobe, breaking the salt bridge between the catalytic lysine residue and the glutamate present in helix αC and distorting the regulatory spine. In this conformation, the catalytic lysine forms a salt bridge with the DFG aspartate instead. The conformation of the activation loop also changes significantly in the CDK-/Src-like inactive conformation. It forms a single or double turn helix, which stabilizes the outwardly directed helix αC through a set of hydrophobic interactions. In this conformation,

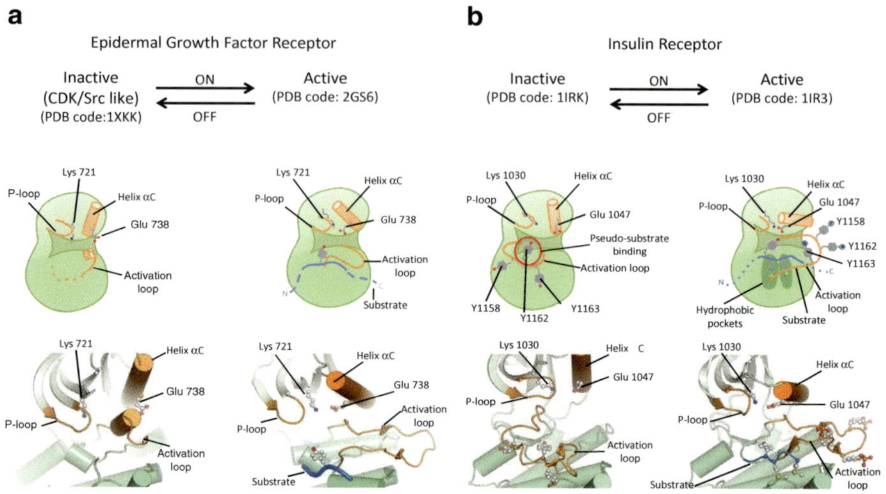

Fig. 9.3 Inactive states of kinases. (**a**) The transition between an inactive (CDK-/Src-like) and an active state of the epidermal growth factor (EGF) receptor is depicted using cartoon representations (*top panel*). The *bottom* panels represent a magnified view of an active site region from the crystal structures of an active (PDB ID: 2GS6) and inactive (PDB ID: 1XKK) EGF receptor kinase that depicts conformational changes associated these different states. The catalytic lysine and glutamate residues and the substrate tyrosine are shown in *ball* and *stick* representation. (**b**) The transition between an inactive and active state of the insulin receptor is controlled by the phosphorylation of the tyrosines in the activation loop (Tyr 1158, Tyr 1162, and Tyr 1163). The *bottom* panels represent a magnified view of the active site region from the crystal structures of an active (PDB ID: 1IR3) and inactive (PDB ID: 1IRK) insulin receptor kinase that depicts a pseudo-substrate inhibition by the unphosphorylated Tyr 1162 in the inactive conformation. The catalytic lysine and glutamate residues, activation loop, and the substrate tyrosines are shown in *ball* and *stick* representation

the substrate-binding site on the activation loop is occluded, and in addition the activation loop physically blocks the entry of the substrate to the active site.

Despite the fact that CDK-/Src-like inactive conformation was observed in the structures of both tyrosine (Src) and the serine/threonine kinases (CDKs), the inactive conformations of kinases had been considered to be diverse between different kinases and relatively uninformative regarding kinase catalysis [11]. However, the CDK-/Src-like inactive conformation was repeatedly observed in the structures of many kinases, including receptor tyrosine kinases such as human epidermal growth factor (EGF) receptor and hepatocyte growth factor receptor (MET) and serine/threonine kinases such as Wnk and NEK2 [12]. The reoccurrence of the same inactive conformation among distantly related kinases raises a possibility that this conformation might be significant for kinase regulation. Because the switch between the active and CDK-/Src-like inactive conformation involves significant change in the N-lobe orientation relative to the C-lobe that allows for movement of helix αC, it has been suggested that the CDK-/Src-like conformation is an intermediate in kinase catalysis that facilitates "DFG motif flipping" [12, 13]. Flipping of the DFG

motif is associated with large crankshaft-like motions of the peptide backbone that result in placement of the DFG phenylalanine in the position of the DFG aspartate. The DFG-aspartate hydrogen bonds with the metal ions and stabilizes ATP binding when a kinase is in the active conformation. The replacement of the DFG aspartate with DFG phenylalanine in the active site results in displacement of the nucleotide, a step in catalysis that has been shown to be rate limiting [14, 15]. Kinases might therefore use the CDK-/Src-like conformation for the regulation of their nucleotide binding and release—an exciting hypothesis that illustrates how seemingly inactive state of a kinase might be essential for catalysis.

The inactive conformations other than the CDK-/Src-like inactive conformation have been observed in numerous kinases. In the structure of the unphosphorylated IR kinase, helix αC is also swung outwardly from the active site, breaking the Lys 1030–Glu 1047 salt bridge; however, the activation loop adopts a conformation, which is significantly different from the one observed in the CDK-/Src-like inactive conformation (Fig. 9.3b). IR kinase has three tyrosine residues in the activation loop (Tyr 1158, Tyr 1162, and Tyr 1163), whose phosphorylation is necessary for kinase activation. In the inactive state, one of the tyrosine residues (Tyr 1162) is bound in the active site in a manner similar to a tyrosine residue in the substrate peptide [16]. Despite the substrate-like binding, Tyr 1162 cannot become phosphorylated because of the distorted active site, which is unable to bind ATP and magnesium. The IR kinase hence uses its own activation loop for blocking the substrate-binding site at the same stabilizing a catalytically incompetent conformation of the kinase active site.

A similar obstruction of ATP and substrate binding is visible in the inactive structure of the receptor tyrosine kinase, muscle-specific kinase (MUSK) where the activation loop adopts a conformation similar to the one seen in the inactive IR kinase [17]. In contrast, the activation loop in the fibroblast growth factor receptor 1 (FGFR1), which has two tyrosine phosphorylation sites, adopts a significantly different conformation [18]. In FGFR1 none of the tyrosine residues are bound in the active site, and the binding of the nucleotide is still possible. The conformation of the activation loop is however not permissive for substrate binding, effectively blocking the phosphorylation of the FGFR1 substrates. These selected examples demonstrate the critical role of the conformation changes in the activation loop for kinase activation and highlight how divergent can be the inactive states that kinases adopt. This diversity likely helps kinases to achieve specificity since very accurate and customized mechanisms had to evolve for catalytic activation of each kinase.

9.3.2 Activation by Phosphorylation of the Activation Loop

The essential role of the activation loop in stabilization of the inactive state of kinases provides an insight into a mechanism by which ligand binding triggers transition from the inactive to active state [16, 18]. In majority of kinases, this transition involves phosphorylation of the highly conserved sites in the activation loop, which include tyrosines or serines and threonines in tyrosine or serine/threonine kinases, respectively. Introduction of the negatively charged phosphate has significant structural consequences on the activation loop and usually breaks tight interactions between the activation loop and the kinase core observed in the inactive kinase structures. In addition, the presence of phosphorylated sites makes the activation loop competent for substrate binding (e.g., in Fig. 9.3b). The phosphorylation of the activation loop is an almost universally shared mechanism of activation among protein kinases including receptor tyrosine kinases such as MET, ephrin B receptor (EphB), FGFR1, IR, and vascular endothelial growth factor receptor (VEGFR) [19, 20]. The critical role of activation loop phosphorylation in releasing the autoinhibitory constraints on the kinase domain is further emphasized by findings that the mutations of an activation loop tyrosine that mimic phosphorylation, such as Y1235D in MET, are found in cancer [21].

The presence of more than one tyrosine residue in the activation loop of several tyrosine kinases suggests a more complex role for activation loop phosphorylation in the kinase activation mechanism. Members of the insulin receptor kinase family including insulin receptor (IR), insulin-like growth factor receptor 1 (IGFR1), and MUSK are regulated by the intermolecular activation at three different sites. In the case of IR and IGFR1, each phosphorylation event results in increase in enzyme turnover number and the decrease in K_M values for ATP and peptide substrate, leading to overall 120-fold increase in catalytic efficiency when all three sites are phosphorylated [22, 23]. While the first phosphorylation event plays a role in destabilizing the inactive conformation, the subsequent ones are important for stabilization of the active state [24]. Sequential phosphorylation events are also critical for the activation of the fibroblast growth factor receptor 1 (FGFR1), which has two tyrosine phosphorylation sites in the activation loop [25]. In the first step, a tyrosine (Tyr 653) in the activation loop becomes transphosphorylated, which results in 50–100-fold increase in kinase activity. This increase is sufficient to trigger autophosphorylation of FGFR1 on it effector-docking sites located in the juxtamembrane segment, kinase insert region, and the C-terminal tail and subsequent recruitment of downstream effectors. Only then, the second tyrosine site in the activation loop (Tyr 654) of the FGFR1 kinase is phosphorylated to enable phosphorylation of the recruited effectors. This carefully orchestrated sequence of events likely plays a role in controlling the fidelity of receptor/effector interactions.

In receptor tyrosine kinases, activation loop phosphorylation is triggered by ligand-induced receptor oligomerization. It is well established that activation loop phosphorylation occurs by an intermolecular mechanism in which kinase domains transphosphorylated each other upon oligomerization. The analysis of the crystal structures of kinases such as IR kinase and MUSK helps to understand why kinases do not autophosphorylate in cis. This is because the activation loop imposes steric constraints on the active site and destabilizes the active conformation, when its own tyrosine residues are presented as substrates. There are some rare exceptions to this rule however, and some kinases can phosphorylate their activation loop residues in an intramolecular fashion either in partially unfolded state (such as dual-specificity tyrosine-regulated (DYR) kinase [26]) or when two kinases are located on one polypeptide chain, such as ribosomal S6 kinase (RSK) and the mitogen and stress-activated protein kinase (MSK), in which the N-terminal kinase is phosphorylated by the C-terminal kinase [26, 27].

9.3.3 Activation by the Activation Loop Phosphorylation-Independent Mechanism: Unique Case of HER Receptors

Human epidermal growth factor receptors (HERs), which include HER1/EGF receptor, HER2, HER3, and HER4, are thus far the only known receptor tyrosine kinases that do not require phosphorylation of the activation loop tyrosine to for catalytic activation. Although these receptors conserve the phosphorylation site in the activation loop, which undergoes rapid phosphorylation upon ligand binding [28], this phosphorylation event is not required for receptor activation [29–31]. Early crystallographic studies showed that the unphosphorylated activation loop of EGF receptor can adopt a fully extended conformation, analogous to a conformation induced by phosphorylation in other kinases, and as such it is compatible with substrate binding and an active catalytic state [31, 32]. These observations led to speculations that EGF receptor does not adopt an autoinhibited state and is always in the active conformation [32]. The subsequently obtained crystal structure of the EGF receptor kinase domain bound to an ATP analog drug, lapatinib, captured EGF receptor kinase in the inactive state for the first time [33]. This inactive conformation was structurally analogous to the CDK-/Src-like inactive conformation (Fig. 9.3a), suggesting that it might be physiologically relevant. Further support for this notion came from finding that the CDK-/Src-like inactive conformation of EGF receptor has been repeatedly observed in several other EGF receptor kinase domain structures, as well as in the structures of its homologues, HER4 and, the catalytically inactive member of the HER family, HER3 [31, 34–36]. The realization that EGF receptor adopts an inactive conformation underscored that there must be an activation mechanism in place to control EGF receptor activation, but it is likely independent from the phosphorylation of the activation loop.

How do HER receptors transition between the CDK-/Src-like inactive and active conformation in a manner independent from phosphorylation of the activation loop? Close analysis of the EGF receptor kinase domain crystal structures in which the kinase adopts an active conformation has revealed a unique mechanism behind catalytic activation of HER family of receptors [31]. This mechanism involves formation of an asymmetric dimer between the kinase domains in which the N-lobe of one kinase domain (denoted as a "receiver" kinase) interacts with the C-lobe of the second kinase domain (denoted as an "activator" kinase) forming a large hydrophobic interface spanning ~2,000 Å^2 (Fig. 9.4a). This interface undergoes significant structural rearrangements in the receiver kinase upon its transition from the inactive to active state (Fig. 9.4b). Once established, the hydrophobic interactions stabilize

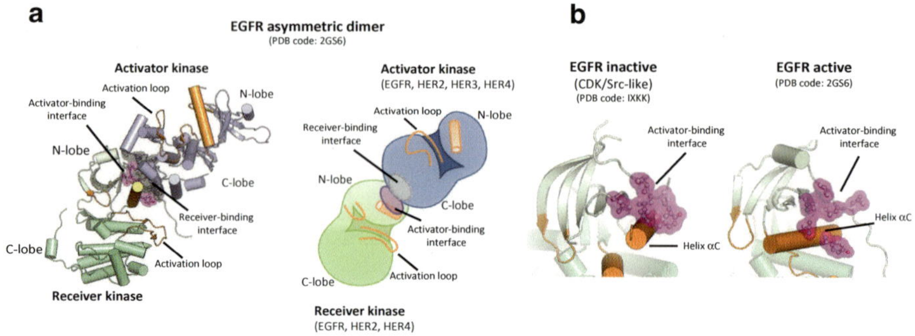

Fig. 9.4 Asymmetric dimerization underlies activation of the EGF receptor kinase domain. (**a**) A crystal structure and the cartoon representation of the head-to-tail asymmetric EGF receptor kinase domain dimer. The relative position of the activator and the receiver kinase, as well as of the catalytically critical components such as helix αC and the activation loop, is shown. The allosteric dimer interface is highlighted using *magenta* and *gray* colors. (**b**) The conformational changes within the hydrophobic activator-binding interface that occur during transition from the inactive CDK-/Src-like conformation to the active state in the EGF receptor are depicted using stick and surface representation and colored in *magenta*. During activation, the hydrophobic residues become exposed enabling binding of the receiver kinase. The PDB IDs for the used EGF receptor structures are 1XKK (inactive) and 2GS6 (active)

few structural features within the N-lobe of the receiver kinase domain that are critical for its catalytic activation. As a result, helix αC in the receiver kinase is positioned in the swung-in conformation enabling formation of the salt bridge between the catalytic lysine (Lys 721 in EGF receptor) and the glutamate contributed by helix αC (Glu 738 in EGF receptor). The receiver kinase therefore becomes catalytically activated by the activator kinase which itself remains catalytically inactive. The asymmetric dimer interface is highly conserved in the HER family of receptors and has been observed in the subsequently obtained crystal structures of the HER4 and HER2 kinase domains [35, 37]. The catalytically inactive HER3 receptor is an exception as its N-lobe interface is highly divergent and prevents HER3 from adopting a position of the receiver kinase [34]. HER3 can only act as an activator kinase and therefore is thought to exclusively signal only in heterodimers with other HER family members (Fig. 9.4a).

The allosteric activation mechanism operative in the HER family of kinases is reminiscent of the way by which cyclin-dependent kinases (CDKs) are activated by their allosteric activators, the cyclins. Although the overall binding between a CDK and a cyclin is quite different from the asymmetric HER kinase dimer interface, it also involves a largely hydrophobic patch and results in stabilization of helix αC in the swung-in conformation [38]. It turns out that similar principle is followed by many other kinases, such as PKA, PDK1, Aurora kinase, and RET receptor tyrosine kinase, in which divergent *cis* or *trans* regulatory elements rely on primarily hydrophobic interactions to position helix αC in the catalytically active conformation [12]. However, although allosteric activation by cyclins is indispensable for activation of CDKs, CDKs and other kinases relying on the hydrophobic patch are still dependent on the activation loop phosphorylation to achieve full catalytic activity [39]. This is in contrast to EGF receptor, where activation loop phosphorylation seems to be largely dispensable for kinase activation [29–31].

9.3.4 Non-catalytic Role of Activation Loop Phosphorylation in HER Receptors

Although not needed for receptor activation, the phosphorylation of the activation loop tyrosine in HER receptors seems to play an important role in their downstream signaling. Mutation of the activation loop tyrosine in the EGF receptor (Tyr 845) to a phenylalanine inhibits EGF-dependent DNA synthesis [28, 30]. The functional consequence of mutating the equivalent tyrosine residue in HER2, Tyr 899, seems to be even more complex. Although mutation of Tyr 899 to phenylalanine in HER2/neu, which carries an activating mutation V664E in the transmembrane domain, does not affect its phosphorylation in vivo, the mutation decreases transforming potential of HER2/neu and the catalytic activity of the immunoprecipitated receptor from cells [40].

One of the possible non-catalytic roles of activation loop tyrosine phosphorylation in HER receptors could be to facilitate substrate binding to the kinase active site, similar to its role in CDK2 and several other kinases [39]. This is however unlikely the case because there are no observed effects of Tyr 845 to phenylalanine mutation in EGF receptor on K_M of binding of the substrate peptides derived from the C-terminal tail of the receptor [41]. It seems that the phosphorylated activation loop more likely acts as a scaffold for recruiting downstream effectors to EGF receptor, such as STAT5 [42], or to mediate EGFR localization to the subcellular compartments [43]. Phosphorylated Tyr 845 recruits cytochrome C oxidase subunit II (CoII) to EGFR, and this association was suggested to be important in redistribution of EGF receptor to mitochondria [43].

As already discussed, the phosphorylation of the activation loop tyrosines is thought to occur intermolecularly when two receptor tyrosine kinases oligomerize in response to ligand binding. The intriguing question is how the activation loop tyrosine becomes phosphorylated in the EGF receptor in which asymmetric geometry of the active dimer keeps the activation loops away from their active sites (Fig. 9.4a). In support of this structural prediction, studies have shown that the wild-type EGF receptor does not efficiently autophosphorylate Tyr 845 and that Tyr 845 is predominantly phosphorylated indirectly, in the Src kinase-dependent manner [28, 41]. Interestingly, EGF receptor variants found in non-small cell lung cancer (NSCLC) patients that carry mutations (Leu 834 to Arg and Leu 837 to Gln) can autophosphorylate Tyr 845 in the ligand-independent manner in cells and also in vitro using the purified kinase domains [44]. These mutations are thought to destabilize the CDK-/Src-like inactive state and increase basal level of kinase activity [45]. This finding demonstrates that under certain conditions the EGF receptor has the physical ability to phosphorylate Tyr 845, which might be relevant for the fine-tuning of its downstream signaling.

9.4 Structural Basis for Substrate Binding and Specificity

The activated receptor tyrosine kinases can phosphorylate both their own tyrosine residues and the tyrosine sites of their bound effectors. The structural insights into how substrate sequences are presented to the active site of kinases in both of these cases are significantly more limited than our knowledge about ATP binding and hydrolysis. This is likely due to transient nature of interactions between substrate sequences and the kinase, which makes it difficult to capture them in complexes suitable for high resolution analysis such as X-ray crystallography. In the in vitro enzymatic assays, kinases usually display low substrate specificity towards the peptides containing the phosphorylation sites. Substrate recognition by kinases in vivo is however very specific, which points to the determinants other than substrate sequence alone that underlie substrate binding and selection. Those factors may include interaction with the domains adjacent to the catalytic domain, cellular co-localization of kinases with their substrates, additional third party cofactors, or possibly posttranslational modifications on either a substrate or a kinase that facilitate binding. Recent structural insights into some of these interactions underscore how diverse and unique these interactions are for each kinase (Fig. 9.5).

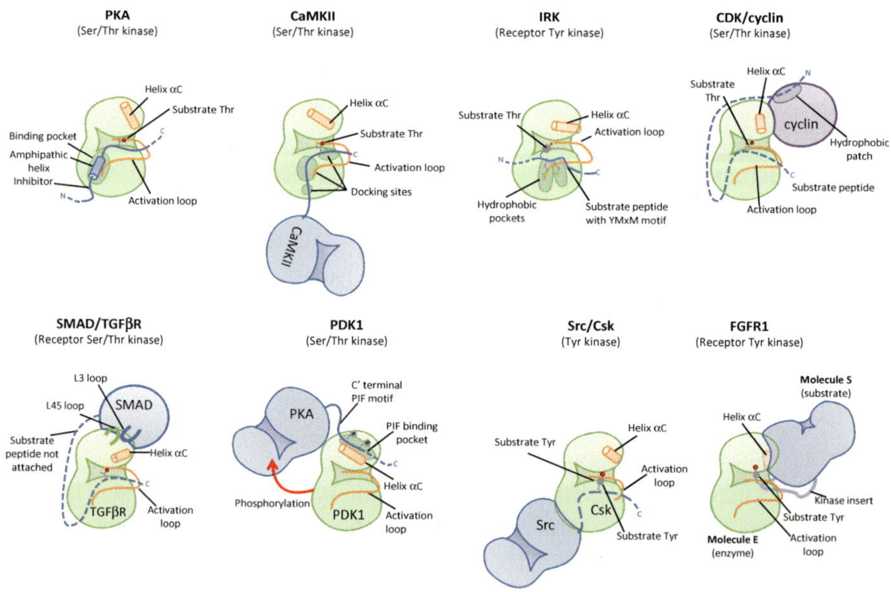

Fig. 9.5 Modes of substrate presentation. Cartoon representation of diversity of the mechanisms by which kinases bind and recognize their substrates, illustrating the examples described in the text. The following PDB IDs were used as a template for the cartoon images: PKA (PDB ID: 1ATP), CamKII (PDB ID: 3KK8), IRK (PDB ID: 1IR3), CDK (PDB ID: 1JSU), TGFβR (PDB ID: 1B6C), PDK1 (Based on the proposed model) [58], Csk (PDB ID: 3D7T), FGFR1 (PDB ID: 3KXX)

9.4.1 Substrate Recognition by Serine/Threonine Kinases

The first structures of PKA already visualized the kinase/substrate interaction in the ternary complex of this serine/threonine kinase with Mg-ATP and a peptide inhibitor [3, 46, 47]. This naturally occurring peptide inhibitor, which competes with the substrate peptide for binding to PKA, uses an amphipathic helix to bind to the bottom of the C-lobe of the kinase between helices αD and αG. The amphipathic helix is followed by an extended region that positions itself in the cleft between the C-lobe and the N-lobe and directs a five-residue consensus sequence towards the active site (Fig. 9.5) [48]. The peptide inhibitor makes several other additional peripheral interactions with the kinase, which are both hydrophobic and hydrophilic. The diversity of these interactions suggested that PKA could tolerate many different substrates, an observation already made based on the in vivo phosphorylation studies.

A similar binding mode was recently exemplified in a structure of another serine/threonine kinase, calcium/calmodulin-dependent kinase II (CaMKII). CaMKII exists as a holoenzyme composed of 12 kinases arranged around a ring-shaped scaffold, six kinases on each side. The scaffold is assembled from the hub domains that connect directly to the kinase domains. In the inactive holoenzyme adjacent kinases position their regulatory regions that contain phosphorylation sites in a daisy-chain arrangement such that they bind a neighboring kinase in the analogous substrate-binding cleft observed in PKA (Fig. 9.5) [49]. Interestingly, in the inactive CaMKII holoenzyme, binding of the regulatory region in the substrate-binding cleft stabilizes the inactive conformation of the CaMKII kinase through facilitating autoinhibitory interactions with the hub domain. Since in such conformation the phosphorylation site in the regulatory region is positioned right in the active site of the kinase, its phosphorylation is predicted to destabilize the inactive conformation between kinase and hub domains and consequently the activation of CaMKII holoenzyme [49]. So CaMKII kinase domain uses a conserved substrate-binding channel for both inhibition and substrate presentation, facilitating the switch between on and off states of the holoenzyme.

9.4.2 Substrate Recognition by Tyrosine Kinases

The structure of the activated insulin receptor kinase in complex with a non-hydrolyzable ATP analog, AMP-PNP, and the 18-residue substrate peptide provided first glimpse into substrate binding in a tyrosine kinase and revealed a slightly different binding mode than observed in PKA (Fig. 9.2) [23]. Only six substrate residues were visualized in the structure, in which they occupy two hydrophobic pockets in the C-lobe formed by helices αEF and αG. This portion of the peptide binds as a short antiparallel β-strand that pairs in the antiparallel fashion with the β11 strand of the C-lobe. The six-residue sequence contains a YMxM motif (where x is any amino acid), which represents a consensus-binding site for the p85 regulatory subunit of phosphoinositide 3-kinase (PI3K) that is specifically phosphorylated by the insulin receptor. The structure revealed how this specificity can be achieved by visualizing recognition of the methionine side chains by two hydrophobic pockets localized in the C-lobe of the insulin receptor. The differences in amino acid composition of these pockets likely play a role in determining substrate specificity in kinases and are sometimes a target of mutations in cancer [50–52].

The insulin receptor kinase structure also shed light on the role of the activation loop in binding and positioning substrate sequences for phosphorylation. In the structure, the activation loop of the receptor is in a fully extended conformation, stabilized by phosphorylation of three tyrosine residues (Y1158, Y1162, Y1163,) in the activation loop. In this conformation, the activation loop creates a binding site for the substrate peptide, which positions the YMxM tyrosine residue directly into the active site. The hydroxyl group of a tyrosine is further stabilized by interaction with the conserved aspartate (D1132) in the HRD motif, and the upstream arginine, R1136. In the inactive IR kinase, the activation loop is not phosphorylated and adopts a more compact conformation that occupies the substrate-binding site [16]. In addition, in the inactive kinase Y1162 in the activation loop plays a role of a pseudo-substrate and takes the place of the YMxM tyrosine, making analogous interactions with D1132 and R1136. In this case, however, the phosphotransfer cannot occur due to the lack of the catalytically important salt bridge between Lys 1039–Glu 1047 resulting in nonproductive ATP coordination. Phosphorylation of the activation loop is therefore important for both releasing the autoinhibitory interactions in the kinase domain and for stabilizing binding of substrates in a conformation that is compatible with phosphotransfer.

How do individual kinases maintain specificity for their substrates? There are certainly at least a few different strategies that have been described. The sequence of the activation loop including the number of phosphorylation sites and the length of the activation loop are relatively divergent often contributing to substrate-binding specificity. The depth of the catalytic cleft between the C-lobe and the N-lobe is a critical determinant whether a kinase is a tyrosine kinase (deeper cleft) or a serine/threonine kinase (shallower cleft) [23]. This rule is more stringent in the case of tyrosine kinases; serine/threonine kinases can sometimes phosphorylate tyrosine residues [53]. Although the interaction between the phosphorylation site of the substrate and the active site of the kinase provides a considerable amount of binding energy and specificity, it is frequently not sufficient. Kinases rely on additional mechanisms to increase binding specificity of their substrates, and the diversity of these mechanisms is remarkable (Fig. 9.2) [54]. An interesting example is provided by cyclin-dependent kinases (CDKs), which become catalytically activated upon binding their allosteric activators, cyclins. In addition to using cyclins for activation, the CDKs also utilize cyclins for help in binding a subset of substrates. This is achieved through a conserved hydrophobic patch localized on cyclins, which is localized ~35 Å away from the kinase active site when cyclin is bound to a CDK. This hydrophobic patch binds a consensus motif RXL present in some CDK substrates and is critical for their phosphorylation by CDKs [55].

Another example comes from the family of the transforming growth factor (TGF) β-receptors, which are transmembrane serine/threonine kinases that phosphorylate the C-terminal tail of its cytoplasmic substrates SMADs. Despite being the direct substrate for the TGFβR receptor kinase, the SMAD's tail does not significantly contribute to receptor binding. On the contrary, SMADs bind TGFβR stronger in the absence of the tail [56]. This is because SMADs utilize a different region to bind to the TGFβR kinase which includes the residues located in their so-called L3 loop. The L3 loop of SMADs binds to the L45 loop located between β4 and β5 strands in the N-lobe kinase domain of TGFβR creating an interface that is sufficient to determine the specificity of binding [57]. This interface also determines the difference between substrate recognition by the TGFβ receptors and their close homologues, bone morphogenetic protein (BMP) receptors in which residue composition of the L45 loop and of the L3 loop in its substrates is different. Another serine/threonine kinase that uses the N-lobe to bind its substrates is PDK1. PDK1 binds and phosphorylates the members of the AGC family of kinases that contain a PDK1-interaction fragment motif (PIF). The PIF motif binds to the hydrophobic pocket in PDK1 located in the N-lobe next to the helix αC [58]. In addition to serving as a substrate-recruiting motif, PIF binding is also necessary for the catalytic activation of PDK1 [59], increasing the specificity with which the catalytic activity of PDK1 is directed towards its substrates.

An interesting example of a substrate docking mechanism that is completely independent from the activation loop is illustrated by the binding between the Csk tyrosine kinase and its substrate Src kinase. In Csk the activation loop is significantly shorter than in other kinases, which results in the loss of the Csk ability to bind substrate peptides via its activation loop. Instead Csk evolved an alternative mode of substrate presentation in which Csk utilizes the side of its C-lobe to dock to the bottom of the C-lobe of Src [60]. This docking mechanism assures high substrate-binding specificity, and Src kinases are the only known substrates of Csk.

For majority of other kinases, including the receptor tyrosine kinases, the substrate-presenting mechanisms remain obscure and still await structural characterization. Only recently, an exciting insight into the mechanism of transphosphorylation of the members of the FGFR family of receptor tyrosine kinases was provided by the analysis of the crystal structures of the FGFR1 and FGFR2 kinase domains [61]. The structure of the FGFR1 kinase revealed an interface formed between the N-lobe of one kinase domain that serves as an enzyme and the C-lobe of another kinase domain that serves as a substrate (Fig. 9.2). In this asymmetric arrangement the substrate kinase presents the phosphorylation site located in the kinase insert region into the active site of the enzyme kinase capturing an active complex between two FGFR1 kinases. This interface is disrupted by mutations in several human diseases in which loss of function in FGFR1 has been reported [62, 63]. In an earlier structure of the FGFR2 kinase domain, a slightly different asymmetric dimer with partial overlap with the interface observed in FGFR1 captures a presentation of another substrate tyrosine for phosphorylation [64]. Since activation of the members of the FGFR family requires sequential phosphorylation of the multiple tyrosine residues [25], these snapshots likely represent different steps of the activation process and demonstrate that each phosphorylation event is dependent on formation of a highly specific molecular complex between the enzyme and the substrate.

9.5 Structural Basis for Regulation of Kinase Activity in Receptor Tyrosine Kinases

Because activation of receptor tyrosine kinases is critical for cellular homeostasis, it is subject to multiple regulatory steps prior and post ligand binding. An essential aspect of this regulation is the maintenance of low basal activity of a kinase in the absence of an activating stimulus. This is challenging in case of membrane receptor kinases, which operate under conditions of high local concentration favoring their oligomerization and subsequent activation even in the absence of a ligand. To limit such spontaneous incidents, receptor tyrosine kinases evolved intricate autoinhibitory mechanisms that keep their basal activity in check. A subset of these mechanisms impinges directly on the kinase domain by restricting it in the inactive state through a set of homo-oligomeric interactions or engagement of the adjacent domains, such as the juxtamembrane segment and the C-terminal tail. Current structural and functional data underscore how diverse are the mechanisms by which receptors are stabilized in the inactive state in the absence of a ligand. Likewise, the mechanisms by which ligand binding induces receptor kinase activity seem to be unique for each receptor subtype and in numerous instances challenge the canonical view of receptor tyrosine kinase activation that states that ligand binding brings two monomers together to form an active dimer. Characterizing the molecular basis behind these differences is necessary to understand how receptor tyrosine kinases achieve precision and specificity of signaling and to advance therapeutic treatments directed against the deregulated states of these kinases in human diseases.

9.5.1 *Receptor Oligomerization*

In general receptor tyrosine kinases are activated by ligand-induced oligomerization, most commonly dimerization, which results in juxtapositioning of the kinase domains of receptor monomers next to each other allowing for transphosphorylation. This canonical mechanism applies to receptors such as FGFR family and PDGFR family of receptor tyrosine kinases (Fig. 9.6a). This simple model does not apply to several other receptor tyrosine kinases that were demonstrated to be pre-organized into higher-order oligomeric states rather than monomeric in the absence of a ligand (Fig. 9.6). These inactive oligomers are thought to provide

Fig. 9.6 (continued) state of the insulin receptor, but leads to a rearrangement of the catalytic domains resulting in activation. (**c**) In the absence of a ligand, EGF receptor is thought to exist as a monomer or a preformed inactive dimer. The activation requires formation of at least EGF receptor dimers, but higher-order EGF receptor oligomers have also been reported. (**d**) HER3, a catalytically inactive member of the HER family of receptors, has been reported to homoligomerize in the ligand-free state. Ligand binding was suggested to cause HER3 dissociation and subsequent heterodimerization with other HER family members. (**e**) The activation of ephrin receptors (Ephs) requires high-order clustering of receptors by binding the clusters of membrane-associated ligands presented by a neighboring cell

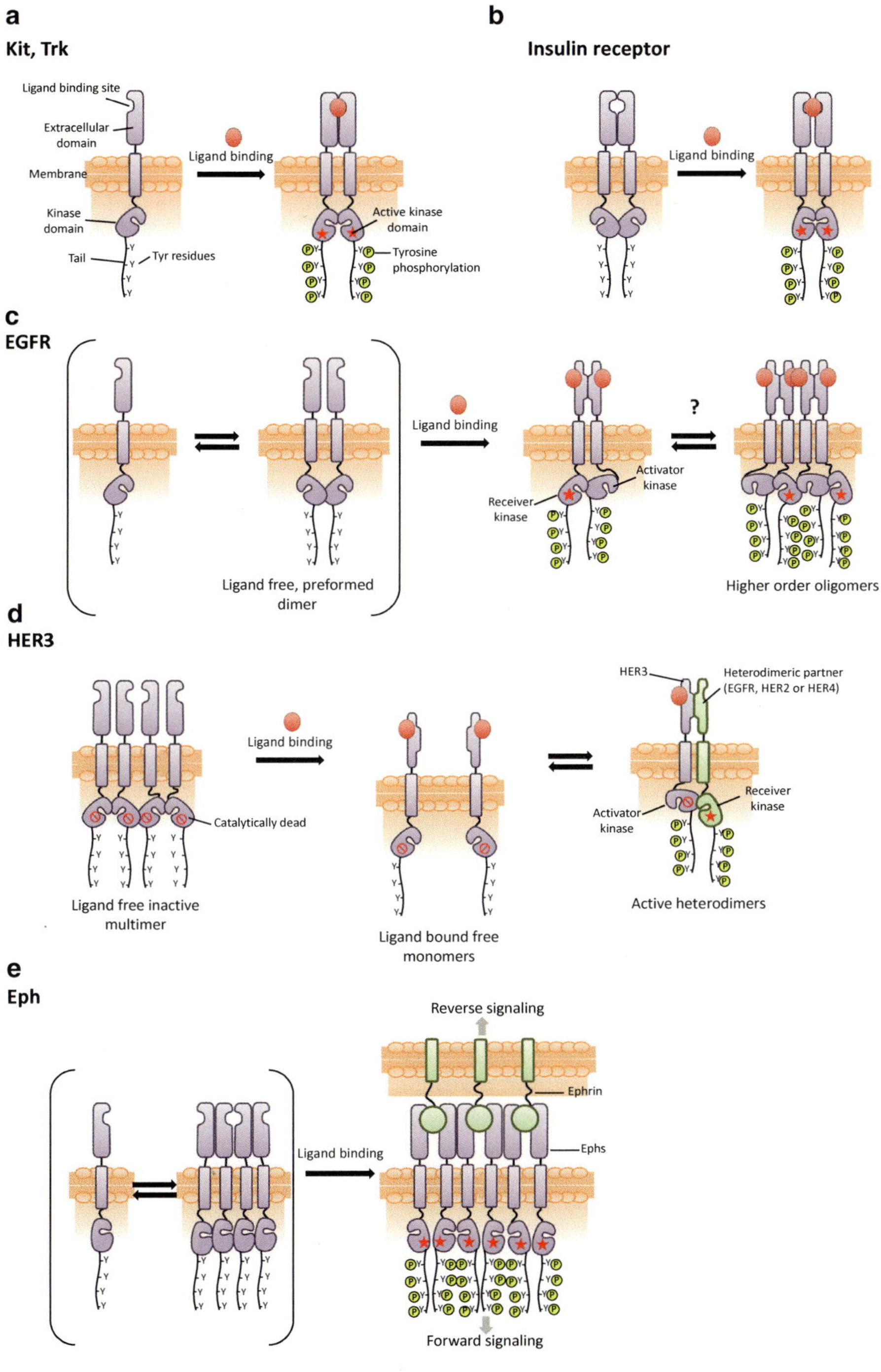

Fig. 9.6 Different activation strategies of receptor tyrosine kinases by oligomerization. (**a**) A canonical mechanism in which a monomeric RTK is activated by ligand-induced dimerization is represented by c-Kit and Trk receptors. (**b**) Ligand binding does not affect the oligomerization

autoinhibitory constraints that lock the receptors in the inactive state and prevent their ligand-independent activation. At the same time pre-organization of receptors into inactive oligomers prior to ligand binding could facilitate their activation once the ligand becomes available, due to local concentration of receptor molecules. This strategy might be particularly important under conditions of low receptor abundance in physiological setting.

The classical example of receptor organization into inactive dimers is provided by the insulin receptor family, whose members exist as disulfide-linked dimers both in the absence and presence of a ligand (Fig. 9.6b) [65]. Ligand binding induces a still not fully understood conformational change in the inactive dimeric receptor complex that results in the catalytic activation of the kinase domains. The presence of preformed ligand-independent inactive oligomeric species of the EGF receptor on the cell surface has also been repeatedly reported (Fig. 9.6c) [66–71]. Single-molecule fluorescence resonance energy transfer (FRET) and diffusion studies indicated that EGF receptor is in the equilibrium between the monomeric and dimeric state in the absence of a ligand, with the dimeric state being more predominant [67, 70]. In contrast to the insulin receptor, the inactive EGF receptor oligomers are most likely not stabilized by covalent interactions, but their exact structure and composition remain to be determined. The structural model of the inactive dimer between the extracellular domains was proposed based on the crystal structure of the inactive (tethered) extracellular domain of EGF receptor; however, so far its significance for receptor autoinhibition was not unequivocally demonstrated [72]. Likewise, the proposed structural models of inactive intracellular dimers of EGF receptor still await their full functional characterization [73, 74]. Due to the limited insight into the nature of inactive EGF receptor oligomers, it is unclear how ligand binding promotes their transition to an active state. In one model, formation of higher oligomeric species, such as tetramers, has been proposed as a mechanism for ligand-induced activation of the preformed inactive dimers of the EGF receptor (Fig. 9.6) [69].

Another HER receptor with a demonstrated propensity to oligomerize prior to ligand binding is the catalytically inactive HER3 [75, 76]. Since the catalytically inactive HER3 only functions in heterodimers and does not homodimerize upon ligand binding [77], its "active" state corresponds to a liganded HER3 monomer, which is competent to dimerize with other HER receptors (Fig. 9.6d). Sequestration of HER3 in inactive oligomers prior to ligand binding would therefore decrease the possibility that HER3 could spontaneously heterodimerize with the active HER family members. In such scenario, ligand binding should result in dissociation of HER3 homo-oligomers to a pool of monomeric HER3 that is now available for heterodimerization, and this is indeed what has been observed [76]. Although it is not yet known what is the molecular architecture of the HER3 homo-oligomers, a structural model for the oligomers formed between the isolated HER3 kinase domains was recently proposed based on the solved crystal structure of the HER3 kinase [34]. In this model HER3 kinase domains oligomerize through the alternating interactions, involving their N-lobes and the C-lobes. The N-lobe interaction interface is built by the helices αC and the C-terminal portions of the juxtamembrane segments of both interacting kinases. The C-lobe interactions are mediated by the C-terminal tails, which interact in trans with the C-lobes of two adjacent HER3 kinase domains. The flexible nature of the C-terminal tail is predicted to permit branching of the HER3 oligomers into a two-dimensional receptor lattice at the plasma membrane and hence formation of a cluster. Interestingly, analogous interactions are observed in the crystal lattices of the HER4 kinase domain crystallized in its inactive conformation. Since HER3 and HER4 are evolutionarily the closest HER receptors, it is possible that they are subject to similar regulatory mechanisms suggested by the structural models.

Perhaps an example, which is the most divergent from the canonical receptor tyrosine kinase model of activation through monomer to dimer transition, is ligand-dependent activation of the largest family of receptor tyrosine kinases—the ephrin receptors (Ephs). Their ligands, ephrins, are membrane-bound molecules and are presented to the Eph receptors on the surface of neighboring cells. In fact ephrins can activate their receptors only when bound to the membrane and soluble monomeric ephrins act as antagonists of Eph receptor signaling [78]. Ephrins localize to microclusters in the plasma membrane and upon binding to Ephs induce their clustering. The minimal signaling unit of the Eph/ephrin system is thought to be a heterotetramer that has been structurally characterized [79]. This unit is proposed to quickly expand by lateral interactions between Eph/ephrin complexes and result in formation of large receptor/ligand clusters on adjacent cells (Fig. 9.6e). These complexes then induce signaling in both interacting cells—so-called forward signaling on the receptor site and reverse signaling on the ligand site. Recent structural insights into the extracellular domains of the Ephs provided a molecular model for formation of their higher-order oligomers [80]. These studies also suggested that unliganded Ephs can be recruited to the initial receptor/ligand clusters hence allowing for spreading of the activation signal downstream from Ephs in a ligand-independent manner. The final size and the lateral mobility of the cluster determine the functional outcome of Eph/ephrin activation [81, 82], and it will be exciting to find out how these parameters are regulated during cell signaling.

9.5.2 *Regulation by the Intracellular Juxtamembrane Segment*

The intracellular juxtamembrane segment in receptor tyrosine kinases corresponds to a linker connecting the kinase domain to the transmembrane domain. The juxtamembrane segment is thought to play a vital role in transducing the conformational changes induced by ligand binding on the outside to the intracellular receptor portions. The length and sequence of the juxtamembrane segment varies significantly between receptors and therefore allows for integration of unique inputs that ultimately distinguish different receptors. In the case of the majority of receptor tyrosine kinases in which the function of the juxtamembrane segment has been characterized structurally, this region was shown to play an essential role in establishing a set of intramolecular interactions that stabilize an inactive conformation of the kinase [83].

For example, the juxtamembrane segment of Ephrin receptors (Ephs) contains a conserved stretch of ten amino acids including two tyrosines that is essential in suppressing basal activity of Ephs. The crystal structure of the EphB2 receptor fragment encompassing the kinase domain and the juxtamembrane segment revealed that in the inactive state the juxtamembrane segment of EphB2 interacts directly with helix αC in the kinase domain, utilizing two tyrosine residues, Tyr Y608 and Y620. This interaction stabilizes helix αC in a conformation that prevents hydrogen bonding between the catalytic lysine (Lys 661) and the glutamate residue (Glu 678) (Fig. 9.7a) [84]. Following ligand binding to the receptor, phosphorylation of the two tyrosine residues within the juxtamembrane segment releases the autoinhibitory interactions and fully activates EphB2.

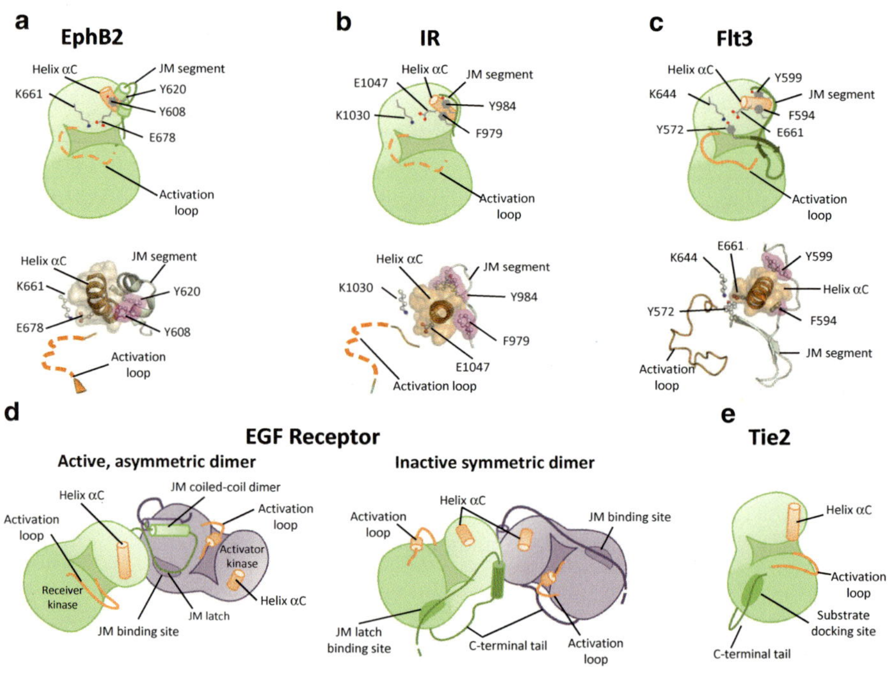

Fig. 9.7 Cis regulation of receptor tyrosine kinase activity. (**a**) In EphB2 binding of two tyrosine residues (Tyr 608 and Tyr 620) located in the juxtamembrane (JM) segment to the N-lobe of the kinase results in distortion of helix αC and misalignment of the catalytic glutamate (cartoon and structure representation based on the PDB ID: 1JPA). (**b**) The inactive conformation of the insulin receptor kinase is stabilized by the hydrophobic interactions between the juxtamembrane segment and helix αC that disrupts the salt bridge between the catalytic lysine (K1030) and glutamate (E1047) (PDB ID: 1P14). (**c**) The juxtamembrane segment of the Flt3 receptor occludes substrate-binding site and results in positioning of the Tyr 572 residue as a pseudo-substrate (PDB ID: 1RJB and 2R4B). (**d**) The juxtamembrane segment in the EGF receptor plays an important role in the stabilization of the active asymmetric dimer (model based on the PDB ID: 3G0P) through the formation of the juxtamembrane latch and the juxtamembrane coiled–coil dimer. In the structure of the inactive EGF receptor dimer, the juxtamembrane latch-binding site is occluded by the C-terminal tail (PDB ID: 3GT8). (**e**) The Tie2 receptor tyrosine kinase is subject to inhibition in cis by the C-terminal tail, which sterically blocks the substrate docking site (PDB ID: 1FVR)

In a similar fashion, the insulin receptor kinase is autoinhibited by the juxtamembrane segment that stabilizes the helix αC in a conformation, which prevents effective hydrogen bonding between the catalytic lysine (Lys 1030) and the glutamate in the helix αC (Glu 1047). In contrast to EphB2, the activation of the insulin receptor occurs in a manner independent from the phosphorylation of the juxtamembrane segment despite the presence of a potential phosphorylation site (Tyr 984) in the juxtamembrane segment (Fig. 9.7b) [85]. Yet another example of the juxtamembrane segment-mediated autoinhibition is operative in the platelet-derived growth factor (PDGF) receptor family that includes PDGFRα and PDGFRβ, Flt3, and c-Kit. The crystal structure of the Flt3 cytoplasmic domain revealed details of this mechanism [86]. In the Flt3 structure the juxtamembrane segment makes extensive contacts with the N-lobe and the C-lobe and contributes to the stabilization of the activation loop in the inactive conformation, which positions the activation loop tyrosine phosphorylation site, Tyr 842, as a pseudo-substrate (Fig. 9.7c). In the case of Flt3, phosphorylation of two tyrosine residues in the juxtamembrane segment is necessary to release the autoinhibitory interactions. The critical importance of these interactions is evidenced by numerous mutations that are observed in the juxtamembrane segments of the PDGF family of receptors in cancer [87, 88].

The juxtamembrane segment does not always play an autoinhibitory role in regulation of receptor tyrosine kinases. In the case of HER receptors, the juxtamembrane segment serves as an activation domain [89–91]. The mechanism by which this occurs was revealed by the crystal structures of the EGF receptor and HER4 receptor portions encompassing the kinase domains and the C-terminal portion of the juxtamembrane segment that is directly adjacent to the kinase domain [74, 92, 93]. In these structures the juxtamembrane segment of the receiver kinase participates in the asymmetric dimer interface providing a latch between the receiver kinase and the activator kinase (Fig. 9.7d). The juxtamembrane latch is conserved in all active HER receptors with the exception of HER3, most likely because HER3 does not function as a receiver kinase. Interestingly in the structures of the EGF receptor and HER4 receptor fragments that visualize the juxtamembrane latch, the kinase domains are in the inactive, CDK-/Src-like conformation. In case of HER4, the inactive conformation is enforced by the presence of an inhibitor [93] and in the EGF receptor by a mutation in the catalytic lysine residue (Lys 721 Met) [92]. The structures therefore suggest that an autoinhibited HER receptor kinase can occupy the receiver position thanks to the latching interaction provided by the juxtamembrane segment. Latching through the juxtamembrane segment could therefore be a first step in HER receptor activation in which two inactive kinases are brought in close proximity to form a dimer.

The function of the remaining N-terminal portion of the juxtamembrane segment of HER receptors is less understood but certainly necessary for HER receptor activation [74, 89, 90]. The N-terminal region of the juxtamembrane segment is indispensable for dimerization of the intracellular portions of EGF receptor in vitro and for activation of the full-length EGF receptor in cells [74, 89]. This segment has been proposed to dimerize through a short coiled–coil interaction and enhance the dimerization affinity of the receptor kinase domains in the active state [74]. The amphipathic nature of the coiled–coil sequences has also led to a suggestion that they additionally regulate EGF receptor activation by reversible interactions with the plasma membrane [94].

9.5.3 Regulation by the C-Terminal Tail

The carboxy-terminal (C-terminal) tail of receptor tyrosine kinases is a highly diverse region that contains tyrosine phosphorylation sites linking activated receptors to their downstream signaling outputs. In several receptors, the C-terminal tail also plays a role of an autoinhibitory domain that restricts basal activity of the receptor in the absence of ligand binding. Structurally, such regulation was first characterized in Tie2, an endothelium specific receptor tyrosine kinase critical for angiogenesis and vasculature maintenance [95, 96]. In Tie2 deletion of the C-terminal tail potentiates kinase activity. In the autoinhibited structure of the Tie2 kinase domain, the C-terminal tail is bound in the substrate-binding pocket effectively preventing substrate phosphorylation (Fig. 9.7e). The binding interface between the C-terminal tail and the substrate pocket involves the hydroxyl groups of two tyrosine residues located in the tail. These tyrosines are phosphorylated in the active state of the receptor and serve to recruit downstream effectors of Tie2. Their phosphorylation is also predicted to be a mechanism by which autoinhibition of Tie2 is released upon ligand binding.

The C-terminal tail has also been implicated in autoinhibition of the EGF receptor and its viral form vErbB in which the deletion of the tail results in receptor activation [97–99]. C-terminal deletions are also found in glioblastoma patients, spanning from small truncations to the truncations that remove almost the entire tail domain [100]. Some insights into the molecular basis for the role of the C-terminal tail in the EGF receptor autoinhibition were provided by a recent crystal structure of the inactive EGF receptor kinase domain dimer, which visualizes a portion of the C-terminal tail region deleted in glioblastoma variants, called vIV [74]. In this dimer both kinases are stabilized in the inactive conformation by the C-terminal tail-mediated interactions in a manner reminiscent of how the SH2 domain linker stabilizes the inactive conformation of the Src kinase domain [101]. The interactions within the inactive EGF receptor dimer also position the C-terminal tails to bind in *cis* to the hinge region of the kinase domains forming a network of electrostatic interactions that lock kinases in the inactive conformation. This interaction resembles the mechanism for autoinhibition of the zeta-chain-associated protein tyrosine kinase 70 (Zap70) [102]. Finally, the most C-terminal portion of the visualized tail fragment tail occludes the site within the C-lobe where the juxtamembrane segment binds during receptor activation (Fig. 9.7d). While the first two of the three described interactions serve to prevent the EGF receptor kinase from spontaneously adopting the active conformation, the occlusion of the juxtamembrane latch-binding site would prevent activation of EGF receptor by ligand-induced dimerization. Although the relevance of this inactive dimeric form for the EGF receptor autoinhibition has still not been unequivocally determined, EGF receptor has been repeatedly reported to exist as a dimer prior to ligand binding [67, 69, 70]. Additional support for the autoinhibitory role of the C-terminal tail for EGF receptor activation comes from a recent study in which deletion of the full-length C-terminal tail was demonstrated to result in ligand-independent dimerization and activation of EGFR [103].

The C-terminal tail interactions seem to be different for each HER receptor. HER2 is most similar to EGFR, and the deletion of its C-terminal tail also results in receptor activation although at present there are not structural insights into the molecular basis of this inhibition [104]. In the inactive structures of the HER4 and HER3 kinases, the C-terminal tails do not make interactions with the kinase domains in *cis* [34, 35]. In contrast, the tail regions in these structures are found to bind to the allosteric activator interface in the C-lobe of an adjacent kinase domain. Such binding physically blocks the activator interface and is predicted to prevent formation of an active receptor dimer and consequently limit signaling by HER3 and HER4. Interestingly, an analogous mechanism is explored by a negative regulator of EGF receptor signaling, Mig6, which binds to the activator interface of EGF receptor post ligand binding and limits receptor activation [105]. HER receptors seem therefore to evolve their unique strategies to regulate kinase activity by the C-terminal tail. The specificity of these mechanisms is supported by an observation that in a chimeric EGF receptor in which the tail was replaced by the tail sequence of HER3, the C-terminal tail is no longer able to impose autoinhibition on the EGF receptor kinase [103].

Several other receptor tyrosine kinases have been shown to utilize their C-terminal tails for autoinhibition prior to ligand binding, although the molecular basis for these interactions is not completely understood. The catalytic activity of close homologues, MET and RON receptor tyrosine kinases, can be significantly increased by deletion of the short C-terminal tail or inhibited by the peptides corresponding to the C-terminal tail [106, 107]. The C-terminal tail of these receptors contains two tyrosine phosphorylation sites, which when substituted with phenylalanines enhance receptor inhibition. This suggests that the C-terminal-mediated autoinhibition of MET and RON receptors might be regulated by phosphorylation. Similarly in the PDGFβ receptor, the deletion of a portion of the C-terminal tail containing a proline and glutamic acid-rich motif enhances receptor activity, while addition of the peptides corresponding to this region has an inhibitory effect on receptor activation [108]. Hence, the PDGF family of receptor tyrosine kinases utilizes both their juxtamembrane and C-terminal segments for the inhibition of receptor kinase activity.

9.6 Conclusions

The crystal structures of kinases have enabled basic understanding of the mechanism of phosphorylation and uncovered unexpected diversity within the regulatory mechanisms that control their function. Despite decades of research, receptor tyrosine kinases still remain quite poorly characterized branch of the human kinome. A detailed understanding of how ligand binding to receptors' extracellular domains is translated into precise set of molecular motions that activate the intracellular kinase domains has largely not been accomplished. This is mostly due to the lack of structural information on the receptor fragments that encompass both extracellular and intracellular portions of the receptors simultaneously. The advancements in experimental approaches that allow such investigation will be a huge step towards understanding the molecular basis for receptor tyrosine kinase signaling. The pioneering steps towards this goal have been recently made through electron microscopy imaging of the detergent preparations of almost full-length EGF/EGF receptor [109] and of a cytokine receptor gp130 in complex with the cytokine interleukin-6, its alpha receptor (IL-6Rα), and Jak1 kinase [110]. Although detergent preparations poorly recapitulate the native membrane environment, alternative approaches such as protein reconstitution in model membranes [111] provide new avenues for structural studies of full-length integral membrane proteins. These studies will shed light on the complex nature of interactions that control catalytic activation of receptor tyrosine kinases and ultimately determine the outcome of their downstream signaling.

References

1. Eswaran J, Knapp S. Insights into protein kinase regulation and inhibition by large scale structural comparison. Biochim Biophys Acta. 2010;1804(3):429–32.
2. Robinson DR, Wu YM, Lin SF. The protein tyrosine kinase family of the human genome. Oncogene. 2000;19(49):5548–57.
3. Knighton DR, Zheng JH, Ten Eyck LF, Ashford VA, Xuong NH, Taylor SS, et al. Crystal structure of the catalytic subunit of cyclic adenosine monophosphate-dependent protein kinase. Science. 1991;253(5018):407–14.
4. Herberg FW, Doyle ML, Cox S, Taylor SS. Dissection of the nucleotide and metal-phosphate binding sites in cAMP-dependent protein kinase. Biochemistry. 1999;38(19):6352–60.
5. Bossemeyer D. The glycine-rich sequence of protein kinases: a multifunctional element. Trends Biochem Sci. 1994;19(5):201–5.
6. Kornev AP, Haste NM, Taylor SS, Eyck LF. Surface comparison of active and inactive protein kinases identifies a conserved activation mechanism. Proc Natl Acad Sci U S A. 2006;103(47):17783–8.
7. Kornev AP, Taylor SS, Ten Eyck LF. A helix scaffold for the assembly of active protein kinases. Proc Natl Acad Sci U S A. 2008;105(38):14377–82.
8. Azam M, Seeliger MA, Gray NS, Kuriyan J, Daley GQ. Activation of tyrosine kinases by mutation of the gatekeeper threonine. Nat Struct Mol Biol. 2008;15(10):1109–18.
9. Knight ZA, Shokat KM. Features of selective kinase inhibitors. Chem Biol. 2005;12(6):621–37.
10. De Bondt HL, Rosenblatt J, Jancarik J, Jones HD, Morgan DO, Kim SH. Crystal structure of cyclin-dependent kinase 2. Nature. 1993;363(6430):595–602.
11. Noble ME, Endicott JA, Johnson LN. Protein kinase inhibitors: insights into drug design from structure. Science. 2004;303(5665):1800–5.
12. Jura N, Zhang X, Endres NF, Seeliger MA, Schindler T, Kuriyan J. Catalytic control in the EGF receptor and its connection to general kinase regulatory mechanisms. Mol Cell. 2011;42(1):9–22.
13. Shan Y, Seeliger MA, Eastwood MP, Frank F, Xu H, Jensen MO, et al. A conserved protonation-dependent switch controls drug binding in the Abl kinase. Proc Natl Acad Sci U S A. 2009;106(1):139–44.
14. Grant BD, Adams JA. Pre-steady-state kinetic analysis of cAMP-dependent protein kinase using rapid quench flow techniques. Biochemistry. 1996;35(6):2022–9.
15. Lew J, Taylor SS, Adams JA. Identification of a partially rate-determining step in the catalytic mechanism of cAMP-dependent protein kinase: a transient kinetic study using stopped-flow fluorescence spectroscopy. Biochemistry. 1997;36(22):6717–24.

16. Hubbard SR, Wei L, Ellis L, Hendrickson WA. Crystal structure of the tyrosine kinase domain of the human insulin receptor. Nature. 1994;372(6508):746–54.
17. Till JH, Becerra M, Watty A, Lu Y, Ma Y, Neubert TA, et al. Crystal structure of the MuSK tyrosine kinase: insights into receptor autoregulation. Structure. 2002;10(9):1187–96.
18. Mohammadi M, Schlessinger J, Hubbard SR. Structure of the FGF receptor tyrosine kinase domain reveals a novel autoinhibitory mechanism. Cell. 1996;86(4):577–87.
19. Huse M, Kuriyan J. The conformational plasticity of protein kinases. Cell. 2002;109(3):275–82.
20. Schlessinger J. Signal transduction. Autoinhibition control. Science. 2003;300(5620):750–2.
21. Rickert KW, Patel SB, Allison TJ, Byrne NJ, Darke PL, Ford RE, et al. Structural basis for selective small molecule kinase inhibition of activated c-Met. J Biol Chem. 2011;286(13):11218–25.
22. Favelyukis S, Till JH, Hubbard SR, Miller WT. Structure and autoregulation of the insulin-like growth factor 1 receptor kinase. Nat Struct Biol. 2001;8(12):1058–63.
23. Hubbard SR. Crystal structure of the activated insulin receptor tyrosine kinase in complex with peptide substrate and ATP analog. EMBO J. 1997;16(18):5572–81.
24. Li W, Miller WT. Role of the activation loop tyrosines in regulation of the insulin-like growth factor I receptor-tyrosine kinase. J Biol Chem. 2006;281(33):23785–91.
25. Furdui CM, Lew ED, Schlessinger J, Anderson KS. Autophosphorylation of FGFR1 kinase is mediated by a sequential and precisely ordered reaction. Mol Cell. 2006;21(5):711–7.
26. Lochhead PA, Sibbet G, Morrice N, Cleghon V. Activation-loop autophosphorylation is mediated by a novel transitional intermediate form of DYRKs. Cell. 2005;121(6):925–36.
27. Pearce LR, Komander D, Alessi DR. The nuts and bolts of AGC protein kinases. Nat Rev Mol Cell Biol. 2010;11(1):9–22.
28. Biscardi JS, Maa MC, Tice DA, Cox ME, Leu TH, Parsons SJ. c-Src-mediated phosphorylation of the epidermal growth factor receptor on Tyr845 and Tyr1101 is associated with modulation of receptor function. J Biol Chem. 1999;274(12):8335–43.
29. Gotoh N, Tojo A, Hino M, Yazaki Y, Shibuya M. A highly conserved tyrosine residue at codon 845 within the kinase domain is not required for the transforming activity of human epidermal growth factor receptor. Biochem Biophys Res Commun. 1992;186(2):768–74.
30. Tice DA, Biscardi JS, Nickles AL, Parsons SJ. Mechanism of biological synergy between cellular Src and epidermal growth factor receptor. Proc Natl Acad Sci U S A. 1999;96(4):1415–20.
31. Zhang X, Gureasko J, Shen K, Cole PA, Kuriyan J. An allosteric mechanism for activation of the kinase domain of epidermal growth factor receptor. Cell. 2006;125(6):1137–49.
32. Stamos J, Sliwkowski MX, Eigenbrot C. Structure of the epidermal growth factor receptor kinase domain alone and in complex with a 4-anilinoquinazoline inhibitor. J Biol Chem. 2002;277(48):46265–72.
33. Wood ER, Truesdale AT, McDonald OB, Yuan D, Hassell A, Dickerson SH, et al. A unique structure for epidermal growth factor receptor bound to GW572016 (Lapatinib): relationships among protein conformation, inhibitor off-rate, and receptor activity in tumor cells. Cancer Res. 2004;64(18):6652–9.
34. Jura N, Shan Y, Cao X, Shaw DE, Kuriyan J. Structural analysis of the catalytically inactive kinase domain of the human EGF receptor 3. Proc Natl Acad Sci U S A. 2009;106(51):21608–13.
35. Qiu C, Tarrant MK, Choi SH, Sathyamurthy A, Bose R, Banjade S, et al. Mechanism of activation and inhibition of the HER4/ErbB4 kinase. Structure. 2008;16(3):460–7.
36. Shi F, Telesco SE, Liu Y, Radhakrishnan R, Lemmon MA. ErbB3/HER3 intracellular domain is competent to bind ATP and catalyze autophosphorylation. Proc Natl Acad Sci U S A. 2010;107(17):7692–7.
37. Aertgeerts K, Skene R, Yano J, Sang BC, Zou H, Snell G, et al. Structural analysis of the mechanism of inhibition and allosteric activation of the kinase domain of HER2 protein. J Biol Chem. 2011;286(21):18756–65.
38. Jeffrey PD, Russo AA, Polyak K, Gibbs E, Hurwitz J, Massague J, et al. Mechanism of CDK activation revealed by the structure of a cyclinA-CDK2 complex. Nature. 1995;376(6538):313–20.

39. Hagopian JC, Kirtley MP, Stevenson LM, Gergis RM, Russo AA, Pavletich NP, et al. Kinetic basis for activation of CDK2/cyclin A by phosphorylation. J Biol Chem. 2001;276(1):275–80.
40. Zhang HT, O'Rourke DM, Zhao H, Murali R, Mikami Y, Davis JG, et al. Absence of autophosphorylation site Y882 in the p185neu oncogene product correlates with a reduction of transforming potential. Oncogene. 1998;16(22):2835–42.
41. Fan YX, Wong L, Deb TB, Johnson GR. Ligand regulates epidermal growth factor receptor kinase specificity: activation increases preference for GAB1 and SHC versus autophosphorylation sites. J Biol Chem. 2004;279(37):38143–50.
42. Kloth MT, Laughlin KK, Biscardi JS, Boerner JL, Parsons SJ, Silva CM. STAT5b, a mediator of synergism between c-Src and the epidermal growth factor receptor. J Biol Chem. 2003;278(3):1671–9.
43. Boerner JL, Demory ML, Silva C, Parsons SJ. Phosphorylation of Y845 on the epidermal growth factor receptor mediates binding to the mitochondrial protein cytochrome c oxidase subunit II. Mol Cell Biol. 2004;24(16):7059–71.
44. Yang S, Park K, Turkson J, Arteaga CL. Ligand-independent phosphorylation of Y869 (Y845) links mutant EGFR signaling to stat-mediated gene expression. Exp Cell Res. 2008;314(2):413–9.
45. Yun CH, Boggon TJ, Li Y, Woo MS, Greulich H, Meyerson M, et al. Structures of lung cancer-derived EGFR mutants and inhibitor complexes: mechanism of activation and insights into differential inhibitor sensitivity. Cancer Cell. 2007;11(3):217–27.
46. Zheng J, Knighton DR, ten Eyck LF, Karlsson R, Xuong N, Taylor SS, et al. Crystal structure of the catalytic subunit of cAMP-dependent protein kinase complexed with MgATP and peptide inhibitor. Biochemistry. 1993;32(9):2154–61.
47. Bossemeyer D, Engh RA, Kinzel V, Ponstingl H, Huber R. Phosphotransferase and substrate binding mechanism of the cAMP-dependent protein kinase catalytic subunit from porcine heart as deduced from the 2.0 A structure of the complex with Mn2+ adenylyl imidodiphosphate and inhibitor peptide PKI(5-24). EMBO J. 1993;12(3):849–59.
48. Knighton DR, Zheng JH, Ten Eyck LF, Xuong NH, Taylor SS, Sowadski JM. Structure of a peptide inhibitor bound to the catalytic subunit of cyclic adenosine monophosphate-dependent protein kinase. Science. 1991;253(5018):414–20.
49. Chao LH, Stratton MM, Lee IH, Rosenberg OS, Levitz J, Mandell DJ, et al. A mechanism for tunable autoinhibition in the structure of a human Ca2+/calmodulin- dependent kinase II holoenzyme. Cell. 2011;146(5):732–45.
50. Hofstra RM, Landsvater RM, Ceccherini I, Stulp RP, Stelwagen T, Luo Y, et al. A mutation in the RET proto-oncogene associated with multiple endocrine neoplasia type 2B and sporadic medullary thyroid carcinoma. Nature. 1994;367(6461):375–6.
51. Songyang Z, Carraway 3rd KL, Eck MJ, Harrison SC, Feldman RA, Mohammadi M, et al. Catalytic specificity of protein-tyrosine kinases is critical for selective signalling. Nature. 1995;373(6514):536–9.
52. Pandit SD, Donis-Keller H, Iwamoto T, Tomich JM, Pike LJ. The multiple endocrine neoplasia type 2B point mutation alters long-term regulation and enhances the transforming capacity of the epidermal growth factor receptor. J Biol Chem. 1996;271(10):5850–8.
53. Zhu H, Klemic JF, Chang S, Bertone P, Casamayor A, Klemic KG, et al. Analysis of yeast protein kinases using protein chips. Nat Genet. 2000;26(3):283–9.
54. Ubersax JA, Ferrell Jr JE. Mechanisms of specificity in protein phosphorylation. Nat Rev Mol Cell Biol. 2007;8(7):530–41.
55. Schulman BA, Lindstrom DL, Harlow E. Substrate recruitment to cyclin-dependent kinase 2 by a multipurpose docking site on cyclin A. Proc Natl Acad Sci U S A. 1998;95(18):10453–8.
56. Lo RS, Chen YG, Shi Y, Pavletich NP, Massague J. The L3 loop: a structural motif determining specific interactions between SMAD proteins and TGF-beta receptors. EMBO J. 1998;17(4):996–1005.
57. Chen YG, Hata A, Lo RS, Wotton D, Shi Y, Pavletich N, et al. Determinants of specificity in TGF-beta signal transduction. Genes Dev. 1998;12(14):2144–52.

58. Biondi RM, Cheung PC, Casamayor A, Deak M, Currie RA, Alessi DR. Identification of a pocket in the PDK1 kinase domain that interacts with PIF and the C-terminal residues of PKA. EMBO J. 2000;19(5):979–88.
59. Biondi RM, Nebreda AR. Signalling specificity of Ser/Thr protein kinases through docking-site-mediated interactions. Biochem J. 2003;372(Pt 1):1–13.
60. Levinson NM, Seeliger MA, Cole PA, Kuriyan J. Structural basis for the recognition of c-Src by its inactivator Csk. Cell. 2008;134(1):124–34.
61. Bae JH, Boggon TJ, Tome F, Mandiyan V, Lax I, Schlessinger J. Asymmetric receptor contact is required for tyrosine autophosphorylation of fibroblast growth factor receptor in living cells. Proc Natl Acad Sci U S A. 2010;107(7):2866–71.
62. Rohmann E, Brunner HG, Kayserili H, Uyguner O, Nurnberg G, Lew ED, et al. Mutations in different components of FGF signaling in LADD syndrome. Nat Genet. 2006;38(4):414–7.
63. Rand V, Huang J, Stockwell T, Ferriera S, Buzko O, Levy S, et al. Sequence survey of receptor tyrosine kinases reveals mutations in glioblastomas. Proc Natl Acad Sci U S A. 2005;102(40):14344–9.
64. Chen H, Xu CF, Ma J, Eliseenkova AV, Li W, Pollock PM, et al. A crystallographic snapshot of tyrosine trans-phosphorylation in action. Proc Natl Acad Sci U S A. 2008; 105(50):19660–5.
65. Hubbard SR, Miller WT. Receptor tyrosine kinases: mechanisms of activation and signaling. Curr Opin Cell Biol. 2007;19(2):117–23.
66. Gadella Jr TW, Jovin TM. Oligomerization of epidermal growth factor receptors on A431 cells studied by time-resolved fluorescence imaging microscopy. A stereochemical model for tyrosine kinase receptor activation. J Cell Biol. 1995;129(6):1543–58.
67. Sako Y, Minoghchi S, Yanagida T. Single-molecule imaging of EGFR signalling on the surface of living cells. Nat Cell Biol. 2000;2(3):168–72.
68. Martin-Fernandez M, Clarke DT, Tobin MJ, Jones SV, Jones GR. Preformed oligomeric epidermal growth factor receptors undergo an ectodomain structure change during signaling. Biophys J. 2002;82(5):2415–27.
69. Clayton AH, Walker F, Orchard SG, Henderson C, Fuchs D, Rothacker J, et al. Ligand-induced dimer-tetramer transition during the activation of the cell surface epidermal growth factor receptor-A multidimensional microscopy analysis. J Biol Chem. 2005;280(34):30392–9.
70. Chung I, Akita R, Vandlen R, Toomre D, Schlessinger J, Mellman I. Spatial control of EGF receptor activation by reversible dimerization on living cells. Nature. 2010;464(7289):783–7.
71. Moriki T, Maruyama H, Maruyama IN. Activation of preformed EGF receptor dimers by ligand-induced rotation of the transmembrane domain. J Mol Biol. 2001;311(5):1011–26.
72. Ferguson KM, Berger MB, Mendrola JM, Cho HS, Leahy DJ, Lemmon MA. EGF activates its receptor by removing interactions that autoinhibit ectodomain dimerization. Mol Cell. 2003;11(2):507–17.
73. Landau M, Fleishman SJ, Ben-Tal N. A putative mechanism for downregulation of the catalytic activity of the EGF receptor via direct contact between its kinase and C-terminal domains. Structure. 2004;12(12):2265–75.
74. Jura N, Endres NF, Engel K, Deindl S, Das R, Lamers MH, et al. Mechanism for activation of the EGF receptor catalytic domain by the juxtamembrane segment. Cell. 2009;137(7):1293–307.
75. Kani K, Park E, Landgraf R. The extracellular domains of ErbB3 retain high ligand binding affinity at endosome pH and in the locked conformation. Biochemistry. 2005;44(48):15842–57.
76. Landgraf R, Eisenberg D. Heregulin reverses the oligomerization of HER3. Biochemistry. 2000;39(29):8503–11.
77. Berger MB, Mendrola JM, Lemmon MA. ErbB3/HER3 does not homodimerize upon neuregulin binding at the cell surface. FEBS Lett. 2004;569(1–3):332–6.
78. Himanen JP, Saha N, Nikolov DB. Cell-cell signaling via Eph receptors and ephrins. Curr Opin Cell Biol. 2007;19(5):534–42.
79. Himanen JP, Rajashankar KR, Lackmann M, Cowan CA, Henkemeyer M, Nikolov DB. Crystal structure of an Eph receptor-ephrin complex. Nature. 2001;414(6866):933–8.

80. Seiradake E, Harlos K, Sutton G, Aricescu AR, Jones EY. An extracellular steric seeding mechanism for Eph-ephrin signaling platform assembly. Nat Struct Mol Biol. 2010; 17(4):398–402.
81. Salaita K, Nair PM, Petit RS, Neve RM, Das D, Gray JW, et al. Restriction of receptor movement alters cellular response: physical force sensing by EphA2. Science. 2010; 327(5971):1380–5.
82. Wimmer-Kleikamp SH, Janes PW, Squire A, Bastiaens PI, Lackmann M. Recruitment of Eph receptors into signaling clusters does not require ephrin contact. J Cell Biol. 2004; 164(5):661–6.
83. Hubbard SR. Juxtamembrane autoinhibition in receptor tyrosine kinases. Nat Rev Mol Cell Biol. 2004;5(6):464–71.
84. Wybenga-Groot LE, Baskin B, Ong SH, Tong J, Pawson T, Sicheri F. Structural basis for autoinhibition of the Ephb2 receptor tyrosine kinase by the unphosphorylated juxtamembrane region. Cell. 2001;106(6):745–57.
85. Li S, Covino ND, Stein EG, Till JH, Hubbard SR. Structural and biochemical evidence for an autoinhibitory role for tyrosine 984 in the juxtamembrane region of the insulin receptor. J Biol Chem. 2003;278(28):26007–14.
86. Griffith J, Black J, Faerman C, Swenson L, Wynn M, Lu F, et al. The structural basis for autoinhibition of FLT3 by the juxtamembrane domain. Mol Cell. 2004;13(2):169–78.
87. Hirota S, Isozaki K, Moriyama Y, Hashimoto K, Nishida T, Ishiguro S, et al. Gain-of-function mutations of c-kit in human gastrointestinal stromal tumors. Science. 1998; 279(5350):577–80.
88. Heinrich MC, Corless CL, Duensing A, McGreevey L, Chen CJ, Joseph N, et al. PDGFRA activating mutations in gastrointestinal stromal tumors. Science. 2003;299(5607):708–10.
89. Thiel KW, Carpenter G. Epidermal growth factor receptor juxtamembrane region regulates allosteric tyrosine kinase activation. Proc Natl Acad Sci U S A. 2007;104(49):19238–43.
90. Aifa S, Aydin J, Nordvall G, Lundstrom I, Svensson SP, Hermanson O. A basic peptide within the juxtamembrane region is required for EGF receptor dimerization. Exp Cell Res. 2005;302(1):108–14.
91. Macdonald-Obermann JL, Pike LJ. The intracellular juxtamembrane domain of the epidermal growth factor (EGF) receptor is responsible for the allosteric regulation of EGF binding. J Biol Chem. 2009;284(20):13570–6.
92. Red Brewer M, Choi SH, Alvarado D, Moravcevic K, Pozzi A, Lemmon MA, et al. The juxtamembrane region of the EGF receptor functions as an activation domain. Mol Cell. 2009;34(6):641–51.
93. Wood ER, Shewchuk LM, Ellis B, Brignola P, Brashear RL, Caferro TR, et al. 6-Ethynylthieno[3,2-d]- and 6-ethynylthieno[2,3-d]pyrimidin-4-anilines as tunable covalent modifiers of ErbB kinases. Proc Natl Acad Sci U S A. 2008;105(8):2773–8.
94. McLaughlin S, Smith SO, Hayman MJ, Murray D. An electrostatic engine model for autoinhibition and activation of the epidermal growth factor receptor (EGFR/ErbB) family. J Gen Physiol. 2005;126(1):41–53.
95. Shewchuk LM, Hassell AM, Ellis B, Holmes WD, Davis R, Horne EL, et al. Structure of the Tie2 RTK domain: self-inhibition by the nucleotide binding loop, activation loop, and C-terminal tail. Structure. 2000;8(11):1105–13.
96. Niu XL, Peters KG, Kontos CD. Deletion of the carboxyl terminus of Tie2 enhances kinase activity, signaling, and function. Evidence for an autoinhibitory mechanism. J Biol Chem. 2002;277(35):31768–73.
97. Gamett DC, Tracy SE, Robinson HL. Differences in sequences encoding the carboxyl-terminal domain of the epidermal growth factor receptor correlate with differences in the disease potential of viral erbB genes. Proc Natl Acad Sci U S A. 1986;83(16):6053–7.
98. Khazaie K, Dull TJ, Graf T, Schlessinger J, Ullrich A, Beug H, et al. Truncation of the human EGF receptor leads to differential transforming potentials in primary avian fibroblasts and erythroblasts. EMBO J. 1988;7(10):3061–71.

99. Pines G, Huang PH, Zwang Y, White FM, Yarden Y. EGFRvIV: a previously uncharacterized oncogenic mutant reveals a kinase autoinhibitory mechanism. Oncogene. 2010;29:5850.
100. Frederick L, Wang XY, Eley G, James CD. Diversity and frequency of epidermal growth factor receptor mutations in human glioblastomas. Cancer Res. 2000;60(5):1383–7.
101. Gonfloni S, Frischknecht F, Way M, Superti-Furga G. Leucine 255 of Src couples intramolecular interactions to inhibition of catalysis. Nat Struct Biol. 1999;6(8):760–4.
102. Deindl S, Kadlecek TA, Brdicka T, Cao X, Weiss A, Kuriyan J. Structural basis for the inhibition of tyrosine kinase activity of ZAP-70. Cell. 2007;129(4):735–46.
103. Bublil EM, Pines G, Patel G, Fruhwirth G, Ng T, Yarden Y. Kinase-mediated quasi-dimers of EGFR. FASEB J. 2010;24:4744.
104. Akiyama T, Matsuda S, Namba Y, Saito T, Toyoshima K, Yamamoto T. The transforming potential of the c-erbB-2 protein is regulated by its autophosphorylation at the carboxyl-terminal domain. Mol Cell Biol. 1991;11(2):833–42.
105. Zhang X, Pickin KA, Bose R, Jura N, Cole PA, Kuriyan J. Inhibition of the EGF receptor by binding of MIG6 to an activating kinase domain interface. Nature. 2007;450(7170):741–4.
106. Yokoyama N, Ischenko I, Hayman MJ, Miller WT. The C terminus of RON tyrosine kinase plays an autoinhibitory role. J Biol Chem. 2005;280(10):8893–900.
107. Bardelli A, Longati P, Williams TA, Benvenuti S, Comoglio PM. A peptide representing the carboxyl-terminal tail of the met receptor inhibits kinase activity and invasive growth. J Biol Chem. 1999;274(41):29274–81.
108. Chiara F, Bishayee S, Heldin CH, Demoulin JB. Autoinhibition of the platelet-derived growth factor beta-receptor tyrosine kinase by its C-terminal tail. J Biol Chem. 2004;279(19):19732–8.
109. Mi LZ, Lu C, Li Z, Nishida N, Walz T, Springer TA. Simultaneous visualization of the extracellular and cytoplasmic domains of the epidermal growth factor receptor. Nat Struct Mol Biol. 2011;18(9):984–9.
110. Lupardus PJ, Skiniotis G, Rice AJ, Thomas C, Fischer S, Walz T, et al. Structural snapshots of full-length Jak1, a transmembrane gp130/IL-6/IL-6Ralpha cytokine receptor complex, and the receptor-Jak1 holocomplex. Structure. 2011;19(1):45–55.
111. Nath A, Atkins WM, Sligar SG. Applications of phospholipid bilayer nanodiscs in the study of membranes and membrane proteins. Biochemistry. 2007;46(8):2059–69.

Chapter 10
Targeting Receptor Tyrosine Kinases in Cancer

Wolfgang J. Köstler and Christoph C. Zielinski

10.1 Mechanisms of Normal RTK Activation and Signaling

The trajectory originating from a single, non-transformed cell to a metastatic and drug-resistant tumor is caused by multiple perturbations of regulators of cell-fate decisions [1]. The 20 families including 58 receptor tyrosine kinases (RTKs) represent one of the most prominent regulators of such cell fate decisions. RTKs typically sense signals emanating from the extracellular environment in the form of bound polypeptide ligands ("growth factors") or by their interaction with other activated receptors. The prototypical ligand-induced activation mechanism entails direct ligand binding to the RTK, thereby the trapping of the RTK's extracellular domain in a conformation, which enables the formation of stable receptor homodimers or heterodimers (i.e., complex formation with other members of the same RTK family). For instance, in the Epidermal Growth Factor Receptor (EGFR, also ErbB) family of receptor tyrosine kinases, receptor dimerization acts by relieving the inhibition that the extracellular domain imposes on the receptor's intracellular kinase domain [2]. In some RTKs, however, activation entails ligand binding to a co-receptor (e.g., co-receptors of the glycosylphosphatidylinositol-anchored GDNF family in case of the RET RTK). Formation of receptor dimers or higher order oligomers may occur between different members of the same RTK family or between other receptors [3–5].

The resulting activation of the tyrosine kinase domain leads to transfer of phosphate moieties to multiple tyrosine residues located in the cytoplasmic receptor portion. Phosphorylated tyrosine residues serve as docking sites for SH2- (SRC

W.J. Köstler, M.D., Ph.D. (✉) • C.C. Zielinski, M.D., Ph.D. h.c.
Clinical Division of Oncology, Department of Medicine I, Medical University of Vienna, Wahringer Guertel 18-20, 1090 Vienna, Austria

Comprehensive Cancer Center, Medical University of Vienna, Wahringer Guertel 18-20, 1090 Vienna, Austria
e-mail: wolfgang.koestler@meduniwien.ac.at; christoph.zielinski@meduniwien.ac.at

D.L. Wheeler, Y. Yarden (eds.), *Receptor Tyrosine Kinases: Structure, Functions and Role in Human Disease*, DOI 10.1007/978-1-4939-2053-2_10

homology 2) or PTB- (phosphotyrosine-binding) domain-containing adaptor and effector proteins [6]. These proteins link activated receptors to multiple intracellular signaling pathways. For instance, in the EGFR family of RTKs these include the mitogen-activated protein kinase (MAPK) pathways, the phosphatidylinositol 3-kinase (PI3K)-AKT-mTOR pathway, adhesion signaling (e.g., integrin, focal adhesion kinase) pathways, and to phospholipase C. Moreover, activated RTKs may directly or indirectly couple to transcription factors such as the signal transducers and activators of transcription (STAT) family transcription factors. The exact composition of engaged signaling effectors and adaptor proteins is encoded by the ligand- and receptor dimer-specific set of phosphorylated tyrosine residues, and by the cell type-specific relative abundance of these effectors and adaptors. Hence, RTKs encode information through multisite receptor tail phosphorylation and combinations of overlapping sets of adaptor proteins [7, 8]. For instance, the EGFR RTK family consists of four receptors—ERBB1 (also EGFR), ERBB2 (also Her-2/*neu*), ERBB3 and ERBB4, that bind 11 distinct stimulatory ligands with distinct specificity and affinity (reviewed in [9]). Those ligands are released through proteolytic cleavage of their transmembrane precursors by metalloproteinases. The later are activated in response to a different types of extracellular signals by a broad set of heterologous receptors, including cytokine receptors, GPCRs (G-protein-coupled receptors), integrins and others—a process termed *receptor transactivation* (reviewed in [10]). For instance, cell motility and proliferation by ErbB receptors are non-redundantly achieved by different phosphorylation sites [11, 12].

In addition to ligand binding, RTK activation by *receptor transmodulation* occurs by means of receptor phosphorylation on critical amino acid residues by other cellular kinases (e.g., by the SRC family kinases) or by inhibition of regulatory receptor phosphatases [13–15]. Moreover, several RTKs, their oncogenic variants reviewed below, and cleaved c-terminal fragments of ligand precursors translocate to the nucleus, where they may act as transcriptional regulators [16–28].

10.2 Oncogenic Insults Resulting in Aberrant RTK Signaling

Oncogenic insults leading to aberrant activation of RTK signaling hardly occur as single hits, but typically comprise orchestrated combinations of aberrations involving RTKs themselves (e.g., by overexpression, amplification of the encoding genomic region, activating mutation, formation of gene fusions) along with aberrations in the both positive and negative regulatory components that normally warrant failsafe RTK signaling. A detailed review of all oncogenic aberrations in all RTKs and their accompanying aberrations in regulators in all human cancers is beyond the limits of this chapter. Hence, the first two parts of this section will give a brief overview on the plenitude of oncogenic aberrations in positive and negative regulators of RTK activity frequently observed in breast and other cancers—with particular emphasis on the ErbB system. The following sections will give an overview on the types of aberrations of RTKs themselves encountered in different human neoplasms.

Some of the best-studied RTK aberrations that served as role models in the development of RTK-targeting therapeutics—specifically kinase domain mutations in the EGFR, the ERBB2 amplicon and ALK fusion kinases—will be reviewed in detail and are supplemented by comprehensive lists of hitherto characterized oncogenic RTK aberrations detected in human cancers. A comprehensive list of RTK aberrations in different neoplasms along with drugs successfully deployed to target these diseases is given in the respective tables. As will be evident from the fact that the same aberration(s) can be observed in a diverse set of malignant diseases, a single genetic aberration does not suffice to define a disease. A prototypical example includes gene fusions of the Anaplastic Lymphoma Kinase (ALK): TPM3-ALK was the first oncogene fusion shown to transform both mesenchymal and lymphoid cells and occurs in a variety of neoplasms of different tissue origins [29]. In the same manner, fusions of RET have occurred in more than one-third of papillary thyroid cancers and in roughly 1 % of non-small cell lung cancers (NSCLC) [30].

10.2.1 Network Insults Resulting in Enhanced Signal Generation

Beyond RTKs themselves, prototypical examples of orchestrated aberrant activation of positive regulatory elements include overexpression of their respective ligands and/or adaptor proteins, and cooccurring mutations downstream in the signaling pathways instigated by RTKs. As an example of such oncogenic hits, the well-studied ERBB2 amplification is reviewed in detail below.

An important feature of oncogenic aberrations involving increased abundance and/or activity of RTKs is that they do not simply represent "*more of the same*." For instance, overexpression of the Her-2/*neu* kinase shifts the activation/inactivation equilibrium towards higher prevalence of active receptors [31], likely by forming heterodimers with other ErbB family members even in the absence of their ligands [32]. Moreover, Her-2/*neu* overexpression is almost universally accompanied by a disproportional shift in the expression of receptor variants (described below) compared to the more abundant wild type receptor.

On the other hand, the stoichiometry and specificity of adaptor binding is also sensitive to changes in the abundance of interaction partners [7], thus linking states of increased RTK activation not only to enhanced, but also to broader signaling. For instance, via phosphorylated tyrosine residues 1068 and 1086 the EGFR recruits the adaptor protein GRB2, which can bind multiple positive (e.g., SOS1) and negative (e.g., CBL) signaling regulators. Hence, the net output in terms of signal specificity, strength, and duration depends upon the abundance and activity of both, RTKs and their downstream pathway components.

Oncogenic aberrations in RTKs also change the way signal attenuation is achieved by the degradative endocytic machinery. For instance, overexpression of RTKs can saturate the binding capacity of both clathrin-coated pits and endosomal adaptors, further prolonging half-life of active receptors [33]. Directly overcoming the initial steps of degradative receptor sorting by ubiquitin ligases such as CBL represents another key event required for RTK induced neoplasms. For instance, increased activity of the kinase SRC, e.g., by Her-2/*neu* which promotes SRC synthesis and stability and physically associates with SRC [34], triggers CBL's autoubiquitination and proteasomal degradation, thereby promoting RTK recycling [35]. Overproduction of growth factors that loose receptor affinity in the mildly acidic early endosomes, such as TGFA, EPGN, and HBEGF, also results in evasion of the regulatory role of CBL: Dissociation of receptor–ligand complexes decreases phosphorylation of endosomal EGFR and its association with CBL, hence resulting in enhanced receptor recycling [36, 37]. Several oncogenic RTK variants are poorly downregulated because they lack docking sites for endocytic adaptors (e.g., the EGFRvIII variant reviewed below exhibits decreased phosphorylation on the CBL docking site at tyrosine 1045) or because they exhibit constitutive association with HSP90 [38]. Unlike EGFR, Her-2/*neu* poorly couples to the degradative endocytic machinery. Hence, states of Her-2/*neu* overexpression or enhanced expression of ligands promoting the formation of Her-2/*neu* containing heterodimers (e.g., EREG and EPGN) enhance the protein levels and prolong the signaling of the respective heterodimerization partners [8, 36, 39–42].

An important feature of tumors driven by mutations of oncogenic RTKs (e.g., lung cancers harboring EGFR kinase domain mutations) or by genomic rearrangements involving RTKs (e.g., ALK-rearranged lung cancer) is that their viability critically depends upon the activity of the respective kinase, whereas shutdown of kinase activity results in cell death. This *oncogene addiction* as been explained by changes in the expression or activity levels of BCL-2 family members (e.g., BIM, PUMA, MCL1) resulting as a consequence of shutdown of MAPK or PI3K pathways activity upon inhibition of the kinase [43–50]. Oncogene addiction, however, also creates novel hubs of network sensitivity. For instance, the protein stability of many oncogenic forms of RTKs (e.g., many of the oncogenic ALK-rearrangements [51] or Her-2/*neu* [52, 53]) critically depends upon the chaperoning functions of the heat-shock protein 90 (HSP90). This Achilles heel creates *non-oncogene addiction* and forms the basis of therapeutic concepts employing synthetic lethality (i.e., where an otherwise relatively harmless pharmacologic assault such as HSP90 inhibition becomes lethal specifically to cells displaying a specific aberration such as ALK-rearrangements).

10.2.1.1 Enhanced Ligand Production

Enhanced ligand production and release by autocrine or paracrine modes frequently accompanies oncogenic RTK aberrations. In case of the ErbB system, enhanced synthesis of the ligands involves autocrine transcriptional loops of ligand-stimulation-mediated transcriptional induction of ligand production , and proteolytic release of ligands (e.g., AREG, HB-EGF, TGFA). Ligand release predominantly occurs through upregulation or mutations of sheddases (e.g., ADAMTS1, MMP1, ADAM17 and ADAM12) and other extracellular matrix-degrading enzymes.

Paracrine sources of ligand secretion also include the tumor microenvironment. For instance, EGF is produced by tumor-associated M2 macrophages following chemoattractant stimuli such as colony-stimulating factor-1 (CSF1) released by tumor cells [59].

Tumor-associated aberrations in ligand production also includes increased production of alternatively spliced variant forms of ligand precursors (e.g., of TGFA and neuregulin), which mediate enhanced transforming ability by increased secretion, differential binding to ErbB family members or by their enhanced nuclear translocation [60].

Similarly, breast tumors also shift the expression of RTKs, particularly ERBB4, towards cleavable isoforms, along with enhanced expression of their sheddase, ADAM17. This results in tumorigenic ERBB4 activity—even in the absence of ligand [26].

10.2.1.2 Adaptor Proteins

Changes in the abundance or activity of adaptor proteins frequently accompany and boost the transforming potential of oncogenic aberrations in RTKs. For instance, the adaptor protein GRB7 (growth factor receptor-bound protein 7) is co-amplified in ERBB2 amplified breast cancers, where it forms a physical complex with HER-2/*neu*. This complex facilitates phosphorylation of the receptor and its interactions with phospholipase C, AKT, the adaptor Shc, and to calmodulin and the focal adhesion kinase [61, 62].

Other examples include the adaptor proteins GRB2 and GAB2: GRB2 (growth factor receptor-bound protein 2) located on chromosome 17q24-q25, is frequently overexpressed in mammary tumors [63]. This adaptor links ErbB receptors to the RAS-MAPK pathway through binding of SHC, GAB1, CBL, TNK2, and SOS, to the PI3K pathway via binding the p85 regulatory subunit of PI3K [64]. Likewise, GAB2 (Grb2-associated binding protein 2), which maps to a region (11q13-14) amplified in approximately 15 % and overexpressed in almost 50 % of mammary tumors [65], links Her-2/neu to the RAS-MAPK pathway. Other, less frequently encountered aberrations of adaptors in breast cancer involve SRC, GAB1, and IRS2.

10.2.1.3 Signaling Pathways Downstream of RTKs

In addition to RTKs, signaling pathways downstream of these receptors often exhibit oncogenic aberrations. For instance, in breast cancer mutations of BRAF, KRAS, several MAP kinases (e.g., MAP3K6, MAPK13), actin organizers (e.g., CDC42-binding protein kinases), regulators and effectors of RHO GTPases (e.g., ARHGEF4, ARHGAP25) and calcium/calmodulin-dependent kinases, as well as mTOR signaling components and ribosomal protein S6 kinases (e.g., RPS6KA3, RPS6KA4, RPS6KC1) have been detected [58, 66, 67]. Also, multiple signaling modules exhibit enhanced expression due to gene amplification at lower frequency

10.2.2.1.2 Perturbations of the Endocytic Machinery

Aberrations in endocytic components that normally regulate in RTK turnover represent another characteristic of RTK driven cancers. For instance, the CDC42 guanine nucleotide exchange factor ARHGEF7 is frequently overexpressed and/or activated by EGFR and SRC in breast cancer [104]. The resulting formation of a CDC42/ARHGEF7/CBL complex, which sequesters CBL from EGFR, diminishes EGFR ubiquitination and downregulation [105]. Similarly, the actin-binding protein cortactin (CTTN) is overexpressed due to amplification of the 11q13 region in about 13 % of primary breast carcinomas [106] and inhibits CBL-mediated downregulation of EGFR [107] via impaired tyrosine phosphorylation of CBL and via activation of CDC42 [107]. Other examples of impaired RTK endocytosis in breast cancer involve LIMK1, the actin adaptor HIP1, several ubiquitin E3 ligases (e.g., WWP1, ITCH, NRDP1, and NEDD4), and the ubiquitin-binding protein TSG101. At this point, the reader is referred to several excellent articles describing the oncogenic role of aberrations of the endocytic machinery [108, 109].

10.2.2.2 Failure to Achieve Delayed Signal Attenuation

In non-transformed cells, delayed signal attenuation is typically induced by sets of signal modulators that transcriptionally induced by RTK signaling. These include autoregulatory loops in which products of forward-driving *immediate early genes* (e.g., FOS) repress the activity of their own promoters [110]. Consecutive transcription of *delayed early genes*, including transcriptional repressors and RNA-binding proteins, confines the temporal activity of *immediate early genes* [56, 110]. *Delayed early genes* also include inhibitors of signaling cascades such as the MAPK phosphatases. Not surprisingly, immediate early and delayed early genes regulating the window of transcriptional activity induced by RTKs rank among the most frequently perturbed genes in human tumors [56]. Another, later wave of transcription instigated by RTKs encodes for regulators that mostly act in a more receptor-proximal way. For instance, LRIG1 associates with the extracellular domains of multiple RTKs and recruits CBL to tag RTKs for degradation [111]. Unsurprisingly, LRIG1 is frequently underexpressed in breast cancer, and its loss enhances tumorigenesis in ERBB2 transgenic mice [112]. ERRFI1, another transcriptionally induced feedback inhibitor of the ErbB network, directly binds to the kinase domains of ErbB receptors, thereby quenching their activation [113]. ERFFI1 under-expression owing to genomic loss of the encoding chromosomal region (1p36), frequently occurs in Her-2/*neu* overexpressing breast tumors, enhances their aggressiveness and promotes their resistance to ErbB-targeted therapies [114]. RTK-induced RAS-MAPK signaling also induces transcription and translocation of Sprouty proteins (SPRY1, SPRY2) to early endosomes and to the plasma membrane, where these proteins inhibit the activation of RAS and RAF1 [115–117]. Decreased expression of Sprouty proteins enhances tumorigenesis in preclinical breast cancer models and is frequently observed in human breast cancer [118].

10.2.3 Oncogenic Mutations of RTKs

Oncogenic mutations of RTKs include point mutations, small duplications, and small insertions/deletions (mostly encompassing the kinase domain or critical amino acid residues involved in dimerization of the extracellular domain), as well as larger deletions. The later typically occur in gene regions encoding for receptor portions that restrain the activity of the respective RTK (e.g., the receptor's extracellular domain rendering the receptor's activity ligand-independent or the receptor's C-terminal tail). A detailed description how each of the known mutations in RTKs results in aberrant activation of the respective kinase is beyond the scope of this section. Rather, the transforming potential of different types of mutations will be exemplified by the most common mutations seen in 3 kinases: KIT, FGFR3 and EGFR and a comprehensive list of common mutations in RTKs along with the drugs approved for treatment of patients are given in Table 10.1.

For instance, the most common primary mutations encountered in gastrointestinal stromal tumors (GIST; a rare mesenchymal neoplasm considered to originate form the intestinal Cayal cells) encompass exon 11 of the KIT (v-kit Hardy-Zuckerman 4 feline sarcoma viral oncogene homolog) RTK (85 % of all KIT mutations and approx. 70 % of all mutations in GIST) or exon 12 of the PDFRA (platelet-derived growth factor receptor, alpha polypeptide) RTK (approx. 5 % of all mutations in GIST), respectively [119]. In both RTKs, these mutations encompass the intracellular juxtamembrane domain, which negatively regulates the receptors' kinase activity [120, 121]. The second most common KIT mutation in GIST is found in exon 9, which encodes for the extracellular dimerization motif thereby enhancing constitutive receptor dimerization.

Similarly, the vast majority of FGFR3 mutations in occur at a single position in the extracellular domain (specifically, S249C). This mutation leads to the formation of an intermolecular cysteine disulphide bridge, which results in constitutive dimerization and receptor activation [122].

Oncogenic mutations occurring in EGFR exhibit a much broader spectrum, which appears distinct for different cancer types. In lung cancer, the pathognomonic point mutations, small duplications and insertions/deletions typically cluster in the catalytic cleft of the receptor's kinase domain. For instance, the most common EGFR mutations observed in lung cancer encompass exon 19 deletions (encoding for the alpha-C-helix of the kinase domain) and the L858R (located in the activation loop oft the kinase domain) mutation [123–126]. These mutations weaken the hydrophobic interactions between the activation loop and the N-lobe of the kinase domain, thereby interfere with packing of the kinase domain in the inactive conformation resulting in constitutive catalysis [127]. Moreover, these mutations also differentially alter the binding affinities for ATP and ATP-mimetic drugs (e.g., the

Table 10.1 Mutations, short duplications and insertions/deletions involving receptor tyrosine kinases that are frequently encountered in human cancers

		Drugs	
Kinase	Disease (frequency within disease)	Tyrosine kinase inhibitors	Antibodies
ALK	Neuroblastoma (6–12 %) [272–275]		
EPH-family	Melanoma (rare): EPHA6, EPHA10, EPHB1, EPHB2, EPHB6 [276], NSCLC (i.e., non-small cell lung cancer; rare): EPHA2, EPHA3, EPHA4 [277]		
ERBB1 (EGFR)	NSCLC (up to 20 %) [278], glioblastoma multiforme (up to 30 %) [128, 129], esophageal and pancreatic cancer (rare) [279]	NSCLC: erlotinib, gefitinib, [196, 280–291], afatinib [292], pancreatic cancer: erlotinib	NSCLC: cetuximab [293]
ERBB2 (Her-2/neu)	NSCLC (4 %) [294–297], ovarian (rare) [298], breast (rare) [299], gastric (rare) [299] and colorectal cancer (rare) [299]		
ERBB3	Colon and gastric cancer (11 %) [300]		
ERBB4	Melanoma (19 %) [276]		
FGFR1	Melanoma (rare) [301]		
FGFR2	Endometrial carcinoma (10–12 %) [302, 303], gastric cancer (7 %) [304]		
FGFR3	Bladder canner [305, 306], multiple myeloma (rare) [307], cervix carcinoma [305], prostate cancer (8 %) [308], seminoma (7 %) [309], oral squamous cell carcinoma (0–62 %) [310, 311], NSCLC (rare) [278], colorectal cancer (5 %) [304]		
FGFR4	Rhabdomyosarcoma (7 %) [312]		
FLT1	Melanoma (rare) [276]		
FLT3	Acute myeloid leukemia (24 %), myelodysplastic syndrome (rare) [313], acute lymphoblastic leukemia (rare)		
KIT	GIST (i.e., gastrointestinal stromal tumor; 85 %) [119], melanoma (2–6 %), germ cell tumors incl. intracranial dysgerminoma, mastocytosis (>90 %) and mast cell leukemia, SCLC,	GIST: imatinib [187, 188, 202, 203, 314–316], sunitinib [204], regorafenib [205], mastocytosis: imatinib [317, 318], masitinib, ponatinib, midostaurin	
MET	NSCLC [277], small-cell lung cancer, (hereditary) papillary renal cell carcinoma [319], hepatocellular carcinoma	Medullary thyroid cancer and prostate cancer: Cabozantinib [320, 321]	
PDGFRA	GIST [322, 323], melanoma [276], GBM (3 %) [324]	GIST: Imatinib, sunitinib, regorafenib	

(continued)

Table 10.1 (continued)

		Drugs	
Kinase	Disease (frequency within disease)	Tyrosine kinase inhibitors	Antibodies
RET	(papillary and) Medullary thyroid carcinoma (sporadic form and hereditary form together with pheochromocytoma in multiple endocrine neoplasia syndrome type 2 (MEN 2) [325, 326]	Thyroid cancer: Vandetanib [327, 328], cabozantinib [321]	
TIE1	Melanoma [276]		

Drugs listed include those that are approved for clinical use for common cancers (e.g., lung cancer, breast cancer, colorectal cancer) and those that have shown activity in phase II clinical trials in rare cancer (sub)types

tyrosine kinase inhibitors, TKIs, reviewed below), which explains in the favorable therapeutic index of TKIs in tumors harboring these aberrations [127].

Point mutations of EGFR appear rare in other cancer types, with the exception of point mutations in the extracellular domain of EGFR observed in malignant gliomas [128, 129]. Indeed, the extracellular domain appears to be a mutational hub in gliomas, which frequently exhibit expression of the EGFRvIII variant, which lacks large parts of the extracellular ligand-binding domain encoded by exons 2–7. The later variant has also been described to occur at lower frequency in breast and other cancers [130, 131]. Owing to the loss of the negative regulatory function the extracellular domain imposes on EGFR, the EGFRvIII variant exhibits constitutive, ligand-independent receptor and PI3K pathway activity [132].

10.2.4 Gene Amplifications of RTKs

While not specific for a certain cancer type, gene amplifications (and other genomic aberrations) of specific RTKs are increasingly being associated with certain histological subtypes and clinical characteristics of malignant tumors. In breast cancer, for instance, high level EGFR amplification may represent the pathognomonic oncogenic insult of basal-like tumors with metaplastic histologic features [133], the ETV6-NTRK3 gene fusion characterizes basal-like carcinomas exhibiting the secretory breast carcinoma phenotype [134], whereas the ALK gene is amplified in most inflammatory (i.e., displaying extensive involvement of the lymph and blood vessels of the skin) breast cancers [135]. The most frequent RTK amplification in breast cancer, however, entails amplification of the ERBB2 gene. This genomic insult serves as a role model how gene amplifications involve not only the RTK gene itself, but also simultaneous derangements in multiple activating and inhibitory regulatory mechanisms involved in RTK signaling. Hence it is reviewed in more details and a comprehensive list of RTKs frequently amplified in human cancers along with the drugs approved for treatment of patients are given in Table 10.2.

Table 10.2 Gene amplifications involving receptor tyrosine kinases that are frequently encountered in human cancers

		Drugs		
Kinase	Disease (frequency within disease)	Tyrosine kinase inhibitors	Antibodies	Other
ALK	Alveolar rhabdomyosarcoma (approx. 20 %) [329], renal cell carcinoma (rare) [330]			
ERBB1 (EGFR)	Colorectal cancer (rare) [331], squamous cell carcinoma of the head and neck (approx. 20 %), NSCLC (approx. 30 %) [277, 278] GBM (up to 40 %) [128], breast cancer (rare)	NSCLC: gefitinib, erlotinib [196, 280–291]	Colon cancer: cetuximab, panitumumab [332–335], squamous cell carcinoma of the head and neck: cetuximab [336–338], NSCLC: cetuximab [293]	
ERBB2 (HER-2/*neu*)	Breast cancer (20–25 %) [136], gastric and gastroesophageal cancer (12–25 %) [339, 340] NSCLC (approx. 20 %) [278, 341, 342], ovarian cancer (7–10 %) [343, 344], salivary duct carcinoma (approx. 50 %) [345], urothelial carcinoma (5 %) [346]	Breast: lapatinib [28, 251, 252, 256], neratinib [347–349]	Breast: trastuzumab [156, 157, 350–353], pertuzumab [253, 254], gastric and gastroesophageal: trastuzumab [340] NSCLC: trastuzumab? [342, 354]	Breast: Trastuzumab-DM1 (immunotoxin) [179], tanespimycin (HSP90 inhibitor) [52]
FGFR1	Breast cancer (10 %) [355–357], ovarian cancer [358], bladder cancer (3 %) [306, 359], rhabdomyosarcoma (3 %) [360], lung squamous cancer (22 %) [361]			
FGFR2	Gastric cancer (3–10 %) [362], breast cancer			

(continued)

Table 10.2 (continued)

		Drugs		
Kinase	Disease (frequency within disease)	Tyrosine kinase inhibitors	Antibodies	Other
FGFR3	Bladder cancer [363], salivary gland andenoid cystic carcinoma (28 %) [364]			
MET	NSCLC (approx. 8 %) [277], colon cancer (rare) [190], papillary renal cell carcinoma			
PDGFRA	Glioblastoma multiforme, diffuse gliomas (5 %) [324, 365]			

Drugs listed include those that are approved for clinical use for common cancers (e.g., lung cancer, breast cancer, colorectal cancer) and those that have shown activity in phase II clinical trials in rare cancer (sub)types

Amplification of a genomic region on chromosome 17q12 containing the ERBB2 gene occurs in up to 25 % of invasive breast tumors [136] and at lower frequency in gastric cancer, NSCLC, and endometrial cancer. Beyond ERBB2, the minimum region of recurrent amplification includes STARD3 (involved in endosomal vesicle trafficking) and the Her-2/*neu* adaptor protein GRB7 (reviewed above). The commonly amplified region contains 5 additional genes, which are overexpressed along with Her-2/*neu* and constitute part of the ERBB2 gene expression signature in breast cancer [137]. These include PPP1R1B (protein phosphatase 1, regulatory subunit 1B), splice variants of which have been associated with resistance to the anti-Her-2/*neu* antibody trastuzumab, and C17ORF37 (chromosome 17 open reading frame 37), which promotes a migratory and invasive phenotype by serving as a positive regulator of AKT phosphorylation. Furthermore, multiple genes promoting the growth of Her-2/*neu* amplified breast cancer, such as topoisomerase II, are frequently co-amplified and overexpressed along with the RTK [69]. Importantly, ERBB2 amplification often occurs on the background of an aneusomy of chromosome 17, leading to deregulation of many genes involved in breast cancer pathogenesis and treatment response (reviewed in [138, 139]).

The gene dosage gain of usually a less than a dozen extra copies of ERBB2, however, may not account for the remarkable increase in mRNA and protein levels observed in ERBB2 amplified breast tumors. Indeed, additional mechanisms are operative that govern the increased, predominantly monoallelic, transcription, from amplified ERBB2 genes [140]. In most Her-2/*neu* overexpressing breast tumors, multiple transcription factors enhancing ERBB2 promoter activity (e.g., the AP-2 family of proteins, factors engaging the AP-2 complex such as Yin Yang 1, various

Ets factor family members, EGR2, ZO-1, and ZONAB) exhibit enhanced expression or activity. In addition, these are accompanied by frequent genomic or functional loss of (trans)repressors of the ERBB2 promoter, including FOXP3, GATA4, RB1, WWOX, and the Ets protein PEA [141, 142].

10.2.5 Oncogene Fusions Involving RTKs

Tyrosine kinase fusion genes can be observed in 5–7 % of hematological, mesenchymal, and epithelial tumors (reviewed in [143]). In analogy to oncogene amplifications, a specific chromosomal rearrangement is not characteristic for a certain type of tumor and conversely, identical appearing tumors may exhibit very distinct oncogenic aberrations (e.g., in NSCLC with adenocarcinoma histology rearrangements of ALK, ROS, and RET may occur in a mutually exclusive fashion with mutations in EGFR and KRAS [30]). The mechanisms underlying the formation of gene fusions are incompletely understood. Exposure to increased levels of ionizing radiation has been convincingly linked to formation fusions involving the RTKs RET, ALK, and NTRK1 in papillary thyroid carcinomas, but the mechanisms underlying the formation of the more abundant sporadic papillary thyroid carcinomas remain unknown [144]. RTK fusion proteins typically contain the kinase domain for the RTK fusion partner and may have lost trans-regulatory sequences (e.g., microRNA binding sites) [145]. The non-RTK fusion partner typically contains protein domains (e.g., coiled-coil or leucine zipper domains) that drive the multimerization of the fusion kinase in the absence of ligand. Moreover, fusion partners often alter the subcellular location of the kinase (e.g., most ALK fusion proteins no longer localize to the plasma membrane), contain 5′ regulatory sequences that drive high-level expression. In addition, the fusion often confers conformational changes that favor the activated state of the kinase (reviewed in [143]). A comprehensive list of RTKs that frequently undergo genomic rearrangements in human cancers along with the drugs approved for treatment of patients is given in Table 10.3.

Probably the most abundant and best-characterized oncogene fusions of RTKs involve the anaplastic lymphoma kinase (ALK; located on chromosome 2). For instance, a translocation involving chromosomes 2 and 5 is found in approximately 60 % anaplastic large-cell lymphomas (ALCLs). The rearrangement fuses the 3′ half of the ALK gene (encoding for the catalytic domain) to the 5′ portion of the nucleophosmin (NPM) gene from chromosome 5. In a smaller fraction of ALCLs, ALK is fused to the 5′ portion of other genes including ALO17, ATIC, CLTC1, MSN, MYH9TFG, TMP3, or TPM4. Another rearrangement involving ALK, the EML4-ALK fusion is responsible for approximately 5 % of NSCLC and predominantly occurs in never or light smokers with adenocarcinoma histology. ALK fusions also occur in more rare cancers. For instance, RTK fusions involving ALK, ROS1, or PDGFRB can be found in virtually all patients with inflammatory myofibroblastic tumor (a rare type of soft tissue sarcoma).

Table 10.3 Translocations resulting in fusion genes involving receptor tyrosine kinases that are frequently encountered in human cancers

Kinase	Fusion partner	Disease (frequency within disease)	Drugs: Tyrosine kinase inhibitors	Drugs: Other
ALK	EML4	NSCLC (5 %) [30, 366, 367]	Crizotinib [147, 216, 368], ceritinib [222]	HSP90 inhibitors? [149, 150]
	TFG			
	KIF5B			
	KLC1			
	STRN			
	NPM	Anaplastic large cell lymphoma (60 %) DLBCL (rare) [369–373]	Crizotinib [368], imatinib [374]	
	TFG			
	CLTC			
	RNF213			
	MSN			
	ATIC			
	TPM3			
	TPM4			
	MYH9			
	EML4	Breast cancer (rare) [375]		
	VCL	(Pediatric) renal cell cancer (<1 %) [330, 376–378]		
	TPM3			
	EML4			
	TPM4	Esophageal cancer [379, 380]		
	TPM3	Inflammatory myofibroblastic tumor (50 %) [29, 373, 381–389]	Crizotinib [368]	
	TPM4			
	LMNA			
	SEC31A			
	TFG			
	RANBP2			
	CLTC			
	EML4			
	CARS			
	ATIC			
	DCTN1			
	PPFIBP1			
	EML4	Colorectal cancer (<1 %) [375, 390]		
	C2orf44			
AXL	MBIP	NSCLC (adenocarcinoma) [391]		

(continued)

Table 10.3 (continued)

Kinase	Fusion partner	Disease (frequency within disease)	Drugs: Tyrosine kinase inhibitors	Drugs: Other
FGFR1	BCR	Myeloid and lymphoid neoplasms with eosinophilia, CML (rare) [392, 393], MPS [394, 395]		
	CNTRL	Stem cell myeloproliferative disorders (rare) [396]		
	ZMYM2	Stem cell leukemia lymphoma syndrome (rare) [397, 398]		
	BAG4	NSCLC (squamous cell type) [399]		
	ERLIN2	Breast cancer [399]		
	TACC1	Glioblastoma multiforme (3 %) [400]		
	TACC3			
FGFR2	CCAR2	NSCLC (squamous cell carcinoma) [399]		
	CIT	NSCLC (adenocarcinoma; rare) [391]		
	AFF3	Breast cancer (rare) [399]		
	CASP7			
	CCDC6			
	OFD1	Thyroid cancer (rare) [399]		
	SLC45A3	Prostate cancer (rare) [399]		
	BICC1	Cholangiocarcinoma (rare) [399]		
FGFR3	TACC1	Glioblastoma multiforme (3 %) [400]		
	TACC3			
	ETV6	Peripheral T-cell lymphoma (rare) [401]		
	IGH	Myeloma (rare) [402, 403]		
	FGFR1OP	Atypical stem cell myeloproliferative disorder (rare) [404]		
	TACC3	Bladder cancer (5–10 %) [399, 405]		
	BAIAP2L1			
	TACC3	Squamous cell carcinoma (head and neck, lung) [399, 406]		

(continued)

Table 10.3 (continued)

Kinase	Fusion partner	Disease (frequency within disease)	Drugs: Tyrosine kinase inhibitors	Drugs: Other
NTRK1	TPM3	Colorectal cancer (rare) [407]		
	TPM3	Papillary thyroid cancer (12 %) [408–411]		
	TPR			
	TFG			
	MPRIP	NSCLC (rare)		
	CD74			
NTRK3	ETV6	Secretory basal-like breast carcinoma (pathognomonic) [134]		
	ETV6	Salivary gland carcinoma (rare) [412]		
	ETV6	Infantile fibrosarcoma, (pathognomonic) [413]		
	ETV6	Mesoblastic nephroma (>80 %) [414]		
PDGFRA	FIP1L1	Hypereosinophilic syndrome (approx. 50 %)	Imatinib [415]	
	FIP1L1	Mastocytosis (rare)	Imatinib [317, 318]	
	SCAF11	NSCLC (adenocarcinoma; rare) [391]		
PDGFRB	COL1A1	Dermatofibrosarcoma protuberans (100 %) [416]	Imatinib [417, 418]	
	EBF1	Acute lymphoblastic leukemia (rare) [419]		
	ETV6	Myeloid and lymphoid neoplasms with eosinophilia (rare) [420–424]		
	HIP1			
	CCDC6			
	RABEP1			
	CEP85L			
	NAB2	Inflammatory myofibroblastic tumor (approx. 25 %) [425]	Crizotinib	

(continued)

Table 10.3 (continued)

Kinase	Fusion partner	Disease (frequency within disease)	Drugs: Tyrosine kinase inhibitors	Drugs: Other
RET	KIF5B	NSCLC (1–2 %) [30, 277, 390, 426–430]	Cabozantinib ? [146], vandetanib ? [431]	
	CCDC6			
	NCOA4			
	TRIM33			
	CCDC6	Papillary thyroid carcinoma (approx. 30 %) [432–443]	Cabozantinib?	
	NCOA4			
	PRKAR1A			
	GOLGA5			
	ERC1			
	MBD1			
	TRIM24			
	KTN1			
	TRIM33			
	TRIM27			
	HOOK3			
ROS1	GOPC	NSCLC (1–2 %) [277, 391] [30, 367, 444], [445] [446]	NSCLC: crizotinib [444, 447, 448]	HSP90 inhibitors? [149, 150]
	SLC34A2			
	CD74			
	SDC4			
	TPM3			
	EZR			
	LRIG3			
	KDELR2			
	CCDC6			
	GOPC	Glioblastoma multiforme (rare) [449]		
	GOPC	Cholangiocellular carcinoma (rare) [450]		
	GOPC	Ovarian cancer (rare) [451]		
	SLC34A2	Gastric cancer (rare) [452]		
	YWHAE	Inflammatory myofibroblastic tumor (approx. 25 %) [425]		
	TFG			

Drugs listed include those that are approved for clinical use for common cancers (e.g., lung cancer, breast cancer, colorectal cancer) and those that have shown activity in phase II clinical trials in rare cancer (sub)types

10.3 How RTK Targets are Identified in Clinical Routine

Once conventional histopathologic examination (e.g., by hematoxylin–eosin staining and immunohistochemistry) has determined the tissue of origin (e.g., lung) and the subtype (e.g., adenocarcinoma) of a specific cancer, depending on the most probable RTK aberration specific sets of investigations are performed.

Point mutations and small duplications and insertions/deletions (e.g., EGFR mutations in lung cancer, KIT mutations in GIST) are typically identified by sequencing (e.g., Sanger sequencing, target amplification and sequencing by mass spectrometry) or mutation-specific real-time PCR techniques. For many tumors, identification of the exact mutation has critical therapeutic consequences. For instance, in GIST patients with tumor harboring KIT exon 9 mutations benefit from higher upfront doses of the TKI imatinib as compared to tumors with exon 11 mutations, for which higher doses only add toxicity. Moreover, in contrast to GISTs with PDGFR exon 12 mutations, those exhibiting exon 18 mutations (particularly the D842V mutation) do not benefit from imatinib at all, but their tumors often exhibit an indolent course even in the absence of treatment.

An interesting variation of this approach is the exclusion of mutations in pathways downstream of the kinase rather than in the targeted kinase itself. For instance, in colorectal cancer (CRC) mutations in KRAS, NRAS, and BRAF all have been associated with intrinsic resistance to treatment with EGFR targeting antibodies (reviewed below). Indeed, in CRC harboring these mutations treatment with EGFR-targeting antibodies has been shown to exhibit a deleterious effect on patient survival, possibly because intercepting with EGFR abrogates some feed-forward inhibition executed by the receptor on RAS family members. Hence, testing for the most common mutations in these genes has become routine practice before EGFR-targeting therapies are considered in CRC.

Amplification of oncogenic RTKs is usually detected by fluorescence in situ hybridization (FISH) or related assays (e.g., chromogenic in situ hybridization CISH, silver in situ hybridization, SISH) that are more readily applicable on paraffin-embedded tissue. Copy number gains of the gene of interest are then identified by an increase in the number of signals above the two copies normally present in all autosomes of somatic cells. For instance, for detection of ERBB2 amplification status in breast cancer, FISH assays commonly employ two probes—one for ERBB2 and one for a centromeric region of the encoding chromosome 17—to distinguish real ERBB2 amplification from polysomy of chromosome 17. This distinction has clinical relevance, because polysomy of chromosome 17 alone (i.e., without accompanying ERBB2 amplification) does not confer increased sensitivity to Her-2/*neu* targeting agents.

Similarly, oncogenic fusion kinases (e.g., fusions involving ALK, ROS1, and RET [146, 147]) are typically identified using break-apart FISH assays. These assays employ two differently colored fluorescent probes that hybridize to DNA sequences on either side of the translocation breakpoint. A chromosomal rearrangement is visualized either by spatial separation of the probes—which otherwise generate a fusion signal—or by loss of the 5′ probe. The presence of a gene fusion and the identity of the fusion partner are then often confirmed by sequencing.

Immunohistochemistry is a readily available, inexpensive, and semiquantitative tool to detect kinases overexpressed due to gene amplification (e.g., Her-2/*neu* overexpression in breast cancer) or kinases that are usually absent in the investigated tissue. For instance, in the absence of specific rearrangements ALK expression is not detectable in adult lung tissue and positive staining for ALK in lung cancers indicates the presence of ALK fusions with high probability [148]. Hence, owing to its high sensitivity but often limited specificity, IHC often serves as a screening tool to direct the most appropriate tests for identifying overexpressed or rearranged oncogenes.

Owing to the absence of specific investigations predicting the benefit of targeted therapies some RTKs, however, are targeted without previous specific testing. For instance, most anti-angiogenic agents are employed without testing for specific molecular aberrations beforehand.

All of these assays, however, are limited by the ever-increasing number of targetable RTK aberrations and the cost and low throughput of performing multiple single gene assays such as Sanger sequencing, IHC, and FISH. Rather, targeted sequencing of all known oncogenic drivers by next generation sequencing techniques will soon replace those techniques to identification of druggable RTK targets. Another important limitation of routine assays employed to date encounters the fact that tissue—and hence an inconvenient biopsy—is needed. The plenitude of mechanisms of resistance to RTK-targeting agents reviewed below indicates that a "one-stop" molecular diagnostic approach at the time of diagnosis may not be appropriate along the whole the course of treatment. In this respect, innovative diagnostic approaches that harvest circulating tumor cells or cell-free DNA from the patient's blood in for molecular analyses using the above assays hold great promise.

10.4 Drugs Targeting Oncogenic RTK Signaling

10.4.1 Drug Classes

Currently licensed pharmaceuticals targeting RTKs can broadly be divided in four classes: monoclonal antibodies, antibody—drug conjugates (i.e., immunotoxins consisting of an antibody linked to a cytotoxic agent), decoy receptors (i.e., recombinant fusion proteins sequestering RTK ligands; "traps"), and small molecule tyrosine kinase inhibitors (TKIs). These drugs are most often not applied as single agents, but as combinations (e.g., combinations of anti-Her-2/*neu* antibodies reviewed below), and in combination with drugs intercepting downstream pathways (e.g., PI3K inhibitors, MEK inhibitors), and with cytotoxic agents. Moreover, is anticipated that the future armamentarium of RTK-targeting agents will also include radioimmunoconjugates, bispecific antibodies, antibody-like constructs with improved tumor diffusion efficiency (e.g., diabodies, minibodies, Fab fragments, single chain Fv domains, and single chain antibodies), along with inhibitors of chaperone proteins (e.g., HSP90 inhibitors in NSCLC with ALK rearrangements [149, 150]), and vaccines targeting wild-type or mutated RTKs.

10.4.2 Monoclonal Antibodies

10.4.2.1 Modes of Action

Antibodies predominantly target RTKs by binding to the extracellular domain. Monoclonal antibodies, which—like decoy receptors described below—bind soluble growth factors to prevent their binding to their cognate RTKs represent an important exception to this rule. A typical example is the monoclonal antibody bevacizumab, which binds and neutralizes the angiogenic growth factor VEGFA. Antibodies targeting the extracellular portion of RTKs exert several distinct biologic actions: First, by binding to the extracellular receptor portion, monoclonal antibodies opsonize tumor cells for recognition and subsequent destruction by the immune system by induction of antibody-dependent cellular cytotoxicity (ADCC) and complement-mediated cytotoxicity (CDC). Indeed, antibodies like the Her-2/*neu* antibody trastuzumab elicit ADCC in vitro [151]. In patients, germline polymorphisms influencing the affinity of activating and inhibitory immunoglobulin receptors expressed on immune cells have been associated with the clinical benefit of the anti-Her-2/*neu* antibody trastuzumab in breast cancer [152] and of the anti-EGFR antibody cetuximab in colorectal cancer [153].

The second main mode of action of monoclonal antibodies entails interference with the activation of oncogenic RTKs. This mode of action becomes particularly important when combinations of antibodies and conventional cytotoxic chemotherapy are employed, as the shutdown of survival signals resulting from blockage of RTKs may sensitize cancer cells to cytotoxic agents. For instance, when used as single-agent, the anti-Her-2/*neu* antibody trastuzumab elicits only moderate clinical activity patients with Her-2/*neu* overexpressing metastatic breast cancer [154, 155]. By contrast combinations of trastuzumab with cytotoxic chemotherapy have shown impressive clinical activity in this clinical setting [156, 157]. Importantly, depending on the exact bound epitope, monoclonal antibodies may disrupt distinct signaling complexes formed by their cognate RTKs. For instance, binding an epitope in subdomain IV of the Her-2/*neu* extracellular domain [158], trastuzumab predominantly disrupts receptor homodimers. Conversely, another anti-Her-2/*neu* antibody, pertuzumab, binds the dimerization loop of Her-2/*neu*, thereby preferentially blocking heterodimers containing Her-2/neu [32, 159]. As will be reviewed below, combinations of these antibodies exhibit synergism in patients with breast cancer and the same may hold true for other types of cancer such as head and neck cancer where EGFR and ERBB4 antibodies exhibit synergistic effects [160].

This principle, however, may not apply for all antibody combinations. For instance, compelling preclinical data have linked an angiogenic switch to resistance to EGFR targeting monoclonal antibodies such as cetuximab in colorectal cancer and to resistance to Her-2/nu targeting therapies such as trastuzumab in breast cancer. Indeed, in preclinical models combinations of cetuximab or trastuzumab with the anti-VEGFA antibody bevacizumab have demonstrated encouraging results in preclinical models curtailing acquired resistance to ErbB targeting emanating from upregulation of VEGF gene expression [161]. In the same vein, addition of bevacizumab to cytotoxic chemotherapy has resulted in moderate improvements in the survival of patients with breast cancer, lung cancer, ovarian cancer, and colorectal cancer [162–165]. By contrast, in clinical trials addition of bevacizumab has not shown any appreciable benefit in breast cancer patients undergoing trastuzumab-based therapy [166] and proved to be even somewhat deleterious in patients with colorectal cancer undergoing cetuximab-based therapy [167, 168]. Likewise, the benefits of adding bevacizumab to the EGFR TKI erlotinib in pancreatic or lung cancer have been modest [169].

Third, combinations of antibodies targeting distinct epitopes on RTKs or RTK oligomers result in formation of large receptor–antibody complexes, which are internalized and experience endocytic routing for degradation, hence resulting in receptor downregulation [170–173].

10.4.2.2 Subcutaneous Formulations, Biosimilars and Mutation-Specific Antibodies

A recent important advancement in pharmaceutical manufacturing of monoclonal antibodies entails the ability to concentrate antibodies in small volumes, so that antibodies can be applied as subcutaneous bolus injection rather than intravenous infusion. For several therapeutic antibodies targeting RTKs, including the anti-Her-2/*neu* antibody trastuzumab, this route of administration has not only been associated with enhanced patient convenience compared to intravenous infusion of antibodies, but also with more favorable pharmacokinetic and safety profiles with at least comparable therapeutic efficacy [174, 175].

The second main development entails the generic production of therapeutic antibodies that are no longer protected by manufacturer's patents. In terms of safety and efficacy, to date it is not clear how similar to the parental antibody these so-called *biosimilars* have to be. For instance, beyond recognizing the same epitope, the clinical relevance of specific posttranslational modifications such as glycosylation patterns remains to be explored. In the same vein, there are approaches under development to improve the efficacy of therapeutic antibodies by modifying their affinity to the various activating and inhibitory immunoglobulin receptors. These approaches can include modifications to the amino acid sequence of the antibody's Fc portion and modifications of the antibody's glycosylation state (reviewed in [176]).

Hitherto developed monoclonal antibodies target receptor portions also contained in the "wild type" protein of RTKs. A third novel development is represented by the generation of antibodies targeting novel epitopes specific to mutated forms of RTKs such as the EGFRvIII variant [177] and truncated Her-2/*neu* receptors [178].

10.4.3 Immunotoxins

Like monoclonal antibodies, antibody–drug conjugates target RTKs by binding to the receptor's extracellular portion. Immunotoxins additionally act by selectively delivering a toxic payload into cells expressing the RTK recognized by the antibody. Importantly, immunotoxins may still exert significant antineoplastic effects even when tumor cells have become resistant to monoclonal antibodies interfering with the activity of the cognate RTKs, but maintain the expression of these receptors. For instance, the immunotoxin T-DM1 (a conjugate of the anti-Her-2/*neu* antibody trastuzumab to the tubulin inhibitor emtansine) has shown impressive clinical efficacy in patients with Her-2/*neu* overexpressing metastatic breast cancer that failed previous therapy with trastuzumab and chemotherapy [179]. The current clinical development of immunotoxins focuses on the employment of these compounds in less heavily pretreated patients as a substitute cytotoxic chemotherapy.

Another exciting development entails the development of immunotoxins with even more selective intratumoral delivery. For instance, the antibody–drug conjugate brentuximab vedotin consists of the monoclonal antibody brentuximab (targeting the cell-membrane protein CD30) linked via a stable peptide-based linker to the antimitotic agent monomethyl auristatin E to target CD30 positive lymphomas. The linker is stable under physiologic conditions, but gets cleaved by cathepsins in the acidic endosomal or peritumoral milieu [180].

10.4.4 Decoy Receptors

Whereas ligand-targeting antibodies such as bevacizumab (which binds VEGFA) exhibit a small spectrum of targets, decoy receptors may bind multiple ligands. For instance, the decoy receptor ziv-aflibercept, consists of VEGF-binding portions from the extracellular domains of human VEGF receptors 1 and 2, which are fused to the Fc portion of the human IgG1 immunoglobulin. Thereby, the decoy receptor binds not only VEGFA, but also VEGF-B, placental growth factor (PGF), and many oncogenic splice variants of these growth factors. To date, ziv-aflibercept is the only decoy receptor targeting oncogenic RTK signaling in clinical use, but its activity appears to be limited to patients with metastatic colorectal cancer as trials employing this agent in other diseases such as ovarian cancer, prostate, lung and pancreatic lung cancer, respectively, did not demonstrate a survival benefit with addition of this agent [181–185]. Sequestration of ligands to intercept oncogenic RTK signaling, however, remains a promising therapeutic concept. Indeed, novel agents, such as decoy receptors sequestering ErbB ligands have shown encouraging efficacy in preclinical models [186].

10.4.5 Tyrosine Kinase Inhibitors (TKIs)

As opposed to monoclonal antibodies and decoy receptors, TKIs represent small molecules, which more readily cross the plasma membrane (and blood–brain barrier) and act on the receptors' intracellular tyrosine kinase domain.

Most small molecular TKIs are 4-anilinoquinazolones and act as ATP competitive inhibitors at the P-loop of the cytoplasmic tyrosine kinase domain of RTKs. Some TKIs reversibly bind the kinase domain (e.g., erlotinib and gefitinib reversibly bind EGFR), whereas others irreversibly (mostly via covalent bonds to cysteine residues) bind their substrate kinase (e.g., afatinib irreversibly binds to EGFR family kinases). Some TKIs exhibit high specificity to a single RTK. For instance, at concentrations used for therapeutic purposes, the TKIs gefitinib and erlotinib almost exclusively bind EGFR. Other TKIs exhibit a more broad-spectrum activity binding a specific RTK family or an even broader set of substrates: For instance, afatinib covalently binds to all EGFR family members (including EGFR, Her-2/*neu*, ERBB3, ERBB4). Imatinib, by contrast, binds RTKs and non-receptor tyrosine kinases belonging to distinct families including ABL, KIT, PDGF-Receptors, SCF, and oncogenic fusion proteins involving these kinases (e.g., BCR-ABL).

10.5 Resistance to Therapies Targeting RTKs in Cancer

10.5.1 Intrinsic (Primary) and Adaptive (Acquired, Secondary) Resistance

From a clinical point of view, a distinction between intrinsic and adaptive resistance to RTK-targeting agents is often made: Tumors that grow rapidly (i.e., within weeks) despite the use of RTK-targeting agents are commonly considered primary resistant, whereas tumors that exhibit an initial period of tumor shrinkage or growth arrest and only later progress despite ongoing RTK-targeting treatment are often considered to have developed secondary resistance. The time period to distinguish intrinsic from acquired resistance, however, largely depends on the disease, treatment and the clinical setting (e.g., disease stage). For instance, advanced ALK-rearranged lung cancers exhibit initial dramatic responses to TKIs such as crizotinib, but then usually progress within much less than a year. By contrast, adjuvant (i.e., postoperative in patients with localized tumors to prevent metastatic relapse) therapy of patients with GIST with the TKI imatinib for a period of 3 years results in significantly better survival than the same treatment for only 1 year [187, 188]. This finding suggests that in early GIST resistance to imatinib—if ever—occurs much later than by 1 year.

From a biological point of view, only very few mechanisms can be strictly divided into those mediating intrinsic and those mediating adaptive resistance. For instance, colorectal cancers exhibiting mutations of NRAS, KRAS, or BRAF exhibit intrinsic resistance to antibodies targeting the upstream RTK EGFR [189]. Acquired KRAS mutations and amplifications, however, have also been identified to occur in >50 % of patients with colorectal cancer that developed secondary resistance to EGFR-targeting antibodies [190, 191].

Similarly, mutations in exon 18—particularly the D842V substitution—represent frequent primary PDGFRA mutations observed in GIST. These mutations trap the PDGFRA kinase domain in a confirmation that is constitutively active, thus unable to bind the majority of PDGFRA TKIs, which bind the kinase in the inactive conformation. Accordingly GIST with the PDGFRA D842V mutation exhibit primary resistance to TKIs such as imatinib and most other available TKIs, but trials employing novel TKIs that bind the activated conformation of the kinase or antibodies targeting PDGFRA are underway [192–195].

For many other treatments, however, intrinsic and acquired mechanisms to targeted therapies may overlap. For instance, most advanced lung cancers harboring EGFR kinase domain mutations exhibit an initial dramatic response to EGFR TKIs. Renewed disease progression, however, typically occurs within a year [196] and is almost universally associated with either appearance of secondary mutations in the kinase domain of the EGFR (e.g., the T790M mutation reviewed below), or with amplification of the MET kinase. Importantly, both the T790M EGFR mutation and the MET amplification have been detected in TKI-naïve EGFR-mutant tumors [197–201]. Hence, the relative allelic frequencies of these alterations may determine whether tumors exhibit intrinsic or acquired resistance to EGFR targeting TKIs.

Another important consideration in resistance to RTK-targeting agents is that several mechanisms are specific for a certain drug. For instance, in GIST the identity of the KIT mutation critically influences response to different TKIs: Advanced GIST with KIT exon 11 mutations exhibit a longer disease control with imatinib than those with KIT exon 9 mutations [202, 203]. The opposite is true for second-line therapy of imatinib-refractory GIST treated with another TIK, sunitinib [204]. When imatinib- and sunitinib-refractory patients with GIST receive third-line therapy with another TKI, regorafenib, no difference in clinical benefit according to mutation type is observed [205]. Metastatic GIST also represents a role model of the heterogeneity of mechanisms of that may mediate resistance within the same patient. Typically, acquired resistance to imatinib in metastatic GIST occurs in a few displaced lesions (i.e., the so-called nodule within a mass or lesion within a lesion progression type), while the majority of metastatic lesions may still remain under prolonged control with imatinib. When several of these resistant lesions escaping the control of imatinib are biopsied, sequencing of their KIT kinase phenotype typically reveals remarked heterogeneity in the identity of secondary kinase mutations [206, 207].

Moreover, as elaborated above, oncogenic aberrations in RTKs hardly ever occur alone, but encompass simultaneous aberrations of many positive and negative regulators of RTK signaling, some of which (e.g., the secondary KRAS mutations in colorectal cancer that were cited above) occur during RTK-targeting therapies. Hence, single therapeutic hits against oncogenic RTKs unlikely achieve sustained control of tumor growth. The following section will, therefore, highlight the mechanisms underlying primary or acquired resistance to RTK-targeting agents and point out clinical lessons we have learned from rational design of drug combinations targeting RTKs.

10.5.2 Mechanisms Underlying Resistance to RTK-Targeting Agents

10.5.2.1 Competition with Drug Binding

A prototypical example of drug resistance emanating from competition with drug binding entails the EGFR T790M mutation. This secondary kinase mutation restores the ATP affinity of the catalytic domain of the EGFR kinase to levels observed in wild type receptors. Moreover, the T790M mutation also lowers the binding affinity of TKIs. Hence, the "ATP-mimetic" TKIs can no longer outcompete ATP, which is highly abundant in the intracellular milieu, for binding to the kinase. The consequence is resistance to reversibly binding TKIs, but tumors harboring the T790M mutation (as well as tumors harboring Her-2/*neu* kinase domain mutation) remain sensitive to irreversibly binding TKIs [208–213].

Increased ligand production represents another mode of competing with drug binding. For instance, ligand production causes resistance to cetuximab—in part by competing with antibody binding to EGFR [214]. Likewise, ligand production in Her-2/*neu* overexpressing breast tumors increases resistance to trastuzumab by formation of ligand-mediated receptor heterodimers which are not inhibited by the antibody [215].

A typical example of the multitude of mechanisms resulting in acquired resistance includes ALK-rearranged lung cancers. Crizotinib exhibits dramatic activity in ALK-rearranged NSCLS, with the majority of patients responding to treatment [216]. Resistance to crizotinib, however, typically occurs within a year. By then, approximately on third of crizotinib-resistant tumors have acquired mutations within the tyrosine kinase domain of ALK, whereas others may exhibit acquired amplification of the fusion gene [217–219]. Similar mechanisms may apply to crizotinib resistance in ROS1-rearranged NSCLC [220].

Several next-generation ALK inhibitors that are more potent and selective (e.g., in terms of cross-reactivity with ROS1 and MET) than crizotinib may exhibit increased upfront efficacy. Moreover, several of these inhibitors exhibit activity against mutant forms of ALK that are resistant to crizotinib. Ceritinib has recently been approved for use in patients with crizotinib-refractory ALK-rearranged

NSCLC [221, 222]. The efficacy of these new ALK inhibitors to overcome resistance in most patients suggests that most crizotinib-resistant tumours remain oncogene addicted to the ALK kinase activity and that subtherapeutic inhibition of ALK kinase activity may have contributed to crizotinib resistance.

10.5.2.2 Masking or Loss of the Drug-Binding Site

Approximately half of Her-2/*neu* overexpressing breast tumors exhibit enhanced expression of truncated Her-2/*neu* receptors lacking most of the extracellular receptor portion. These variant receptors are generated by alternative initiation of translation or by metalloprotease-mediated cleavage of their extracellular domain [22, 178, 223–228] and lack the trastuzumab- or pertuzumab-binding epitopes, but remain sensitive to the TKI lapatinib [229, 230]. Moreover, the cleaved extracellular domain binds and neutralizes trastuzumab, resulting in sub-therapeutic trastuzumab serum levels in patients presenting very high extracellular domain serum levels [231, 232]. Conversely, trastuzumab-binding inhibits proteolytic cleavage of Her-2/*neu* [233], embedding trastuzumab and these receptor variants in an interesting network of mutual pharmacokinetic and pharmacodynamic interactions [234, 235]. In a similar manner, an EGFR ectodomain mutation (S492R), which prevents cetuximab binding, has been shown to confer resistance to the antibody in patients with advanced colorectal cancer [236].

Masking of the drug-binding site represents another mechanisms to escape RTK-targeting pharmacologic insults. For instance, the glycoprotein MUC4 associates with Her-2/*neu* at the plasma membrane and prevents binding of trastuzumab [237, 238].

Acquired resistance to TKIs often involves development of a secondary mutation that interferes with drug binding (e.g., by steric hindrance). For instance, in GIST acquired imatinib resistance is predominantly associated acquired secondary KIT mutations (e.g., the V654A substitution) that interfere with drug binding [239–241]. Some secondary mutations, however, only somewhat decrease the binding affinity of imatinib in agreement with the frequently observed clinical benefit to imatinib dose-escalation in patients with GISTs that have become resistant to standard doses of imatinib [242].

10.5.2.3 Exploiting Network Plasticity, Feed-Forward and Feedback Regulation

The RTK network-intrinsic mechanisms of feedback regulation are also often exploited by tumors to escape pharmacological assaults. For instance, incomplete inhibition of EGFR and/or Her-2/*neu* kinase activity by TKIs may by buffered by tumor cells via a shift in the phosphorylation—dephosphorylation equilibrium of ERBB3: The uncoupling of an AKT-dependent negative regulatory feedback loop enhances the translocation of ERBB3 from the cytoplasm to the plasma membrane, where it is available to form heterodimers with incompletely suppressed RTKs. The later mechanism may be prevented by irreversible kinase inhibitors and by combining inhibitors of RTKs and AKT [32, 243].

Receptor trans-modulation by SRC and other kinases represents another means of escaping pharmacological assaults by taking advantage of network plasticity. For instance, transmodulation-induced recruitment of EGFR into signaling complexes, independent of its kinase activity, may account for resistance of EGFR overexpressing mammary tumors to EGFR TKIs [244]. Recent studies also implicate the EGFR-SRC axis in resistance to pathway inhibitors downstream of the RTK: For instance, rapamycin induces EGFR activation via SRC and this activation has been implicated in resistance to mTOR inhibitors [245].

Another example entails a specific HER-2 variant, HER2Δ16, that is generated by alternative mRNA splicing and lacks a small part of the extracellular domain. This variant accounts for up to 9 % of Her-2/*neu* transcripts in mammary tumors [246–249]. HER2Δ16 potently couples to downstream signaling pathways—in particular to SRC. Even though HER2Δ16 binds trastuzumab, the feed-forward resulting in SRC kinase activity interferes with plasma membrane recruitment and activation of PTEN required for trastuzumab efficacy [84]. Tumors expressing HER2Δ16, however, remain exquisitely sensitive to lapatinib and SRC inhibitors [247, 250].

Network plasticity, however, also occurs at the level of RTKs. For instance, when overexpressed, Her-2/*neu* may engage in homodimers or heterodimers (preferentially with ERBB3). Thereby, tumor cells can escape pharmacological assaults, which inhibit only homodimers (e.g., trastuzumab) or heterodimers (e.g., pertuzumab). Moreover, antibody-binding may downregulate receptors from the plasma membrane, whereas TKIs may stabilize Her-2/*neu* at the plasma membrane. A series of elegant clinical studies have demonstrated that targeting Her-2/*neu* overexpressing breast cancers by double hits (e.g., combinations of trastuzumab and pertuzumab or combinations of trastuzumab and lapatinib) exhibits impressive clinical activity even when each single-agent alone has failed [159, 229, 251–258].

10.5.2.4 Kinase and Pathway Switching

Another mode of acquired resistance to pharmacological assaults targeting specific RTKs or RTK families entails switching to alternative kinases or alternative cellular differentiation programs. For instance, switching to estrogen receptor signaling occurs in a proportion of breast cancers that express both Her-2/*neu* and the hormone receptor in resistance to Her-2/*neu* targeting therapies [259]. Indeed, in these tumors dual targeting of estrogen receptor and ErbB signaling may prevent or postpone the resistance occurring to either target alone [260–262].

Switching to another, often amplified, kinase represents another mechanisms of acquired resistance. For instance, overexpression of AXL has been incriminated in secondary lapatinib resistance in breast cancer [263]. Similarly, MET and ERBB2 amplification have been identified as potential secondary resistance mechanism in colorectal cancers that become resistant to EGFR targeting antibodies in the absence of acquired KRAS mutations [190, 214]. In the same manner, occurrence of MET amplification has been identified as a major mechanism underlying acquired resistance to EGFR TKIs in lung cancer [197].

A more drastic change underlies the epithelial-to-mesenchymal transition and acquisition of a stem cell-like phenotype that is observed in some EGFR mutated lung cancers that become resistant to TKIs targeting the EGFR [264].

10.5.2.5 Pharmacodynamic and Pharmacokinetic Mechanisms

Beyond kinases, pharmacodynamic resistance to TKIs has also been attributed to the activity of mechanisms that mediate drug uptake into (e.g., low activity of the transporter SLC22A1 in imatinib resistance [265, 266]) or drug efflux out of tumor cells (e.g., increased activity of P-glycoprotein transporters [267–269]).

Pharmacokinetic mechanisms may likely play another, underappreciated role in the development of resistance to RTK-targeting therapies. The relevance of shed Her-2/neu extracellular domain on trastuzumab levels has been elaborated above. Similarly, imatinib plasma levels have been associated with the size of benefit from imatinib in advanced GIST [270]. Decreased imatinib levels have been observed in patients that had undergone surgical gastrectomy (e.g., for GIST), but also increased drug metabolism over time has been identified as source of resistance to imatinib in GIST [271].

10.6 Conclusions

Over the last two decades, tremendous advances have been made in the detection and clinical targeting of oncogenic RTKs and many aberrations likely remain to be discovered—particularly in rare tumor types. We envisage that future advances will depend on several routes that are already being explored. First, the development of more potent and more selective inhibitors, or combinations thereof, is expected to improve the therapeutic application of RTK-targeting agents for prolonged time periods. Second, the combination of RTK-targeting agents with inhibitors of the pathways downstream of RTKs in order to avoid resistance emanating from network plasticity holds great promise. Third, identification and simultaneous targeting of the novel hubs of network sensitivity conferred oncogenic aberrations, i.e., by approaches employing synthetic lethality, will further foster the field. Lastly, monitoring oncogenic aberrations in RTKs and resistance mechanisms in real-time—predominantly by blood sampling and next generation sequencing—and rapid adaption of treatment according to the analytic results appears another promising road to take.

References

1. Hanahan D, Weinberg RA. The hallmarks of cancer. Cell. 2000;100(1):57–70.
2. Zhang X, Gureasko J, Shen K, Cole PA, Kuriyan J. An allosteric mechanism for activation of the kinase domain of epidermal growth factor receptor. Cell. 2006;125(6):1137–49.
3. Balana ME, Labriola L, Salatino M, Movsichoff F, Peters G, Charreau EH, et al. Activation of ErbB-2 via a hierarchical interaction between ErbB-2 and type I insulin-like growth factor receptor in mammary tumor cells. Oncogene. 2001;20(1):34–47.
4. Liu D, Aguirre Ghiso J, Estrada Y, Ossowski L. EGFR is a transducer of the urokinase receptor initiated signal that is required for in vivo growth of a human carcinoma. Cancer Cell. 2002;1(5):445–57.
5. Brantley-Sieders DM, Zhuang G, Hicks D, Fang WB, Hwang Y, Cates JM, et al. The receptor tyrosine kinase EphA2 promotes mammary adenocarcinoma tumorigenesis and metastatic progression in mice by amplifying ErbB2 signaling. J Clin Invest. 2008;118(1):64–78.
6. Pawson T. Specificity in signal transduction: from phosphotyrosine-SH2 domain interactions to complex cellular systems. Cell. 2004;116(2):191–203.
7. Jones RB, Gordus A, Krall JA, MacBeath G. A quantitative protein interaction network for the ErbB receptors using protein microarrays. Nature. 2006;439(7073):168–74.
8. Pinkas-Kramarski R, Soussan L, Waterman H, Levkowitz G, Alroy I, Klapper L, et al. Diversification of Neu differentiation factor and epidermal growth factor signaling by combinatorial receptor interactions. EMBO J. 1996;15(10):2452–67.
9. Yarden Y, Sliwkowski MX. Untangling the ErbB signalling network. Nat Rev Mol Cell Biol. 2001;2(2):127–37.
10. Fischer OM, Hart S, Gschwind A, Ullrich A. EGFR signal transactivation in cancer cells. Biochem Soc Trans. 2003;31(Pt 6):1203–8.
11. Chen P, Gupta K, Wells A. Cell movement elicited by epidermal growth factor receptor requires kinase and autophosphorylation but is separable from mitogenesis. J Cell Biol. 1994;124(4):547–55.
12. Ursini-Siegel J, Schade B, Cardiff RD, Muller WJ. Insights from transgenic mouse models of ERBB2-induced breast cancer. Nat Rev Cancer. 2007;7(5):389–97.
13. Yamauchi T, Ueki K, Tobe K, Tamemoto H, Sekine N, Wada M, et al. Tyrosine phosphorylation of the EGF receptor by the kinase Jak2 is induced by growth hormone. Nature. 1997;390(6655):91–6.
14. Frank SJ. Mechanistic aspects of crosstalk between GH and PRL and ErbB receptor family signaling. J Mammary Gland Biol Neoplasia. 2008;13(1):119–29.
15. Zwang Y, Yarden Y. p38 MAP kinase mediates stress-induced internalization of EGFR: implications for cancer chemotherapy. EMBO J. 2006;25(18):4195–206.
16. Xie Y, Hung MC. Nuclear localization of p185neu tyrosine kinase and its association with transcriptional transactivation. Biochem Biophys Res Commun. 1994;203(3):1589–98.
17. Lin SY, Makino K, Xia W, Matin A, Wen Y, Kwong KY, et al. Nuclear localization of EGF receptor and its potential new role as a transcription factor. Nat Cell Biol. 2001;3(9):802–8.
18. Ni CY, Murphy MP, Golde TE, Carpenter G. Gamma-Secretase cleavage and nuclear localization of ErbB-4 receptor tyrosine kinase. Science. 2001;294(5549):2179–81.
19. Offterdinger M, Schofer C, Weipoltshammer K, Grunt TW. c-erbB-3: a nuclear protein in mammary epithelial cells. J Cell Biol. 2002;157(6):929–39.
20. Wang SC, Lien HC, Xia W, Chen IF, Lo HW, Wang Z, et al. Binding at and transactivation of the COX-2 promoter by nuclear tyrosine kinase receptor ErbB-2. Cancer Cell. 2004;6(3):251–61.
21. Bao J, Lin H, Ouyang Y, Lei D, Osman A, Kim TW, et al. Activity-dependent transcription regulation of PSD-95 by neuregulin-1 and Eos. Nat Neurosci. 2004;7(11):1250–8.
22. Scaltriti M, Rojo F, Ocana A, Anido J, Guzman M, Cortes J, et al. Expression of p95HER2, a truncated form of the HER2 receptor, and response to anti-HER2 therapies in breast cancer. J Natl Cancer Inst. 2007;99(8):628–38.

23. Grasl-Kraupp B, Schausberger E, Hufnagl K, Gerner C, Low-Baselli A, Rossmanith W, et al. A novel mechanism for mitogenic signaling via pro-transforming growth factor alpha within hepatocyte nuclei. Hepatology. 2002;35(6):1372–80.
24. Lo HW, Xia W, Wei Y, Ali-Seyed M, Huang SF, Hung MC. Novel prognostic value of nuclear epidermal growth factor receptor in breast cancer. Cancer Res. 2005;65(1):338–48.
25. Wang SC, Nakajima Y, Yu YL, Xia W, Chen CT, Yang CC, et al. Tyrosine phosphorylation controls PCNA function through protein stability. Nat Cell Biol. 2006;8(12):1359–68.
26. Hollmen M, Maatta JA, Bald L, Sliwkowski MX, Elenius K. Suppression of breast cancer cell growth by a monoclonal antibody targeting cleavable ErbB4 isoforms. Oncogene. 2009;28(10):1309–19.
27. Kim HP, Yoon YK, Kim JW, Han SW, Hur HS, Park J, et al. Lapatinib, a dual EGFR and HER2 tyrosine kinase inhibitor, downregulates thymidylate synthase by inhibiting the nuclear translocation of EGFR and HER2. PLoS One. 2009;4(6):e5933.
28. Geyer CE, Forster J, Lindquist D, Chan S, Romieu CG, Pienkowski T, et al. Lapatinib plus capecitabine for HER2-positive advanced breast cancer. N Engl J Med. 2006;355(26):2733–43.
29. Lawrence B, Perez-Atayde A, Hibbard MK, Rubin BP, Dal Cin P, Pinkus JL, et al. TPM3-ALK and TPM4-ALK oncogenes in inflammatory myofibroblastic tumors. Am J Pathol. 2000;157(2):377–84.
30. Takeuchi K, Soda M, Togashi Y, Suzuki R, Sakata S, Hatano S, et al. RET, ROS1 and ALK fusions in lung cancer. Nat Med. 2012;18(3):378–81.
31. Yuste L, Montero JC, Esparis-Ogando A, Pandiella A. Activation of ErbB2 by overexpression or by transmembrane neuregulin results in differential signaling and sensitivity to herceptin. Cancer Res. 2005;65(15):6801–10.
32. Junttila TT, Akita RW, Parsons K, Fields C, Lewis Phillips GD, Friedman LS, et al. Ligand-independent HER2/HER3/PI3K complex is disrupted by trastuzumab and is effectively inhibited by the PI3K inhibitor GDC-0941. Cancer Cell. 2009;15(5):429–40.
33. French AR, Sudlow GP, Wiley HS, Lauffenburger DA. Postendocytic trafficking of epidermal growth factor-receptor complexes is mediated through saturable and specific endosomal interactions. J Biol Chem. 1994;269(22):15749–55.
34. Kim H, Chan R, Dankort DL, Zuo D, Najoukas M, Park M, et al. The c-Src tyrosine kinase associates with the catalytic domain of ErbB-2: implications for ErbB-2 mediated signaling and transformation. Oncogene. 2005;24(51):7599–607.
35. Bao J, Gur G, Yarden Y. Src promotes destruction of c-Cbl: implications for oncogenic synergy between Src and growth factor receptors. Proc Natl Acad Sci U S A. 2003; 100(5):2438–43.
36. Kochupurakkal BS, Harari D, Di-Segni A, Maik-Rachline G, Lyass L, Gur G, et al. Epigen, the last ligand of ErbB receptors, reveals intricate relationships between affinity and mitogenicity. J Biol Chem. 2005;280(9):8503–12.
37. French AR, Tadaki DK, Niyogi SK, Lauffenburger DA. Intracellular trafficking of epidermal growth factor family ligands is directly influenced by the pH sensitivity of the receptor/ligand interaction. J Biol Chem. 1995;270(9):4334–40.
38. Grandal MV, Zandi R, Pedersen MW, Willumsen BM, van Deurs B, Poulsen HS. EGFRvIII escapes down-regulation due to impaired internalization and sorting to lysosomes. Carcinogenesis. 2007;28(7):1408–17.
39. Offterdinger M, Bastiaens PI. Prolonged EGFR Signaling by ERBB2-Mediated Sequestration at the Plasma Membrane. Traffic. 2008;9(1):147–55.
40. Baulida J, Kraus MH, Alimandi M, Di Fiore PP, Carpenter G. All ErbB receptors other than the epidermal growth factor receptor are endocytosis impaired. J Biol Chem. 1996;271(9):5251–7.
41. Lenferink AE, Pinkas-Kramarski R, van de Poll ML, van Vugt MJ, Klapper LN, Tzahar E, et al. Differential endocytic routing of homo- and hetero-dimeric ErbB tyrosine kinases confers signaling superiority to receptor heterodimers. EMBO J. 1998;17(12):3385–97.
42. Shelly M, Pinkas-Kramarski R, Guarino BC, Waterman H, Wang LM, Lyass L, et al. Epiregulin is a potent pan-ErbB ligand that preferentially activates heteromeric receptor complexes. J Biol Chem. 1998;273(17):10496–505.

43. Bean GR, Ganesan YT, Dong Y, Takeda S, Liu H, Chan PM, et al. PUMA and BIM are required for oncogene inactivation-induced apoptosis. Sci Signal. 2013;6(268):ra20.
44. Costa DB, Halmos B, Kumar A, Schumer ST, Huberman MS, Boggon TJ, et al. BIM mediates EGFR tyrosine kinase inhibitor-induced apoptosis in lung cancers with oncogenic EGFR mutations. PLoS Med. 2007;4(10):1669–79. discussion 80.
45. Cragg MS, Kuroda J, Puthalakath H, Huang DC, Strasser A. Gefitinib-induced killing of NSCLC cell lines expressing mutant EGFR requires BIM and can be enhanced by BH3 mimetics. PLoS Med. 2007;4(10):1681–9. discussion 90.
46. Faber AC, Li D, Song Y, Liang MC, Yeap BY, Bronson RT, et al. Differential induction of apoptosis in HER2 and EGFR addicted cancers following PI3K inhibition. Proc Natl Acad Sci U S A. 2009;106(46):19503–8.
47. Luo J, Solimini NL, Elledge SJ. Principles of cancer therapy: oncogene and non-oncogene addiction. Cell. 2009;136(5):823–37.
48. Ng KP, Hillmer AM, Chuah CT, Juan WC, Ko TK, Teo AS, et al. A common BIM deletion polymorphism mediates intrinsic resistance and inferior responses to tyrosine kinase inhibitors in cancer. Nat Med. 2012;18(4):521–8.
49. Sharma SV, Settleman J. Oncogene addiction: setting the stage for molecularly targeted cancer therapy. Genes Dev. 2007;21(24):3214–31.
50. Takezawa K, Okamoto I, Nishio K, Janne PA, Nakagawa K. Role of ERK-BIM and STAT3-survivin signaling pathways in ALK inhibitor-induced apoptosis in EML4-ALK-positive lung cancer. Clin Cancer Res. 2011;17(8):2140–8.
51. Sang J, Acquaviva J, Friedland JC, Smith DL, Sequeira M, Zhang C, et al. Targeted inhibition of the molecular chaperone Hsp90 overcomes ALK inhibitor resistance in non-small cell lung cancer. Cancer Discov. 2013;3(4):430–43.
52. Modi S, Stopeck A, Linden H, Solit D, Chandarlapaty S, Rosen N, et al. HSP90 inhibition is effective in breast cancer: a phase II trial of tanespimycin (17-AAG) plus trastuzumab in patients with HER2-positive metastatic breast cancer progressing on trastuzumab. Clin Cancer Res. 2011;17(15):5132–9.
53. Modi S, Stopeck AT, Gordon MS, Mendelson D, Solit DB, Bagatell R, et al. Combination of trastuzumab and tanespimycin (17-AAG, KOS-953) is safe and active in trastuzumab-refractory HER-2 overexpressing breast cancer: a phase I dose-escalation study. J Clin Oncol. 2007;25(34):5410–7.
54. Earp HS, Austin KS, Blaisdell J, Rubin RA, Nelson KG, Lee LW, et al. Epidermal growth factor (EGF) stimulates EGF receptor synthesis. J Biol Chem. 1986;261(11):4777–80.
55. Nagashima T, Shimodaira H, Ide K, Nakakuki T, Tani Y, Takahashi K, et al. Quantitative transcriptional control of ErbB receptor signaling undergoes graded to biphasic response for cell differentiation. J Biol Chem. 2007;282(6):4045–56.
56. Amit I, Citri A, Shay T, Lu Y, Katz M, Zhang F, et al. A module of negative feedback regulators defines growth factor signaling. Nat Genet. 2007;39(4):503–12.
57. Borrell-Pages M, Rojo F, Albanell J, Baselga J, Arribas J. TACE is required for the activation of the EGFR by TGF-alpha in tumors. EMBO J. 2003;22(5):1114–24.
58. Greenman C, Stephens P, Smith R, Dalgliesh GL, Hunter C, Bignell G, et al. Patterns of somatic mutation in human cancer genomes. Nature. 2007;446(7132):153–8.
59. Wyckoff J, Wang W, Lin EY, Wang Y, Pixley F, Stanley ER, et al. A paracrine loop between tumor cells and macrophages is required for tumor cell migration in mammary tumors. Cancer Res. 2004;64(19):7022–9.
60. Hayes NV, Gullick WJ. The neuregulin family of genes and their multiple splice variants in breast cancer. J Mammary Gland Biol Neoplasia. 2008;13(2):205–14.
61. Bai T, Luoh SW. GRB-7 facilitates HER-2/Neu-mediated signal transduction and tumor formation. Carcinogenesis. 2008;29(3):473–9.
62. Stein D, Wu J, Fuqua SA, Roonprapunt C, Yajnik V, D'Eustachio P, et al. The SH2 domain protein GRB-7 is co-amplified, overexpressed and in a tight complex with HER2 in breast cancer. EMBO J. 1994;13(6):1331–40.
63. Daly RJ, Binder MD, Sutherland RL. Overexpression of the Grb2 gene in human breast cancer cell lines. Oncogene. 1994;9(9):2723–7.

64. Lowenstein EJ, Daly RJ, Batzer AG, Li W, Margolis B, Lammers R, et al. The SH2 and SH3 domain-containing protein GRB2 links receptor tyrosine kinases to ras signaling. Cell. 1992;70(3):431–42.
65. Bentires-Alj M, Gil SG, Chan R, Wang ZC, Wang Y, Imanaka N, et al. A role for the scaffolding adapter GAB2 in breast cancer. Nat Med. 2006;12(1):114–21.
66. Sjoblom T, Jones S, Wood LD, Parsons DW, Lin J, Barber TD, et al. The consensus coding sequences of human breast and colorectal cancers. Science. 2006;314(5797):268–74.
67. Wood LD, Parsons DW, Jones S, Lin J, Sjoblom T, Leary RJ, et al. The genomic landscapes of human breast and colorectal cancers. Science. 2007;318(5853):1108–13.
68. Bellacosa A, de Feo D, Godwin AK, Bell DW, Cheng JQ, Altomare DA, et al. Molecular alterations of the AKT2 oncogene in ovarian and breast carcinomas. Int J Cancer. 1995;64(4):280–5.
69. Neve RM, Chin K, Fridlyand J, Yeh J, Baehner FL, Fevr T, et al. A collection of breast cancer cell lines for the study of functionally distinct cancer subtypes. Cancer Cell. 2006;10(6):515–27.
70. Xiang B, Chatti K, Qiu H, Lakshmi B, Krasnitz A, Hicks J, et al. Brk is coamplified with ErbB2 to promote proliferation in breast cancer. Proc Natl Acad Sci U S A. 2008;105(34):12463–8.
71. Samuels Y, Wang Z, Bardelli A, Silliman N, Ptak J, Szabo S, et al. High frequency of mutations of the PIK3CA gene in human cancers. Science. 2004;304(5670):554.
72. Hanker AB, Pfefferle AD, Balko JM, Kuba MG, Young CD, Sanchez V, et al. Mutant PIK3CA accelerates HER2-driven transgenic mammary tumors and induces resistance to combinations of anti-HER2 therapies. Proc Natl Acad Sci U S A. 2013;110(35):14372–7.
73. Eichhorn PJ, Gili M, Scaltriti M, Serra V, Guzman M, Nijkamp W, et al. Phosphatidylinositol 3-kinase hyperactivation results in lapatinib resistance that is reversed by the mTOR/phosphatidylinositol 3-kinase inhibitor NVP-BEZ235. Cancer Res. 2008;68(22):9221–30.
74. Tseng PH, Wang YC, Weng SC, Weng JR, Chen CS, Brueggemeier RW, et al. Overcoming trastuzumab resistance in HER2-overexpressing breast cancer cells by using a novel celecoxib-derived phosphoinositide-dependent kinase-1 inhibitor. Mol Pharmacol. 2006;70(5):1534–41.
75. Carpten JD, Faber AL, Horn C, Donoho GP, Briggs SL, Robbins CM, et al. A transforming mutation in the pleckstrin homology domain of AKT1 in cancer. Nature. 2007;448(7152):439–44.
76. Stemke-Hale K, Gonzalez-Angulo AM, Lluch A, Neve RM, Kuo WL, Davies M, et al. An integrative genomic and proteomic analysis of PIK3CA, PTEN, and AKT mutations in breast cancer. Cancer Res. 2008;68(15):6084–91.
77. Liu H, Radisky DC, Nelson CM, Zhang H, Fata JE, Roth RA, et al. Mechanism of Akt1 inhibition of breast cancer cell invasion reveals a protumorigenic role for TSC2. Proc Natl Acad Sci U S A. 2006;103(11):4134–9.
78. Hutchinson JN, Jin J, Cardiff RD, Woodgett JR, Muller WJ. Activation of Akt-1 (PKB-alpha) can accelerate ErbB-2-mediated mammary tumorigenesis but suppresses tumor invasion. Cancer Res. 2004;64(9):3171–8.
79. Irie HY, Pearline RV, Grueneberg D, Hsia M, Ravichandran P, Kothari N, et al. Distinct roles of Akt1 and Akt2 in regulating cell migration and epithelial-mesenchymal transition. J Cell Biol. 2005;171(6):1023–34.
80. Cizkova M, Dujaric ME, Lehmann-Che J, Scott V, Tembo O, Asselain B, et al. Outcome impact of PIK3CA mutations in HER2-positive breast cancer patients treated with trastuzumab. Br J Cancer. 2013;108(9):1807–9.
81. Jensen JD, Knoop A, Laenkholm AV, Grauslund M, Jensen MB, Santoni-Rugiu E, et al. PIK3CA mutations, PTEN, and pHER2 expression and impact on outcome in HER2-positive early-stage breast cancer patients treated with adjuvant chemotherapy and trastuzumab. Ann Oncol. 2012;23(8):2034–42.
82. Dave B, Migliaccio I, Gutierrez MC, Wu MF, Chamness GC, Wong H, et al. Loss of phosphatase and tensin homolog or phosphoinositol-3 kinase activation and response to trastuzumab or lapatinib in human epidermal growth factor receptor 2-overexpressing locally advanced breast cancers. J Clin Oncol. 2011;29(2):166–73.

83. Wang L, Zhang Q, Zhang J, Sun S, Guo H, Jia Z, et al. PI3K pathway activation results in low efficacy of both trastuzumab and lapatinib. BMC Cancer. 2011;11:248.
84. Nagata Y, Lan KH, Zhou X, Tan M, Esteva FJ, Sahin AA, et al. PTEN activation contributes to tumor inhibition by trastuzumab, and loss of PTEN predicts trastuzumab resistance in patients. Cancer Cell. 2004;6(2):117–27.
85. Peschard P, Fournier TM, Lamorte L, Naujokas MA, Band H, Langdon WY, et al. Mutation of the c-Cbl TKB domain binding site on the Met receptor tyrosine kinase converts it into a transforming protein. Mol Cell. 2001;8(5):995–1004.
86. Ostman A, Hellberg C, Bohmer FD. Protein-tyrosine phosphatases and cancer. Nat Rev Cancer. 2006;6(4):307–20.
87. Sun T, Aceto N, Meerbrey KL, Kessler JD, Zhou C, Migliaccio I, et al. Activation of multiple proto-oncogenic tyrosine kinases in breast cancer via loss of the PTPN12 phosphatase. Cell. 2011;144(5):703–18.
88. Vermeer PD, Bell M, Lee K, Vermeer DW, Wieking BG, Bilal E, et al. ErbB2, EphrinB1, Src kinase and PTPN13 signaling complex regulates MAP kinase signaling in human cancers. PLoS One. 2012;7(1):e30447.
89. Prasad NK, Tandon M, Badve S, Snyder PW, Nakshatri H. Phosphoinositol phosphatase SHIP2 promotes cancer development and metastasis coupled with alterations in EGF receptor turnover. Carcinogenesis. 2008;29(1):25–34.
90. Bianco R, Shin I, Ritter CA, Yakes FM, Basso A, Rosen N, et al. Loss of PTEN/MMAC1/TEP in EGF receptor-expressing tumor cells counteracts the antitumor action of EGFR tyrosine kinase inhibitors. Oncogene. 2003;22(18):2812–22.
91. She QB, Solit D, Basso A, Moasser MM. Resistance to gefitinib in PTEN-null HER-overexpressing tumor cells can be overcome through restoration of PTEN function or pharmacologic modulation of constitutive phosphatidylinositol 3′-kinase/Akt pathway signaling. Clin Cancer Res. 2003;9(12):4340–6.
92. Robinson AG, Turbin D, Thomson T, Yorida E, Ellard S, Bajdik C, et al. Molecular predictive factors in patients receiving trastuzumab-based chemotherapy for metastatic disease. Clin Breast Cancer. 2006;7(3):254–61.
93. Miller TW, Perez-Torres M, Narasanna A, Guix M, Stal O, Perez-Tenorio G, et al. Loss of Phosphatase and Tensin homologue deleted on chromosome 10 engages ErbB3 and insulin-like growth factor-I receptor signaling to promote antiestrogen resistance in breast cancer. Cancer Res. 2009;69(10):4192–201.
94. Johnston S, Trudeau M, Kaufman B, Boussen H, Blackwell K, LoRusso P, et al. Phase II study of predictive biomarker profiles for response targeting human epidermal growth factor receptor 2 (HER-2) in advanced inflammatory breast cancer with lapatinib monotherapy. J Clin Oncol. 2008;26(7):1066–72.
95. Xia W, Husain I, Liu L, Bacus S, Saini S, Spohn J, et al. Lapatinib antitumor activity is not dependent upon phosphatase and tensin homologue deleted on chromosome 10 in ErbB2-overexpressing breast cancers. Cancer Res. 2007;67(3):1170–5.
96. Faratian D, Goltsov A, Lebedeva G, Sorokin A, Moodie S, Mullen P, et al. Systems biology reveals new strategies for personalizing cancer medicine and confirms the role of PTEN in resistance to trastuzumab. Cancer Res. 2009;69(16):6713–20.
97. Stambolic V, Suzuki A, de la Pompa JL, Brothers GM, Mirtsos C, Sasaki T, et al. Negative regulation of PKB/Akt-dependent cell survival by the tumor suppressor PTEN. Cell. 1998;95(1):29–39.
98. Yin Y, Shen WH. PTEN: a new guardian of the genome. Oncogene. 2008;27(41):5443–53.
99. Muraoka RS, Lenferink AE, Law B, Hamilton E, Brantley DM, Roebuck LR, et al. ErbB2/Neu-induced, cyclin D1-dependent transformation is accelerated in p27-haploinsufficient mammary epithelial cells but impaired in p27-null cells. Mol Cell Biol. 2002;22(7):2204–19.
100. Lane HA, Beuvink I, Motoyama AB, Daly JM, Neve RM, Hynes NE. ErbB2 potentiates breast tumor proliferation through modulation of p27(Kip1)-Cdk2 complex formation: receptor overexpression does not determine growth dependency. Mol Cell Biol. 2000;20(9):3210–23.

101. Yakes FM, Chinratanalab W, Ritter CA, King W, Seelig S, Arteaga CL. Herceptin-induced inhibition of phosphatidylinositol-3 kinase and Akt Is required for antibody-mediated effects on p27, cyclin D1, and antitumor action. Cancer Res. 2002;62(14):4132–41.
102. Nahta R, Takahashi T, Ueno NT, Hung MC, Esteva FJ. P27(kip1) down-regulation is associated with trastuzumab resistance in breast cancer cells. Cancer Res. 2004;64(11):3981–6.
103. Scaltriti M, Eichhorn PJ, Cortes J, Prudkin L, Aura C, Jimenez J, et al. Cyclin E amplification/overexpression is a mechanism of trastuzumab resistance in HER2+ breast cancer patients. Proc Natl Acad Sci U S A. 2011;108(9):3761–6.
104. Feng Q, Baird D, Peng X, Wang J, Ly T, Guan JL, et al. Cool-1 functions as an essential regulatory node for EGF receptor- and Src-mediated cell growth. Nat Cell Biol. 2006;8(9): 945–56.
105. Wu WJ, Tu S, Cerione RA. Activated Cdc42 sequesters c-Cbl and prevents EGF receptor degradation. Cell. 2003;114(6):715–25.
106. Hui R, Campbell DH, Lee CS, McCaul K, Horsfall DJ, Musgrove EA, et al. EMS1 amplification can occur independently of CCND1 or INT-2 amplification at 11q13 and may identify different phenotypes in primary breast cancer. Oncogene. 1997;15(13):1617–23.
107. Timpson P, Lynch DK, Schramek D, Walker F, Daly RJ. Cortactin overexpression inhibits ligand-induced down-regulation of the epidermal growth factor receptor. Cancer Res. 2005;65(8):3273–80.
108. Mosesson Y, Mills GB, Yarden Y. Derailed endocytosis: an emerging feature of cancer. Nat Rev Cancer. 2008;8(11):835–50.
109. Hu CT, Wu JR, Wu WS. The role of endosomal signaling triggered by metastatic growth factors in tumor progression. Cell Signal. 2013;25(7):1539–45.
110. Sassone-Corsi P, Sisson JC, Verma IM. Transcriptional autoregulation of the proto-oncogene fos. Nature. 1988;334(6180):314–9.
111. Gur G, Rubin C, Katz M, Amit I, Citri A, Nilsson J, et al. LRIG1 restricts growth factor signaling by enhancing receptor ubiquitylation and degradation. EMBO J. 2004;23(16):3270–81.
112. Miller JK, Shattuck DL, Ingalla EQ, Yen L, Borowsky AD, Young LJ, et al. Suppression of the negative regulator LRIG1 contributes to ErbB2 overexpression in breast cancer. Cancer Res. 2008;68(20):8286–94.
113. Zhang X, Pickin KA, Bose R, Jura N, Cole PA, Kuriyan J. Inhibition of the EGF receptor by binding of MIG6 to an activating kinase domain interface. Nature. 2007;450(7170):741–4.
114. Anastasi S, Sala G, Huiping C, Caprini E, Russo G, Iacovelli S, et al. Loss of RALT/MIG-6 expression in ERBB2-amplified breast carcinomas enhances ErbB-2 oncogenic potency and favors resistance to Herceptin. Oncogene. 2005;24(28):4540–8.
115. Kim HJ, Taylor LJ, Bar-Sagi D. Spatial regulation of EGFR signaling by Sprouty2. Curr Biol. 2007;17(5):455–61.
116. Rubin C, Zwang Y, Vaisman N, Ron D, Yarden Y. Phosphorylation of carboxyl-terminal tyrosines modulates the specificity of Sprouty-2 inhibition of different signaling pathways. J Biol Chem. 2005;280(10):9735–44.
117. Hanafusa H, Torii S, Yasunaga T, Nishida E. Sprouty1 and Sprouty2 provide a control mechanism for the Ras/MAPK signalling pathway. Nat Cell Biol. 2002;4(11):850–8.
118. Raff JP, Rajdev L, Malik U, Novik Y, Manalo JM, Negassa A, et al. Phase II study of weekly docetaxel alone or in combination with trastuzumab in patients with metastatic breast cancer. Clin Breast Cancer. 2004;4(6):420–7.
119. Hirota S, Isozaki K, Moriyama Y, Hashimoto K, Nishida T, Ishiguro S, et al. Gain-of-function mutations of c-kit in human gastrointestinal stromal tumors. Science. 1998;279(5350):577–80.
120. Chan PM, Ilangumaran S, La Rose J, Chakrabartty A, Rottapel R. Autoinhibition of the kit receptor tyrosine kinase by the cytosolic juxtamembrane region. Mol Cell Biol. 2003;23(9):3067–78.
121. Yuzawa S, Opatowsky Y, Zhang Z, Mandiyan V, Lax I, Schlessinger J. Structural basis for activation of the receptor tyrosine kinase KIT by stem cell factor. Cell. 2007;130(2):323–34.
122. Naski MC, Wang Q, Xu J, Ornitz DM. Graded activation of fibroblast growth factor receptor 3 by mutations causing achondroplasia and thanatophoric dysplasia. Nat Genet. 1996;13(2): 233–7.

123. Yasuda H, Kobayashi S, Costa DB. EGFR exon 20 insertion mutations in non-small-cell lung cancer: preclinical data and clinical implications. Lancet Oncol. 2012;13(1):e23–31.
124. Wu JY, Wu SG, Yang CH, Gow CH, Chang YL, Yu CJ, et al. Lung cancer with epidermal growth factor receptor exon 20 mutations is associated with poor gefitinib treatment response. Clin Cancer Res. 2008;14(15):4877–82.
125. Greulich H, Chen TH, Feng W, Janne PA, Alvarez JV, Zappaterra M, et al. Oncogenic transformation by inhibitor-sensitive and -resistant EGFR mutants. PLoS Med. 2005;2(11):e313.
126. Pao W, Chmielecki J. Rational, biologically based treatment of EGFR-mutant non-small-cell lung cancer. Nat Rev Cancer. 2010;10(11):760–74.
127. Sharma SV, Bell DW, Settleman J, Haber DA. Epidermal growth factor receptor mutations in lung cancer. Nat Rev Cancer. 2007;7(3):169–81.
128. Lee JC, Vivanco I, Beroukhim R, Huang JH, Feng WL, DeBiasi RM, et al. Epidermal growth factor receptor activation in glioblastoma through novel missense mutations in the extracellular domain. PLoS Med. 2006;3(12):e485.
129. Vivanco I, Robins HI, Rohle D, Campos C, Grommes C, Nghiemphu PL, et al. Differential sensitivity of glioma- versus lung cancer-specific EGFR mutations to EGFR kinase inhibitors. Cancer Discov. 2012;2(5):458–71.
130. Moscatello DK, Holgado-Madruga M, Godwin AK, Ramirez G, Gunn G, Zoltick PW, et al. Frequent expression of a mutant epidermal growth factor receptor in multiple human tumors. Cancer Res. 1995;55(23):5536–9.
131. Nieto Y, Nawaz F, Jones RB, Shpall EJ, Nawaz S. Prognostic significance of overexpression and phosphorylation of epidermal growth factor receptor (EGFR) and the presence of truncated EGFRvIII in locoregionally advanced breast cancer. J Clin Oncol. 2007;25(28):4405–13.
132. Moscatello DK, Holgado-Madruga M, Emlet DR, Montgomery RB, Wong AJ. Constitutive activation of phosphatidylinositol 3-kinase by a naturally occurring mutant epidermal growth factor receptor. J Biol Chem. 1998;273(1):200–6.
133. Reis-Filho JS, Milanezi F, Carvalho S, Simpson PT, Steele D, Savage K, et al. Metaplastic breast carcinomas exhibit EGFR, but not HER2, gene amplification and overexpression: immunohistochemical and chromogenic in situ hybridization analysis. Breast Cancer Res. 2005;7(6):R1028–35.
134. Tognon C, Knezevich SR, Huntsman D, Roskelley CD, Melnyk N, Mathers JA, et al. Expression of the ETV6-NTRK3 gene fusion as a primary event in human secretory breast carcinoma. Cancer Cell. 2002;2(5):367–76.
135. Tuma RS. ALK gene amplified in most inflammatory breast cancers. J Natl Cancer Inst. 2012;104(2):87–8.
136. Slamon DJ, Clark GM, Wong SG, Levin WJ, Ullrich A, McGuire WL. Human breast cancer: correlation of relapse and survival with amplification of the HER-2/neu oncogene. Science. 1987;235(4785):177–82.
137. van de Vijver M, van de Bersselaar R, Devilee P, Cornelisse C, Peterse J, Nusse R. Amplification of the neu (c-erbB-2) oncogene in human mammmary tumors is relatively frequent and is often accompanied by amplification of the linked c-erbA oncogene. Mol Cell Biol. 1987;7(5):2019–23.
138. Reinholz MM, Bruzek AK, Visscher DW, Lingle WL, Schroeder MJ, Perez EA, et al. Breast cancer and aneusomy 17: implications for carcinogenesis and therapeutic response. Lancet Oncol. 2009;10(3):267–77.
139. Menendez JA, Lupu R. Fatty acid synthase and the lipogenic phenotype in cancer pathogenesis. Nat Rev Cancer. 2007;7(10):763–77.
140. Kraus MH, Popescu NC, Amsbaugh SC, King CR. Overexpression of the EGF receptor-related proto-oncogene erbB-2 in human mammary tumor cell lines by different molecular mechanisms. EMBO J. 1987;6(3):605–10.
141. Zuo T, Wang L, Morrison C, Chang X, Zhang H, Li W, et al. FOXP3 is an X-linked breast cancer suppressor gene and an important repressor of the HER-2/ErbB2 oncogene. Cell. 2007;129(7):1275–86.

142. Xing X, Wang SC, Xia W, Zou Y, Shao R, Kwong KY, et al. The ets protein PEA3 suppresses HER-2/neu overexpression and inhibits tumorigenesis. Nat Med. 2000;6(2):189–95.
143. Shaw AT, Hsu PP, Awad MM, Engelman JA. Tyrosine kinase gene rearrangements in epithelial malignancies. Nat Rev Cancer. 2013;13(11):772–87.
144. Rabes HM, Demidchik EP, Sidorow JD, Lengfelder E, Beimfohr C, Hoelzel D, et al. Pattern of radiation-induced RET and NTRK1 rearrangements in 191 post-chernobyl papillary thyroid carcinomas: biological, phenotypic, and clinical implications. Clin Cancer Res. 2000;6(3):1093–103.
145. Parker BC, Annala MJ, Cogdell DE, Granberg KJ, Sun Y, Ji P, et al. The tumorigenic FGFR3-TACC3 gene fusion escapes miR-99a regulation in glioblastoma. J Clin Invest. 2013;123(2):855–65.
146. Drilon A, Wang L, Hasanovic A, Suehara Y, Lipson D, Stephens P, et al. Response to Cabozantinib in patients with RET fusion-positive lung adenocarcinomas. Cancer Discov. 2013;3(6):630–5.
147. Kwak EL, Bang YJ, Camidge DR, Shaw AT, Solomon B, Maki RG, et al. Anaplastic lymphoma kinase inhibition in non-small-cell lung cancer. N Engl J Med. 2010;363(18):1693–703.
148. Mino-Kenudson M, Chirieac LR, Law K, Hornick JL, Lindeman N, Mark EJ, et al. A novel, highly sensitive antibody allows for the routine detection of ALK-rearranged lung adenocarcinomas by standard immunohistochemistry. Clin Cancer Res. 2010;16(5):1561–71.
149. Sequist LV, Gettinger S, Senzer NN, Martins RG, Janne PA, Lilenbaum R, et al. Activity of IPI-504, a novel heat-shock protein 90 inhibitor, in patients with molecularly defined non-small-cell lung cancer. J Clin Oncol. 2010;28(33):4953–60.
150. Socinski MA, Goldman J, El-Hariry I, Koczywas M, Vukovic V, Horn L, et al. A multicenter phase II study of ganetespib monotherapy in patients with genotypically defined advanced non-small cell lung cancer. Clin Cancer Res. 2013;19(11):3068–77.
151. Kostler WJ, Tomek S, Brodowicz T, Budinsky AC, Flamm M, Hejna M, et al. Soluble ICAM-1 in breast cancer: clinical significance and biological implications. Cancer Immunol Immunother. 2001;50(9):483–90.
152. Musolino A, Naldi N, Bortesi B, Pezzuolo D, Capelletti M, Missale G, et al. Immunoglobulin G fragment C receptor polymorphisms and clinical efficacy of trastuzumab-based therapy in patients with HER-2/neu-positive metastatic breast cancer. J Clin Oncol. 2008;26(11):1789–96.
153. Zhang W, Gordon M, Schultheis AM, Yang DY, Nagashima F, Azuma M, et al. FCGR2A and FCGR3A polymorphisms associated with clinical outcome of epidermal growth factor receptor expressing metastatic colorectal cancer patients treated with single-agent cetuximab. J Clin Oncol. 2007;25(24):3712–8.
154. Vogel CL, Cobleigh MA, Tripathy D, Gutheil JC, Harris LN, Fehrenbacher L, et al. First-line Herceptin monotherapy in metastatic breast cancer. Oncology. 2001;61 Suppl 2:37–42.
155. Vogel CL, Cobleigh MA, Tripathy D, Gutheil JC, Harris LN, Fehrenbacher L, et al. Efficacy and safety of trastuzumab as a single agent in first-line treatment of HER2-overexpressing metastatic breast cancer. J Clin Oncol. 2002;20(3):719–26.
156. Slamon DJ, Leyland-Jones B, Shak S, Fuchs H, Paton V, Bajamonde A, et al. Use of chemotherapy plus a monoclonal antibody against HER2 for metastatic breast cancer that overexpresses HER2. N Engl J Med. 2001;344(11):783–92.
157. Marty M, Cognetti F, Maraninchi D, Snyder R, Mauriac L, Tubiana-Hulin M, et al. Randomized phase II trial of the efficacy and safety of trastuzumab combined with docetaxel in patients with human epidermal growth factor receptor 2-positive metastatic breast cancer administered as first-line treatment: the M77001 study group. J Clin Oncol. 2005;23(19):4265–74.
158. Cho HS, Mason K, Ramyar KX, Stanley AM, Gabelli SB, Denney Jr DW, et al. Structure of the extracellular region of HER2 alone and in complex with the Herceptin Fab. Nature. 2003;421(6924):756–60.
159. Agus DB, Akita RW, Fox WD, Lewis GD, Higgins B, Pisacane PI, et al. Targeting ligand-activated ErbB2 signaling inhibits breast and prostate tumor growth. Cancer Cell. 2002;2(2):127–37.

160. Barnea I, Haif S, Keshet R, Karaush V, Lev-Ari S, Khafif A, et al. Targeting ErbB-1 and ErbB-4 in irradiated head and neck cancer: results of in vitro and in vivo studies. Head Neck. 2013;35(3):399–407.
161. Viloria-Petit A, Crombet T, Jothy S, Hicklin D, Bohlen P, Schlaeppi JM, et al. Acquired resistance to the antitumor effect of epidermal growth factor receptor-blocking antibodies in vivo: a role for altered tumor angiogenesis. Cancer Res. 2001;61(13):5090–101.
162. Sandler A, Gray R, Perry MC, Brahmer J, Schiller JH, Dowlati A, et al. Paclitaxel-carboplatin alone or with bevacizumab for non-small-cell lung cancer. N Engl J Med. 2006;355(24):2542–50.
163. Sandler A, Yi J, Dahlberg S, Kolb MM, Wang L, Hambleton J, et al. Treatment outcomes by tumor histology in Eastern Cooperative Group Study E4599 of bevacizumab with paclitaxel/carboplatin for advanced non-small cell lung cancer. J Thorac Oncol. 2010;5(9):1416–23.
164. Reck M, von Pawel J, Zatloukal P, Ramlau R, Gorbounova V, Hirsh V, et al. Overall survival with cisplatin-gemcitabine and bevacizumab or placebo as first-line therapy for nonsquamous non-small-cell lung cancer: results from a randomised phase III trial (AVAiL). Ann Oncol. 2010;21(9):1804–9.
165. Reck M, von Pawel J, Zatloukal P, Ramlau R, Gorbounova V, Hirsh V, et al. Phase III trial of cisplatin plus gemcitabine with either placebo or bevacizumab as first-line therapy for non-squamous non-small-cell lung cancer: AVAil. J Clin Oncol. 2009;27(8):1227–34.
166. Gianni L, Romieu GH, Lichinitser M, Serrano SV, Mansutti M, Pivot X, et al. AVEREL: a randomized phase III Trial evaluating bevacizumab in combination with docetaxel and trastuzumab as first-line therapy for HER2-positive locally recurrent/metastatic breast cancer. J Clin Oncol. 2013;31(14):1719–25.
167. Tol J, Koopman M, Cats A, Rodenburg CJ, Creemers GJ, Schrama JG, et al. Chemotherapy, bevacizumab, and cetuximab in metastatic colorectal cancer. N Engl J Med. 2009;360(6):563–72.
168. Saltz L, Badarinath S, Dakhil S, Bienvenu B, Harker WG, Birchfield G, et al. Phase III trial of cetuximab, bevacizumab, and 5-fluorouracil/leucovorin vs. FOLFOX-bevacizumab in colorectal cancer. Clin Colorectal Cancer. 2012;11(2):101–11.
169. Herbst RS, Ansari R, Bustin F, Flynn P, Hart L, Otterson GA, et al. Efficacy of bevacizumab plus erlotinib versus erlotinib alone in advanced non-small-cell lung cancer after failure of standard first-line chemotherapy (BeTa): a double-blind, placebo-controlled, phase 3 trial. Lancet. 2011;377(9780):1846–54.
170. Spangler JB, Neil JR, Abramovitch S, Yarden Y, White FM, Lauffenburger DA, et al. Combination antibody treatment down-regulates epidermal growth factor receptor by inhibiting endosomal recycling. Proc Natl Acad Sci U S A. 2010;107(30):13252–7.
171. Maron R, Schechter B, Mancini M, Mahlknecht G, Yarden Y, Sela M. Inhibition of pancreatic carcinoma by homo- and heterocombinations of antibodies against EGF-receptor and its kin HER2/ErbB-2. Proc Natl Acad Sci U S A. 2013;110(38):15389–94.
172. Ferraro DA, Gaborit N, Maron R, Cohen-Dvashi H, Porat Z, Pareja F, et al. Inhibition of triple-negative breast cancer models by combinations of antibodies to EGFR. Proc Natl Acad Sci U S A. 2013;110(5):1815–20.
173. Ben-Kasus T, Schechter B, Lavi S, Yarden Y, Sela M. Persistent elimination of ErbB-2/HER2-overexpressing tumors using combinations of monoclonal antibodies: relevance of receptor endocytosis. Proc Natl Acad Sci U S A. 2009;106(9):3294–9.
174. Ismael G, Hegg R, Muehlbauer S, Heinzmann D, Lum B, Kim SB, et al. Subcutaneous versus intravenous administration of (neo)adjuvant trastuzumab in patients with HER2-positive, clinical stage I-III breast cancer (HannaH study): a phase 3, open-label, multicentre, randomised trial. Lancet Oncol. 2012;13(9):869–78.
175. Pivot X, Gligorov J, Muller V, Barrett-Lee P, Verma S, Knoop A, et al. Preference for subcutaneous or intravenous administration of trastuzumab in patients with HER2-positive early breast cancer (PrefHer): an open-label randomised study. Lancet Oncol. 2013;14(10):962–70.
176. Presta LG. Molecular engineering and design of therapeutic antibodies. Curr Opin Immunol. 2008;20(4):460–70.

177. Scott AM, Lee FT, Tebbutt N, Herbertson R, Gill SS, Liu Z, et al. A phase I clinical trial with monoclonal antibody ch806 targeting transitional state and mutant epidermal growth factor receptors. Proc Natl Acad Sci U S A. 2007;104(10):4071–6.
178. Sperinde J, Jin X, Banerjee J, Penuel E, Saha A, Diedrich G, et al. Quantitation of p95HER2 in paraffin sections by using a p95-specific antibody and correlation with outcome in a cohort of trastuzumab-treated breast cancer patients. Clin Cancer Res. 2010;16(16):4226–35.
179. Verma S, Miles D, Gianni L, Krop IE, Welslau M, Baselga J, et al. Trastuzumab emtansine for HER2-positive advanced breast cancer. N Engl J Med. 2012;367(19):1783–91.
180. Vaklavas C, Forero-Torres A. Safety and efficacy of brentuximab vedotin in patients with Hodgkin lymphoma or systemic anaplastic large cell lymphoma. Ther Adv Hematol. 2012;3(4):209–25.
181. Van Cutsem E, Tabernero J, Lakomy R, Prenen H, Prausova J, Macarulla T, et al. Addition of aflibercept to fluorouracil, leucovorin, and irinotecan improves survival in a phase III randomized trial in patients with metastatic colorectal cancer previously treated with an oxaliplatin-based regimen. J Clin Oncol. 2012;30(28):3499–506.
182. Tew WP, Colombo N, Ray-Coquard I, Del Campo JM, Oza A, Pereira D, et al. Intravenous aflibercept in patients with platinum-resistant, advanced ovarian cancer: results of a randomized, double-blind, phase 2, parallel-arm study. Cancer. 2014;120(3):335–43.
183. Tannock IF, Fizazi K, Ivanov S, Karlsson CT, Flechon A, Skoneczna I, et al. Aflibercept versus placebo in combination with docetaxel and prednisone for treatment of men with metastatic castration-resistant prostate cancer (VENICE): a phase 3, double-blind randomised trial. Lancet Oncol. 2013;14(8):760–8.
184. Rougier P, Riess H, Manges R, Karasek P, Humblet Y, Barone C, et al. Randomised, placebo-controlled, double-blind, parallel-group phase III study evaluating aflibercept in patients receiving first-line treatment with gemcitabine for metastatic pancreatic cancer. Eur J Cancer. 2013;49(12):2633–42.
185. Ramlau R, Gorbunova V, Ciuleanu TE, Novello S, Ozguroglu M, Goksel T, et al. Aflibercept and Docetaxel versus Docetaxel alone after platinum failure in patients with advanced or metastatic non-small-cell lung cancer: a randomized, controlled phase III trial. J Clin Oncol. 2012;30(29):3640–7.
186. Lindzen M, Carvalho S, Starr A, Ben-Chetrit N, Pradeep CR, Kostler WJ, et al. A recombinant decoy comprising EGFR and ErbB-4 inhibits tumor growth and metastasis. Oncogene. 2012;31(30):3505–15.
187. Dematteo RP, Ballman KV, Antonescu CR, Maki RG, Pisters PW, Demetri GD, et al. Adjuvant imatinib mesylate after resection of localised, primary gastrointestinal stromal tumour: a randomised, double-blind, placebo-controlled trial. Lancet. 2009;373(9669):1097–104.
188. Joensuu H, Eriksson M, Sundby Hall K, Hartmann JT, Pink D, Schutte J, et al. One vs three years of adjuvant imatinib for operable gastrointestinal stromal tumor: a randomized trial. JAMA. 2012;307(12):1265–72.
189. Benvenuti S, Sartore-Bianchi A, Di Nicolantonio F, Zanon C, Moroni M, Veronese S, et al. Oncogenic activation of the RAS/RAF signaling pathway impairs the response of metastatic colorectal cancers to anti-epidermal growth factor receptor antibody therapies. Cancer Res. 2007;67(6):2643–8.
190. Bardelli A, Corso S, Bertotti A, Hobor S, Valtorta E, Siravegna G, et al. Amplification of the MET receptor drives resistance to anti-EGFR therapies in colorectal cancer. Cancer Discov. 2013;3(6):658–73.
191. Misale S, Yaeger R, Hobor S, Scala E, Janakiraman M, Liska D, et al. Emergence of KRAS mutations and acquired resistance to anti-EGFR therapy in colorectal cancer. Nature. 2012;486(7404):532–6.
192. Heinrich MC, Griffith D, McKinley A, Patterson J, Presnell A, Ramachandran A, et al. Crenolanib inhibits the drug-resistant PDGFRA D842V mutation associated with imatinib-resistant gastrointestinal stromal tumors. Clin Cancer Res. 2012;18(16):4375–84.
193. Cassier PA, Fumagalli E, Rutkowski P, Schoffski P, Van Glabbeke M, Debiec-Rychter M, et al. Outcome of patients with platelet-derived growth factor receptor alpha-mutated

gastrointestinal stromal tumors in the tyrosine kinase inhibitor era. Clin Cancer Res. 2012;18(16):4458–64.
194. Corless CL, Schroeder A, Griffith D, Town A, McGreevey L, Harrell P, et al. PDGFRA mutations in gastrointestinal stromal tumors: frequency, spectrum and in vitro sensitivity to imatinib. J Clin Oncol. 2005;23(23):5357–64.
195. Weisberg E, Wright RD, Jiang J, Ray A, Moreno D, Manley PW, et al. Effects of PKC412, nilotinib, and imatinib against GIST-associated PDGFRA mutants with differential imatinib sensitivity. Gastroenterology. 2006;131(6):1734–42.
196. Rosell R, Carcereny E, Gervais R, Vergnenegre A, Massuti B, Felip E, et al. Erlotinib versus standard chemotherapy as first-line treatment for European patients with advanced EGFR mutation-positive non-small-cell lung cancer (EURTAC): a multicentre, open-label, randomised phase 3 trial. Lancet Oncol. 2012;13(3):239–46.
197. Engelman JA, Zejnullahu K, Mitsudomi T, Song Y, Hyland C, Park JO, et al. MET amplification leads to gefitinib resistance in lung cancer by activating ERBB3 signaling. Science. 2007;316(5827):1039–43.
198. Bean J, Brennan C, Shih JY, Riely G, Viale A, Wang L, et al. MET amplification occurs with or without T790M mutations in EGFR mutant lung tumors with acquired resistance to gefitinib or erlotinib. Proc Natl Acad Sci U S A. 2007;104(52):20932–7.
199. Inukai M, Toyooka S, Ito S, Asano H, Ichihara S, Soh J, et al. Presence of epidermal growth factor receptor gene T790M mutation as a minor clone in non-small cell lung cancer. Cancer Res. 2006;66(16):7854–8.
200. Sequist LV, Martins RG, Spigel D, Grunberg SM, Spira A, Janne PA, et al. First-line gefitinib in patients with advanced non-small-cell lung cancer harboring somatic EGFR mutations. J Clin Oncol. 2008;26(15):2442–9.
201. Turke AB, Zejnullahu K, Wu YL, Song Y, Dias-Santagata D, Lifshits E, et al. Preexistence and clonal selection of MET amplification in EGFR mutant NSCLC. Cancer Cell. 2010;17(1):77–88.
202. Heinrich MC, Owzar K, Corless CL, Hollis D, Borden EC, Fletcher CD, et al. Correlation of kinase genotype and clinical outcome in the North American Intergroup Phase III Trial of imatinib mesylate for treatment of advanced gastrointestinal stromal tumor: CALGB 150105 Study by Cancer and Leukemia Group B and Southwest Oncology Group. J Clin Oncol. 2008;26(33):5360–7.
203. Gastrointestinal Stromal Tumor Meta-Analysis Group (MetaGIST). Comparison of two doses of imatinib for the treatment of unresectable or metastatic gastrointestinal stromal tumors: a meta-analysis of 1,640 patients. J Clin Oncol. 2010;28(7):1247–53.
204. Demetri GD, van Oosterom AT, Garrett CR, Blackstein ME, Shah MH, Verweij J, et al. Efficacy and safety of sunitinib in patients with advanced gastrointestinal stromal tumour after failure of imatinib: a randomised controlled trial. Lancet. 2006;368(9544):1329–38.
205. Demetri GD, Reichardt P, Kang YK, Blay JY, Rutkowski P, Gelderblom H, et al. Efficacy and safety of regorafenib for advanced gastrointestinal stromal tumours after failure of imatinib and sunitinib (GRID): an international, multicentre, randomised, placebo-controlled, phase 3 trial. Lancet. 2013;381(9863):295–302.
206. Wardelmann E, Thomas N, Merkelbach-Bruse S, Pauls K, Speidel N, Buttner R, et al. Acquired resistance to imatinib in gastrointestinal stromal tumours caused by multiple KIT mutations. Lancet Oncol. 2005;6(4):249–51.
207. Liegl B, Kepten I, Le C, Zhu M, Demetri GD, Heinrich MC, et al. Heterogeneity of kinase inhibitor resistance mechanisms in GIST. J Pathol. 2008;216(1):64–74.
208. Pao W, Miller VA, Politi KA, Riely GJ, Somwar R, Zakowski MF, et al. Acquired resistance of lung adenocarcinomas to gefitinib or erlotinib is associated with a second mutation in the EGFR kinase domain. PLoS Med. 2005;2(3):e73.
209. Yun CH, Mengwasser KE, Toms AV, Woo MS, Greulich H, Wong KK, et al. The T790M mutation in EGFR kinase causes drug resistance by increasing the affinity for ATP. Proc Natl Acad Sci U S A. 2008;105:2070.

210. Kobayashi S, Ji H, Yuza Y, Meyerson M, Wong KK, Tenen DG, et al. An alternative inhibitor overcomes resistance caused by a mutation of the epidermal growth factor receptor. Cancer Res. 2005;65(16):7096–101.
211. Engelman JA, Zejnullahu K, Gale CM, Lifshits E, Gonzales AJ, Shimamura T, et al. PF00299804, an irreversible pan-ERBB inhibitor, is effective in lung cancer models with EGFR and ERBB2 mutations that are resistant to gefitinib. Cancer Res. 2007;67(24):11924–32.
212. Minami Y, Shimamura T, Shah K, LaFramboise T, Glatt KA, Liniker E, et al. The major lung cancer-derived mutants of ERBB2 are oncogenic and are associated with sensitivity to the irreversible EGFR/ERBB2 inhibitor HKI-272. Oncogene. 2007;26(34):5023–7.
213. Zhou W, Ercan D, Chen L, Yun CH, Li D, Capelletti M, et al. Novel mutant-selective EGFR kinase inhibitors against EGFR T790M. Nature. 2009;462(7276):1070–4.
214. Yonesaka K, Zejnullahu K, Okamoto I, Satoh T, Cappuzzo F, Souglakos J, et al. Activation of ERBB2 signaling causes resistance to the EGFR-directed therapeutic antibody cetuximab. Sci Transl Med. 2011;3(99):99ra86.
215. Motoyama AB, Hynes NE, Lane HA. The efficacy of ErbB receptor-targeted anticancer therapeutics is influenced by the availability of epidermal growth factor-related peptides. Cancer Res. 2002;62(11):3151–8.
216. Shaw AT, Kim DW, Nakagawa K, Seto T, Crino L, Ahn MJ, et al. Crizotinib versus chemotherapy in advanced ALK-positive lung cancer. N Engl J Med. 2013;368(25):2385–94.
217. Doebele RC, Pilling AB, Aisner DL, Kutateladze TG, Le AT, Weickhardt AJ, et al. Mechanisms of resistance to crizotinib in patients with ALK gene rearranged non-small cell lung cancer. Clin Cancer Res. 2012;18(5):1472–82.
218. Katayama R, Shaw AT, Khan TM, Mino-Kenudson M, Solomon BJ, Halmos B, et al. Mechanisms of acquired crizotinib resistance in ALK-rearranged lung Cancers. Sci Transl Med. 2012;4(120):120ra17.
219. Katayama R, Khan TM, Benes C, Lifshits E, Ebi H, Rivera VM, et al. Therapeutic strategies to overcome crizotinib resistance in non-small cell lung cancers harboring the fusion oncogene EML4-ALK. Proc Natl Acad Sci U S A. 2011;108(18):7535–40.
220. Awad MM, Katayama R, McTigue M, Liu W, Deng YL, Brooun A, et al. Acquired resistance to crizotinib from a mutation in CD74-ROS1. N Engl J Med. 2013;368(25):2395–401.
221. Perez CA, Velez M, Raez LE, Santos ES. Overcoming the resistance to crizotinib in patients with non-small cell lung cancer harboring EML4/ALK translocation. Lung Cancer. 2014;84(2):110–5.
222. Shaw AT, Engelman JA. Ceritinib in ALK-rearranged non-small-cell lung cancer. N Engl J Med. 2014;370(26):2537–9.
223. Anido J, Scaltriti M, Bech Serra JJ, Santiago Josefat B, Todo FR, Baselga J, et al. Biosynthesis of tumorigenic HER2 C-terminal fragments by alternative initiation of translation. EMBO J. 2006;25(13):3234–44.
224. Codony-Servat J, Albanell J, Lopez-Talavera JC, Arribas J, Baselga J. Cleavage of the HER2 ectodomain is a pervanadate-activable process that is inhibited by the tissue inhibitor of metalloproteases-1 in breast cancer cells. Cancer Res. 1999;59(6):1196–201.
225. Pedersen K, Angelini PD, Laos S, Bach-Faig A, Cunningham MP, Ferrer-Ramon C, et al. A naturally occurring HER2 carboxy-terminal fragment promotes mammary tumor growth and metastasis. Mol Cell Biol. 2009;29(12):3319–31.
226. Xia W, Liu LH, Ho P, Spector NL. Truncated ErbB2 receptor (p95ErbB2) is regulated by heregulin through heterodimer formation with ErbB3 yet remains sensitive to the dual EGFR/ErbB2 kinase inhibitor GW572016. Oncogene. 2004;23(3):646–53.
227. Saez R, Molina MA, Ramsey EE, Rojo F, Keenan EJ, Albanell J, et al. p95HER-2 predicts worse outcome in patients with HER-2-positive breast cancer. Clin Cancer Res. 2006;12(2):424–31.
228. Hudelist G, Kostler WJ, Attems J, Czerwenka K, Muller R, Manavi M, et al. Her-2/neu-triggered intracellular tyrosine kinase activation: in vivo relevance of ligand-independent

activation mechanisms and impact upon the efficacy of trastuzumab-based treatment. Br J Cancer. 2003;89(6):983–91.
229. Scaltriti M, Verma C, Guzman M, Jimenez J, Parra JL, Pedersen K, et al. Lapatinib, a HER2 tyrosine kinase inhibitor, induces stabilization and accumulation of HER2 and potentiates trastuzumab-dependent cell cytotoxicity. Oncogene. 2009;28(6):803–14.
230. Scaltriti M, Chandarlapaty S, Prudkin L, Aura C, Jimenez J, Angelini PD, et al. Clinical benefit of lapatinib-based therapy in patients with human epidermal growth factor receptor 2-positive breast tumors coexpressing the truncated p95HER2 receptor. Clin Cancer Res. 2010;16(9):2688–95.
231. Pegram MD, Lipton A, Hayes DF, Weber BL, Baselga JM, Tripathy D, et al. Phase II study of receptor-enhanced chemosensitivity using recombinant humanized anti-p185HER2/neu monoclonal antibody plus cisplatin in patients with HER2/neu-overexpressing metastatic breast cancer refractory to chemotherapy treatment. J Clin Oncol. 1998;16(8):2659–71.
232. Brodowicz T, Wiltschke C, Budinsky AC, Krainer M, Steger GG, Zielinski CC. Soluble HER-2/neu neutralizes biologic effects of anti-HER-2/neu antibody on breast cancer cells in vitro. Int J Cancer. 1997;73(6):875–9.
233. Molina MA, Codony-Servat J, Albanell J, Rojo F, Arribas J, Baselga J. Trastuzumab (herceptin), a humanized anti-Her2 receptor monoclonal antibody, inhibits basal and activated Her2 ectodomain cleavage in breast cancer cells. Cancer Res. 2001;61(12):4744–9.
234. Ali SM, Carney WP, Esteva FJ, Fornier M, Harris L, Kostler WJ, et al. Serum HER-2/neu and relative resistance to trastuzumab-based therapy in patients with metastatic breast cancer. Cancer. 2008;113(6):1294–301.
235. Kostler WJ, Schwab B, Singer CF, Neumann R, Rucklinger E, Brodowicz T, et al. Monitoring of serum Her-2/neu predicts response and progression-free survival to trastuzumab-based treatment in patients with metastatic breast cancer. Clin Cancer Res. 2004;10(5):1618–24.
236. Montagut C, Dalmases A, Bellosillo B, Crespo M, Pairet S, Iglesias M, et al. Identification of a mutation in the extracellular domain of the Epidermal Growth Factor Receptor conferring cetuximab resistance in colorectal cancer. Nat Med. 2012;18(2):221–3.
237. Nagy P, Friedlander E, Tanner M, Kapanen AI, Carraway KL, Isola J, et al. Decreased accessibility and lack of activation of ErbB2 in JIMT-1, a herceptin-resistant, MUC4-expressing breast cancer cell line. Cancer Res. 2005;65(2):473–82.
238. Price-Schiavi SA, Jepson S, Li P, Arango M, Rudland PS, Yee L, et al. Rat Muc4 (sialomucin complex) reduces binding of anti-ErbB2 antibodies to tumor cell surfaces, a potential mechanism for herceptin resistance. Int J Cancer. 2002;99(6):783–91.
239. Chen LL, Trent JC, Wu EF, Fuller GN, Ramdas L, Zhang W, et al. A missense mutation in KIT kinase domain 1 correlates with imatinib resistance in gastrointestinal stromal tumors. Cancer Res. 2004;64(17):5913–9.
240. Antonescu CR, Besmer P, Guo T, Arkun K, Hom G, Koryotowski B, et al. Acquired resistance to imatinib in gastrointestinal stromal tumor occurs through secondary gene mutation. Clin Cancer Res. 2005;11(11):4182–90.
241. Heinrich MC, Corless CL, Blanke CD, Demetri GD, Joensuu H, Roberts PJ, et al. Molecular correlates of imatinib resistance in gastrointestinal stromal tumors. J Clin Oncol. 2006;24(29):4764–74.
242. Hislop J, Mowatt G, Sharma P, Fraser C, Elders A, Jenkinson D, et al. Systematic review of escalated imatinib doses compared with sunitinib or best supportive care, for the treatment of people with unresectable/metastatic gastrointestinal stromal tumours whose disease has progressed on the standard imatinib dose. J Gastrointest Cancer. 2012;43(2):168–76.
243. Sergina NV, Rausch M, Wang D, Blair J, Hann B, Shokat KM, et al. Escape from HER-family tyrosine kinase inhibitor therapy by the kinase-inactive HER3. Nature. 2007;445(7126):437–41.
244. Mueller KL, Hunter LA, Ethier SP, Boerner JL. Met and c-Src cooperate to compensate for loss of epidermal growth factor receptor kinase activity in breast cancer cells. Cancer Res. 2008;68(9):3314–22.
245. Chaturvedi D, Gao X, Cohen MS, Taunton J, Patel TB. Rapamycin induces transactivation of the EGFR and increases cell survival. Oncogene. 2009;28(9):1187–96.

246. Siegel PM, Ryan ED, Cardiff RD, Muller WJ. Elevated expression of activated forms of Neu/ErbB-2 and ErbB-3 are involved in the induction of mammary tumors in transgenic mice: implications for human breast cancer. EMBO J. 1999;18(8):2149–64.
247. Kwong KY, Hung MC. A novel splice variant of HER2 with increased transformation activity. Mol Carcinog. 1998;23(2):62–8.
248. Castiglioni F, Tagliabue E, Campiglio M, Pupa SM, Balsari A, Menard S. Role of exon-16-deleted HER2 in breast carcinomas. Endocr Relat Cancer. 2006;13(1):221–32.
249. Alajati A, Sausgruber N, Aceto N, Duss S, Sarret S, Voshol H, et al. Mammary tumor formation and metastasis evoked by a HER2 splice variant. Cancer Res. 2013;73(17):5320–7.
250. Mitra D, Brumlik MJ, Okamgba SU, Zhu Y, Duplessis TT, Parvani JG, et al. An oncogenic isoform of HER2 associated with locally disseminated breast cancer and trastuzumab resistance. Mol Cancer Ther. 2009;8(8):2152–62.
251. Baselga J, Bradbury I, Eidtmann H, Di Cosimo S, de Azambuja E, Aura C, et al. Lapatinib with trastuzumab for HER2-positive early breast cancer (NeoALTTO): a randomised, open-label, multicentre, phase 3 trial. Lancet. 2012;379(9816):633–40.
252. Guarneri V, Frassoldati A, Bottini A, Cagossi K, Bisagni G, Sarti S, et al. Preoperative chemotherapy plus trastuzumab, lapatinib, or both in human epidermal growth factor receptor 2-positive operable breast cancer: results of the randomized phase II CHER-LOB study. J Clin Oncol. 2012;30(16):1989–95.
253. Gianni L, Pienkowski T, Im YH, Roman L, Tseng LM, Liu MC, et al. Efficacy and safety of neoadjuvant pertuzumab and trastuzumab in women with locally advanced, inflammatory, or early HER2-positive breast cancer (NeoSphere): a randomised multicentre, open-label, phase 2 trial. Lancet Oncol. 2012;13(1):25–32.
254. Baselga J, Cortes J, Kim SB, Im SA, Hegg R, Im YH, et al. Pertuzumab plus trastuzumab plus docetaxel for metastatic breast cancer. N Engl J Med. 2012;366(2):109–19.
255. Cortes J, Fumoleau P, Bianchi GV, Petrella TM, Gelmon K, Pivot X, et al. Pertuzumab monotherapy after trastuzumab-based treatment and subsequent reintroduction of trastuzumab: activity and tolerability in patients with advanced human epidermal growth factor receptor 2-positive breast cancer. J Clin Oncol. 2012;30(14):1594–600.
256. Blackwell KL, Burstein HJ, Storniolo AM, Rugo HS, Sledge G, Aktan G, et al. Overall survival benefit with lapatinib in combination with trastuzumab for patients with human epidermal growth factor receptor 2-positive metastatic breast cancer: final results from the EGF104900 Study. J Clin Oncol. 2012;30(21):2585–92.
257. Cortés J, Baselga J, Petrella T, Gelmon K, Fumoleau P, Verma S, et al. Pertuzumab monotherapy following trastuzumab-based treatment: activity and tolerability in patients with advanced HER2- positive breast cancer. J Clin Oncol. 2009;27(15S):abstract 1022.
258. Franklin MC, Carey KD, Vajdos FF, Leahy DJ, de Vos AM, Sliwkowski MX. Insights into ErbB signaling from the structure of the ErbB2-pertuzumab complex. Cancer Cell. 2004;5(4):317–28.
259. Xia W, Bacus S, Hegde P, Husain I, Strum J, Liu L, et al. A model of acquired autoresistance to a potent ErbB2 tyrosine kinase inhibitor and a therapeutic strategy to prevent its onset in breast cancer. Proc Natl Acad Sci U S A. 2006;103(20):7795–800.
260. Rimawi MF, Mayer IA, Forero A, Nanda R, Goetz MP, Rodriguez AA, et al. Multicenter phase II study of neoadjuvant lapatinib and trastuzumab with hormonal therapy and without chemotherapy in patients with human epidermal growth factor receptor 2-overexpressing breast cancer: TBCRC 006. J Clin Oncol. 2013;31(14):1726–31.
261. Johnston S, Pegram M, Press M, Pippen J, Pivot X, Gomez H, et al. Lapatinib combined with letrozole vs. letrozole alone for front line postmenopausal hormone receptor positive (HR+) metastatic breast cancer (MBC): first results from the EGF30008 Trial. Cancer research. 2008;69(2 (suppl)):74s (abstr 46).
262. Kaufman B, Mackey JR, Clemens MR, Bapsy PP, Vaid A, Wardley A, et al. Trastuzumab plus anastrozole versus anastrozole alone for the treatment of postmenopausal women with human epidermal growth factor receptor 2-positive, hormone receptor-positive metastatic breast cancer: results from the randomized phase III TAnDEM study. J Clin Oncol. 2009;27(33):5529–37.

263. Liu L, Greger J, Shi H, Liu Y, Greshock J, Annan R, et al. Novel mechanism of lapatinib resistance in HER2-positive breast tumor cells: activation of AXL. Cancer Res. 2009;69(17):6871–8.
264. Shien K, Toyooka S, Yamamoto H, Soh J, Jida M, Thu KL, et al. Acquired resistance to EGFR inhibitors is associated with a manifestation of stem cell-like properties in cancer cells. Cancer Res. 2013;73(10):3051–61.
265. White DL, Saunders VA, Dang P, Engler J, Venables A, Zrim S, et al. Most CML patients who have a suboptimal response to imatinib have low OCT-1 activity: higher doses of imatinib may overcome the negative impact of low OCT-1 activity. Blood. 2007;110(12):4064–72.
266. White DL, Saunders VA, Dang P, Engler J, Zannettino AC, Cambareri AC, et al. OCT-1-mediated influx is a key determinant of the intracellular uptake of imatinib but not nilotinib (AMN107): reduced OCT-1 activity is the cause of low in vitro sensitivity to imatinib. Blood. 2006;108(2):697–704.
267. Illmer T, Schaich M, Platzbecker U, Freiberg-Richter J, Oelschlagel U, von Bonin M, et al. P-glycoprotein-mediated drug efflux is a resistance mechanism of chronic myelogenous leukemia cells to treatment with imatinib mesylate. Leukemia. 2004;18(3):401–8.
268. Burger H, van Tol H, Brok M, Wiemer EA, de Bruijn EA, Guetens G, et al. Chronic imatinib mesylate exposure leads to reduced intracellular drug accumulation by induction of the ABCG2 (BCRP) and ABCB1 (MDR1) drug transport pumps. Cancer Biol Ther. 2005;4(7):747–52.
269. Zhang WW, Cortes JE, Yao H, Zhang L, Reddy NG, Jabbour E, et al. Predictors of primary imatinib resistance in chronic myelogenous leukemia are distinct from those in secondary imatinib resistance. J Clin Oncol. 2009;27(22):3642–9.
270. Demetri GD, Wang Y, Wehrle E, Racine A, Nikolova Z, Blanke CD, et al. Imatinib plasma levels are correlated with clinical benefit in patients with unresectable/metastatic gastrointestinal stromal tumors. J Clin Oncol. 2009;27(19):3141–7.
271. Eechoute K, Fransson MN, Reyners AK, de Jong FA, Sparreboom A, van der Graaf WT, et al. A long-term prospective population pharmacokinetic study on imatinib plasma concentrations in GIST patients. Clin Cancer Res. 2012;18(20):5780–7.
272. Mosse YP, Laudenslager M, Longo L, Cole KA, Wood A, Attiyeh EF, et al. Identification of ALK as a major familial neuroblastoma predisposition gene. Nature. 2008;455(7215):930–5.
273. Chen Y, Takita J, Choi YL, Kato M, Ohira M, Sanada M, et al. Oncogenic mutations of ALK kinase in neuroblastoma. Nature. 2008;455(7215):971–4.
274. George RE, Sanda T, Hanna M, Frohling S, Luther 2nd W, Zhang J, et al. Activating mutations in ALK provide a therapeutic target in neuroblastoma. Nature. 2008;455(7215):975–8.
275. Janoueix-Lerosey I, Lequin D, Brugieres L, Ribeiro A, de Pontual L, Combaret V, et al. Somatic and germline activating mutations of the ALK kinase receptor in neuroblastoma. Nature. 2008;455(7215):967–70.
276. Prickett TD, Agrawal NS, Wei X, Yates KE, Lin JC, Wunderlich JR, et al. Analysis of the tyrosine kinome in melanoma reveals recurrent mutations in ERBB4. Nat Genet. 2009;41(10):1127–32.
277. Govindan R, Ding L, Griffith M, Subramanian J, Dees ND, Kanchi KL, et al. Genomic landscape of non-small cell lung cancer in smokers and never-smokers. Cell. 2012;150(6):1121–34.
278. Imielinski M, Berger AH, Hammerman PS, Hernandez B, Pugh TJ, Hodis E, et al. Mapping the hallmarks of lung adenocarcinoma with massively parallel sequencing. Cell. 2012;150(6):1107–20.
279. Kwak EL, Jankowski J, Thayer SP, Lauwers GY, Brannigan BW, Harris PL, et al. Epidermal growth factor receptor kinase domain mutations in esophageal and pancreatic adenocarcinomas. Clin Cancer Res. 2006;12(14 Pt 1):4283–7.
280. Lynch TJ, Bell DW, Sordella R, Gurubhagavatula S, Okimoto RA, Brannigan BW, et al. Activating mutations in the epidermal growth factor receptor underlying responsiveness of non-small-cell lung cancer to gefitinib. N Engl J Med. 2004;350(21):2129–39.

281. Paez JG, Janne PA, Lee JC, Tracy S, Greulich H, Gabriel S, et al. EGFR mutations in lung cancer: correlation with clinical response to gefitinib therapy. Science. 2004;304(5676):1497–500.
282. Cappuzzo F, Ciuleanu T, Stelmakh L, Cicenas S, Szczesna A, Juhasz E, et al. Erlotinib as maintenance treatment in advanced non-small-cell lung cancer: a multicentre, randomised, placebo-controlled phase 3 study. Lancet Oncol. 2010;11(6):521–9.
283. Perol M, Chouaid C, Perol D, Barlesi F, Gervais R, Westeel V, et al. Randomized, phase III study of gemcitabine or erlotinib maintenance therapy versus observation, with predefined second-line treatment, after cisplatin-gemcitabine induction chemotherapy in advanced non-small-cell lung cancer. J Clin Oncol. 2012;30(28):3516–24.
284. Ku GY, Haaland BA, De Lima Lopes Jr G. Gefitinib vs chemotherapy as first-line therapy in advanced non-small cell lung cancer: meta-analysis of phase III trials. Lung Cancer. 2011;74(3):469–73.
285. Lee CK, Brown C, Gralla RJ, Hirsh V, Thongprasert S, Tsai CM, et al. Impact of EGFR inhibitor in non-small cell lung cancer on progression-free and overall survival: a meta-analysis. J Natl Cancer Inst. 2013;105(9):595–605.
286. Gaafar RM, Surmont VF, Scagliotti GV, Van Klaveren RJ, Papamichael D, Welch JJ, et al. A double-blind, randomised, placebo-controlled phase III intergroup study of gefitinib in patients with advanced NSCLC, non-progressing after first line platinum-based chemotherapy (EORTC 08021/ILCP 01/03). Eur J Cancer. 2011;47(15):2331–40.
287. Zhang L, Ma S, Song X, Han B, Cheng Y, Huang C, et al. Gefitinib versus placebo as maintenance therapy in patients with locally advanced or metastatic non-small-cell lung cancer (INFORM; C-TONG 0804): a multicentre, double-blind randomised phase 3 trial. Lancet Oncol. 2012;13(5):466–75.
288. Mok TS, Wu YL, Thongprasert S, Yang CH, Chu DT, Saijo N, et al. Gefitinib or carboplatin-paclitaxel in pulmonary adenocarcinoma. N Engl J Med. 2009;361(10):947–57.
289. Mitsudomi T, Morita S, Yatabe Y, Negoro S, Okamoto I, Tsurutani J, et al. Gefitinib versus cisplatin plus docetaxel in patients with non-small-cell lung cancer harbouring mutations of the epidermal growth factor receptor (WJTOG3405): an open label, randomised phase 3 trial. Lancet Oncol. 2010;11(2):121–8.
290. Maemondo M, Inoue A, Kobayashi K, Sugawara S, Oizumi S, Isobe H, et al. Gefitinib or chemotherapy for non-small-cell lung cancer with mutated EGFR. N Engl J Med. 2010;362(25):2380–8.
291. Zhou C, Wu YL, Chen G, Feng J, Liu XQ, Wang C, et al. Erlotinib versus chemotherapy as first-line treatment for patients with advanced EGFR mutation-positive non-small-cell lung cancer (OPTIMAL, CTONG-0802): a multicentre, open-label, randomised, phase 3 study. Lancet Oncol. 2011;12(8):735–42.
292. Miller VA, Hirsh V, Cadranel J, Chen YM, Park K, Kim SW, et al. Afatinib versus placebo for patients with advanced, metastatic non-small-cell lung cancer after failure of erlotinib, gefitinib, or both, and one or two lines of chemotherapy (LUX-Lung 1): a phase 2b/3 randomised trial. Lancet Oncol. 2012;13(5):528–38.
293. Pirker R, Pereira JR, Szczesna A, von Pawel J, Krzakowski M, Ramlau R, et al. Cetuximab plus chemotherapy in patients with advanced non-small-cell lung cancer (FLEX): an open-label randomised phase III trial. Lancet. 2009;373(9674):1525–31.
294. Stephens P, Hunter C, Bignell G, Edkins S, Davies H, Teague J, et al. Lung cancer: intragenic ERBB2 kinase mutations in tumours. Nature. 2004;431(7008):525–6.
295. Shigematsu H, Takahashi T, Nomura M, Majmudar K, Suzuki M, Lee H, et al. Somatic mutations of the HER2 kinase domain in lung adenocarcinomas. Cancer Res. 2005;65(5):1642–6.
296. Buttitta F, Barassi F, Fresu G, Felicioni L, Chella A, Paolizzi D, et al. Mutational analysis of the HER2 gene in lung tumors from Caucasian patients: mutations are mainly present in adenocarcinomas with bronchioloalveolar features. Int J Cancer. 2006;119(11):2586–91.
297. Arcila ME, Chaft JE, Nafa K, Roy-Chowdhuri S, Lau C, Zaidinski M, et al. Prevalence, clinicopathologic associations, and molecular spectrum of ERBB2 (HER2) tyrosine kinase mutations in lung adenocarcinomas. Clin Cancer Res. 2012;18(18):4910–8.

298. Anglesio MS, Arnold JM, George J, Tinker AV, Tothill R, Waddell N, et al. Mutation of ERBB2 provides a novel alternative mechanism for the ubiquitous activation of RAS-MAPK in ovarian serous low malignant potential tumors. Mol Cancer Res. 2008;6(11):1678–90.
299. Lee JW, Soung YH, Seo SH, Kim SY, Park CH, Wang YP, et al. Somatic mutations of ERBB2 kinase domain in gastric, colorectal, and breast carcinomas. Clin Cancer Res. 2006;12(1):57–61.
300. Jaiswal BS, Kljavin NM, Stawiski EW, Chan E, Parikh C, Durinck S, et al. Oncogenic ERBB3 mutations in human cancers. Cancer Cell. 2013;23(5):603–17.
301. Lin WM, Baker AC, Beroukhim R, Winckler W, Feng W, Marmion JM, et al. Modeling genomic diversity and tumor dependency in malignant melanoma. Cancer Res. 2008;68(3): 664–73.
302. Dutt A, Salvesen HB, Chen TH, Ramos AH, Onofrio RC, Hatton C, et al. Drug-sensitive FGFR2 mutations in endometrial carcinoma. Proc Natl Acad Sci U S A. 2008;105(25): 8713–7.
303. Pollock PM, Gartside MG, Dejeza LC, Powell MA, Mallon MA, Davies H, et al. Frequent activating FGFR2 mutations in endometrial carcinomas parallel germline mutations associated with craniosynostosis and skeletal dysplasia syndromes. Oncogene. 2007;26(50):7158–62.
304. Jang JH, Shin KH, Park JG. Mutations in fibroblast growth factor receptor 2 and fibroblast growth factor receptor 3 genes associated with human gastric and colorectal cancers. Cancer Res. 2001;61(9):3541–3.
305. Cappellen D, De Oliveira C, Ricol D, de Medina S, Bourdin J, Sastre-Garau X, et al. Frequent activating mutations of FGFR3 in human bladder and cervix carcinomas. Nat Genet. 1999;23(1):18–20.
306. Iyer G, Al-Ahmadie H, Schultz N, Hanrahan AJ, Ostrovnaya I, Balar AV, et al. Prevalence and co-occurrence of actionable genomic alterations in high-grade bladder cancer. J Clin Oncol. 2013;31:3133.
307. Onwuazor ON, Wen XY, Wang DY, Zhuang L, Masih-Khan E, Claudio J, et al. Mutation, SNP, and isoform analysis of fibroblast growth factor receptor 3 (FGFR3) in 150 newly diagnosed multiple myeloma patients. Blood. 2003;102(2):772–3.
308. Hernandez S, de Muga S, Agell L, Juanpere N, Esgueva R, Lorente JA, et al. FGFR3 mutations in prostate cancer: association with low-grade tumors. Mod Pathol. 2009;22(6):848–56.
309. Goriely A, Hansen RM, Taylor IB, Olesen IA, Jacobsen GK, McGowan SJ, et al. Activating mutations in FGFR3 and HRAS reveal a shared genetic origin for congenital disorders and testicular tumors. Nat Genet. 2009;41(11):1247–52.
310. Zhang Y, Hiraishi Y, Wang H, Sumi KS, Hayashido Y, Toratani S, et al. Constitutive activating mutation of the FGFR3b in oral squamous cell carcinomas. Int J Cancer. 2005;117(1): 166–8.
311. Aubertin J, Tourpin S, Janot F, Ahomadegbe JC, Radvanyi F. Analysis of fibroblast growth factor receptor 3 G697C mutation in oral squamous cell carcinomas. Int J Cancer. 2007;120(9):2058–9. author reply 60.
312. Taylor JG, Cheuk AT, Tsang PS, Chung JY, Song YK, Desai K, et al. Identification of FGFR4-activating mutations in human rhabdomyosarcomas that promote metastasis in xenotransplanted models. J Clin Invest. 2009;119(11):3395–407.
313. Yokota S, Kiyoi H, Nakao M, Iwai T, Misawa S, Okuda T, et al. Internal tandem duplication of the FLT3 gene is preferentially seen in acute myeloid leukemia and myelodysplastic syndrome among various hematological malignancies. A study on a large series of patients and cell lines. Leukemia. 1997;11(10):1605–9.
314. Joensuu H, Roberts PJ, Sarlomo-Rikala M, Andersson LC, Tervahartiala P, Tuveson D, et al. Effect of the tyrosine kinase inhibitor STI571 in a patient with a metastatic gastrointestinal stromal tumor. N Engl J Med. 2001;344(14):1052–6.
315. Demetri GD, von Mehren M, Blanke CD, Van den Abbeele AD, Eisenberg B, Roberts PJ, et al. Efficacy and safety of imatinib mesylate in advanced gastrointestinal stromal tumors. N Engl J Med. 2002;347(7):472–80.

316. Heinrich MC, Corless CL, Demetri GD, Blanke CD, von Mehren M, Joensuu H, et al. Kinase mutations and imatinib response in patients with metastatic gastrointestinal stromal tumor. J Clin Oncol. 2003;21(23):4342–9.
317. Droogendijk HJ, Kluin-Nelemans HJ, van Doormaal JJ, Oranje AP, van de Loosdrecht AA, van Daele PL. Imatinib mesylate in the treatment of systemic mastocytosis: a phase II trial. Cancer. 2006;107(2):345–51.
318. Vega-Ruiz A, Cortes JE, Sever M, Manshouri T, Quintas-Cardama A, Luthra R, et al. Phase II study of imatinib mesylate as therapy for patients with systemic mastocytosis. Leuk Res. 2009;33(11):1481–4.
319. Schmidt L, Duh FM, Chen F, Kishida T, Glenn G, Choyke P, et al. Germline and somatic mutations in the tyrosine kinase domain of the MET proto-oncogene in papillary renal carcinomas. Nat Genet. 1997;16(1):68–73.
320. Smith DC, Smith MR, Sweeney C, Elfiky AA, Logothetis C, Corn PG, et al. Cabozantinib in patients with advanced prostate cancer: results of a phase II randomized discontinuation trial. J Clin Oncol. 2013;31(4):412–9.
321. Elisei R, Schlumberger MJ, Muller SP, Schoffski P, Brose MS, Shah MH, et al. Cabozantinib in progressive medullary thyroid cancer. J Clin Oncol. 2013;31(29):3639–46.
322. Heinrich MC, Corless CL, Duensing A, McGreevey L, Chen CJ, Joseph N, et al. PDGFRA activating mutations in gastrointestinal stromal tumors. Science. 2003;299(5607):708–10.
323. Hirota S, Ohashi A, Nishida T, Isozaki K, Kinoshita K, Shinomura Y, et al. Gain-of-function mutations of platelet-derived growth factor receptor alpha gene in gastrointestinal stromal tumors. Gastroenterology. 2003;125(3):660–7.
324. Alentorn A, Marie Y, Carpentier C, Boisselier B, Giry M, Labussiere M, et al. Prevalence, clinico-pathological value, and co-occurrence of PDGFRA abnormalities in diffuse gliomas. Neuro Oncol. 2012;14(11):1393–403.
325. Mulligan LM, Kwok JB, Healey CS, Elsdon MJ, Eng C, Gardner E, et al. Germ-line mutations of the RET proto-oncogene in multiple endocrine neoplasia type 2A. Nature. 1993;363(6428):458–60.
326. Hofstra RM, Landsvater RM, Ceccherini I, Stulp RP, Stelwagen T, Luo Y, et al. A mutation in the RET proto-oncogene associated with multiple endocrine neoplasia type 2B and sporadic medullary thyroid carcinoma. Nature. 1994;367(6461):375–6.
327. Leboulleux S, Bastholt L, Krause T, de la Fouchardiere C, Tennvall J, Awada A, et al. Vandetanib in locally advanced or metastatic differentiated thyroid cancer: a randomised, double-blind, phase 2 trial. Lancet Oncol. 2012;13(9):897–905.
328. Wells Jr SA, Robinson BG, Gagel RF, Dralle H, Fagin JA, Santoro M, et al. Vandetanib in patients with locally advanced or metastatic medullary thyroid cancer: a randomized, double-blind phase III trial. J Clin Oncol. 2012;30(2):134–41.
329. Williamson D, Missiaglia E, de Reynies A, Pierron G, Thuille B, Palenzuela G, et al. Fusion gene-negative alveolar rhabdomyosarcoma is clinically and molecularly indistinguishable from embryonal rhabdomyosarcoma. J Clin Oncol. 2010;28(13):2151–8.
330. Sukov WR, Hodge JC, Lohse CM, Akre MK, Leibovich BC, Thompson RH, et al. ALK alterations in adult renal cell carcinoma: frequency, clinicopathologic features and outcome in a large series of consecutively treated patients. Mod Pathol. 2012;25(11):1516–25.
331. Sartore-Bianchi A, Moroni M, Veronese S, Carnaghi C, Bajetta E, Luppi G, et al. Epidermal growth factor receptor gene copy number and clinical outcome of metastatic colorectal cancer treated with panitumumab. J Clin Oncol. 2007;25(22):3238–45.
332. Cunningham D, Humblet Y, Siena S, Khayat D, Bleiberg H, Santoro A, et al. Cetuximab monotherapy and cetuximab plus irinotecan in irinotecan-refractory metastatic colorectal cancer. N Engl J Med. 2004;351(4):337–45.
333. Van Cutsem E, Kohne CH, Hitre E, Zaluski J, Chang Chien CR, Makhson A, et al. Cetuximab and chemotherapy as initial treatment for metastatic colorectal cancer. N Engl J Med. 2009;360(14):1408–17.
334. Jonker DJ, O'Callaghan CJ, Karapetis CS, Zalcberg JR, Tu D, Au HJ, et al. Cetuximab for the treatment of colorectal cancer. N Engl J Med. 2007;357(20):2040–8.

335. Bokemeyer C, Bondarenko I, Hartmann JT, de Braud F, Schuch G, Zubel A, et al. Efficacy according to biomarker status of cetuximab plus FOLFOX-4 as first-line treatment for metastatic colorectal cancer: the OPUS study. Ann Oncol. 2011;22(7):1535–46.
336. Bonner JA, Harari PM, Giralt J, Azarnia N, Shin DM, Cohen RB, et al. Radiotherapy plus cetuximab for squamous-cell carcinoma of the head and neck. N Engl J Med. 2006;354(6): 567–78.
337. Bonner JA, Harari PM, Giralt J, Cohen RB, Jones CU, Sur RK, et al. Radiotherapy plus cetuximab for locoregionally advanced head and neck cancer: 5-year survival data from a phase 3 randomised trial, and relation between cetuximab-induced rash and survival. Lancet Oncol. 2010;11(1):21–8.
338. Vermorken JB, Mesia R, Rivera F, Remenar E, Kawecki A, Rottey S, et al. Platinum-based chemotherapy plus cetuximab in head and neck cancer. N Engl J Med. 2008;359(11): 1116–27.
339. Tanner M, Hollmen M, Junttila TT, Kapanen AI, Tommola S, Soini Y, et al. Amplification of HER-2 in gastric carcinoma: association with Topoisomerase IIalpha gene amplification, intestinal type, poor prognosis and sensitivity to trastuzumab. Ann Oncol. 2005;16(2):273–8.
340. Bang YJ, Van Cutsem E, Feyereislova A, Chung HC, Shen L, Sawaki A, et al. Trastuzumab in combination with chemotherapy versus chemotherapy alone for treatment of HER2-positive advanced gastric or gastro-oesophageal junction cancer (ToGA): a phase 3, open-label, randomised controlled trial. Lancet. 2010;376(9742):687–97.
341. Pellegrini C, Falleni M, Marchetti A, Cassani B, Miozzo M, Buttitta F, et al. HER-2/Neu alterations in non-small cell lung cancer: a comprehensive evaluation by real time reverse transcription-PCR, fluorescence in situ hybridization, and immunohistochemistry. Clin Cancer Res. 2003;9(10 Pt 1):3645–52.
342. Langer CJ, Stephenson P, Thor A, Vangel M, Johnson DH. Trastuzumab in the treatment of advanced non-small-cell lung cancer: is there a role? Focus on Eastern Cooperative Oncology Group study 2598. J Clin Oncol. 2004;22(7):1180–7.
343. Tuefferd M, Couturier J, Penault-Llorca F, Vincent-Salomon A, Broet P, Guastalla JP, et al. HER2 status in ovarian carcinomas: a multicenter GINECO study of 320 patients. PLoS One. 2007;2(11):e1138.
344. Vermeij J, Teugels E, Bourgain C, Xiangming J, in 't Veld P, Ghislain V, et al. Genomic activation of the EGFR and HER2-neu genes in a significant proportion of invasive epithelial ovarian cancers. BMC Cancer. 2008;8:3.
345. Cornolti G, Ungari M, Morassi ML, Facchetti F, Rossi E, Lombardi D, et al. Amplification and overexpression of HER2/neu gene and HER2/neu protein in salivary duct carcinoma of the parotid gland. Arch Otolaryngol Head Neck Surg. 2007;133(10):1031–6.
346. Lae M, Couturier J, Oudard S, Radvanyi F, Beuzeboc P, Vieillefond A. Assessing HER2 gene amplification as a potential target for therapy in invasive urothelial bladder cancer with a standardized methodology: results in 1005 patients. Ann Oncol. 2010;21(4):815–9.
347. Burstein HJ, Sun Y, Dirix LY, Jiang Z, Paridaens R, Tan AR, et al. Neratinib, an irreversible ErbB receptor tyrosine kinase inhibitor, in patients with advanced ErbB2-positive breast cancer. J Clin Oncol. 2010;28(8):1301–7.
348. Awada A, Dirix L, Manso Sanchez L, Xu B, Luu T, Dieras V, et al. Safety and efficacy of neratinib (HKI-272) plus vinorelbine in the treatment of patients with ErbB2-positive metastatic breast cancer pretreated with anti-HER2 therapy. Ann Oncol. 2013;24(1):109–16.
349. Untch M, Loibl S, Bischoff J, Eidtmann H, Kaufmann M, Blohmer JU, et al. Lapatinib versus trastuzumab in combination with neoadjuvant anthracycline-taxane-based chemotherapy (GeparQuinto, GBG 44): a randomised phase 3 trial. Lancet Oncol. 2012;13(2):135–44.
350. Gianni L, Dafni U, Gelber RD, Azambuja E, Muehlbauer S, Goldhirsch A, et al. Treatment with trastuzumab for 1 year after adjuvant chemotherapy in patients with HER2-positive early breast cancer: a 4-year follow-up of a randomised controlled trial. Lancet Oncol. 2011;12(3):236–44.
351. Romond EH, Perez EA, Bryant J, Suman VJ, Geyer Jr CE, Davidson NE, et al. Trastuzumab plus adjuvant chemotherapy for operable HER2-positive breast cancer. N Engl J Med. 2005;353(16):1673–84.

352. Joensuu H, Kellokumpu-Lehtinen PL, Bono P, Alanko T, Kataja V, Asola R, et al. Adjuvant docetaxel or vinorelbine with or without trastuzumab for breast cancer. N Engl J Med. 2006;354(8):809–20.
353. Gianni L, Eiermann W, Semiglazov V, Manikhas A, Lluch A, Tjulandin S, et al. Neoadjuvant chemotherapy with trastuzumab followed by adjuvant trastuzumab versus neoadjuvant chemotherapy alone, in patients with HER2-positive locally advanced breast cancer (the NOAH trial): a randomised controlled superiority trial with a parallel HER2-negative cohort. Lancet. 2010;375(9712):377–84.
354. Gatzemeier U, Groth G, Butts C, Van Zandwijk N, Shepherd F, Ardizzoni A, et al. Randomized phase II trial of gemcitabine-cisplatin with or without trastuzumab in HER2-positive non-small-cell lung cancer. Ann Oncol. 2004;15(1):19–27.
355. Courjal F, Cuny M, Simony-Lafontaine J, Louason G, Speiser P, Zeillinger R, et al. Mapping of DNA amplifications at 15 chromosomal localizations in 1875 breast tumors: definition of phenotypic groups. Cancer Res. 1997;57(19):4360–7.
356. Jacquemier J, Adelaide J, Parc P, Penault-Llorca F, Planche J, de Lapeyriere O, et al. Expression of the FGFR1 gene in human breast-carcinoma cells. Int J Cancer. 1994; 59(3):373–8.
357. Chin K, DeVries S, Fridlyand J, Spellman PT, Roydasgupta R, Kuo WL, et al. Genomic and transcriptional aberrations linked to breast cancer pathophysiologies. Cancer Cell. 2006; 10(6):529–41.
358. Gorringe KL, Jacobs S, Thompson ER, Sridhar A, Qiu W, Choong DY, et al. High-resolution single nucleotide polymorphism array analysis of epithelial ovarian cancer reveals numerous microdeletions and amplifications. Clin Cancer Res. 2007;13(16):4731–9.
359. Simon R, Richter J, Wagner U, Fijan A, Bruderer J, Schmid U, et al. High-throughput tissue microarray analysis of 3p25 (RAF1) and 8p12 (FGFR1) copy number alterations in urinary bladder cancer. Cancer Res. 2001;61(11):4514–9.
360. Missiaglia E, Selfe J, Hamdi M, Williamson D, Schaaf G, Fang C, et al. Genomic imbalances in rhabdomyosarcoma cell lines affect expression of genes frequently altered in primary tumors: an approach to identify candidate genes involved in tumor development. Genes Chromosomes Cancer. 2009;48(6):455–67.
361. Weiss J, Sos ML, Seidel D, Peifer M, Zander T, Heuckmann JM, et al. Frequent and focal FGFR1 amplification associates with therapeutically tractable FGFR1 dependency in squamous cell lung cancer. Sci Transl Med. 2010;2(62):6293.
362. Kunii K, Davis L, Gorenstein J, Hatch H, Yashiro M, Di Bacco A, et al. FGFR2-amplified gastric cancer cell lines require FGFR2 and Erbb3 signaling for growth and survival. Cancer Res. 2008;68(7):2340–8.
363. Nord H, Segersten U, Sandgren J, Wester K, Busch C, Menzel U, et al. Focal amplifications are associated with high grade and recurrences in stage Ta bladder carcinoma. Int J Cancer. 2010;126(6):1390–402.
364. Vekony H, Ylstra B, Wilting SM, Meijer GA, van de Wiel MA, Leemans CR, et al. DNA copy number gains at loci of growth factors and their receptors in salivary gland adenoid cystic carcinoma. Clin Cancer Res. 2007;13(11):3133–9.
365. Zarghooni M, Bartels U, Lee E, Buczkowicz P, Morrison A, Huang A, et al. Whole-genome profiling of pediatric diffuse intrinsic pontine gliomas highlights platelet-derived growth factor receptor alpha and poly (ADP-ribose) polymerase as potential therapeutic targets. J Clin Oncol. 2010;28(8):1337–44.
366. Soda M, Choi YL, Enomoto M, Takada S, Yamashita Y, Ishikawa S, et al. Identification of the transforming EML4-ALK fusion gene in non-small-cell lung cancer. Nature. 2007; 448(7153):561–6.
367. Rikova K, Guo A, Zeng Q, Possemato A, Yu J, Haack H, et al. Global survey of phosphotyrosine signaling identifies oncogenic kinases in lung cancer. Cell. 2007;131(6):1190–203.
368. Mosse YP, Lim MS, Voss SD, Wilner K, Ruffner K, Laliberte J, et al. Safety and activity of crizotinib for paediatric patients with refractory solid tumours or anaplastic large-cell lymphoma: a Children's Oncology Group phase 1 consortium study. Lancet Oncol. 2013; 14(6):472–80.

369. Morris SW, Kirstein MN, Valentine MB, Dittmer KG, Shapiro DN, Saltman DL, et al. Fusion of a kinase gene, ALK, to a nucleolar protein gene, NPM, in non-Hodgkin's lymphoma. Science. 1994;263(5151):1281–4.
370. Hernandez L, Pinyol M, Hernandez S, Bea S, Pulford K, Rosenwald A, et al. TRK-fused gene (TFG) is a new partner of ALK in anaplastic large cell lymphoma producing two structurally different TFG-ALK translocations. Blood. 1999;94(9):3265–8.
371. Gascoyne RD, Lamant L, Martin-Subero JI, Lestou VS, Harris NL, Muller-Hermelink HK, et al. ALK-positive diffuse large B-cell lymphoma is associated with Clathrin-ALK rearrangements: report of 6 cases. Blood. 2003;102(7):2568–73.
372. De Paepe P, Baens M, van Krieken H, Verhasselt B, Stul M, Simons A, et al. ALK activation by the CLTC-ALK fusion is a recurrent event in large B-cell lymphoma. Blood. 2003; 102(7):2638–41.
373. Cools J, Wlodarska I, Somers R, Mentens N, Pedeutour F, Maes B, et al. Identification of novel fusion partners of ALK, the anaplastic lymphoma kinase, in anaplastic large-cell lymphoma and inflammatory myofibroblastic tumor. Genes Chromosomes Cancer. 2002;34(4): 354–62.
374. Laimer D, Dolznig H, Kollmann K, Vesely PW, Schlederer M, Merkel O, et al. PDGFR blockade is a rational and effective therapy for NPM-ALK-driven lymphomas. Nat Med. 2012;18(11):1699–704.
375. Lin E, Li L, Guan Y, Soriano R, Rivers CS, Mohan S, et al. Exon array profiling detects EML4-ALK fusion in breast, colorectal, and non-small cell lung cancers. Mol Cancer Res. 2009;7(9):1466–76.
376. Debelenko LV, Raimondi SC, Daw N, Shivakumar BR, Huang D, Nelson M, et al. Renal cell carcinoma with novel VCL-ALK fusion: new representative of ALK-associated tumor spectrum. Mod Pathol. 2011;24(3):430–42.
377. Sugawara E, Togashi Y, Kuroda N, Sakata S, Hatano S, Asaka R, et al. Identification of anaplastic lymphoma kinase fusions in renal cancer: large-scale immunohistochemical screening by the intercalated antibody-enhanced polymer method. Cancer. 2012;118(18):4427–36.
378. Marino-Enriquez A, Ou WB, Weldon CB, Fletcher JA, Perez-Atayde AR. ALK rearrangement in sickle cell trait-associated renal medullary carcinoma. Genes Chromosomes Cancer. 2011;50(3):146–53.
379. Jazii FR, Najafi Z, Malekzadeh R, Conrads TP, Ziaee AA, Abnet C, et al. Identification of squamous cell carcinoma associated proteins by proteomics and loss of beta tropomyosin expression in esophageal cancer. World J Gastroenterol. 2006;12(44):7104–12.
380. Du XL, Hu H, Lin DC, Xia SH, Shen XM, Zhang Y, et al. Proteomic profiling of proteins dysregulted in Chinese esophageal squamous cell carcinoma. J Mol Med. 2007;85(8):863–75.
381. Patel AS, Murphy KM, Hawkins AL, Cohen JS, Long PP, Perlman EJ, et al. RANBP2 and CLTC are involved in ALK rearrangements in inflammatory myofibroblastic tumors. Cancer Genet Cytogenet. 2007;176(2):107–14.
382. Debelenko LV, Arthur DC, Pack SD, Helman LJ, Schrump DS, Tsokos M. Identification of CARS-ALK fusion in primary and metastatic lesions of an inflammatory myofibroblastic tumor. Lab Invest. 2003;83(9):1255–65.
383. Griffin CA, Hawkins AL, Dvorak C, Henkle C, Ellingham T, Perlman EJ. Recurrent involvement of 2p23 in inflammatory myofibroblastic tumors. Cancer Res. 1999;59(12):2776–80.
384. Debiec-Rychter M, Marynen P, Hagemeijer A, Pauwels P. ALK-ATIC fusion in urinary bladder inflammatory myofibroblastic tumor. Genes Chromosomes Cancer. 2003;38(2):187–90.
385. Ma Z, Hill DA, Collins MH, Morris SW, Sumegi J, Zhou M, et al. Fusion of ALK to the Ran-binding protein 2 (RANBP2) gene in inflammatory myofibroblastic tumor. Genes Chromosomes Cancer. 2003;37(1):98–105.
386. Bridge JA, Kanamori M, Ma Z, Pickering D, Hill DA, Lydiatt W, et al. Fusion of the ALK gene to the clathrin heavy chain gene, CLTC, in inflammatory myofibroblastic tumor. Am J Pathol. 2001;159(2):411–5.
387. Panagopoulos I, Nilsson T, Domanski HA, Isaksson M, Lindblom P, Mertens F, et al. Fusion of the SEC31L1 and ALK genes in an inflammatory myofibroblastic tumor. Int J Cancer. 2006;118(5):1181–6.

388. Wang X, Krishnan C, Nguyen EP, Meyer KJ, Oliveira JL, Yang P, et al. Fusion of dynactin 1 to anaplastic lymphoma kinase in inflammatory myofibroblastic tumor. Hum Pathol. 2012;43(11):2047–52.
389. Takeuchi K, Soda M, Togashi Y, Sugawara E, Hatano S, Asaka R, et al. Pulmonary inflammatory myofibroblastic tumor expressing a novel fusion, PPFIBP1-ALK: reappraisal of anti-ALK immunohistochemistry as a tool for novel ALK fusion identification. Clin Cancer Res. 2011;17(10):3341–8.
390. Lipson D, Capelletti M, Yelensky R, Otto G, Parker A, Jarosz M, et al. Identification of new ALK and RET gene fusions from colorectal and lung cancer biopsies. Nat Med. 2012;18(3):382–4.
391. Seo JS, Ju YS, Lee WC, Shin JY, Lee JK, Bleazard T, et al. The transcriptional landscape and mutational profile of lung adenocarcinoma. Genome Res. 2012;22(11):2109–19.
392. Roumiantsev S, Krause DS, Neumann CA, Dimitri CA, Asiedu F, Cross NC, et al. Distinct stem cell myeloproliferative/T lymphoma syndromes induced by ZNF198-FGFR1 and BCR-FGFR1 fusion genes from 8p11 translocations. Cancer Cell. 2004;5(3):287–98.
393. Demiroglu A, Steer EJ, Heath C, Taylor K, Bentley M, Allen SL, et al. The t(8;22) in chronic myeloid leukemia fuses BCR to FGFR1: transforming activity and specific inhibition of FGFR1 fusion proteins. Blood. 2001;98(13):3778–83.
394. Fioretos T, Panagopoulos I, Lassen C, Swedin A, Billstrom R, Isaksson M, et al. Fusion of the BCR and the fibroblast growth factor receptor-1 (FGFR1) genes as a result of t(8;22)(p11;q11) in a myeloproliferative disorder: the first fusion gene involving BCR but not ABL. Genes Chromosomes Cancer. 2001;32(4):302–10.
395. Mugneret F, Chaffanet M, Maynadie M, Guasch G, Favre B, Casasnovas O, et al. The 8p12 myeloproliferative disorder. t(8;19)(p12;q13.3): a novel translocation involving the FGFR1 gene. Br J Haematol. 2000;111(2):647–9.
396. Guasch G, Mack GJ, Popovici C, Dastugue N, Birnbaum D, Rattner JB, et al. FGFR1 is fused to the centrosome-associated protein CEP110 in the 8p12 stem cell myeloproliferative disorder with t(8;9)(p12;q33). Blood. 2000;95(5):1788–96.
397. Xiao S, Nalabolu SR, Aster JC, Ma J, Abruzzo L, Jaffe ES, et al. FGFR1 is fused with a novel zinc-finger gene, ZNF198, in the t(8;13) leukaemia/lymphoma syndrome. Nat Genet. 1998;18(1):84–7.
398. Popovici C, Adelaide J, Ollendorff V, Chaffanet M, Guasch G, Jacrot M, et al. Fibroblast growth factor receptor 1 is fused to FIM in stem-cell myeloproliferative disorder with t(8;13). Proc Natl Acad Sci U S A. 1998;95(10):5712–7.
399. Wu YM, Su F, Kalyana-Sundaram S, Khazanov N, Ateeq B, Cao X, et al. Identification of targetable FGFR gene fusions in diverse cancers. Cancer Discov. 2013;3(6):636–47.
400. Singh D, Chan JM, Zoppoli P, Niola F, Sullivan R, Castano A, et al. Transforming fusions of FGFR and TACC genes in human glioblastoma. Science. 2012;337(6099):1231–5.
401. Yagasaki F, Wakao D, Yokoyama Y, Uchida Y, Murohashi I, Kayano H, et al. Fusion of ETV6 to fibroblast growth factor receptor 3 in peripheral T-cell lymphoma with a t(4;12)(p16;p13) chromosomal translocation. Cancer Res. 2001;61(23):8371–4.
402. Avet-Loiseau H, Li JY, Facon T, Brigaudeau C, Morineau N, Maloisel F, et al. High incidence of translocations t(11;14)(q13;q32) and t(4;14)(p16;q32) in patients with plasma cell malignancies. Cancer Res. 1998;58(24):5640–5.
403. Chesi M, Nardini E, Brents LA, Schrock E, Ried T, Kuehl WM, et al. Frequent translocation t(4;14)(p16.3;q32.3) in multiple myeloma is associated with increased expression and activating mutations of fibroblast growth factor receptor 3. Nat Genet. 1997;16(3):260–4.
404. Popovici C, Zhang B, Gregoire MJ, Jonveaux P, Lafage-Pochitaloff M, Birnbaum D, et al. The t(6;8)(q27;p11) translocation in a stem cell myeloproliferative disorder fuses a novel gene, FOP, to fibroblast growth factor receptor 1. Blood. 1999;93(4):1381–9.
405. Williams SV, Hurst CD, Knowles MA. Oncogenic FGFR3 gene fusions in bladder cancer. Hum Mol Genet. 2013;22(4):795–803.
406. Majewski IJ, Mittempergher L, Davidson NM, Bosma A, Willems SM, Horlings HM, et al. Identification of recurrent FGFR3 fusion genes in lung cancer through kinome-centred RNA sequencing. J Pathol. 2013;230(3):270–6.

407. Martin-Zanca D, Hughes SH, Barbacid M. A human oncogene formed by the fusion of truncated tropomyosin and protein tyrosine kinase sequences. Nature. 1986;319(6056):743–8.
408. Butti MG, Bongarzone I, Ferraresi G, Mondellini P, Borrello MG, Pierotti MA. A sequence analysis of the genomic regions involved in the rearrangements between TPM3 and NTRK1 genes producing TRK oncogenes in papillary thyroid carcinomas. Genomics. 1995;28(1):15–24.
409. Bongarzone I, Vigneri P, Mariani L, Collini P, Pilotti S, Pierotti MA. RET/NTRK1 rearrangements in thyroid gland tumors of the papillary carcinoma family: correlation with clinicopathological features. Clin Cancer Res. 1998;4(1):223–8.
410. Greco A, Pierotti MA, Bongarzone I, Pagliardini S, Lanzi C, Della PG. TRK-T1 is a novel oncogene formed by the fusion of TPR and TRK genes in human papillary thyroid carcinomas. Oncogene. 1992;7(2):237–42.
411. Greco A, Mariani C, Miranda C, Lupas A, Pagliardini S, Pomati M, et al. The DNA rearrangement that generates the TRK-T3 oncogene involves a novel gene on chromosome 3 whose product has a potential coiled-coil domain. Mol Cell Biol. 1995;15(11):6118–27.
412. Skalova A, Vanecek T, Sima R, Laco J, Weinreb I, Perez-Ordonez B, et al. Mammary analogue secretory carcinoma of salivary glands, containing the ETV6-NTRK3 fusion gene: a hitherto undescribed salivary gland tumor entity. Am J Surg Pathol. 2010;34(5):599–608.
413. Knezevich SR, McFadden DE, Tao W, Lim JF, Sorensen PH. A novel ETV6-NTRK3 gene fusion in congenital fibrosarcoma. Nat Genet. 1998;18(2):184–7.
414. Rubin BP, Chen CJ, Morgan TW, Xiao S, Grier HE, Kozakewich HP, et al. Congenital mesoblastic nephroma t(12;15) is associated with ETV6-NTRK3 gene fusion: cytogenetic and molecular relationship to congenital (infantile) fibrosarcoma. Am J Pathol. 1998; 153(5):1451–8.
415. Cools J, DeAngelo DJ, Gotlib J, Stover EH, Legare RD, Cortes J, et al. A tyrosine kinase created by fusion of the PDGFRA and FIP1L1 genes as a therapeutic target of imatinib in idiopathic hypereosinophilic syndrome. N Engl J Med. 2003;348(13):1201–14.
416. Simon MP, Pedeutour F, Sirvent N, Grosgeorge J, Minoletti F, Coindre JM, et al. Deregulation of the platelet-derived growth factor B-chain gene via fusion with collagen gene COL1A1 in dermatofibrosarcoma protuberans and giant-cell fibroblastoma. Nat Genet. 1997;15(1):95–8.
417. McArthur GA, Demetri GD, van Oosterom A, Heinrich MC, Debiec-Rychter M, Corless CL, et al. Molecular and clinical analysis of locally advanced dermatofibrosarcoma protuberans treated with imatinib: Imatinib Target Exploration Consortium Study B2225. J Clin Oncol. 2005;23(4):866–73.
418. Rutkowski P, Van Glabbeke M, Rankin CJ, Ruka W, Rubin BP, Debiec-Rychter M, et al. Imatinib mesylate in advanced dermatofibrosarcoma protuberans: pooled analysis of two phase II clinical trials. J Clin Oncol. 2010;28(10):1772–9.
419. Loh ML, Mullighan CG. Advances in the genetics of high-risk childhood B-progenitor acute lymphoblastic leukemia and juvenile myelomonocytic leukemia: implications for therapy. Clin Cancer Res. 2012;18(10):2754–67.
420. Golub TR, Barker GF, Lovett M, Gilliland DG. Fusion of PDGF receptor beta to a novel ets-like gene, tel, in chronic myelomonocytic leukemia with t(5;12) chromosomal translocation. Cell. 1994;77(2):307–16.
421. Ross TS, Bernard OA, Berger R, Gilliland DG. Fusion of Huntingtin interacting protein 1 to platelet-derived growth factor beta receptor (PDGFbetaR) in chronic myelomonocytic leukemia with t(5;7)(q33;q11.2). Blood. 1998;91(12):4419–26.
422. Kulkarni S, Heath C, Parker S, Chase A, Iqbal S, Pocock CF, et al. Fusion of H4/D10S170 to the platelet-derived growth factor receptor beta in BCR-ABL-negative myeloproliferative disorders with a t(5;10)(q33;q21). Cancer Res. 2000;60(13):3592–8.
423. Schwaller J, Anastasiadou E, Cain D, Kutok J, Wojiski S, Williams IR, et al. H4(D10S170), a gene frequently rearranged in papillary thyroid carcinoma, is fused to the platelet-derived growth factor receptor beta gene in atypical chronic myeloid leukemia with t(5;10)(q33;q22). Blood. 2001;97(12):3910–8.

424. Magnusson MK, Meade KE, Brown KE, Arthur DC, Krueger LA, Barrett AJ, et al. Rabaptin-5 is a novel fusion partner to platelet-derived growth factor beta receptor in chronic myelomonocytic leukemia. Blood. 2001;98(8):2518–25.
425. Lovly CM, Gupta A, Lipson D, Otto G, Brennan T, Chung CT, et al. Inflammatory Myofibroblastic Tumors harbor multiple potentially actionable kinase fusions. Cancer Discov. 2014;4:889.
426. Kohno T, Ichikawa H, Totoki Y, Yasuda K, Hiramoto M, Nammo T, et al. KIF5B-RET fusions in lung adenocarcinoma. Nat Med. 2012;18(3):375–7.
427. Wang R, Hu H, Pan Y, Li Y, Ye T, Li C, et al. RET fusions define a unique molecular and clinicopathologic subtype of non-small-cell lung cancer. J Clin Oncol. 2012;30(35):4352–9.
428. Matsubara D, Kanai Y, Ishikawa S, Ohara S, Yoshimoto T, Sakatani T, et al. Identification of CCDC6-RET fusion in the human lung adenocarcinoma cell line, LC-2/ad. J Thorac Oncol. 2012;7(12):1872–6.
429. Ju YS, Lee WC, Shin JY, Lee S, Bleazard T, Won JK, et al. A transforming KIF5B and RET gene fusion in lung adenocarcinoma revealed from whole-genome and transcriptome sequencing. Genome Res. 2012;22(3):436–45.
430. Suehara Y, Arcila M, Wang L, Hasanovic A, Ang D, Ito T, et al. Identification of KIF5B-RET and GOPC-ROS1 fusions in lung adenocarcinomas through a comprehensive mRNA-based screen for tyrosine kinase fusions. Clin Cancer Res. 2012;18(24):6599–608.
431. Gautschi O, Zander T, Keller FA, Strobel K, Hirschmann A, Aebi S, et al. A patient with lung adenocarcinoma and RET fusion treated with vandetanib. J Thorac Oncol. 2013;8(5):e43–4.
432. Grieco M, Cerrato A, Santoro M, Fusco A, Melillo RM, Vecchio G. Cloning and characterization of H4 (D10S170), a gene involved in RET rearrangements in vivo. Oncogene. 1994; 9(9):2531–5.
433. Santoro M, Dathan NA, Berlingieri MT, Bongarzone I, Paulin C, Grieco M, et al. Molecular characterization of RET/PTC3; a novel rearranged version of the RETproto-oncogene in a human thyroid papillary carcinoma. Oncogene. 1994;9(2):509–16.
434. Bongarzone I, Butti MG, Coronelli S, Borrello MG, Santoro M, Mondellini P, et al. Frequent activation of ret protooncogene by fusion with a new activating gene in papillary thyroid carcinomas. Cancer Res. 1994;54(11):2979–85.
435. Bongarzone I, Monzini N, Borrello MG, Carcano C, Ferraresi G, Arighi E, et al. Molecular characterization of a thyroid tumor-specific transforming sequence formed by the fusion of ret tyrosine kinase and the regulatory subunit RI alpha of cyclic AMP-dependent protein kinase A. Mol Cell Biol. 1993;13(1):358–66.
436. Ciampi R, Giordano TJ, Wikenheiser-Brokamp K, Koenig RJ, Nikiforov YE. HOOK3-RET: a novel type of RET/PTC rearrangement in papillary thyroid carcinoma. Endocr Relat Cancer. 2007;14(2):445–52.
437. Corvi R, Berger N, Balczon R, Romeo G. RET/PCM-1: a novel fusion gene in papillary thyroid carcinoma. Oncogene. 2000;19(37):4236–42.
438. Grieco M, Santoro M, Berlingieri MT, Melillo RM, Donghi R, Bongarzone I, et al. PTC is a novel rearranged form of the ret proto-oncogene and is frequently detected in vivo in human thyroid papillary carcinomas. Cell. 1990;60(4):557–63.
439. Klugbauer S, Demidchik EP, Lengfelder E, Rabes HM. Detection of a novel type of RET rearrangement (PTC5) in thyroid carcinomas after Chernobyl and analysis of the involved RET-fused gene RFG5. Cancer Res. 1998;58(2):198–203.
440. Klugbauer S, Rabes HM. The transcription coactivator HTIF1 and a related protein are fused to the RET receptor tyrosine kinase in childhood papillary thyroid carcinomas. Oncogene. 1999;18(30):4388–93.
441. Nakata T, Kitamura Y, Shimizu K, Tanaka S, Fujimori M, Yokoyama S, et al. Fusion of a novel gene, ELKS, to RET due to translocation t(10;12)(q11;p13) in a papillary thyroid carcinoma. Genes Chromosomes Cancer. 1999;25(2):97–103.
442. Saenko V, Rogounovitch T, Shimizu-Yoshida Y, Abrosimov A, Lushnikov E, Roumiantsev P, et al. Novel tumorigenic rearrangement, Delta rfp/ret, in a papillary thyroid carcinoma from externally irradiated patient. Mutat Res. 2003;527(1–2):81–90.

443. Salassidis K, Bruch J, Zitzelsberger H, Lengfelder E, Kellerer AM, Bauchinger M. Translocation t(10;14)(q11.2:q22.1) fusing the kinetin to the RET gene creates a novel rearranged form (PTC8) of the RET proto-oncogene in radiation-induced childhood papillary thyroid carcinoma. Cancer Res. 2000;60(11):2786–9.
444. Bergethon K, Shaw AT, Ou SH, Katayama R, Lovly CM, McDonald NT, et al. ROS1 rearrangements define a unique molecular class of lung cancers. J Clin Oncol. 2012;30(8): 863–70.
445. Rimkunas VM, Crosby KE, Li D, Hu Y, Kelly ME, Gu TL, et al. Analysis of receptor tyrosine kinase ROS1-positive tumors in non-small cell lung cancer: identification of a FIG-ROS1 fusion. Clin Cancer Res. 2012;18(16):4449–57.
446. Davies KD, Le AT, Theodoro MF, Skokan MC, Aisner DL, Berge EM, et al. Identifying and targeting ROS1 gene fusions in non-small cell lung cancer. Clin Cancer Res. 2012;18(17): 4570–9.
447. Bos M, Gardizi M, Schildhaus HU, Heukamp LC, Geist T, Kaminsky B, et al. Complete metabolic response in a patient with repeatedly relapsed non-small cell lung cancer harboring ROS1 gene rearrangement after treatment with crizotinib. Lung Cancer. 2013;81:142.
448. Komiya T, Thomas A, Khozin S, Rajan A, Wang Y, Giaccone G. Response to crizotinib in ROS1-rearranged non-small-cell lung cancer. J Clin Oncol. 2012;30(27):3425–6. author reply 6.
449. Charest A, Lane K, McMahon K, Park J, Preisinger E, Conroy H, et al. Fusion of FIG to the receptor tyrosine kinase ROS in a glioblastoma with an interstitial del(6)(q21q21). Genes Chromosomes Cancer. 2003;37(1):58–71.
450. Gu TL, Deng X, Huang F, Tucker M, Crosby K, Rimkunas V, et al. Survey of tyrosine kinase signaling reveals ROS kinase fusions in human cholangiocarcinoma. PLoS One. 2011;6(1): e15640.
451. Birch AH, Arcand SL, Oros KK, Rahimi K, Watters AK, Provencher D, et al. Chromosome 3 anomalies investigated by genome wide SNP analysis of benign, low malignant potential and low grade ovarian serous tumours. PLoS One. 2011;6(12):e28250.
452. Lee J, Lee SE, Kang SY, Do IG, Lee S, Ha SY, et al. Identification of ROS1 rearrangement in gastric adenocarcinoma. Cancer. 2013;119(9):1627–35.

Chapter 11
Mouse Models of Receptor Tyrosine Kinases

Martin Holcmann, Nicole Amberg, Barbara Drobits, Elisabeth Glitzner, Karin Komposch, Jonathan Robson, Fabio Savarese, Sriram Srivatsa, Gabriel Stulnig, and Maria Sibilia

11.1 Alk Receptor Family

11.1.1 Overview About Family Members

Anaplastic lymphoma kinase (Alk) and the closely related leukocyte receptor tyrosine kinase (Ltk) show high homology and constitute their own subgroup within the insulin receptor superfamily of receptor tyrosine kinases [1]. Ltk is expressed in hematopoietic cells, the brain and placenta [2]. However, a ligand for Ltk has not been identified and the physiologic role of Ltk remains elusive to date [1, 2]. Alk was discovered as a part of the oncogenic fusion protein nucleophosmin-anaplastic lymphoma kinase (NPM-ALK) in non-Hodgkin's lymphoma [3]. Subsequently, Alk has been implicated in tumorigenesis as a product of numerous chromosomal translocations which potentially result in the expression of chimeric Alk oncogenes [4]. Alk oncogene formation leads to constitutive activation of the Alk tyrosine kinase and increases proliferation and cell survival [5]. Two possible ligands for Alk, the growth factors pleiotrophin (PTN) and midkine (MK) have been identified [2, 6]. Alk plays an important role in the development of the brain and nervous system, metabolic homeostasis, and tissue regeneration [6].

RTKs: Alk, Ltk
Alk ligands: pleiotrophin (PTN), midkine (MK)
Ltk ligands: currently unknown

M. Holcmann (✉) • N. Amberg • B. Drobits • E. Glitzner • K. Komposch
J. Robson • F. Savarese • S. Srivatsa • G. Stulnig • M. Sibilia (✉)
Department of Medicine I, Comprehensive Cancer Center, Institute of Cancer Research, Medical University of Vienna, Vienna, Austria
e-mail: martin.holcmann@meduniwien.ac.at; maria.sibilia@meduniwien.ac.at

D.L. Wheeler, Y. Yarden (eds.), *Receptor Tyrosine Kinases: Structure, Functions and Role in Human Disease*, DOI 10.1007/978-1-4939-2053-2_11

11.1.2 Alk

11.1.2.1 Constitutive Alk Knockout Strains

11.1.2.1.1 Alk$^{-/-}$ (Bisland)

Generation of knockout strain: This strain was generated via conventional insertional mutagenesis of an IRES-*β-gal* cassette, resulting in a 35-bp deletion in exon 22 of the *Alk* gene locus resulting in deletion of the transmembrane and kinase domains of Alk. The linearized construct was electroporated into 129/OlaHsd ES-cells. Southern blot positive clones were injected into C57BL/6J blastocysts [4].

Phenotype: These mice display no obvious phenotypes and can be maintained via homozygous breeding. However, in adult mice brain behavioral defects could be observed. Analysis demonstrates an increase in basal dopaminergic signaling selectively within the frontal cortex and an age-dependent increase in basal hippocampal progenitor proliferation and alterations in behavioral tests [4].

11.1.2.1.2 Alk$^{-/-}$ (Lasek)

Generation of knockout strain: Mice were generated by targeting exons 20-21 of the *Alk* genomic locus, which encodes the juxtamembrane and N-terminal portion of the Alk tyrosine kinase domain [7].

Phenotype: *Alk*$^{-/-}$ mice are more sensitive to ethanol exposure. Mice sedate longer in response to high doses of ethanol and consume more ethanol than wild-type mice [7].

11.1.2.1.3 Alk$^{-/-}$ (Morris)

An *Alk* knockout mouse has been generated in the group of Morris, however, details about this strain have not been published [8]. Reportedly, these animals show no obvious phenotype [8–11].

11.1.2.2 Transgenic Mouse Lines

11.1.2.2.1 ALK^{F1174L}

Generation of a conditional knockout strain: *ALK*^{F1174L} is the most frequent and most aggressive transforming mutation found in neuroblastoma. To proof that this mutation indeed is causative for neuroblastoma formation inducible transgenic mice were generated. Under the control of the chicken β-actin promoter a floxed STOP cassette followed by *ALK*^{F1174L} cDNA is expressed [12]. Transgenic founder mice were generated via pronucleus injection of the vector. The *ALK*^{F1174L} transgene was only expressed upon deletion of a stop cassette by Cre recombinase. Thus, tissue specific transgene expression was achieved by mating the *ALK*^{F1174L} transgenic mice with DBHiCre [13] and TH-IRES-Cre [14] strains driving recombination mainly in neural crest cells and its precursors.

Phenotype: Independently of the Cre line used in both models transgene expression resulted in neuroblastoma like tumor development in ~40 % of transgenic mice. These tumors recapitulated the genetic signature and metastatic behavior of human neuroblastoma. Targeted ALK^{F1174L} and Mycn co-expression revealed a strong synergism [12].

11.1.2.2.2 tTA-NPM-ALK; tTA TPM3-ALK

Generation of transgenic mice: Nucleophosmin 1-ALK (NPM-ALK) and nonmuscular tropomyosin 3-ALK (TPM3-ALK) transgenic founder mice were generated by insertion of human *NPM-ALK* and *TPM3-ALK* cDNA downstream of a Tet-regulated cytomegalovirus mini promoter [15]. Introduction of the vector-construct into the pronucleus of FVB/N mouse oocytes resulted in transgenic mice. To induce transgene expression specifically in T- and B-lymphocytes NPM-ALK or TPM3-ALK mice were bread to tet-off mice expressing tTA protein under the control of the EμSRα promoter/enhancer [15].

Phenotype: Transgenic expression of either of the ALK fusion proteins resulted in development of an ALK-positive B-cell lympho-proliferative disorder and kerato-acanthoma in the skin. Inactivation of ALK oncogenes upon doxycycline treatment in the lymphoid linage leads to regression of hematopoietic tumors and skin lesions [15].

11.1.2.2.3 Cd4-NPM-ALK

Generation of transgenic mice: Human *NPM-ALK* cDNA was cloned under the control of the murine *Cd4* promoter/enhancer [5]. The vector was injected into Swiss-Webster blastocysts and transgenic NPM-ALK founder mice were back-crossed to balb/c and C57BL/6 strains [5].

Phenotype: *Cd4 NPM-ALK* transgenic mice were born with normal lymphoid organs, and a normal number and proportions of T-cells. All mice developed malignant lympho-proliferative disorders with a mean survival of 18 weeks. A subset of the NPM-ALK transgenic mice also developed clonal B-cell plasma cell neoplasms. These tumors arose in peripheral lymphoid organs (plasmacytomas) or within the bone marrow and often led to peripheral neuropathies and limb paralysis [5].

11.1.2.2.4 Vav-NPM-ALK

Generation of transgenic mice: Human *NPM-ALK* cDNA was cloned under the control of the *vav* promoter for expression in all cells of the hematopoietic lineage. The construct was microinjected into the pronuclei of eggs from C57BL/6J × CBA F1 mice [16].

Phenotype: Two transgenic lines developed distinct B-cells lymphomas with aberrant myeloid gene expression in lymph nodes, spleen, and liver [16]. Transgene copy number correlated to aggressiveness of the disease.

11.1.2.2.5 CD2-NPM-ALK

Generation of transgenic mice: Human *NPM-ALK* cDNA was cloned behind the CD2 promoter to induce expression in T-cells and the linearized vector was microinjected into pronuclei of eggs from C57BL/6J × CBA F1 mice. Transfer of microinjected eggs into ovarian ducts of recipients resulted in the birth of NPM-ALK transgenic mice [17].

Phenotype: Development of B-cell lymphomas but not T-cell lymphoma [17].

11.1.2.2.6 Lck-NPM-ALK

Generation of transgenic mice: Human *NPM-ALK* cDNA was cloned between the *lck* promoter, 3′ exon–intron structure and pA signal of the human growth hormone gene and the locus control region of human CD2 to induce expression in early T-cells. The linearized vector was microinjected into pronuclei of eggs from C57BL/6J × DBA2 F1 mice. Transfer of micro-injected eggs into ovarian ducts of recipients resulted in the birth of NPM-ALK transgenic mice [18].

Phenotype: Development of large cell lymphoblastic lymphomas mainly in the thymus, lymph nodes, and extranodal organs. Transgenic mice on the one hand showed a strongly reduced life expectancy but on the other hand also increased body size [18].

11.1.2.2.7 SPC-EML4-ALK

Generation of transgenic mice: FLAG tagged *Eml4-Alk* cDNA was cloned in a construct under the surfactant protein C (*Spc)* promoter that drives expression exclusively in lung alveolar type-II epithelial cells. EML4-ALK is a fusion-type oncogene that is generated in human non-small-cell lung cancer due to recurrent chromosomal inversion. The vector was injected into pronuclei of C57BL/6J mice. Southern blot analysis of the progeny confirmed the successful generation of *EML4-ALK* transgenic mice [19].

Phenotype: All of the transgenic mice examined developed hundreds of adenocarcinoma nodules in the lung within a few weeks after birth [19]. Treatment with small molecule ALK inhibitors lead to rapid disappearance of the nodules.

11.1.3 Ltk

11.1.3.1 Transgenic Mouse Line

11.1.3.1.1 β-Actin-Ltk

Generation of transgenic mice: Human leukocyte tyrosine kinase (*LTK)* cDNA was cloned into a vector under the control of cytomegalovirus enhancer and a chicken β-actin promoter [20]. Microinjection into pronuclei of eggs from C57BL/6XDBA/2F2 mice and thereafter embryonal transfer resulted in Ltk transgenic mice [20].

Phenotype: Mice appeared normal at birth but were growth retarded soon after birth and died at young age. The different transgenic lines generated were shown distinct expression patterns. Two transgenic lines showing LTK phosphorylation specifically in the heart were analyzed in more details. Echocardiography showed concentric hypertrophy, which resulted in reduced cardiac output, low blood pressure, and increased heart rate. Histological examination of the heart exhibited focal degeneration of cardiomyocytes [20].

11.2 TAM Receptor Family

11.2.1 Overview About Family Members

The TAM family of receptor tyrosine kinases includes Tyro-3, Axl, and Mer. These three receptors share the vitamin K-dependent ligands Gas6 and Protein S. Upon ligand binding the receptors dimerize and subsequently get trans-autophosphorylated at tyrosine residues within the intracellular domain, resulting in activation of further downstream signaling cascades including PI3K, Akt, Ras, ERK, and PLC pathways [21].

Although being a small family of kinases it regulates a large variety of processes such as innate and adaptive immunity, apoptosis, and phagocytosis as well as cell survival and proliferation [22, 23].

11.2.2 Tyro-3

Box: list of all mouse strains

Tyro3$^{-/-}$	Lu 1999 [24]

11.2.2.1 Constitutive Knockout Strain and Phenotype

11.2.2.1.1 Tyro-3$^{-/-}$

Generation of knockout strain: The *Tyro-3* receptor gene exon 9 and parts of exon 10 were disrupted by insertion of a G418 (Neo) resistance cassette by homologous recombination in mouse embryonic stem cells [24].

Phenotype: Mice homozygous for Tyro-3 mutation are viable and fertile and display no gross anatomical defects. They are obtained at Mendelian frequency and viable for a long time, indicating that the receptor is not essential during embryonic development. Application of bleeding, thrombosis, and ischemia models demonstrates delayed clot retraction and milder or no thrombosis symptoms as well as reduced infarct size and edema. Additionally, inefficient platelet aggregation is reported, being a consequence of impaired tyrosine phosphorylation of the β3 integrin and thus for outside-in $\alpha_{IIb}\beta3$ signaling via PI3K activation [25]. Moreover, activity induced seizures can occur in mutant mice over the age of 7 months. Tyro3 mediates ProteinS (PS) induced vasculoprotection. After Tyro3 ligation, PS activated sphingosine 1-phosphate receptor resulting in Rac1-dependent blood–brain barrier protection [26].

11.2.3 *Axl*

Box: List of all mouse strains

Axl$^{-/-}$	Lu 1999 [24]
Axl^{lacZ}	Ruan 2012 [28]
GCSFR-*axl*^{tg}	Augustine 1999 [29]
ApoE soluble *axl*^{tg}	Augustine 1999 [29]

11.2.3.1 Constitutive Axl Knockout Strain

11.2.3.1.1 Axl$^{-/-}$

Generation of knockout strain: The Axl receptor gene exon 9 (comprising the second last exon of the extracellular domain) was disrupted by insertion of a Neo resistance cassette by homologous recombination in mouse embryonic stem cells [24].

Phenotype: Similar to *Tyro-3*$^{-/-}$ mice, animals homozygous for *Axl* mutation are viable, fertile, do not display gross anatomical defects and are obtained at Mendelian frequency and viable for a long time. Bleeding and thrombosis phenotypes are similar to *Tyro-3*$^{-/-}$ mice. Upon induction of vascular remodeling *Axl*$^{-/-}$ mice develop significantly smaller intima and media thickness and exhibit fivefold increased apoptosis as compared to WT mice, due to reduced levels of phosphorylated Akt. A cuprizone toxicity study suggests that *Axl*$^{-/-}$ mice have delayed clearance of apoptotic oligodendrocytes and myelin debris resulting in prolonged axonal damage and recovery from cuprizone toxicity [27].

11.2.3.2 Constitutive Axl Knock in Strain

11.2.3.2.1 Axl^{lacZ}

Generation of knockout strain: The targeted mutation was achieved by inserting a LacO-SA-IRES-LacZ-Neo555G/Kan cassette into the *Axl* gene in E14 ES-cells.

Phenotype: There are no significant differences detected in the homozygous mutant mice when compared with age- and gender-matched WT control mice. Both homozygous mutant males and females are fertile. Their progeny is viable until weaning (DeltagenInc, MGI Ref ID J:101679). LacZ expression could be detected in every tissue investigated.

Axl is required for VEGF-A-dependent activation of PI3K/Akt and thereby involved in the regulation of migration, tube formation, vascular permeability and corneal neovascularization by. It is also noticed that lung lysates and lung endothelial cells from *Axl*$^{lacZ/lacZ}$ mice display increased basal Akt activation [28].

11.2.3.3 Transgenic Mouse Lines of Axl

11.2.3.3.1 Gcsfr-AXL

Generation of transgenic strain: The murine *gcsfr* promoter and the human *AXL* cDNA were ligated to the lacZ-sv40 fragment and injected into fertilized BDF1 oocytes that were then transferred into CD1 females [29].

Phenotype: Transgenic Axl expression is detected in Gr-1 and Mac-1 positive hematopoietic cells as well as in B220 positive early B cells. F2 transgenic males exhibit non-insulin dependent diabetes mellitus but normal serum triglyceride and corticosterone levels. Peripheral blood analysis indicates a threefold increase in serum TNFα levels as compared to WT animals in response to minimal LPS stimulation.

11.2.3.3.2 ApoE-axl^{S}

Generation of transgenic strain: After adding a stop codon, the cDNA fragment of the extracellular domain of human axl was cloned into a vector containing a fragment of the apolipoproteinE (ApoE) promoter, ApoE intron, and an sv40polyA site. This line is not existing anymore, as all nine genomic positive transgenic animals were unsuccessful in producing progeny.

Phenotype: The testes of transgenic males reveal significant lesions or were absent. Similar to GCSFR-axl transgenic mice, *ApoE-axl*^{S} mice exhibited an obese-diabetes phenotype. Peripheral blood analysis also showed highly elevated levels of TNFα [29].

11.2.4 Mer

Box: List of all mouse strains

Mer^{TK}	Lu 1999 [24]
Mer^{kd}	Camenisch 1999 [31]
Mer^{tg}	Keating 2006 [34]
Mer^{nmf12}	Jackson labs MGI Ref. ID: 82238, 2003

11.2.4.1 Constitutive Mer Knockout Strain

11.2.4.1.1 Mer^{TK}

Generation of knockout strain: Exon 18 of *Mer* (comprising the last exon encoding the 3′ end of the kinase domain) was disrupted by insertion of a Neo resistance cassette by homologous recombination in mouse embryonic stem cells [24].

Phenotype: Mice homozygous for the targeted *Mer* allele are viable and fertile and display no gross anatomical defects. They are obtained at the expected Mendelian frequency and viable for a long time, indicating that also this receptor is not essential during embryonic development. However, mutant Sertoli cells show a 35 % decrease in phagocytosis of apoptotic spermatogenic cells.

No neurological or hematological disorders have been reported for Mer^{TK} mice, with the exception that as adults, Mer$^{-/-}$ mice are blind as a result of photoreceptor death. This defect is non-cell autonomous, since it results from impaired phagocytotic capacity of retinal pigment epithelium cells [30].

Moreover, apoptotic cell accumulation in post-lactational mammary glands is observed, going in line with increased vascularization, fibrosis, and epithelial hyperproliferation. This effect is due to failed efferocytosis after lactation, but does not manifest in nulliparous mice. Mer also seems to play a role in vaginal development, as Mer$^{-/-}$ mice exhibit a high incidence of distal vaginal atresia. Additionally, Mer$^{-/-}$ mice display a twofold increase in spleen size and a higher susceptibility to autoimmune disease development due to anomalously high levels of uncleared apoptotic cells. Prolonged apoptotic cell accumulation in germinal centers of Mer$^{-/-}$ mice causes elevated B cell and CD4^{+} T-helper cell responses leading to autoantibody production.

11.2.4.1.2 Mer^{kd}

Generation of knockout strain: The mer gene was replaced by homologous recombination of the Neo cassette with the last exon of the 3′ kinase domain end of the gene, resulting in loss of 53 residues in the activation-loop domain including the autophosphorylation sites on the cytoplasmic tail of Mer (Tyr749, Typr753, and Tyr 754) [31].

Phenotype: Injection of toxic levels of dexamethasone to induce apoptosis in cortical thymocytes or direct injection of apoptotic cells reduces the rate of apoptotic cell clearance by *Mer*$^{-/-}$ macrophages. This effect is not due to deficiency in binding to apoptotic cells, but in defective engulfment and internalization [32].

Challenging mutant mice with lipopolysaccharide leads to higher susceptibility to lethal septic shock as compared to WT mice, since loss of Mer function results in elevated NFκB and TNFα levels. Usually Mer activation stimulates PI3K-Akt signaling, which inhibits NFκB activation and subsequent TNFα production [31].

11.2.4.2 Transgenic Mouse Lines of Mer

11.2.4.2.1 Vav-Mer

Generation of transgenic strain: A full-length *Mer* cDNA was cloned into the HS21/45 vector containing a fragment of the *vav* promoter and microinjected into oocytes.

Phenotype: In *Vav-Mer* mice, Mer is expressed almost exclusively in hematopoietic cells, particularly in all CD4/CD8 lymphocytes and in all thymocyte developmental subsets [33]. Transgenic animals show fivefold increased rates of lymphoblastic leukemia/lymphoma, resulting from activation of anti-apoptotic pathways, including Akt and ERK1/2 [34].

11.2.4.3 ENU Mutant Strain of Mer

11.2.4.3.1 Mer^{nmf12}

Generation of mutant strain: Chemically induced mutation by ENU treatment in the neuroscience mutagenesis facility 12 from Jax (nmf12) lead to a G→A missense mutation at base pair 2237, resulting in a His→Arg conversion at amino acid position 716 of the protein. G3 descendants were selected for investigation of neurological impairment [35].

Phenotype: Standard pathology work-up on 3 mutants at the age of 266, 325, and 360 days showed thinned peripheral retina in both eyes with white intraretinal patches (indicating loss of pigmentation in the retina).

11.2.4.4 Double and Triple Knockout Strains

11.2.4.4.1 Tyro-3$^{-/-}$ Axl$^{-/-}$

Mice are obtained in the expected Mendelian ratios and are viable and superficially healthy. Moreover, as young adults, they are fertile, although some minor effects on fertility and testicular size, mass and histology can be observed [24].

11.2.4.4.2 Tyro-3$^{-/-}$ Mer$^{-/-}$

Mice are obtained in the expected Mendelian ratios and are viable and superficially healthy. Male mice display a reduced testicular mass relative to WT and only 20 % of them sire offspring. Spermatogenesis seems to be disordered and mature sperm show reduced epididymal numbers. However, since the two genes, Tyro-3 and Mer, are linked in the mouse chromosome, these mice are used to generate knockouts for all three receptor genes [24].

11.2.4.4.3 Axl$^{-/-}$ Mer$^{-/-}$

Mice are obtained in the expected Mendelian ratios and are viable and superficially healthy. Moreover, as young adults, they are fertile. However, retinal self-antigen induces a predominantly Th1 effector response in DKO mice [24].

11.2.4.4.4 Tyro-3$^{-/-}$ Axl$^{-/-}$Mer$^{-/-}$

Triple knockout mice are viable for up to 1 year. Nevertheless, they display multiple major organ defects, neurological abnormalities, and physiological deficits. Altered histology is observed in the hippocampus, cerebellum and neocortex of the brain, in the epithelium of the prostate, in the parenchyma of the liver, and in the walls of blood vessels. As Mer$^{-/-}$ mice, adult TKO mice are blind and their enlarged spleen is populated with apoptotic cells, being consistent with aberrant homeostatsis of those lymphoid populations in which expression of the three receptors is prominent [24].

Ovaries of TKO females are histologically abnormal with cell death evident in the granulose cells associated with growing ovarian follicles, but several of the females are fertile, although they show more pronounced incidence of distal vaginal atresia (fourfold increase as compared to Mer$^{-/-}$ mice). In contrast, TKO males are infertile due to drastic size reduction of their testes (only 1/3 the size of a WT). Normally more than half of the spermatogenic population dies by apoptosis during each cycle of spermatogenesis. In the absence of TAM receptors, the ability of Sertoli cells to phagocytose dead cells is almost completely abrogated. Therefore, death of all germ cells in TAM TKO mice is a non-cell autonomous mechanism [24].

In the immune system, TAM receptors are mainly expressed by macrophages and dendritic cells (DCs), but also by natural killer (NK) cells. Similar to Sertoli cells and RPE cells, macrophages and DCs display reduced phagocytic capacities in TKO mice. As bacterial and viral infections generate a number of apoptotic cell corpses, which need to be cleared by macrophages and DCs, the elevated steady-state number of apoptotic cells in TAM TKO mice is likely to contribute to the development of autoimmune disease. This is accompanied by a second effect of TAM receptor TKO in the immune system: failure in the inhibition of TLR-induced inflammation via NFκB—which is highly elevated, since unremoved apoptotic cell debris presents danger signals and autoantigenes [24].

For a special subset of DCs, the Langerhans cells (LCs), it is shown that TGF-β1-induced Axl expression enhances apoptotic cell uptake and blocks proinflammatory cytokine production, demonstrating that constitutive Axl expression is required for tolerogenic barrier immunity in the epidermis [36].

In NK cells, it is shown that TAM receptor signaling drives end-stage differentiation, as TKO NK cells have normal levels of lytic molecules perforin and granzymeB, but lack the full complement of activating and inhibitory receptors that are expressed by cytotoxic NK cells. They are also unable to secrete IFNγ in response to stimulation and exhibit a tenfold reduction of killing ability against target cells than WT NK cells [37].

11.3 DDR Receptor Family

11.3.1 Overview About Family Members

Discoidin domain receptors (DDRs) are widely expressed RTKs that bind to and are activated by collagen. By acting as collagen sensors, DDRs control cell and tissue homeostasis. Tumors often exploit the DDR pathway to disrupt normal cell-matrix communication and initiate promigratory and proinvasive programs. The DDR family consists of DDR1 (DDR, TrkE, CAK, RTK-6 or MCK-10) and DDR2 (CCK-2, TKT, Tyro 10). Alternative splicing can give rise to five distinct DDR1 isoforms, DDR1a–e, whereas only DDR1a, DDR1b, and DDR1c constitute fully functional receptors, and DDR1d and DDR1e encode truncated or kinase inactive receptors. DDRs play a critical role in bone development and growth [38]. In humans, mutations of DDR2 result in growth disorders characterized by disproportionate short stature and short limbs [39, 40]

Murine receptors: Ddr1, Ddr2
Ligands: various types of collagen

Box: List of all mouse strains

Ddr1$^{-/-}$	Vogel et al. 2001 [41]
Ddr 2$^{-/-}$	Labrador et al. 2001 [42]
Ddr 2^{mut}*(slie)*	Kano et al. 2008 [43]

11.3.2 Ddr 1

11.3.2.1 Constitutive Ddr1 Knockout Strain

11.3.2.1.1 Ddr 1$^{-/-}$

Generation of knockout strain: The *Ddr1* gene was targeted via homologous recombination in R1 ES-cells using a 129/Sv library. Thereby, the first 12 exons were deleted, and a neo cassette was inserted. Successful recombination was confirmed by Southern blot, deletion of the DDR1 protein was confirmed by Western blot [41].

Phenotype: *Ddr1*$^{-/-}$ mice were viable, albeit smaller in size. The majority of mutant females were infertile due to improper blastocyst implantation. When *Ddr1*$^{-/-}$ females produced offspring, they were unable to lactate due to defectuous mammary gland morphogenesis [41]. *Ddr1*$^{-/-}$ mice exhibit perturbed kidney structure resulting in proteinuria [44] and impaired primary mesangial cell adhesion and hyperproliferation [45]. *Ddr1*$^{-/-}$ mice are also unable to control their ear movements [41], accompanied by a loss of auditory function [46].

11.3.3 Ddr 2

11.3.3.1 Constitutive Ddr2 Knockout Strain

11.3.3.1.1 Ddr 2$^{-/-}$

Generation of knockout strain: The *Ddr2* gene was disrupted in R1 ES by replacing the exon K1, coding for the ATP binding region of the kinase domain, with a neomycin resistance cassette via homologous recombination. Correct integration was confirmed by Southern blot. Full loss of DDR2 protein in *Ddr2*$^{-/-}$ mice was demonstrated by Western blot [42].

Phenotype: *Ddr2*-deficient mice exhibited dwarfism, shortened long bones, and delayed dermal wound healing [42], as well as increased inflammatory responses from macrophages and hepatic stellar cells in chronic liver inflammation [47], even though DDR2 promotes hepatic stellar cell activation and fibrosis in acute liver injury [48]. Furthermore, *Ddr2*$^{-/-}$ mice are more susceptible to metastasis of colon carcinoma cells to the liver [49].

11.3.3.2 Spontaneous Mutant Strain of Ddr2

11.3.3.2.1 Ddr 2^{mut} (slie)

Generation of mutated strain: The smallie (slie) mutation is a spontaneous, autosomal-recessive loss-of-function mutation of DDR2, officially called *BKS (HRS)-Ddr2*$_{slie}$*/JngJ*. It was mapped to chromosome 1 and to *Ddr2* by real-time RT PCR. It constitutes a 150 kb deletion [43].

Phenotype: *Smallie (slie) Ddr2*^{mut} mice are born viable, but exhibit dwarfism, mild craniofacial deformities, and are often sterile due to peripheral dysregulation of the endocrine system [43, 50, 51]. This phenotype is similar to human dwarfism Levi type [52].

11.4 ErbB Kinases

11.4.1 Overview About Family Members

The members of the EGFR family are transmembrane receptors that, apart from ErbB2, are activated by extracellular ligand binding and transmit signals into the cell by homo- or heterodimerisation. As such they are critical mediators of basic cellular functions like proliferation, differentiation, and survival. Knockout studies revealed that all ErbB receptors are essential during embryonic development of organs like heart, nervous system, and lung. EGFR (ErbB1) and ErbB2 are furthermore critical for the maintenance of homeostasis and alterations are frequently associated with cancer.

Murine receptors:	erbB1 (egfr), erbB2 (neu), erbB3, erbB4
Ligands:	
for EGFR only:	epidermal growth factor (egf), amphiregulin (areg), transforming growth factor-a (tgfα), and epigen (epgn)
for EGFR and ErbB4:	epiregulin (ereg), heparin-binding EGF-like growth factor (hb egf), betacellulin (btc)
for ErbB3:	neuregulin 1, 2
for ErbB4:	neuregulin 3, 4

11.4.2 ErbB1, EGFR

Box: List of knockout, knockin and mutated mouse strains

egfr$^{-/-}$ *(SIB)*	Sibilia et al. 1995 [53]
egfr$^{-/-}$ *(Dwt)*	Threadgill et al. 1995 [54]
egfr$^{-/-}$ *(Mie)*	Miettinen et al. 1995 [55]
egfr$^{fl/fl}$ *(SIB)*	Natarajan et al. 2007 [56]
egfr$^{fl/fl}$ *(Dwt)*	Lee and Threadgill 2009 [57]
egfr$^{fl/fl}$ *(Diet)*	Dietrich et al. 2000 [58]
hEGFR^{KI} *(SIB)*	Sibilia et al. 2003 [59]
egfr^{wa2}	Luetteke et al. 1994 [60]
egfr^{wa5}	Lee et al. 2004 [61]
egfr^{dsk5}	Fitch et al. 2003 [62]
egfr^{vel}	Du et al. 2004 [63]
egfr$^{tm1.1(L858R)Tyj}$	jax.org, strain 018473

11.4.2.1 Constitutive Knockout Strains and Phenotype

11.4.2.1.1 egfr$^{-/-}$ (Sibilia)

Generation of knockout strain: The gene encoding the *egfr* was inactivated by replacement of part of the first exon including the *egfr* translation initiation site and parts of the leader peptide with an *Escherichia coli lacZ* reporter gene. The linearized construct was injected into R1 and D3 ES-cells [53].

11.4.2.1.2 egfr$^{-/-}$ (Threadgill)

Generation of knockout strain: The *egfr* locus was targeted by homologous recombination to replace 155 bp around at the splice acceptor site of exon 2 with a *neo* cassette. The linearized construct was injected into D3 ES-cells [54].

11.4.2.1.3 egfr$^{-/-}$ (Miettinen)

Generation of knockout strain: The *egfr* locus was targeted by homologous recombination to replace exon 2 with a *PGK-neo*-pA cassette. The linearized construct was injected into JM-1 ES-cells (129SV/J) [55].

Phenotype: Mutant fetuses are growth retarded and die at mid-gestation in a 129/Sv genetic background, whereas in a 129/Sv × C57BL/6 or a CF-1 background some survive until birth and even to postnatal day 20 in a 129/Sv × C57BL/6 × MF1 or a CD-1 background [53, 54]. Death in utero most probably results from a defect in the spongiotrophoblast layer of the placenta. Newborn mutant mice have open eyes, rudimentary whiskers, immature lungs, and defects in the epidermis. Surviving mutant mice were growth retarded, showed wasting disease and were dehydrated [53–55]. Cells of the hematopoietic system were not affected by lack of *egfr* [53]. After birth, defects in skin differentiation and hair development became obvious [54, 55]. Alteration in the differentiation and proliferation of epithelial cells results in a block in the anagen phase of hair follicle morphogenesis [59, 64]. However, egfr signaling is not essential for estrogen-induced epithelial growth in the uterus and vagina [65]. Whereas Threadgill reported only minor alterations of the intestinal epithelium at p18 Miettinen observed massive disintegration already at p8. Defects in kidney function are indicated at p0 by increased blood urea levels accompanied by cystic dilatation of the tubules [54].

Around birth, all mutant mice develop a progressive neurodegeneration in the frontal cortex, olfactory bulb, and thalamus [54, 66]. Cerebral cortices from *egfr* mutants contain lower numbers of gfap positive astrocytes, which display reduced proliferation in vitro [66]. Egfr signaling controls the cortical neuronal degeneration by regulating cortical astrocyte apoptosis [67]. Astrocyte differentiation might be disturbed by an irresponsiveness of *egfr*$^{-/-}$ progenitors to leukemia inhibitory factor (LIF) [68]. In the peripheral nervous system hyper-innervation of the skin was observed in *egfr*$^{-/-}$ but not in *egfr*$^{fl/fl}$ × *K14*-Cre mice indicating a cell autonomous role for egfr in the dorsal root ganglion [69].

11.4.2.2 Conditional Knockout Strains and Phenotype

11.4.2.2.1 egfr$^{fl/fl}$ (SIB)

Generation of knockout strain: The targeting construct for homologous recombination in ES-cells harbors a neomycin (*neo*) resistance cassette flanked by frt sites for positive and a diphtheria toxin cassette for negative selection. The linearized construct was electroporated into HM-1 ES-cells for homologous recombination and generation of a floxed allele (*egfr*^{fl}). Cre-mediated recombination results in the deletion of Exon 1 of the *egfr* gene. Crossing to *Mox2*-Cre mice resulted in germ-line deletion and a phenotype similar to *egfr*$^{-/-}$ mice [56, 70].

Phenotype:

Strain name	Organ	Phenotype
Alfp-Cre egfr$^{fl/fl}$ *Mx1-Cre egfr$^{fl/fl}$* *LysM-Cre egfr$^{fl/fl}$*	For selective deletion in hepatocytes Cre recombinase is expressed under the control of the liver-specific albumin promoter and albumin and α-feto-protein enhancers [71]. The *mx1* promoter is inducible by IFNα and Cre is expressed in parenchymal and non-parenchymal cells of the liver but also of several other organs [72] The LysozymeM-Cre (LysM-Cre) line deletes in monocytes, mature macrophages and neutrophil granulocytes [107].	Reduced liver regeneration and increased mortality after partial hepatectomy in Alfp-Cre *egfr$^{fl/fl}$* and *Mx1*-Cre *egfr$^{fl/fl}$* mice [56]. Lack of *egfr* in hepatocytes rescues from Fas ligand induced fulminant hepatitis, mediated by a timp3/adam17/egfr ligand axis [73]. Tumor growth is increased in mice lacking EGFR in hepatocytes only whereas it is reduced when deleted in macrophages [731].
K5-CreER egfr$^{fl/fl}$	Tamoxifen inducible deletion of *Egfr* in basal keratinocytes [74].	Egfr signaling synergizes with vegfr signaling in promoting epithelial cancer development [75]. Egfr cooperation with hedgehog signaling is necessary during basal cell carcinoma development [76].
K5-Cre egfr$^{fl/fl}$	Constitutive deletion of *Egfr* in basal keratinocytes around E14.5 [77].	These mice resemble very much *egfr$^{-/-}$* mice showing similar skin and hair phenotype and premature death. Egfr deficiency in keratinocytes results in massive production of proinflammatory cytokines, reduced production of antimicrobial peptides and barrier defect mimicking the situation in patients treated with EGFR inhibitors [78]. These alterations are independent of the adaptive immune system.
ElasK-RasV12 egfr$^{fl/fl}$	*Elas-tTA/tetO*-Cre transgenic mice have been crossed to mice inducible for expression of RasV12 [79] and *egfr$^{fl/fl}$* mice.	RasV12 pancreatic tumors are strictly dependent on egfr unless they are also lacking p53 [80].
Ptf1a-Cre egfr$^{fl/fl}$ *KrasG12D;* *p53KO (SIB) (Dtg)*	Pancreas transcription factor 1 subunit alpha (*Ptf1a*)-Cre is active at early stages in the progenitors of pancreatic ducts, exocrine and endocrine cells [81]. Tumor growth is induced by recombination of the *KrasLSL-G12D/+*transgene [82].	Upregulation of *egfr* and increased activity in response to kras in tumors. Reduced formation of pancreatic ductal carcinoma [83].
Vsmc-Cre egfr$^{fl/fl}$	Expression of the Cre recombinase under the smooth muscle cell specific SM22 promoter with the highest levels observed in aorta, intestine and uterus [84].	*Egfr* $^{\Delta VSMC}$ mice show aortic dilatation and altered Ca2+ signaling [85]. Mice have a reduced life span and display a dilated vascular phenotype with minor signs of fibrosis and inflammation as well as progressive cardiac hypertrophy [86].
CD4-Cre egfr$^{fl/fl}$	*CD4*-Cre deletes egfr in all mature T-cells.	Areg induced egfr signals are necessary for efficient regulatory T-cell function in mouse models of colitis and tumorigenesis [87].

11.4.2.2.2 egfr$^{fl/fl}$ (Dwt)

Generation of knockout strain: A conditional allele of Egfr was generated by flanking exon 3 that encodes for residues 57-117 in the ligand binding region of egfr with loxP sites. Homologous recombination was performed in TL1 ES-cells from 129S6/SvEvTAC (129S6) mice [57]. To test for efficient deletion of Exon 3 egfr fl/fl mice were crossed to EIIa-Cre mice that express Cre during preimplantation development [88]. Embryos and placenta showed alterations similar to that observed in *egfr*$^{-/-}$ mice.

Phenotype:

Strain name	Organ	Phenotype
K14-Cre egfr$^{fl/fl}$	Constitutive deletion of *egfr* in basal keratinocytes around E14.5 [89].	In a mixed background of 129S1/SvImJ and FVB K14-Cre *egfr*$^{fl/fl}$ mice show a mild phenotype with wavy hair [57]. These mice were resistant to alopecia induced by cyclophosphamide treatment [90]. However, in a SV129/CL57BL6 background mice developed a phenotype similar to [78]. Mice show a strong skin inflammation, barrier defect and reduced transglutaminase levels in the skin [91].
K5-Cre egfr$^{fl/fl}$	Constitutive deletion of *egfr* in basal keratinocytes around E14.5 [92].	In keratinocytes TNFα synergizes with egfr signaling in the production of gm-csf under inflammatory conditions [93] *Egfr* deficiency in keratinocytes results in massive production of proinflammatory cytokines, reduced production of antimicrobial peptides and barrier defect mimicking the situation in patients treated with EGFR inhibitors. Topical ablation of macrophages reduces local inflammation [94].
Podocin-rtTA-Tet-O-Cre egfr$^{fl/fl}$	Mice carrying the reverse tetracycline transactivator protein under control of the podocin promoter (*Pod-rtTa*) were bred with mice carrying the *Tet-O*-Cre transgene and with mice carrying a *loxP*-flanked *egfr* allele [57, 95, 96]. *EGFR* deletion in podocytes of glomeruli was induced by Dox feeding 2 weeks before experiments.	Conditional deletion of *egfr* from mice alleviates the severity of rapid progressing glomerulonephritis after injection of nephrotoxic serum [97].

(continued)

(continued)

Strain name	Organ	Phenotype
EgfrptKO	Mice deficient for *egfr* in the proximal tubules were generated by crossing γ-glutamyl transpeptidase-Cre (γGT-Cre [98]) mice with *egfr*$^{fl/fl}$ mice.	In a mouse model of acute kidney injury lack of egfr in the proximal tubules delayed recovery [99].
Hoxb7-Cre egfr$^{fl/fl}$	*Hoxb7*-Cre is expressed in the fetal ureteric bud and the adult proximal tubule [100].	Targeted inactivation of *egfr* in the proximal tubule resulted in a thinner medulla, polyuria and reduced urine concentrating ability [101].
Col3.6-Cre egfr$^{fl/fl}$ *Col3.6-Cre egfr*$^{wa5/fl}$	*Col3.6*-Cre *egfr*$^{fl/fl}$ were asymptomatic, therefore *Col3.6*-Cre egfrwa5/fl were breed [102] to delete egfr in (pre-) osteoblasts.	Col3.6-Cre *egfr*$^{wa5/fl}$ show reduced numbers of osteoblasts, bone mineralization and tibial trabecular bone mass [103].
Col2a1-Cre egfr$^{fl/fl}$ *Col2a1-Cre egfr*$^{wa5/fl}$	In neonatal mice Col2a1-Cre is active chondrocytes of the long bones [104]. In Col2a1-Cre egfrfl/fl mice recombination was incomplete.	Whereas in *egfr*$^{wa5/fl}$ mice the hypertrophic chondrocytes zone was only modestly increased in *Col2a1*-Cre *egfr*$^{wa5/fl}$ it was increased up to twofold showing responsibility of egfr in chondrocytes for cartilage remodeling into bone [105]. These mice showed a delay in formation of secondary ossification centers, decreased mineralization and apoptosis of hypertrophic chondrocytes [106].
LysM-Cre egfr$^{fl/fl}$	Cre recombinase is active in monocytes, mature macrophages and neutrophils granulocytes [107].	Mice lacking egfr in macrophages show ameliorated response to DSS induced colitis by increasing IL-10 and reducing TNFα production [108].

11.4.2.2.3 egfr$^{fl/fl}$ (Die)

Generation of knockout strain: *Egfr* exons 9 and 10 were replaced by a targeting cassette containing a floxed *neo* cassette and a CMV promoter and a *lacZ* expression cassette down- and upstream of the *loxP* sites, respectively. *LacZ* expression is induced after Cre mediated excision of the neo cassette. A *HS-1-Cre* line was generated by cloning the heat-shock promoter represented by 536 bp of 58 flanking region and the first 119 (of 225) bp of the 58 non-coding region of the mouse *Hsp70-1* gene into the *pMC*-Cre plasmid [58].

Phenotype: *Hs-Cre1* line behaved as a deleter recombining during early embryogenesis with an efficiency of 70–85 %. Cre recombination after heat shock in adult mice was only partially effective [58]. No data about the phenotype of the floxed allele are available.

11.4.2.3 Knockin Mouse Lines

11.4.2.3.1 hEGFR^{KI}

Generation of knockin strain: A human *EGFR* cDNA cassette flanked by *loxP* sites was inserted into a targeting vector to replace exon 1 of egfr. The construct was linearized and electroporated into GS-1 ES-cells.

Phenotype: Homozygous *hEGFR*$^{KI/KI}$ mice are viable and live up to 6 months. However, these mice are growth retarded and show skin and hair defects similar to *egfr*$^{-/-}$ mutants. *hEGFR*$^{KI/KI}$ mice show severe heart hypertrophy with semilunar valve abnormalities and defects in chondrocytes and osteoblast differentiation. The phenotype correlates with reduced *hEGFR* expression levels in skin and bone [59].

11.4.2.3.2 egfr$^{tm1.1(L858R)Tyj}$

Generation of knockin strain: A construct containing an *FRT-neo-FRT* cassette in intron 20 and a T→G point mutation in exon 21 was targeted to the *egfr* locus in C57BL/6-derived v26.2 ES-cells. The ES-cells were then retargeted to place a *loxP-Stop-loxP* cassette in intron 1 of *egfr*. Mice were crossed to C57BL/6 and an Flpe deleter line to remove the *FRT-neo-FRT* cassette (information at jax.org, strain 018473).

Phenotype: The *loxP-Stop-loxP* cassette (LSL) located in intron 1 of Egfr blocks expression of the gene therefore homozygous floxed mice express no EGFR protein and are not viable. Upon Cre-mediated excision of the floxed stop cassette, the L858R mutant form of the protein is expressed, albeit at levels lower than those found in wild-type mice. Adeno-cre or lentiviral-cre driven excision of the LSL-cassette in lung epithelial cells fails to induce lung tumor development.

If the stop cassette is deleted in the germ line, mice develop a wavy coat characteristic of other Egfr mutations and are also reduced in body size. These mice also develop low grade liver lesions at a late time point. Mutant mice lacking the stop cassette are viable as homozygotes.

11.4.2.4 Mutated Mouse Lines

11.4.2.4.1 egfr^{wa2}

Generation of mutated strain: The *waved 2* mutation is a spontaneous, recessive mutation first reported by Clyde Keeler in 1935. The defect is caused by a point mutation causing a substitution of a glycine for a valine at residue 743 in the protein tyrosine kinase domain of Egfr [60].

Phenotype: The reduced tyrosine kinase activity of egfr causes abnormal hair keratinization, curly vibrissae, and wavy hair. Later it was recognized that these mice also show abnormal semi lunar valve development [109]. This might result in reduced life span depending on the genetic background [110].

In this *egfr* hypomorphic background tumor formation driven by mutations of the RAS pathway is reduced indicating that the egfr provides a survival signal to tumor cells [111]. In the gut $egfr^{wa2}$ mice show increased susceptibility to DSS induced colitis [112]. In the APC^{min} model $egfr^{wa2}$ mice develop reduced number of polyps but normal numbers and pathology of microadenomas [113].

11.4.2.4.2 egfr^{wa5}

Generation of mutated strain: Wa5 is a new allele of Egfr ($egfr^{wa5}$) containing a missense mutation within the coding region for the highly conserved DFG motif of the tyrosine kinase domain exhibiting wavy coat phenotype. The allele was identified by screening mice for dominant, visible mutations induced by ENU mutagenesis [61].

Phenotype: Except for wavy coat $egfr^{wa5}$ mice have open eyes at birth. The mutation encodes a kinase dead egfr and homozygous egfrwa5/wa5 die during midgestation with phenotypes indistinguishable from egfr–/– mice [61]. $Egfr^{wa5/wa5}$ mice show reduced susceptibility to chemically induced multistage carcinogenesis in the skin [114].

11.4.2.4.3 egfr^{vel}

Generation of mutated strain: The mutation was generated by using ENU mutagenesis of male C57BL/6J mice. Chemical mutagenesis induced an adenine-to-guanine transition, leading to a D to G amino acid change at position 833 of the protein [63].

Phenotype: Mice homozygous for this mutation have an embryonic lethal phenotype, failing to develop past embryonic day 13.5 most likely due to placental defects. Heterozygous mice are viable, fertile, do not display any behavioral abnormalities and are born with open eyelids and curly vibrissae. MEFs (mouse embryonic fibroblasts) exhibit reduced migratory ability. Adult heterozygotes often have small eyes and corneal opacity with excessive secretions at the eyelid edges. These mice are also more susceptible to DSS induced colitis [115].

11.4.2.4.4 egfr^{dsk5}

Generation of mutated strain: The mutation was generated by using ENU mutagenesis. Chemical mutagenesis induced a Leu863Gln missense mutation [62]. Whereas mRNA levels in hepatocytes are normal in double mutant mice protein levels decrease up to 90 % in a dose dependent manner, indicating a posttranslational defect. Homozygotes are viable and fertile and the skin phenotype is comparable in homozygous and heterozygous mice.

Phenotype: *Dsk 5* mice show hyper pigmentation on their footpads that occurs in adult mice. The wavy coat observed in 3-week-old animals became less apparent with age. At 2 months of age, epidermal hyper proliferation can be observed whereas obvious pigment accumulation and melanocytosis can be observed at 6 months [62]. Hyperkeratosis and hyper pigmentation are reduced in $EGFR^{dsk5/wa2}$ mice indicating that ^{dsk5} is rather a gain of function mutation.

11.4.2.5 Transgenic Mouse Lines

11.4.2.5.1 K5 HERCD533

Generation of transgenic strain: The cDNA of truncated EGFR was cloned into a plasmid containing the bovine keratin 5 promoter sequences. Linearized DNA was injected into micronuclei of C57BL/6×DBA2 F2 embryos.

Phenotype: The transgene encodes for a dominant negative EGFR in basal keratinocytes and the outer root sheet of the hair follicle. Transgenic mice display wavy hair and curly whiskers with some mice proceeding to severe alopecia. Hair follicles fail to enter catagen stage and these alterations are accompanied by strong inflammation [116].

11.4.2.5.2 Egfr-tr

Generation of transgenic strain: A C-terminal truncated mouse *egfr* cDNA was generated by inserting introducing a stop codon (TGA) after the L690 codon into the SacI site of the mouse *egfr* cDNA sequence [117]. The transgene is expressed under the control of the *MMTV* long terminal repeat. The expressed C-terminally truncated protein acts as a dominant negative protein and inhibits EGFR autophosphorylation.

Phenotype: EGFR signaling is inhibited during mammary duct development, when wt EGFR levels are low. However, during pregnancy endogenous EGFR levels increased and normal presecretory alveoli developed. Postpartum, normal lactation occurred. No phenotype in other tissues observed [117].

11.4.2.5.3 GH-Egfr-tr, PRL- Egfr-tr, TetRE-EGFR-tr

Generation of transgenic strain: Expression of a dominant negative allele in the pituitary gland is either under the control of the rat prolactin promoter (*PRL-Egfr-tr*) or under the control of the rat growth hormone promoter (*GH-Egfr-tr*) that drives expression already during embryogenesis. The transgene directly drives expression of cytoplasmic tyrosine kinase- deleted mutant form of EGFR (*EGFR-tr*) in PRL-producing pituitary cells. Mice to inducibly express *EGFR-tr* under the control of the tetracycline responsive element (TetRE) have also been generated (*TetRE-EGFR-tr*) and crossed with *PRL-tTA*- and *GH-rtTA* mice, respectively, where the transgene drives expression of tetracycline-controlled transactivator (*tTA*) or reverse tetracycline-controlled transactivator (*rtTA*) in PRL- or GH-producing cells in the pituitary, respectively [118].

Phenotype: Egfr-tr overexpression in GH-producing cells during embryogenesis resulted in dwarf mice with pituitary hypoplasia, whereas later expression did not result in an obvious phenotype [118].

11.4.2.5.4 CNPase-hEGFR

Generation of transgenic strain: To express human EGFR in Schwann cells of peripheral nerves a 2′3′cyclic nucleotide 3′ phosphodiesterase promoter was used.

Phenotype: Transgenic expression of EGFR in mouse Schwann cells elicited features of neurofibromas like Schwann cell hyperplasia, excess collagen, mast cell accumulation, and progressive dissociation of non-myelin-forming Schwann cells from axons. Heterozygosity of *Nf1+/-p53+/−* resulted in improved survival [119].

11.4.2.5.5 INS-CA-EGFR

Generation of transgenic strain: Insulin (*INS)-rtTA* mice [120] were bred with *TetOP-EGFR*^{L858R} mice [121] to induce expression in pancreatic islet cells.

Phenotype: Induction of EGFR^{L858R} expression resulted in beta cell proliferation only after doxycycline treatment around birth but not in adult mice. Transgene expression improved glycemia and beta cell apoptosis [122].

11.4.2.5.6 Transgenic Lung Cancer Models

Strain name	Organ	Phenotype
CSSP- *TetO-EGFR*^{wt} *TetO-EGFR*^{L858R} *TetOEGFR*$^{\Delta L747–S752}$	Clara-cell specific tetracycline transactivator (CCSP-rtTA) is active in type II alveolar epithelial cells [123]. L858R and ΔL747–S752 mutations were introduced into wt EGFR cDNA before cloning into tet-plasmid. Single- and bi-transgenic mice were analyzed [121].	*EGFR*^{L858R} mice develop diffuse lung cancer within 2 weeks. *EGFR*$^{\Delta L747–S752}$ mice develop multifocal tumors with longer latency. Withdrawal of doxycycline of mice carrying one allele results in rapid tumor regression [121].

(continued)

(continued)

Strain name	Organ	Phenotype
SP-C EGFR $^{\Delta L748-S752}$	The surfactant protein C (*SP-C*) promoter is active in type II alveolar cells [124].	Mice developed multifocal adenocarcinomas at 5–6 weeks of age and died 2 months later. Treatment with tyrosine kinase inhibitor (TKI) gefitinib abrogated disease within 1 week and prolonged survival [125].
SP-C EGFR ^{L858R}	The *SP-C* promoter is active in type II alveolar cells [124].	Mice developed multifocal adenocarcinomas at 7 weeks of age. Adenomatous hyperplasia is preceding and accompanied by ErbB2 and ErbB3 activity. Treatment with tyrosine kinase inhibitor gefitinib completely abrogated disease [126].
Tet-op-EGFR $^{TL/CCSP\text{-}rtTA}$	Transgenic mice carrying both, the L858R and the T790M mutation, in an inducible manner were generated by site directed mutagenesis and crossed to CCSP-rtTA mice [123].	Expression of mutated egfr led to development of adenocarcinomas in bronchi and alveoli. Only alveolar tumors regressed after treatment with an irreversible TKI alone whereas addition of Rapamycin resulted in significant regression of both tumor types [127].
Tet-op-EGFR^{vIII}/ *CCSP-rtTA*	The constitutively active *EGFR*^{vIII} allele is inducibly expressed in Clara cells of the lung [123].	Tissue-specific expression of EGFRvIII in the murine lung led to the development of NSCLC. Most importantly, these lung tumors depend on EGFRvIII expression for maintenance [128].
Tet-op-EGFR TD $^{\Delta L746-S752/T790M/}$ *CCSP-rtTA*	Transgenic mice carrying both, the deletion mutation in exon 19 and the T790M mutation in an inducible manner were generated by site directed mutagenesis and crossed to CCSP-rtTA mice [123].	Novel irreversible TKI with 30–100 fold increased affinity to T790M mutant EGFR were characterized and proved efficient in the in vivo model [129].

11.4.2.5.7 Transgenic Brain Tumor Models

Strain name	Organ	Phenotype
GFAP-EGFR^{vIII} *GFAP-EGFR*^{wt}	Expression of a constitutively active EGFR in astrocytes in the brain is driven by the human glial fibrillary acidic protein (*GFAP*) promoter.	Mice expressing *GFAP-V12Ha-ras* develop gliomas. Survival is drastically reduced when they co-express the constitutively active *EGFRvIII* transgene but not the *EGFRwt* transgene [130].

(continued)

(continued)

Strain name	Organ	Phenotype
CAGGS^{LSL}*EGFR*^{wt} *CAGGS*^{LSL}*EGFR*^{vIII}	Minigenes consisting of a floxed transcriptional/translational stop cassette inserted between a strong ubiquitous promoter (CAGGS) and the *EGFR* cDNAs (either WT or vIII) were targeted to the mouse Col1a1 locus. Activation of EGFR expression was induced by stereotactic injection of Adenovirus-Cre.	Overexpression of EGFRwt or EGFRvIII together with ablation of Ink4A/Arf and Pten results in rapid development of high grade glioma [131].
TV-A EGFR^{w}	An avian retroviral vector was used to transfer a mutant *EGFR* gene to glial precursors and astrocytes in transgenic mice expressing tv-a, a gene encoding the retrovirus receptor from the *GFAP* promoter.	The mutant EGFR is lacking an extracellular (EGFR^{vIII}) and an intracellular regulatory domain and is constitutively active. Infection with mutant EGFR is only resulting in glioma formation when additional defects like deletions of ink4A/arf or cdk4 are present [132].

11.4.3 ErbB2, HER2, neu

Box: List of knockout, knockin and mutated mouse strains

erbB2−/− (Bir)	Meyer and Birchmeier 1995 [133]
erbB2−/− (Gen)	Erickson et al. 1997 [134]
erbB2−/− (Hau)	Lee et al. 1995 [135]
erbB2−/− (Var)	Park et al. 2001 [136]
erbB2lacZ (Bir)	Britsch et al. 1998 [137]
erbB2fl/fl (Bir)	Garrett et al. 2000 [138]
erbB2fl/fl (Lee)	Crone et al. 2002 [139]
erbB2fl/fl (Mue)	Leu et al. 2003 [140]
erbB2fl/fl (Mul)	Andrechek et al. 2002 [141]
erbB2K757M (Mul)	Chan et al. 2002 [142]
erbB2Y1028F *erbB2Y1144F* *erbB2Y1227F*	Chan et al. 2004 [143]
erbB2 neuNT	Andrechek et al. 2000 [144]

11.4.3.1 Knockout and Knockin Mouse Strains

11.4.3.1.1 erbB2$^{-/-}$ (Bir)

Generation of knockout strain: Exon 7-9 encoding the egf domain of *erbB2* were replaced by a neo cassette by homologous recombination in E14.1 ES-cells [133].

Phenotype: *ErbB2*$^{-/-}$ mice die around midgestation. At day 10.5 mutants display heart malformations. Cranial ganglia are morphologically aberrant and embryos show an altered pattern of erbB3 staining [133].

11.4.3.1.2 erbB2$^{-/-}$ (Gen)

Generation of knockout strain: To target *erbB2*, a 4-kb SmaI-SmaI deletion was made containing the coding region for amino acids 256-304 in the first cysteine-rich domain. Homologous recombination was performed in D3-C12 ES-cells.

Phenotype: Mutant mice die around E10.5 due to lack of cardiomyocyte differentiation resulting in a thinner ventricular wall. Additionally *erbB2*$^{-/-}$ mice displayed defects of cranial ganglia and Schwann cells [134].

11.4.3.1.3 erbB2$^{-/-}$ (Hau)

Generation of knockout strain: To generate *erbB2*$^{-/-}$ mice a targeting construct was designed to replace exon 1 of *erbB2* by homologous recombination. The linearized construct was electroporated into J1 ES-cells.

Phenotype: Mutant mice display defects in heart development, cranial ganglia, and motor neurons. *ErbB2*$^{-/-}$ embryos die before E11 [135]. There are pre- and postsynaptic defects of developing neuromuscular junctions in *erbB2*-deficient embryos. The presynaptic defects include defasciculation and degeneration of the motor nerves, and an absence of Schwann cells. The postsynaptic defect features an impairment of junctional folds at the neuromuscular synapse in the mutants [145].

11.4.3.1.4 erbB2$^{-/-}$ (Var)

Generation of knockout strain: To construct a targeting vector, the mouse *erbB2* gene was cloned from a 129/SvJ genomic library and the PGK-1-driven neomycin resistance cassette was inserted to replace a 2–3-kb fragment containing the exon encoding the essential transmembrane domain of erbB2, as well as flanking sequences encoding portions of the extracellular and cytosolic domains.

Phenotype: *ErbB2*$^{-/-}$ mice die before day 11 as previously reported. ErbB2 is required for transmission of pro-oligoblasts to oligodendrocytes whereas neurite outgrowth and astrocyte development appear normal [136].

11.4.3.1.5 erbB2^{lacZ} (Bir)

Generation of knockin strain: A *lacZ/neo* cassette was inserted into the genomic *erbB2* locus by homologous recombination. *LacZ/neo* was cloned in frame with exon h and results in a fusion protein lacking the tyrosine kinase domain [137].

Phenotype: Homozygous mice die at E10.5 with defects in heart trabeculae, cranial ganglia, and Schwann cell precursors [137].

11.4.3.1.6 erbB2$^{fl/fl}$ (Bir)

Generation of knockout strain: To generate a conditional allele one loxP site was inserted upstream of exon p and a floxed neomycin cassette downstream of exon n. The linearized vector was electroporated into E14.1 ES-cells. To proof efficient recombination homozygous mice were crossed to deleter mice [146]. Mutant mice died before E11.5. To study the role of erbB2 in Schwann cells floxed mice were crossed to krox20 Cre mice [147]. The Krox20 gene encodes a transcription factor expressed in promyelinating and myelinating Schwann cells and, thus, appears from E16 in peripheral nerves. Furthermore, Schwann cell precursors that reside in the spinal roots express Krox20 earlier, starting E11.

Phenotype: Mutant mice show abnormally thin myelin sheets with fewer myelin wraps and the spinal roots are not properly established. Thus, these mice have a widespread neuropathy [138].

11.4.3.1.7 erbB2$^{fl/fl}$ (Lee)

Generation of knockout strain: To generate *erbB2* conditional mutant mice two loxP sites were introduced into the *erbb2* locus to flank a 3-kb region containing the first exon and 1.5 kb of the *erbb2* promoter. Recombination in cardiac muscle cells was achieved by crossing to *MLC2v-Cre* mice and *MCK-Cre* mice [148, 149].

Phenotype: *MCK-Cre erbB2*$^{fl/fl}$ and *MLC2v-Cre erbB2*$^{fl/fl}$ mice develop dilating cardiomyopathy, *MCK-Cre erbB2*$^{fl/fl}$ by 6 weeks of age and *MLC2v-Cre erbB2*$^{fl/fl}$ around 6 months of age. Disease can partially be rescued by bcl-xL expression [139].

11.4.3.1.8 erbB2$^{fl/fl}$ (Mue)

Generation of knockout strain: A loxPneo-IRES-tk-loxP cassette was inserted into exon 1 to delete the signal sequence of *erbB2*. To delete *erbB2* in developing striated muscle HAS-Cre transgenic mice were generated by using the human skeletal alpha actin (HAS) promoter sequence. Expression can be observed around E9.5 [140].

Phenotype: Although neuromuscular synapses form in HAS-Cre *erbB2*$^{fl/fl}$ mice they contain less acetylcholine receptors. Abnormal spindle development causes proprioceptive defects [140].

11.4.3.1.9 erbB2$^{fl/fl}$ (Mul)

Generation of knockout strain: A targeting construct was designed containing a loxP flanked rat-*neuN* cDNA along with a PGK neo cassette. Exon 1 of *erbB2* was replaced by homologous recombination in R1 ES-cells. Recombination in skeletal muscle was achieved by crossing to *MCK* Cre mice. Partial recombination resulted in a penetrance of the phenotype in 80 % of mice.

Phenotype: Lack of erbB2 results in loss of muscle spindles and a progressive defect in proprioception. Induction of muscle damage revealed a defect in regeneration and excessive apoptosis in primary myoblasts differentiating to myofibers [141].

11.4.3.1.10 erbB2^{K757M} (Mul)

Generation of knockin line: Oligonucleotides containing the desired mutations were cloned into wt *erbB2* cDNA. Wt or mutated cDNA encoding for a kinase dead mutant receptor was inserted into exon 1 of *erbB2* by homologous recombination in R1 ES-cells.

Phenotype: Mice homozygous for the kinase-dead *erbB-2* allele die at midgestation and display the same spectrum of embryonic defects seen in *erbB-2* knockout mutants [142].

11.4.3.1.11 erbB2^{Y1028F}, erbB2^{Y1144F}, erbB2^{Y1227F}

Generation of knockin line: Mutated cDNAs of rat *erbB2* were isolated and cloned into a targeting vector to replace exon 1 of mouse *erbB2* as described in [142].

Phenotype: Although mice derived from the different knockin alleles were viable, examination of the levels of erbB2 expression revealed that the knockin strains expressed only 10 % of the expected erbB2 protein. Crossing to *erbB2*$^{-/-}$ mice resulted in perinatal lethality, only Y1028F mutants were viable and showed normal innervations of the diaphragm most likely because of higher expression of erbB2 protein [143].

11.4.3.1.12 erbB2 ^{neuNT}

Generation of knockin line: A targeting vector was designed to replace exon 1 of *erbB2* by a cDNA encoding an oncogenic variant of erbB2 (neuNT). These mice were crossed to Mmtv-Cre mice inducing strong expression of Cre recombinase mainly in mammary epithelium [144]. To achieve germ-line deletion heterozygous embryos were injected with DNA encoding *chicken-β-actin* Cre.

Phenotype: *MMTV-Cre* expression resulted in expression of *neuNT* and development of focal mammary tumors that did however not develop to carcinomas. Interestingly amplification of the neuNT allele relative to the endogenous allele was observed.

Mice bearing germ-line activation of the identical *erbB2* allele on one chromosome were viable whereas homozygous mutants died before E13.5 [150]. Heterozygous mice were completely resistant to mammary tumor development, in contrast to the mammary tumor-prone phenotype of the conditional activated strains. These observations suggest that the time point of expression of *erbB2* is critical for mammary tumor development [150].

11.4.3.2 Transgenic Mouse Lines

Strain name	Organ	Phenotype
MMTV-^{neu} (Mul)	The *MMTV*-LTR was used to express an activated form of the c-neu protein [151].	*MMTV*/c-neu transgenic mice develop tumors uniformly in all glands of the mouse mamma. Hyperplasia can be observed in parotid gland and epididymis [152]. Coexpression of tgfα under the same promoter resulted in synergistic effects [153]
MMTV-^{neuN} (Mul)	Inactivated neu cDNA (neuN) was expressed under the control of the *MMTV* promoter.	Mice developed mammary tumors at around 4 months of age. ErbB activity in tumor cells was increased. Some mice overexpressing erbB2 developed lung metastases [154].
MMTV-^{neu} (Jol)	An activated form of rat neu is expressed under the *MMTV* LTR.	In transgenic mice multiple tumors arose independently at 5–10 months of age. Tumors progressed to adenocarcinomas with necrosis and calcifications [155].
balb/c-^{neu}	Transgenic expression of neu under the control of the *MMTV* promoter. The original strain was backcrossed to balb/c for more than ten generations.	These mice have a different developmental pattern of mammary tumors compared to [152]. Tumors develop multifocal in a multistep process resembling that in human tumors [156]. Treatment with interleukin-12 delayed tumor onset accompanied by immune cell infiltrate into mammary glands [157].
MMTV-^{neuYD,} $_{YB,\ NYPD}$	cDNAs encoding for protein lacking two of the five known autophosphorylation sites of erbB2 (YB, YD) or for all five of the sites (NYPD) were cloned to the *MMTV* promoter.	NYPD transgenic mice developed focal tumors with low penetrance after long latency. In contrast, all YB and YD transgenic females developed mammary tumors with YB tumors being more likely to metastasize [158]. Overexpression of *vegfa* in tumor cells increased tumor development [159].

(continued)

(continued)

Strain name	Organ	Phenotype
MMTV $^{EC\text{-}HER}$	Expression of a truncated form of *erbB2* lacking the cytoplasmic domain.	Many transgenic lines generated developed mammary gland tumors. These tumors showed reduced apoptosis compared to those generated by full length *erbB2* transgene [160].
K5-act.erbB2 *K5-erbB2*	Activated or wt rat *erbB2* cDNAs were expressed under the control of the bovine *keratin 5* promoter	Overexpression of the activated form of *erbB2* resulted in strong epithelial multi-organ-hyperplasia with strongly reduced survival [161]. Overexpression of the wt form shows a milder phenotype, survive longer but spontaneously develop papillomas with potential to develop into squamous cell carcinoma [162].
K14- erbB2 *K14-rtTA/* *TetRE-ErbB2*	The reversible tetracycline-responsive transcriptional activator (rtTA) was used to drive active rat *erbB2* expression or the human keratin 14 promoter [163].	*K14-erbB2* mice are neonatal lethal [163]. Inducible expression of the active rat *erbB2* upon doxycycline treatment resulted in hyperplasia in epidermis, hair follicle, eyelid, and tongue accompanied by increased tgfα levels [164].
WAP-hERBB2	Human full length *ERBB2* cDNA is expressed under the control of the whey acidic protein promoter active in mammary epithelial cells upon hormone stimulation but also in the brain and other organs.	The expression of *ERBB2* resulted in tolerance of injected tumor cells that are highly expressing erbB2 [165]
Prx-1 7 dnErbB2	The 1.4-kb human *dnErbB2*/ Herstatin was cloned into the 2.4-kb mouse Prx1 promoter construct containing the enhancer elements that direct expression to the limbs in the developing mesoderm [166].	DnErbB2 is a soluble inhibitory receptor. Expression results in shortened limbs accompanied by reduced proliferation of chondrocytes and periosteal cells [167].
α-MHC-erbB2	Rat *erbB2* cDNA was cloned behind the myosin heavy chain promoter that drives expression in striated muscle.	Transgenic mice develop hypertrophic cardiomyopathy. Cardiomyocytes are resistant against apoptosis and upregulate a number of related signaling pathways [168].
MMTV-NDL (1) *MMTV-NDL (2)*	Two neu cDNAs possessing activating mutations were placed under the transcriptional control of the *MMTV*-LTR. The NDL1 and NDL2 transgenes contain in-frame deletions in the extracellular domain of neu [169].	Females of all sublines developed mammary gland tumors that were frequently metastasizing to the lung [169].

11.4.4 ErbB3, HER3

Box: List of knockout, knockin and mutated mouse strains

erbB3−/− erbB3$^{\Delta/\Delta}$(Bir)	Riethmacher et al. 1997 [170]
erbB3$^{-/-}$ (Gen)	Erickson et al. 1997 [134]
erbB3^{GIAS} (Qu)	Qu et al. 2006 [171]
erbB3$^{fl/fl}$ (Dwt)	Lee et al. 2009 [172]
erbB3$^{fl/fl}$ (Bir)	Brinkmann et al. 2008 [173]
erbB3$^{\Delta 85}$ (Mul)	Lahlou et al. 2012 [174]
Erbb3^{mspI}	Buac et al. 2008 [175]

11.4.4.1 Knockout and Knockin Mouse Strains

11.4.4.1.1 erbB3$^{-/-}$ (Bir)

Generation of knockout strain: Mouse strains carrying two distinct mutant erbB3 alleles were generated by gene targeting. A genomic region containing seven exons was replaced by a mutated neo cassette. Homologous recombination was performed in E14.1 ES-cells. A heterozygous clone was targeted with the second construct containing a wt neo cassette (erbB3Δ/Δ). In these mice a truncated protein is produced [170].

A second targeting vector contained the neo cassette fused to an exon thus introducing a stop signal (erbB3-). In these mice, no protein is produced.

Phenotype: Most embryos died between E11.5 and E13.5. Homozygous pups were born at 21 % of the expected Mendelian frequency. Homozygous pups born were cyanotic and did neither breathe nor did they react to tactile stimulation. Homozygous erbB3 mutant embryos lack Schwann-cell precursors and Schwann cells at E12.5. Most motor neurons and sensory neurons undergo cell death. In Schwann cells erbB3 acts in a cell autonomous manner [170].

11.4.4.1.2 erbB3$^{-/-}$ (Gen)

Generation of knockout strain: For ErbB3 targeting, a 2-kb deletion was made containing the coding region for amino acids 73-106 to remove the putative ligand binding site. Homologous recombination was performed in D3-C12 ES-cells.

Phenotype: ErbB3$^{-/-}$ embryos survived until E13.5 and displayed differentiation defects of the cerebellar plate, midbrain alterations, defects of cranial ganglia that were less severe than those in erbb2−/− and reduced Schwann cell numbers [134].

11.4.4.1.3 erbB3^{GIAS} (Qu)

Generation of knockout strain: To conditionally target erbB3 a vector containing a flirted neo cassette and an inverted floxed IRESEGFP + spice acceptor (SA) site was flanked by regions homologous to intron 1. Recombination results in inversion of the IRESEGFR cassette (*erbB3*^{SA}) and a fusion protein of exon 1 and IRESEGFR is produced. Deletion was confirmed in EIIa-Cre mice [171].

Phenotype: *ErbB3*$^{SA/SA}$ are embryonic lethal (before E13.5). Histological analysis revealed lack of Schwann cells and their precursors in the dorsal root ganglia. Crossing *erbB3*$^{AS/AS}$ mice with MMTV-Cre mice results in reduction of *erbB3* expression, however, incomplete recombination was observed. Mammary glands from *erbB3*$^{AS/AS}$ × MMTV-Cre mice have lower ductal density, fewer branches, and fewer terminal end buds [171]. Loss of erbB3 in these mice also reduces premalignant alterations in MMTV-Her2 and MMTV-neu transgenic mice [176]. Loss in luminal epithelial cells reduces akt and MAPK signaling and regulates differentiation [177].

11.4.4.1.4 erbB3$^{fl/fl}$ (Dwt)

Generation of knockout strain: To generate a conditional *erbB3* allele loxP sites flanking exon 2 of *erbB3*. In parallel an *erbB3* null allele was generated by a similar strategy that resulted in a similar phenotype as observed in [134]. *ErbB3*$^{fl/fl}$ mice were crossed to *Vil-Cre* mice (*ErbB3*$^{\Delta INT}$) expressing Cre recombinase in intestinal epithelial cells [178].

Phenotype: *ErbB3*$^{\Delta INT}$ mice are more susceptible to DSS induced colitis. However, loss of erbB3 in intestinal epithelial cells reduces the tumor burden in the APC^{MIN} model accompanied by increased numbers of apoptotic cells in erbB3 negative tumors [172].

11.4.4.1.5 erbB3$^{fl/fl}$ (Bir)

Generation of knockout strain: not described

Phenotype: Conditional deletion of *erbB3* either alone or together with *erbB4* in oligodendrocytes by crossing to *CNP-Cre* mice [179] resulted in death during the second postnatal week with severe defects of myelination of the peripheral nervous system, whereas myelination in the central nervous system was independent of erbB3 and 4 [173].

11.4.4.1.6 ERBB3$^{\Delta 85}$ (Mul)

Generation of knockin strain: A cDNA of a mutated human *ERBB3*, lacking the known PI3K binding sites was cloned in frame with exon 10 of the mouse *erbB3* gene and inserted by homologous recombination in E14.1 ES-cells. Thereby exons 11-14 of the endogenous receptor were deleted. Mice were backcrossed to the FVB/N background.

Phenotype: Mice carrying the *Δ85* allele were not born at the expected Mendelian ratio indicating some embryonic lethality. Homozygous mice were growth retarded and females showed impaired mammary gland development. Interestingly, in an *MMTV-NDL2-5* model all females developed metastatic mammary tumors [174].

11.4.4.1.7 *Erbb3^{msp1}*

Generation of knockin strain: The msp1 allele was identified in an ENU mutagenesis screen screening for rescue of neural crest defects. One of the four alleles identified is a mutation resulting in a single amino acid substitution in the extracellular domain of the receptor.

Phenotype: Although protein expression is normal the homozygous mutation leads to embryonic lethality. ErbB3 phosphorylation is strongly reduced [175].

11.4.5 ErbB4, HER4

Box: List of knockout, knockin and transgenic mouse strains

erbB4$^{-/-}$	Gassmann et al. 1995 [180]
erbB4$^{fl/fl}$	Golub et al. 2004 [181–183]
R26ERBB4	Veikkolainen et al. 2012 [184]
erbb4-$^{2A\text{-}CreERT2}$	Madisen et al. 2010 [185]
HER4^{heart}	Tidcombe et al. 2003 [186]

11.4.5.1 Knockout and Knockin Mouse Strains

11.4.5.1.1 erbB4$^{-/-}$

Generation of knockout strain: A targeting vector was designed to replace exon 2 of the *erbB4* gene by a neo selection cassette. For homologous recombination linearized DNA was electroporated into F28-22 ES-cells [180].

Phenotype: Mice lacking *erbB4* die at midgestation from a heart defect. Development of myocardial trabeculae is aborted [180]. Furthermore, innervation of the hindbrain is altered most likely due to aberrant migration of neural crest cells [187].

11.4.5.1.2 erbB4$^{fl/fl}$

Generation of knockout strain: A targeting vector was designed to introduce loxP sites flanking exon 2 of the *erbB4* gene. For homologous recombination linearized DNA was electroporated into R1 ES-cells. ES-cell injected blastocysts were transferred into pseudopregnant CD1 females [181]. *ErbB4*$^{fl/fl}$ mice were crossed with *Nestin (Nes)-Cre* mice to induce recombination in neuronal precursor cells [188].

Phenotype: Conditional neuronal mutants showed lower levels of spontaneous motor neuron activity and reduced grip strength compared to wild type. However, a gene dosage effect could be observed also in learning and memory task tests [181]. Deletion of *erbB4* in luminal mammary epithelial cells by crossing to *MMTV-Cre* mice did not alter tumor development in an MMTV-neu tumor model [182]. Cre-lox mediated deletions of both *Erbb4* alleles within the developing mammary gland (*ErbB4*$^{fl/fl}$ × *Wap-Cre*) fail to accumulate lobuloalveoli or successfully engage lactation at parturition owing, in part, to impaired epithelial proliferation [189]. Conditional inactivation of *erbB4* in ventricular muscle cells led to a severe dilated cardiomyopathy, characterized by thinned ventricular walls with eccentric hypertrophy, reduced contractility, and delayed conduction [183].

11.4.5.1.3 R26ERBB4

Generation of knockout strain: Human *ERBB4 JM-a CYT-2* cDNA encoding an isoform of *ERBB4* detectable in the kidney was cloned into a *rosa26* locus targeting vector containing a floxed transcriptional stop cassette. Kidney specific expression was induced by crossing to *Pax8-Cre* mice [190].

Phenotype: Overexpression of *ERBB4 JM-a CYT2* enhanced proliferation in the collecting ducts, reduced the size of epithelial duct lumens, and promoted formation of cortical tubular cysts. These defects were associated with changes in the subcellular distribution of markers of epithelial cell polarity whereas conditional deletion of *erbB4* in the same cells resulted in larger duct lumens and epithelial cell mispolarization.

11.4.5.1.4 erbb4-$^{2A\text{-}CreERT2}$

Generation of knockin strain: The *2A-CreERT2* sequence was inserted in frame to the 3′ end of the coding sequence immediately upstream of the stop codon of *erbb4*, which is predominantly expressed in interneurons [185].

Phenotype: *Erbb4*-$^{2A\text{-}CreERT2}$ mice showed Cre recombination in a subpopulation of GABAergic interneurons in both cortex and hippocampus only after tamoxifen induction. The reporter-labeled cells colocalized with most, but not all cells that were Erbb4 positive [185].

11.4.5.1.5 HER4^{heart}

Generation of transgenic line: A construct consisting of the human *ERBB4* cDNA under the control of the α-myosin heavy chain promoter was injected into the pronucleus of zygotes [186].

Phenotype: Transgenic mice were born at the expected Mendelian ratio. Transgene expression in wt mice did not induce any overt phenotype. However, transgene expression rescues the heart defects observed in *erb4*$^{-/-}$ mice, however, *erb4*$^{-/-}$ HER4heart mice were still not born at the expected Mendelian ratio and ~80 % of pups died before day 10, whereas only 3 % of *erb4*$^{+/-}$ *HER4*^{heart} pups died. Surviving females showed alterations in mammary gland organization [186] and the cranial defects of *erb4*$^{-/-}$ mice are not rescued.

11.5 EphA Receptor Family

11.5.1 Overview About Family Members

The Ephrin receptor subfamily is one the largest family members of the receptor tyrosine kinases. The interaction between the Ephrin receptors and their ligands, the ephrins, play important roles in development and are crucial for the correct migration of many cell types. Ephrin receptors, or erythropoietin-producing human hepatocellular receptors, were originally identified in 1997 [191]. There are two groups of Ephrin receptors called EphA and EphB which are separated by sequence homology [192]. The EphA and EphB receptors bind to cell-surface associated ephrin ligands which are similarly split into two groups. The ephrin-A ligands are anchored to the cell surface while ephrin-B ligands are transmembrane and, bar a few exceptions, bind to the similarly named receptor. In total there are 16 Eph receptors and 9 ephrins, each bound to cell surface and interactions between them are mediated by cell–cell contact. One of the remarkable features of ephrin-Eph interactions is that the signaling is bidirectional, with both the Eph receptor and ephrin eliciting downstream signaling in opposite directions on the two different cells. The binding of ephrins to Eph receptors results in phosphorylation of the tyrosine residues in the cytoplasmic domain. Interestingly, it was observed that following deletion of the cytoplasmic domain of *EphB2* (*EphB2*^{lacZ} mice) the ephrin containing cell was still able to signal as noted by β-gal activity [193]. Furthermore, it was identified that unlike the majority of receptor tyrosine kinases, whose signaling cascades target the nucleus, ephrin-Eph signaling regulates the cytoskeleton of the cell [194] and thus migration and movement [195, 196]. Specifically, the ephrin-Eph signaling cascades are crucial in the development of the vascular system and the nervous system, particularly in axon guidance [197–199]. Furthermore, the ephrin-Eph signaling cascade has been shown to be involved in tumor metastasis and neovascularization [200, 201]. Collectively, through the use of Eph receptor murine models ephrin-Eph signaling has been revealed as an important regulator of development and cancer metastasis.

11.5.2 EphA Receptors (EphA)

EphA receptors are preferentially activated by the ephrin-A class of ligands which are glycosylphospatidylinositol-linked [202]. Binding of ephrins to the EphA

Table 11.1 *EphA1* constitutive knockout models

Strain	Organ	Phenotype
EphA1$^{-/-}$	Constitutive loss of EphA1 signaling. IRES-human placental alkaline phosphatase inserted into exon 2 of the *EphA1* gene [203].	Abnormal tail development (kinky tail), some improper uterovaginal development (18 % females); these mice are infertile but no change to mortality. Otherwise mice survive with no additional defects.

receptors results in receptor autophosphorylation of tyrosine residues and subsequent activation of the receptors endogenous kinase domain. Currently there are 10 known EphA receptors, 9 of which have been modeled through the use of murine genetics.

11.5.2.1 EphA1

The Ephrin receptor A1 (EphA1/ESK) was the first Ephrin receptor to be identified and is the receptor for the ephrin-A1 ligand [202]. To date one murine model is routinely used to study the effects loss of EphA1 activity has in tissue patterning [203].

11.5.2.1.1 Constitutive EphA1 Knockout Strain (Table 11.1)

11.5.2.2 EphA2

EphA2 (mEcK/Myk2/Sek2) is the primary receptor for ligands ephrin-A3, -A1, -A5, and -A4 [202]. A number of EphA2 mutant mice have been generated to study the various effects that loss of extracellular signaling of EphA2 has on murine development and disease. The first model was created in 1996 using a gene trap retrovirus system disrupting the extracellular domain of *EphA2* [204] followed by a number of subsequent *EphA2* null models (Table 11.2). The use of these *EphA2* null murine models has elucidated the important role of EphA2 in axonal guidance, endothelial cell migration, skin carcinogenesis, and cataract formation [205–207].

11.5.2.2.1 Constitutive EphA2 Knockout Strains (Table 11.2)

11.5.2.3 EphA3

The EphA3 receptor (Mek4) is the receptor for ephrin ligands -A1, -A2, -A3, and -A5 and has shown to play a role in maintaining the tumorigenicity of glioblastoma multiforme [202, 209].

Table 11.2 *EphA2* constitutive knockout models

Strain name	Organ	Phenotype
EphA2$^{-/-}$(Chen)	Constitutive loss of EphA2 signaling. Insertion of a gene trap retrovirus approximately 8 kb upstream of the 5′ end [204].	No discernible phenotype.
EphA2$^{-/-}$ (Brantley-Sieders)	Constitutive loss of EphA2 signaling. A neomycin cassette was inserted at nucleotide 1372, within the sequence encoding the extracellular domain [205].	Homozygous mutant mice are viable, fertile, and display no overt developmental or behavioral abnormalities but do show diminished angiogenesis. Impaired tumor microenvironment formation [200]. Inflammation of respiratory airways [208].
EphA2$^{-/-}$ (Guo)	Constitutive loss of EphA2 signaling. A gene trap vector system was used to truncate exons 6–17 encoding the second fibronectin type III repeat in the extracellular domain [206].	Increased susceptibility to skin carcinogenesis.
EphA2$^{-/-}$ (Guo)/ *β-gal* reporter	Constitutive loss of EphA2 signaling. As per [206] with a β-galactosidase reporter cassette inserted in the ectodomain [207].	Increased frequency of cataract formation.

11.5.2.3.1 Constitutive EphA3 Knockout Strain

11.5.2.3.1.1 EphA3$^{-/-}$

Generation of knockout strain: *EphA3$^{-/-}$* mice have been generated via deletion of part of exon 1 encoding the extracellular domain [210].

Phenotype: Homozygous null mice have a high mortality rate, mostly as a result of pulmonary edema, but interestingly do not show any axon guidance defects (common for many EphA receptor null mice).

11.5.2.4 EphA4

The EphA4 receptor (Sek1) is one of the most studied Ephrin receptors as it is a critical regulator of axon guidance, similar to EphA2 [211–213]. EphA4 is activated by large number of ephrins including ephrin-A1, -A2, -A3, A-5 and also by ephrins-B2 and -B3 [202]. A number of different mouse models have been created to help study the effect loss of *EphA4* has on axon guidance as well as motor coordination, limb development and motor neuron function. The first *EphA4* null mouse model was generated by homologous recombination of the third exon resulting in fertile

Table 11.3 *EphA4* constitutive knockout models

Strain name	Organ	Phenotype
EphA4$^{-/-}$ *(Dottori)*	Constitutive loss of EphA4 signaling. Homologous recombination of a neomycin cassette into the third exon of the *EphA4* gene [213].	Null mice are viable and fertile but show a loss of motor coordination and disruptions of the corticospinal tract.
EphA4$^{-/-}$ [215]	Constitutive loss of EphA4 signaling. Frame shift mutation in exon 3 and homologous recombination of exon 1 and part of intron 1 with a *lacZ-PGK* neomycin cassette [215].	*EphA4* null mice are viable and fertile but show a hopping gait. Defects observed in the corticospinal tract and anterior commissure
EphA4$^{KD/KD}$ *(Kullander)*	Lysine K653 substitution with methionine in the kinase-dead *EphA4KD* allele [214].	*Kinase dead EphA4* null mice are viable and fertile but show a hopping gait.
EphA4^{2F}	Two juxtamembrane tyrosine residues Y596 and Y602 in the *EphA4*^{2F} allele were replaced with phenylalanines [214].	*EphA4*^{2F} knockout mice are viable and fertile but show a hopping gait.
EphA4$^{\Delta SAM}$	The *EphA4*$^{\Delta SAM}$ allele encompassing amino acid residues 905–974 was deleted [214].	*EphA4*$^{\Delta SAM}$ mice showed normal gait and walking patterns

mice that show a coordination defect commonly known as a hopping gait [213]. Subsequent knockout models have attempted to better target specific domains of *EphA4* (Table 11.3). Mice with a specific deletion of amino acids 905-974 encoding the SAM domain had ablated EphA4 activity but showed normal walking patterns indicating that the SAM domain of EphA4 is important in regulating murine posture and gait [214].

11.5.2.4.1 Constitutive EphA4 Knockout and Knockin Strains

See Table 11.3.

11.5.2.4.2 Gene Trap/Transgenic Mouse Lines of EphA4

Many models of Ephrin receptor deletion have been generated through the use of gene trapping, a technique whereby genetrap vectors are inserted into introns downstream of genes of interest and upstream of the promoter that drives that gene. The vectors themselves contain a promoterless reporter gene (DNA) and a stop codon and so when expressed by the promoter of the gene of interest the reporter gene is expressed but the gene of interest is not thereby simultaneously inactivating and reporting the expression of the "trapped" gene. This method was used to generate one *EphA4* null murine model that expresses β-galactosidase in the axons of the anterior commissure where EphA4 activity was lost [216]. Models utilizing green fluorescent protein or cyan fluorescent protein have also been generated where the cDNA encoding for these transgenes were inserted in and around the third exon

Table 11.4 EphA4 gene trap models

Strain name	Organ	Phenotype
EphA4$^{PLAP/\beta\text{-}geo}$ transgenic mouse	Central nervous system. Specifically the anterior commissure. Gene trap insertion of *PLAP/β-geo* vector [216].	Mice show a hopping gait and have axon guidance defects in the anterior commissure.
EphA4^{EGFP} *transgenic mice*	Central nervous system, cDNA containing the extracellular region of *EphA4* fused to *eGFP* was inserted in frame into exon III of *EphA4* as per Kullander et al. [214, 217].	Homozygous *EphA4*$^{EGFP/EGFP}$ mice are viable and moderately fertile but show the hopping gait seen in EphA4 null animals.
EphA4^{mCFP} *transgenic mouse*	Central nervous system. Generated *EphA4* null allele marked with a knock in *mCFP* reporter gene [218].	*EphA4*^{mCFP} mice show a hopping gait with 100 % penetrance.

creating a truncated EphA4 protein that expresses GFP or CFP [217, 218]. In both of these transgenic models the mice display phenotypes similar to the *EphA4* null mice (Table 11.4).

11.5.2.4.3 Conditional EphA4 Knockout Strains (Table 11.5)

11.5.2.5 EphA5

The EphA5 receptor [222] is activated by ligands ephrin-A1, -A2, -A3, -A4, and -A5 [202].

11.5.2.5.1 Constitutive EphA5 Knockout Strain (Table 11.6)

11.5.2.6 EphA6

The EphA6 receptor (mEhk2) has preferences for ephrin ligands -A2, -A3, and -A5 [202].

11.5.2.6.1 Constitutive EphA6 Knockout Strain (Table 11.7)

11.5.2.7 EphA7

The EphA7 receptor (Mdk1/Ebk) is activated by ephrin ligands -A1, -A2, and -A3 [202].

11.5.2.7.1 Constitutive EphA7 Knockout Strain (Table 11.8)

Table 11.5 Cre-lines used with EphA4 conditional knockout mice

Strain name	Organ	Phenotype
EphA4$^{lox/lox}$	Inserted *loxP* sites flanking exon 3 resulting in a splicing of exon 2 and 4 and a frameshift mutation. A floxed *mCFP-neo* reporter gene precedes the *floxed exon 3* [218].	*EphA4*$^{mCFPNeo/Flox}$ mice carrying one floxed and one deleted allele show a hopping gait suggesting the floxedallel to be a loss of function allele. However, *EphA4*$^{lox/lox}$ *exhibited a normal gait.*
EphA4$^{lox/lox}$*Emx1-cre*	Cerebral cortex and hippocampal neurons [219].	Mice viable and fertile but have diminished learning capabilities following mechanical brain injury [219, 220]. EphA4 signaling is an important modulator of glutamate transporter activity in hippocampal astrocytes
EphA4$^{lox/lox}$*Prx1-Cre*	Hind limb [218].	Mice viable and fertile with no change to lateral motor neurons [221].
EphA4$^{lox/lox}$*Nestin-cre*	Central nervous system including the spinal cord [218].	Mice viable and fertile with significant increase of lateral motor neurons. Thus fewer LMC axon projection errors than the ubiquitous deletion of *EphA4* [221].

Table 11.6 EphA5 constitutive knockout models

Strain name	Organ	Phenotype
EphA5$^{LacZ/lacZ}$	Constitutive loss of EphA5 signaling. Homologous recombination of *lacZ* into the intracellular domain [223]. The generated fusion protein lacks the entire tyrosine kinase and C-terminal domains of *EphA5* but has the extracellular, transmembrane, and juxtamembrane domains of *EphA5* linked to β-galactosidase.	*EphA5*$^{lacZ/lacZ}$ mice survive, are fertile and show no gross morphological defects other than some evidence of increased body weight and increased aggression.

Table 11.7 EphA6 constitutive knockout models

Strain name	Phenotype	Organ
EphA6$^{-/-}$	No distinguishable phenotypic differences; some cognitive impairments.	Constitutive loss of EphA6 signaling. Substitution of exon 1 with a neomycin cassette by homologous recombination results in the removal of the translation initiation codon [224].

11.5.2.8 EphA8

The EphA8 receptor (mEek) is activated by ephrin ligands -A2, -A3, and -A5 [202].

11.5.2.8.1 Constitutive EphA8 Knockout Strain (Table 11.9)

Table 11.8 EphA7 constitutive knockout models

Strain name	Organ	Phenotype
EphA7$^{-/-}$	Constitutive loss of EphA7 signaling. Homologous recombination deletes bases 1–330 of the *EphA7*cDNA, as well as part of the upstream sequence (–601 to –1 bp) containing the start codon [225].	No gross morphological changes however the retinocollicular map was disrupted with nasal and temporal axons forming additional or extended termination zones.

Table 11.9 EphA8 constitutive knockout models

Strain name	Organ	Phenotype
EphA8$^{-/-}$	Constitutive loss of EphA8 signaling. The extracellular domain was replaced with transmembrane domain of the human *NTRK1* receptor and *lacZ* genes resulting in a chimeric, nonfunctional EphA8 protein [226].	*EphA8* null mice develop to term, are fertile and do not display obvious anatomical or physiological defects but display an abnormal ipsilateral axonal tract.

11.5.2.9 EphA10

The EphA10 receptor is a relatively new member to the Ephrin receptor family discovered in 2005 [227]. To our knowledge to date no murine models exist.

11.5.3 EphB Receptors (EphB)

EphB receptors are preferentially activated by the Ephrin-B class of transmembrane ligands [202]. Binding of ligands to the EphB receptors results in receptor auto-phosphorylation of tyrosine residues and in activation of the kinase domain similar to EphA receptors. There are five known EphB1 receptors.

11.5.3.1 EphB1

EphB1 is preferentially activate by ephrin ligands -B1, -B2, and -A3 [202]. EphB1 is found exclusively in regions of retina that give rise to the ipsilateral projection and is important for optic chiasm formation. *EphB1* null mice were generated via homologous recombination of exon 3 with a neomycin cassette and display reduced ipsilateral projections and irregular pain behavior following pain induction [228]. Another*EphB1* null reporter mouse was also generated via splicing of exons 1-2 with an IRES-*βgeo* cassette and displays the same phenotype as the first *EphB1* null mice (Table 11.10) [228]. In addition, *EphB1*^{lacZ} mice were generated [229]. This model differs from the null mutants in that the resulting truncated EphB-β-gal protein retains its extracellular and transmembrane domains which allows for the

Table 11.10 EphB1 constitutive knockout models

Strain name	Organ	Phenotype
EphB1$^{-/-}$	Constitutive loss of EphB1 signaling. Insertion of a neomycin resistance gene into Exon 3 causes a frame shift mutation in the coding sequence [228].	*EphB1*$^{-/-}$ mice show reduced ipsilateral projection and pain behavior following pain induction.
EphB1^{lacZ}	Constitutive loss of EphB1 signaling. *EphB1* exons 1 and 2 are spliced with an IRES-*βgeo* cassette resulting in the expression of *βgeo* instead of functional *EphB1* [228].	Phenotype is similar to *EphB1*$^{-/-}$.
EphB1T^{lacZ}	Constitutive loss of EphB1 signaling. *LacZ* inserted into the ninth exon of *EphB1* between the codons encoding the transmembrane domain and the juxtamembrane segment [229].	*EphB1T*$^{lacZ/lacZ}$ show phenotypes similar to *EphB1*$^{-/-}$ mice with failure to form the ipsilateral projecting subpopulation of retinal ganglion cell axons.

mutated protein to traffic to the plasma membrane as well as for ligands to stimulate forward and reverse signaling. Homozygous *EphB1*^{lacZ} mice fail to form the ipsilateral projecting subpopulation of retinal ganglion cell axons (same as *EphB1* nulls). This indicates that reverse signaling through the extracellular domain of EphB1 is not required for proper axon pathfinding of retinal axons at the optic chiasm [229].

11.5.3.1.1 Constitutive EphB1 Knockout Strain

See Table 11.10.

11.5.3.2 EphB2

EphB2 receptor (Nuk/Sek3) is the primary receptor for ephrin ligands -B1, -B2, and -B3 [202]. The EphB2 null mice were generated via insertion of a PGK-neomycin resistance cassette into the 5′ region of the *EphB2* coding sequence [193]. These mice have normal phenotypes but have abnormal axon guidance and show a failure of the cortical neurons to cross the anterior commissure (Table 11.11). *EphB2*null mice were one of the first *Eph* receptor knockout models generated and played an important role in identifying ephrin-Eph bidirectional signaling. Thakar et al. discovered that following deletion of the cytoplasmic domain in *EphB2* of *EphB2*^{lacZ} mice (Table 11.11) the Eph receptors could actually signal through the extracellular domain via interactions with the ephrin ligands [193, 230]. Since then a number of other models have been generated whereby selective sections of the *EphB2* receptor cDNA were deleted to better understand the bidirectional signaling of the EphB2–ephrin interaction (Table 11.11). Overall, the majority of *EphB2* knockout mice display abnormal axon morphology but some models do show normal axonal guidance [193] and/or anterior commissure formation [230] (Table 11.11).

Table 11.11 EphB2 constitutive knockout models

Strain name	Organ	Phenotype
EphB2$^{-/-}$(*Nuk1*$^{-/-}$)	Constitutive loss of EphB2 signaling. Insertion of a PGK-neomycin resistance cassette displacing amino acids 29-50 (1.4 kb) [193].	Normal lifespan and fertile. Abnormal axon guidance. Failure of cortical neurons to link the two temporal lobes (defect in the anterior commissure). Abnormal eye and optic tract morphology [228]. Circling behavior in approximately 60 % of homozygote knockout mice [231].
EphB2^{lacZ}	Constitutive loss of EphB2 signaling. Insertion of a lacZ-neomycin resistance cassette in-frame at codon 622 [193].	Normal lifespan and fertile. Many homozygous *EphB2*^{lacZ}mice have normal axon guidance (different from *Nuk1* germ-line knockout mice)
EphB2$^{\Delta PDZ}$	Homologous recombination of codons 992-994 in exon 17 with an FRT flanked hygro cassette [232].	Abnormal axon morphology
EphB2^{K661R}	Point mutation at exon 12 resulting in an amino acid substitution (K661R) [232].	Abnormal axon morphology
EphB2^{lacZ}	Cytoplasmic domain of *EphB2* replaced with β-gal reporter [230].	Normal lifespan and fertile. Mice homozygous for *EphB2*^{lacZ} showed no defects in the anterior commissure.

11.5.3.2.1 Constitutive EphB2 Knockout Strain

See Table 11.11.

11.5.3.2.2 EphB2 Knock in Strain

11.5.3.2.2.1 EphB2^{F620D}

Generation of knockout mice: An EphB2 targeting vector was produced by homologous recombination-mediated retrieval of an F620D-containing fragment from pBeloBAC11.EphB2.F620D. The linearized vector was electroporated into R1 ES-cells. In this mouse model the conserved phenylalanine codon for residue 620 in *EphB2* was replaced with an aspartic acid codon (F620D) resulting in a constitutive, overexpressing kinase that does not require interactions with ephrin ligands to transduce forward signals [233].

Phenotype: More proliferative cells in the intestine of *EphB2*^{F620D} but no signs of migration defects within the intestinal epithelium of adult mice could be detected.

11.5.3.2.3 Conditional EphB2 Knockout Strain

11.5.3.2.3.1 EphB2^{lx}-CamKIIα-Cre

Generation of knockout mice: Exon 2 of *Eph2B* was flanked by 2 loxP sites. Cam-Cre results in removal of EphB2 specifically from the forebrain [217].

Phenotype: *EphrinB2-CamKcre* mice did not show any gross alterations of the neuronal architecture of the hippocampus nor obvious behavioral abnormalities.

11.5.3.3 EphB3

The EphB3 receptor, or Sek4/Mdk5 receptor, is preferentially activated by ephrin ligands -B1, -B2, and -B3, similar to EphB2 [202].

11.5.3.3.1 Constitutive EphB3 Knockout Strain (Table 11.12)

11.5.3.4 EphB4

The EphB4 receptor (Myk1/Mdk2) is a receptor for ephrin ligands -B1 and -B2 [202].

11.5.3.4.1 Constitutive EphB4 Knockout Strain (Table 11.13)

11.5.3.5 EphB6

The EphB6 (Mep) receptor is one of the more recent EphB receptors to be identified and little is known about its activating ligands. *EphB6* is normally expressed in the brain and thymus and is a unique member of the Ephrin receptor family in that it lacks intrinsic kinase activity, although there is evidence for its phosphorylation following cross talking with other EphB receptors and ephrin-B1 [235, 236].

Table 11.12 EphB3 constitutive knockout models

Strain name	Organ	Phenotype
EphB3$^{-/-}$	Constitutive loss of EphB3 signaling. Homologous recombination of exons encoding part of the kinase domain with a PGK-neomycin resistance cassette [212]	Normal lifespan and have no circling behavior and normal gait [231] but defective axon pruning and some evidence for elongated infra-pyramidal bundle axons [234].

Table 11.13 EphB4 constitutive knockout models

Strain name	Organ	Phenotype
EphB4tau^{lacZ}	Constitutive loss of EphB4 signaling. Homologous recombination of a tau-LacZ-PGK-neomycin cassette in exon 1 targeting the translation start codon [198].	Embryonic lethal. Growth retardation, arrested heart development and lack of blood flow by E9.5–E10. Degeneration and necrosis by E10.5.

Table 11.14 EphB6 constitutive knockout models

Strain name	Organ	Phenotype
EphB6$^{-/-}$ *(Luo)*	Constitutive loss of EphB6 signaling. Homologous recombination of the *ATG* sequence in exon 3 through to the 5′ end of intron 4 with a GFP-neomycin reporter cassette [237].	Mice are viable with no abnormal gross phenotype but show decreased T-cell proliferation.
EphB6$^{-/-}$ *(Shimoyama)*	Constitutive loss of EphB6 signaling. Homologous recombination of a portion of exon 1-3 with a floxed neomycin cassette [238].	Mice developed normally, are fertile and revealed no abnormal gross morphology.

11.5.3.5.1 Constitutive EphB6 Knockout Strain

See Table 11.14.

11.5.4 Compound EphA/B Receptor Mouse Models

In addition to single receptor knockout mouse models a number of compound mutant mice have been used to investigate the combined loss of ephrin receptors in murine development. Crossing of *EphA3/EphA4* null mice gave rise to progeny with no noticeable change to gross morphology but did show axonal misrouting (Table 11.15). Crossing of *EphA3*$^{-/-}$ *mice* with *EphA4*^{fl} *Olig2-cre* mice (*pMNΔFlox/ΔKinase*; Table 11.15) resulted in a severe defect in epaxial sensory pathway formation while unaffecting hypaxial pathways [239]. The *EphA3*$^{-/-}$*;A4*$^{\Delta Kinase}$ mouse displays deficient protein signaling but expression of its extracellular segment is preserved on epaxial motor axons. *EphA4/A3;A4ΔKinase* mice survive and have normal formation of the majority of epaxial sensory projections and thus avoiding the severe epaxial sensory projection defects observed in *EphA3/A3;A4pMNΔFlox* embryos. Combined, these models established the importance of EphA3/4 signaling in peripheral nerve assembly. Specifically, Wang et al. were able to show that EphA3/4 receptors signal forward with a repulsive signaling cue to segregate epaxial motor axons from proximal sensory pathways [240] and also "backwards" via activation of ephrin-A's on sensory growth cones to couple sensory projections towards epaxial motor pathways [239].

Table 11.15 Compound EphA and EphB constitutive knockout models

Strain name	Organ	Phenotype
EphA3$^{-/-}$;*EphA4*$^{-/-}$	Murine crosses [239].	Mice survive but display misrouting of epaxial motor axons. Epaxial sensory projections
EphA3$^{-/-}$;*EphA4*$^{fl/fl}$ *Olig2-cre (ΔpMNΔFlox)*	Murine crosses result in inactivation of EphA3 and 4 in motor neuron progenitors [239].	Severely defective formation of epaxial sensory pathways while leaving unaffecting hypaxial pathways.
EphA3$^{-/-}$*EphA4*$^{\Delta Kinase}$	Homologous recombination of the intracellular domain of *EphA4* with *eGFP* and crossed with *EphA3* knockout mice [239].	Mice survive and have normal formation of the majority of epaxial sensory projections (in contrast to *EphA3/A4* nulls).
EphB2$^{-/-}$; *EphB3*$^{-/-}$ *EphB2*^{lacZ}; *EphB3*$^{-/-}$	Murine crosses [193, 212].	Mice exhibit circling behavior and have vestibular dysfunction because of defects in the endolymph producing dark cells of the inner ear epithelium [231].

In addition to EphA receptor compound mice, *EphB2/B3*null compound mice have been generated through controlled matings. Progeny with homozygous *EphB2/B3* deletions have severe vestibular dysfunction and exhibit a circling behavior as well as defects in the anterior commissure (Table 11.15) [231]. Through the use of compound *EphB2/B3* mice Cowen et al. removed the redundancy of EphB signaling in the vestibular system when either *EphB2* or *EphB3* are knockout alone however it is worth noting that, like so many other compound mutant matings, the results are different depending on the background strain of mice used [231].

Constitutive and conditional Eph double knockout strains

11.6 The FGF Receptor Family

11.6.1 Overview About Family Members

Fibroblast growth factor receptor family (FGFR) consists of four members which are high affinity transmembrane receptors mediating signals of at least 22 ligands [241]. FGFR1, 2, 3 have two major isoforms (IIIb, and IIIc) generated through alternative splicing [242]. FGF signaling is essential for the embryonic development and adult homeostasis of almost all the tissues/organs [241]. Signaling pathways triggered by FGFR activation are involved in almost all kinds of cellular events including cell fate specification, proliferation, differentiation, migration, survival, and apoptosis. Missense mutations in FGFs and FGFRs were found clinically to cause multiple congenital skeleton diseases including chondrodysplasia, craniosynostosis or syndromes with dysregulated phosphate metabolism [243] many of which are mimicked by mouse models lacking or overexpressing variants of fgfrs. FGFR activating mutations and overexpression are closely correlated with the development of various cancers, including breast cancer, bladder cancer, multiple myeloma, liver cancer, and renal cell carcinoma, as well as angiogenesis around the tumor and therefore clinical reagents that specifically target the FGFs or FGF receptors are being developed [244].

Murine receptors: Fgfr1-4
Ligands: fgf1-22

Box: List of the most important constitutive and conditional knockout strains

Fgfr1$^{-/-}$	*Deng 1994* [245]
Fgfr1$^{-/-}$	*Yamaguchi 1994* [248]
Fgfr1$^{fl/fl}$	*Trokovic 2003* [250]
Fgfr1$^{fl/fl}$	*Xu 2002* [259]
Fgfr2	Arman 1998 [275]
Fgfr2$^{\Delta 2}$	Arman 1999 [277]
Fgfr2$^{-/-}$	*Xu 1998* [278]
Fgfr2$^{fl/fl}$	*Yu 2003* [279]
Fgfr2$^{fl/fl}$	*Blak 2007* [288]
Fgfr3$^{-/-}$	*Colvin 1996* [307]
Fgfr3^{fl}	*Su 2010* [312]
Fgfr4$^{-/-}$	*Weinstein 1998* [319]

11.6.2 *Fgfr1*

11.6.2.1 Constitutive Fgfr1 Knockout Strains

11.6.2.1.1 Fgfr1$^{-/-}$ (Deng)

Generation of knockout strain: By homologous recombination a PGKneo cassette was inserted to replace exon 4 of *fgfr1* in J1 ES-cells.

Phenotype: *Fgfr1$^{-/-}$* is embryonic lethal during early postimplantation stage. *Fgfr1* mutant embryos appear growth retarded and cell migration through the primitive streak is impaired [245]. Analysis of chimeric embryos consisting of a mixture of wild-type and *fgfr1* mutant cells has revealed a cell autonomous function for fgfr1 in the mesodermal and endodermal cell movements through the primitive streak [246]. Fgfr1 orchestrates the epithelial to mesenchymal transition and morphogenesis of mesoderm at the primitive streak by controlling snail and e-cadherin expression [247].

11.6.2.1.2 Fgfr1$^{-/-}$(Yamaguchi)

Generation of knockout strain: By homologous recombination a PGKneo cassette was inserted to replace exon 8-14 of *fgfr1* in R1 ES-cells.

Phenotype: In the absence of FGFR1 signaling, embryos displayed early growth defects; however, they remained capable of gastrulating and generating mesoderm. The nascent mesoderm of *fgfr1* homozygous mutant embryos differentiated into diverse mesodermal subtypes, but mesodermal patterning was aberrant. Somites were never generated and axial mesoderm was greatly expanded at the expense of paraxial mesoderm [248].

11.6.2.2 Conditional Knockout Strain

11.6.2.2.1 Fgfr1$^{fl/fl}$ (Trokovic)

Generation of knockout strain: The cells carrying the *fgfr1^{l15}* allele were electroporated with the targeting vector pPNTLoxP-IIIbSTOP introducing a second *loxP* site into intron 7 [249]. The resulting *fgfr1^{l7l15}* (*fgfr1^{fl}*) allele was electroporated into R1 ES-cells. Positive clones were injected into embryos of the ICR strain.

Phenotype: *Fgfr1$^{\Delta/fl}$* mice are normal whereas homozygous mice show defects similar to *fgfr1$^{-/-}$*mice. Crosses with *En1-Cre* and *Wnt1-Cre* mice allowed complete inactivation of *fgfr1* transcription by E9.5 in the midbrain and anterior hindbrain. *Fgfr1* mutants suggest that after establishment of the isthmic organizer, FGFR1 is needed for continued response to the isthmic signals, and that it has direct functions on both sides of the organizer. In addition, FGFR1 appears to regulate expression of specific cell-adhesion molecules [250].

Strain name	Organ	Phenotype
fgfr1$^{fl/fl}$foxG1-Cre	Brain [251]	85 % Reduction of auditory hair cells most likely due to reduced proliferation of precursor cells in the early cochlear duct [252]. Morphological defects at the anterior end of the telencephalon. The olfactory bulbs do not form normally because of excessive proliferation in anterior telencephalic cells supporting a model for olfactory bulb formation in which an FGF-dependent decrease in proliferation is required for initial bulb evagination [253].

(continued)

(continued)

Strain name	Organ	Phenotype
fgfr1$^{fl/fl}$Col2-Cre or Col1a-Cre	Bone/osteo-chondrocyte [104]/differentiated osteoblasts [254]	Inactivation of *fgfr1* in osteo-chondro-progenitor cells delays osteoblast differentiation whereas in differentiated osteoblasts differentiation is accelerated. Both show increased bone mass in adults [255].
fgfr1^{fl}hGFAP-Cre	Embryonic brain [256]	Lack of FGFR1 throughout the embryonic dorsal telencephalon; decrease in dividing radial glial progenitor cells in the hippocampal ventricular zone and dental gyrus; no FGF2+ stem cells. In adults hippocampal pyramidal neurons and dental gyrus granule cells 50 % reduced. FGFR1 in GFAP+ cells is required for neuronal recovery after neonatal hypoxic injury in subventricular zone and olfactory bulb [257, 258].

11.6.2.2.2 Fgfr1$^{fl/fl}$ (Xu)

Generation of knockout strain: *LoxP* sites have been inserted into introns flanking exons 8 and 15 by homologous recombination in C-1 ES-cells [259].

Phenotype:

Strain name	Organ	Phenotype
fgfr1$^{fl/fl}$Col2a1-Cre	Bone/osteo-chondrocyte [104]	Homozygous *fgfr$^{\Delta/\Delta}$* mice reveal a phenotype similar to *fgfr1$^{-/-}$* mice (data not shown). *Col2a1-Cre* used to show principal function of the floxed allele. No phenotype described [259].
fgfr1$^{fl/fl}$ Ap2-Cre	Frontonasal process, limbs [260]	Abnormal anterior digit formation and only mild impact on posterior digit formation [261].
fgfr1fl/fl Hoxb6-Cre	Embryonic mesoderm from E8.5 on [262]	Abnormal hindlimb formation. Abnormal formation of the apical ectodermal ridge. Diminished *shh* expression and absence of the majority of the autopod skeletal elements [261].
fgfr1fl/fl Hes7-Cre	Presomitic mesoderm [263]	Lack of FGFR1 specifically in the presomitic mesoderm. Reduced expression of *sprouty4* and h*es7* results in malformed somites and defect in segmentation [263].

(continued)

(continued)

Strain name	Organ	Phenotype
fgfr1fl/flT-Cre	Recombination in most primitive streak descendants including somites, presomitic mesoderm, and tail bud.	Aborted FGF signaling results in arrest of the dynamic cyclic of gene expression in the presomitic mesoderm and ultimately leads to an arrest of segmentation [264].
fgfr1fl/fl Col2a1-CreERT2	Inducible recombination in articular chondrocytes of adult mice [265].	Resistance to proteoglycan loss induced by AIA and attenuated development of cartilage destruction after surgically induced destabilization of the knee joint. The chondroprotective effect of FGFR1 inhibition was largely associated with decreased expression of matrix metalloproteinase 13 and up-regulation of FGFR-3 [266]

11.6.2.3 Fgfr1 Knockin Mouse Lines

Generation of knockout strain: *Fgfr1*$^{n7,-n15}$ alleles: A floxed neo cassette sites was introduced into the respective intron (7, 15) by homologous recombination. Recombined alleles are called *fgfr1*^{l7}, *fgfr*^{l15}, respectively. *Fgfr1*$^{IIIcn,-IIIbn}$ alleles: A STOP codon was inserted into exon 7 or exon 6, respectively, and a floxed neo cassette between exon 7 and 8 by homologous recombination. *Fgfr1*^{n15YF} allele: A silent mutation creating a shift from Y to F in exon 17 and a floxed neo cassette between exon 16 and 17 were inserted by homologous recombination [249].

Phenotype:

Strain name	Organ	Phenotype
Fgfr1^{n7} *Fgfr1*^{l7}	Development	Heterozygous neo containing allele (*n7*) behaves like a hypomorph, homozygous alleles caused neonatal lethality, defects in craniofacial and limb patterning, and abnormal development of the A-P axis [249].
Fgfr1^{IIIcn} *Fgfr1*^{IIIcl}	Development	Dominant splice variant,phenotype like *fgfr1*$^{-/-}$ [249]
Fgfr1^{IIIbn} *Fgfr1*^{IIIbl}	Development	Heterozygous neo containing allele (*IIIbn*) behaves like a hypomorph, homozygous *IIIbl*-mice are viable and fertile [249]
Fgfr1^{n15} *Fgfr1*^{l15}	Development	Heterozygous neo containing allele (*n15*) behaves like a hypomorph, homozygous alleles caused neonatal lethality, defects in craniofacial and limb patterning, and abnormal development of the A-P axis [249].

(continued)

(continued)

Strain name	Organ	Phenotype
Fgfr1^{n15YF}	Development	Heterozygous neo containing allele (*n15YF*) behaves like a hypomorph, homozygous alleles caused neonatal lethality, defects in craniofacial and limb patterning, and abnormal development of the A-P axis. *Y766Fl* homozygotes are viable and fertile but show vertebral translocations [249]

11.6.2.4 Transgenic Mouse Lines of Fgfr1

Strain name	Generation/organ	Phenotype
K10-DN FGFR1 (IIIc)	A kinase negative form of *Fgfr1(IIIc)* DNA was cloned into a plasmid containing the bovine Keratin 10 promoter. Linearized DNA was injected into fertilized CBYB6F1/J zygotes to direct expression to suprabasal epidermis.	Disorganisation of basal and suprabasal layers with epidermal hyperthickening and increased numbers of K5/K14 positive layers in the mouse tail [267].
Mmtv-DN Fgfr1 (IIIc)	A kinase negative form of *Fgfr1(IIIc)* DNA was constructed by ligating cDNA encoding the three Ig-loops, transmembrane and juxtamembrane domain into pMMTV. Linearized DNA was injected into F1 C57BL×CBA zygotes to direct expression to mammary gland epithelium.	No significant differences to the equivalent control glands during pregnancy and lactation [268].
NSE-DN Fgfr1	A kinase-lacking cDNA of *fgfr1* was cloned downstream of a neuron specific enolase promoter to drive expression in neurins. Linearized DNA was injected into C57/Bl6×CBA(F2) fertilized oocytes.	Neurons isolated from mutant p2-p5 cerebella do not show neurite outgrowth in response to FGF2, NCAM, L1 or n-Cadherin [269].
aA-tR1	A truncated *fgfr1* cDNA encoding the extracellular, transmembrane and a small portion of the cytoplasmic juxtamembrane domains under the lens specific 369 bp aA crystallin promoter was microinjected into pronuclei of FVB/N embryos generating a dominant negative allel of *fgfr1*.	Transgenic families expressing the truncated FGFR exhibited lens defects ranging from cataracts to severe microphthalmia. Lens epithelial cells were reduced in number and the lens fiber cells displayed characteristics consistent with the induction of apoptosis [270].

(continued)

(continued)

Strain name	Generation/organ	Phenotype
th(tk–)/th(tk–)	DNA of a tyrosine kinase negative version of FGFR1 [271] was cloned behind the rat *Th* gene promoter that drives expression in dopamine neurons [272]. DNA was used for pronuclear injection into BCF1 embryos collected from the mating of male C57BL/10J with female C3H/HeJ mice	Reduced number of dopamine neurons in substantia nigra and ventral tegmentum at birth. However, in other brain area, increased dopamine levels were detected. Dopamine antagonists could reverse impaired repulse inhibition similar to schizophrenic patients [273]. Further experiments suggest that midbrain 5-HT2A-controlling systems are important loci of therapeutic action [274].

11.6.3 Fgfr2

11.6.3.1 Constitutive Fgfr2 Knockout Strain

11.6.3.1.1 Fgfr2 (Arman)

Generation of knockout strain: The targeting construct was prepared from a 129SvJ genomic l phage library. Targeting inserted a reverse neo cassette into exon 9. Additionally exons 10-12 were deleted. The construct was electroporated into R1 ES-cells. Two homologous recombinant ES-cell lines were aggregated with eight-cell embryos [275].

Phenotype: Peri-implantation lethality at E4.5-E5.5. Mutant blastocysts hatched, adhered, and formed a layer of trophoblast giant cells in vitro, but after prolonged culture, the growth of the inner cell mass stopped, no visceral endoderm formed, and finally the egg cylinder disintegrated.

11.6.3.1.2 Fgfr2$^{\Delta 2}$ (Arman)

Generation of knockout strain: The construct was prepared from a 129SvJ genomic phage library. Targeting inserted a reverse neo cassette into exon 9. Additionally exon 10 was deleted. The construct was electroporated into R1 ES-cells. Preimplantation lethality was rescued by tetraploid aggregation [276].

Phenotype: Tetraploid aggregation chimeras allowed survival until term and revealed that FGFR2 is required for both limb outgrowth and branching lung morphogenesis. The use of fusion chimeras demonstrated that early lethality was indeed because of trophectoderm defects and indicated that in the embryonic cell lineages Fgfr2 activity manifests in limb and lung development [277].

11.6.3.1.3 Fgfr2$^{-/-}$ (Xu)

Generation of knockout strain: The targeting construct generated from SV129 DNA was used to delete exons 7, 8, and 9 of the *fgfr2* gene. Linearized DNA was electroporated into J1 and TC1 ES-cells to obtain homologous recombination. Positive clones were injected into C57BL/6J blastocysts.

Phenotype: Embryos with a homozygous hypomorphic allele die around E10.5 with no limb buds and a defective placenta. About one third of the mutants failed to form the chorioallantoic fusion junction and the remaining mutants did not have the labyrinthine portion of the placenta [278].

11.6.3.2 Conditional Knockout Strain

11.6.3.2.1 Fgfr2$^{fl/fl}$ (Yu)

Generation of knockout strain: For generation of *fgfr*^{fl} mice 2 loxP sites were introduced into the genomic sequence of *fgfr2*, flanking exons 8-10 containing region *IIIb, IIIc* and the transmembrane region, by homologous recombination into RW4 ES-cells [279].

Phenotype: Recombination of the floxed allel by *β-actin cre* results in a phenotype similar to knockout mice. *Fgfr2*$^{Δ/Δ}$ embryos, which died between E10 and E11 developed no limb buds and failed to form a functional placenta [279].

Strain name	Organ	Phenotype
Fgfr2$^{fl/fl}$ *dermo1-Cre*	*Dermo1-Cre* mice were generated by homologous recombination. Expression occurs at E9.5 at the surface and in the mesoderm and later is mainly found in chondrocytes and osteoblasts [279].	*Fgfr2*^{fl} *Dermo1-Cre* mice display skeletal dwarfism and decreased bone density. Although differentiation of the osteoblast lineage was not disturbed, the proliferation of osteoprogenitors and the anabolic function of mature osteoblasts were severely affected [279].
Fgfr2 $^{fl/fl}$ *tbx18-Cre*	*Tbx18-Cre* conditionally deletes in peri-Wolffian duct stroma [732].	*Fgfr2ST2/2* mice exhibit improper ureteral insertion sites into the bladder and increased vesicoureteral reflux [280].
Fgfr2 $^{fl/fl}$ *msx2-Cre*	*Msx2-cre* is active in the matrix of the postnatal hair follicle and in the dorsal skin during embryonic development [281].	Long, shiny and brittle hair after second week. Increased expression of *shh* and downstream mediators like *gli* whereas the normal downregulation is abolished [282].
Fgfr2$^{fl/fl}$ *cnp1-Cre*	*Cnp1-Cre* mice [179] were crossed with fgfr2fl mice (Yu et al. [279]) to inactivate FGFR2 in oligodendrocytes.	No obvious phenotype was detected in conditional *fgfr2* knockout mice. However, mice homozygous for cre and therefore also *Cnp1*$^{-/-}$ displayed a dramatic hyperactive behavior starting at 2 weeks of age [283].

(continued)

(continued)

Strain name	Organ	Phenotype
Fgfr2^fl/fl^ Sftpc-Cre	Surfactant protein C-cre (*Sftpc-cre)* line, in which Cre recombinase is expressed under the control of *sftpc* promoter elements [284].	Inactivation of FGFR2 in lung epithelium around E12.5 resulted in disruption of lobes and small epithelial outgrowths that arose arbitrarily along the main bronchi [285] .
Fgfr2^fl/fl^ nkx3.1-Cre	*Nkx3.1-Cre* mice direct Cre recombinase activity to all prostate lobes, seminal vesicle, and ductus deferens [733].	FGFR2 is important for prostate organogenesis and for the prostate to develop into a strictly androgen-dependent organ with respect to tissue homeostasis but not to the secretory function [286].

11.6.3.2.2 Fgfr2$^{fl/fl}$ (Blak)

Generation of knockout strain: A conditional allele with a floxed exon 5 of *fgfr2* was generated by homologous recombination.

Phenotype: Ubiquitous recombination by *Pgk-Cre* results in p5 lethality. Recombination in mid- and hindbrain by crossing to *En1 Cre*/+mice [287] resulted in no obvious phenotype [288].

11.6.3.3 Fgfr2 Knockin Mouse Lines

Strain name	Organ	Phenotype
Fgfr2c; *FgfrLR;* *FgfrCLR*	To generate a Crouzon-like point mutation in the extracellular domain Cys-342 was replaced by a Tyr residue encoded in exon 9 of the *fgfr2(IIc)*. A second mutation in cis was made in the juxtamembrane domain. Two amino acids encoded in exon 10, Leu-424 and Arg-426 (LR), were replaced by Ala residues in 129SvJ genomic DNA. Constructs were electroporated into W9.5 ES-cells. Expression occurs at E9.5 at the surface and in the mesoderm and later is mainly found in chondrocytes and osteoblasts [279].	*Fgfr2c* mice display Crouzon-like craniosynostosis induced by the dominant mutation. The premature suture fusion is rescued by mutation of the FRS2 docking sites of FGFR2 in *fgfrCLR* mice due to attenuation of FGFR signaling [289].

(continued)

(continued)

Strain name	Organ	Phenotype
Fgfr-2(IIIb)	A targeting construct using SV129 DNA was constructed containing exon *IIIb* flanked by loxP sites using and HSV-neo cassette. GK129 ES-cells were electroporated and recombination was induced by transient transfection with a pMC-Cre plasmid.	Viable only until birth, because they have no lungs. Mice lack the anterior pituitary gland, show several skeletal abnormalities like no fore and hindlimbs and a thinner skin [290]. Severe dysgenesis of the cochleovestibular membraneous labyrinth and sensory patches and the cochleovestibular ganglion remain at a rudimentary stage [291].
K5-R2IIIb mice	Keratin 5-Cre transgenic mice [292] were used to delete FGFR2 in *Fgfr-2(IIIb)* mice (De Moerlooze et al. [290]).	*K5-R2IIIb* mice display hair and skin alterations with epidermal hyperthickening and increased skin inflammation. Furthermore these mice show increased susceptibility to chemically induced skin carcinogenesis [293].
FgfR2-IIIb$^{-/-}$lacZ	An *IRES-LacZ* gene was introduced behind a translational stop into exon *IIIb* of a 129SvJ-mouse genomic DNA. The linearized plasmid was electroporation into 129P2/OlaHsd ES-cells.	Mice survive to term but show dysgenesis of the kidneys, salivary glands, adrenal glands, thymus, pancreas, skin, otic vesicles, glandular stomach, and hair follicles, and agenesis of the lungs, anterior pituitary, thyroid, teeth, and limbs. There FGFR2 seems to regulate expression of *fgf4* and *shh* [294].
Mmtv-DN Fgfr2(IIIb)	To generate a kinase deleted form of FGFR2 (IIIb) a cDNA encoding the two Ig-loop variant of the receptor was then inserted into expression plasmid pMMTV. Linearized DNA was injected into F1 C57BL×CBA zygotes.	Impairment of lobuloalveolar growth apparent by midpregnancy [268].
SPC-DN FGFR2(IIIb)	The human *FGFR2-IIIb* receptor cDNA was truncated in the intracellular juxtamembrane region removing both kinase domains and the kinase insert. The cDNA was cloned into a plasmid containing the human SPC promoter directing expression to the airway epithelium. Linearized DNA was injected into fertilized eggs of B6SJLF1/J mice.	Mosaic expression with 50 % transmission rate. Affected mice die at birth due to defective branching and alveolar differentiation [295].

(continued)

(continued)

Strain name	Organ	Phenotype
K14-DN FGFR2(IIIb)	A truncated human *FGFR2-lIlb* cDNA from which the tyrosine kinase domains were deleted was fused downstream of the human keratin 14 promoter and upstream of the regulatory elements of the human growth hormone gene.	Epidermal atrophy, abnormalities in the hair follicles, and dermal hyperthickening. Upon skin injury, substantially delayed reepithelialization of the wound and reduced proliferation rate of epidermal keratinocytes [296].
PB-fgfrIIIb (KDNR)	The minimal rat PB promoter was used to target a truncated FGFR2IIIb receptor to prostatic epithelium.	Disorganized epithelium in many of the prostatic ducts containing rounded, cytokeratin-positive cells that were not tightly associated with the basement membrane. The stroma was disorganized and did not form a tight layer of smooth muscle around the epithelial ducts [297].
$Fgfr2^{+/Y394C}$	An 1181 A to G resulting in an Y394C protein mutation in exon 10 was introduced into the genomic locus of *fgfr2* by homologous recombination.	Mice mimic Beare-Stevenson cutis gyrata syndrome. Ligand-independent phosphorylation of FGFR2 in mutant skin and calvarial tissues is p38 dependent [298].
$Fgfr2IIIc^{-/-}$	To inactivate exon 9, the *IIIc*-specific exon of *fgfr2*, dGTP was inserted at codon 333, which created a stop codon five nucleotides downstream from the site of insertion by homologous recombination in R1 ES-cells.	FGFR2IIIc is a positive regulator of ossification affecting the osteoblast and the chondrocyte lineages. Delayed onset of ossification and premature loss of growth results in synostosis at the skull and dwarfism in the long bones and axial skeleton [299].
$FGFR2IIIc^{S252W}$; *sFGFR2IIIc-Ap*	A fragment of the human *FGFR2IIIc* cDNA containing the S252W mutation was cloned into a CMV expression vector. Similarly, a fragment of the human *FGFR2IIIc -Ap* cDNA lacking the transmembrane and cytoplasmic domains (sFGFR2IIIc-Ap) was cloned. DNA was microinjected into the pro-nuclei of fertilized eggs from C57BL/6J mice.	Osteoblasts expressing *FGFR2IIIc-Ap* proliferate and differentiate via highly activated MEK, ERK, and p38 pathways, while these pathways are suppressed in osteoblasts expressing *sFGFR2IIIc-Ap* [300].
$FGFR2^{+/P253R}$	The 758C>G, P253R mutation was introduced into exon IIIa of *fgfr2* by homologous recombination.	*Fgfr2* +/P253R mutation resulted in mice with cranial features that resemble those of the *Fgfr2+/S252W* mice and human Apert syndrome like shortened skull length at p0 and synostosis of sutures [301].

(continued)

(continued)

Strain name	Organ	Phenotype
Fgfr2IIIc	LoxP sites flanking exon 9 of *fgfr2* were introduced into the genomic locus by homologous recombination. 129 ES-cells were injected into blastocysts of C57BL mice to generate chimeras	Heterozygotic abrogation of *fgfr2 IIIc* causes a splicing switch, The gain-of-function mutation results in neonatal growth retardation and death, coronal synostosis, ocular proptosis, precocious sternal fusion, and abnormalities in secondary branching in several organ. Strong parallels to Apert's and Pfeiffer's syndrome patients [302].

11.6.3.4 Fgfr2 Transgenic Mouse Lines

Strain name	Organ	Phenotype
MT-dnFGFR-Tm, *MT-FGFR-HFc*	Expression of transgenes was driven by the mouse MT promoter. *MT-dnFGFR-Tm* is truncated behind the transmembrane domain whereas *MT-FGFR-HFc* encodes for a dimer consisting of two extracellular domains. *MT-dnFGFR-Tm* mice and *MT-FGFR-HFc* embryos were generated by micro-injection into FVB/N zygotes.	Transgenic expression of soluble receptor causes agenesis or severe dysgenesis of kidney, lung, specific cutaneous structures, exocrine and endocrine glands, and craniofacial and limb abnormalities leads to embryonic lethality. In contrast, transgenic mice expressing a membrane-tethered kinase-deficient FGFR were viable [303].
Tie2-fgfr2	A constitutively active form of FGFR2 mutant was caused to be overexpressed using the *Tie2* promoter.	In transgenic mice infarct size, vessel formation and blood perfusion were significantly improved [304]. Hematopoietic cells of *Tie2- Fgfr2* mice are normal except for increased mast of the thymus. The antiapoptotic effect of FGFR2 signaling facilitates the hematopoiesis from c-kit+Sca1+ bone marrow of *Fgfr2* Tg mice [305].

11.6.3.5 Fgfr2 Reporter Mouse Lines

11.6.3.5.1 pGIIIcI2 Reporter, pGIIIcI$^{\Delta\Delta}$ Reporter

Generation of mouse strain: To generate the *pGIIIcI2* splicing reporter, the *IIIc* exon and flanking introns, including intronic activating sequence 2 (IAS2) and the intronic silencing and activating region (ISAR) were inserted within the EGFP reading frame. Skipping of *IIIc* leads to fusion of the EGFP reading frame and expression of EGFP. *pGIIIcIΔΔ*, harbors deletions in IAS2 and ISAR, which abrogates cell-type-specific exon IIIc skipping resulting in almost complete inclusion of exon *IIIc*. The *pGIIIcI2* and *pGIIIcIDD* constructs were inserted by homologous recombination into the ROSA26 locus of A129S4JaeSor embryonic stem (ES) cells. ES-cells were injected into C57BL/6J blastocysts.

Phenotype: Alternative splicing reporters mark epithelial cells in vivo and reveal that FGFR2 isoform switching is not required for keratinocyte migration during wound healing [306].

11.6.4 Fgfr3

11.6.4.1 Constitutive Fgfr3 Knockout Strain

11.6.4.1.1 Fgfr3$^{-/-}$

Generation of mouse strain: The *fgfr3* genomic locus was targeted to delete part of region IgII and IgIII which prevents ligand binding by homologous recombination in SM-1 ES-cells [307].

Phenotype: Fgfr3$^{-/-}$ mice exhibit skeletal dysplasia of the long bones and inner ear defects including failure of pillar cell differentiation and tunnel of Corti formation [307]. FGFR3 appears to regulate endochondral ossification by an essentially negative mechanism, limiting rather than promoting osteogenesis [308]. In FGFR3 knockout mice, an increase in the proliferation of pancreatic epithelial cells in neonates was observed, leading to a marked increase in islet areas in adults along with increase in the proportion of potential islet progenitor cells [309]. Treatment with FGF18 resulted in a significant increase in peri-implant bone formation in both *Fgfr3*$^{+/+}$ and *Fgfr3*$^{-/-}$ mice, with the peri-implant fibrous tissue frequently seen in *Fgfr3*$^{-/-}$ mice being largely replaced by bone [310]. Brains of fgfr3$^{-/-}$ mice show a defect in development of the caudal telencephalon with some effect on brain projections (Fgfr3 regulates development of the caudal telencephalon) [311].

11.6.4.2 Conditional Fgfr3 Knockout Strain and Phenotype

11.6.4.2.1 Fgfr3^{fl}

Generation of mouse strain: The targeting vector was constructed using *Fgfr3* genomic DNA introducing *loxP* sites flanking exon 9 and 10 of *fgfr3*. The targeting construct was electroporated into TC1 ES-cells. Positive clones were injected into C57BL/6 blastocysts.

Phenotype: Crossing with *EIIa-Cre* transgenic mice that express Cre recombinase in germline results in increased length of long bone and decreased bone mineral density similar to *fgfr3*$^{-/-}$ mice. Crossing with Col2a1-Cre mice showed principal function of the conditional allel in vivo [312].

11.6.4.2.2 *Fgfr3*^{Gly374Arg}

Generation of mouse strain: Conditional knockin of a PCR mutated targeting construct.

Phenotype: Crossing to *EIIa-Cre* transgenic mice that express Cre recombinase in germline resulted in mutant mice. Mice exhibited small size, short tail, macrocephaly and had dome-shaped heads, the epiphyseal growth plates of mutant mice were narrower and the hypertrophic chondrocyte zone was also obviously decreased. The majority of female mice were infertile, and the uterus, ovary and mammal gland in mutant female mice were also smaller and underdeveloped [313].

11.6.4.2.3 Fgfr3^{Gly374Arg}

Generation of mouse strain: G to A mutation at codon 374 was introduced into exon 10 of *fgfr3* in SV129 phage DNA. The targeting construct was electroporated into R1 ES-cells. A positive clone was aggregated with 8-cell MF1 blastocysts.

Phenotype: Homozygotes carrying the neo cassette had a phenotype similar to *fgfr3*$^{-/-}$ mice, exhibiting bone overgrowth. After deletion of the neo cassette mice display dwarfism due to achondroplasia with small size, shortened craniofacial area, and hypoplasia of the midface [314]. Expression of MEK1 in the chondrocytes of *fgfr3*^{G374A} mice inhibited skeletal overgrowth and led to shortening of tibia to lengths [315].

11.6.4.2.4 Fgfr3^{G369C}

Generation of mouse strain: The Gly369Cys mutation was introduced into exon 10 of the mouse *Fgfr3* gene. The targeting construct was electroporated into TC1 ES-cells.

Phenotype: *Fgfr3*^{G369C} mice mimic achondrodysplasia displaying dwarfism, macrocephaly, and shortened limbs due to retarded endochondral bone growth and premature closure of cranial base synchondroses [316].

11.6.4.2.5 Fgfr3b; Fgfr3c

Generation of mouse strain: *Fgfr3b* and *Fgfr3c* isoforms were generated by introducing point mutations into exon 8 or exon 9, respectively, by homologous recombination in W9.5 ES-cells. Positive clones were injected into C57BL/6 blastocysts.

Phenotype: No apparent phenotype was detected in *Fgfr3b*$^{-/-}$ mice. *Fgf3c*$^{-/-}$ mice displayed increased chondrocyte proliferation in the growth plates. In addition, *Fgfr3c*$^{-/-}$ mice exhibited decreased bone mineral density in the cortical and trabecular bone [317].

11.6.4.3 Retrovirally Generated Chimeric Strain of Fgfr3

11.6.4.3.1 FGFR3 TDII

Generation of mouse strain: Full-length *FGFR3* wild-type and *TDII(K650E)* cDNAs were individually subcloned into retroviral MSCV-Gateway-neoEB and MSCV2.2-Gateway-IRESGFP. WT bone marrow cells were transduced with retroviral supernatant and injected into irradiated balb/c mice.

Phenotype: Mice that received transplants of *FGFR3 TDII* constructs succumbed to a fatal pre-B-cell lymphoma. Phenylalanine substitution of Y760, essential for phospholipase C binding and activation, significantly attenuated FGFR3 TDII-mediated PLC activation, as well as transformation in Ba/F3 cells and a murine bone marrow transplant leukemia model [318].

11.6.5 Fgfr4

11.6.5.1 Constitutive Fgfr4 Knockout Strain

11.6.5.1.1 Fgfr4$^{-/-}$

Generation of mouse strain: *Fgfr4* was disrupted by interrupting exon 6, which encodes the IgII domain, with the *Pgk-neo* gene. TC1 ES-cells were transfected and positive clones were microinjected into C57BL/6 blastocysts [319].

Phenotype: In the initial publication mice homozygous for a targeted mutation of fgfr-4 exhibited no overt abnormalities [319]. However, it has later been shown that FGFR4 is important in the liver. Mice lacking FGFR4 exhibited normal morphology and regenerated normally in response to partial hepatectomy. However, the FGFR4$^{-/-}$ mice exhibited depleted gallbladders, an elevated bile acid pool and elevated excretion of bile acids [320]. Lack of *fgfr4* inhibits development of hepatocellular carcinoma in FGF19 transgenic mice [321]. FGFR4 is required for the reduction in hepatic expression of CYP7A1, PON1, and plasma HDL cholesterol levels caused by dietary bile acid [322]. Following acute CCl4 exposure, *fgfr4*$^{-/-}$ mice exhibited accelerated liver injury, a significant increase in liver mass and delayed hepatolobular repair. Chronic CCl4 exposure resulted in severe fibrosis in livers of FGFR4-deficient mice compared to normal mice [323].

11.6.6 Double and Triple Conditionally Mutated Strains

Strain name	Organ	Phenotype
Fgfr1/2$^{fl/fl}$ *hoxb7-Cre (fgfr2*$^{UB-/-)}$	*Hoxb7creEGFP* transgenic mice were generated and used to recombine *fgfr1*$^{fl/fl}$ [250] and *fgfr2*$^{fl/fl}$ [279] at E11.5 in ureteric bud derived tissue and in the hindbrain.	*Fgfr2*$^{UB-/-}$ mice have aberrant ureteric bud branching, thin ureteric bud stalks, and fewer ureteric bud tips due to apoptosis and reduced proliferation. Kidneys are small and abnormally shaped or are hydronephrotic. *Fgfr2*$^{UB-/-}$ mice show no obvious phenotype and *Fgfr1/2*$^{UB-/-}$ mice show a phenotype similar tof–/– mice. These defects occur independently of FRS2 [324].
fgfr1/2fl/flMlc2v-cre	Conditional deletion was carried out using the *Mlc2v-cre* transgene. Cre recombinase is expressed specifically in the embryonic ventricle beginning at E10.5 [325]. *Fgfr1*^{fl} and *fgfr2*^{fl} mice used from Pirvola et al. [252] and Yu et al. [279].	Redundant roles for FGFR1 and 2 during heart development. Endocardial and epicardial derived FGF signals regulate myocardial proliferation during midgestation heart development. Impaired coronary vascular development similar to that of *Fgf9–/–* hearts [326]. VEGF-A, B, C and ANG2 expression is regulated in myocardial cells via *shh* [327].
fgfr1/2fl foxG1-Cre fgfr3-/-	Floxed alleles of *fgfr1* and 2 (Trokovic et al. [250]; Yu et al. [279]) were used to generate knock outs when crossed to *Foxg1Cre* or *CAGG-CreER* mice [253, 328] in an *Fgfr3*$^{-/-}$ background [308].	Simultaneous deletion of FGFR1 and FGFR3 specifically in the telencephalon results in the loss of differentiated ventromedial cells and in the *Fgfr1;Fgfr2* double mutant, ventral precursor cells are lost which is not rescued by *Gli3* mutation [329]. Triple *fgfr*-mutant mice exhibit a complete loss of the telencephalon, except the dorsal midline. FGFs promote telencephalic character and are strictly required to keep telencephalic cells alive [330]. FGF signaling, through cooperation between FGFR1 and FGFR2 but not FGFR3, is required for the initial generation of OLPs in the mouse ventral forebrain, with Fgfr1 being a stronger inducer than Fgfr2 [331].

(continued)

(continued)

Strain name	Organ	Phenotype
fgfr1/2flPax3-Cre (Fgfr1/2Mes$^{-/-}$)	*Pax3-Cre* knockin mice [332] were used to conditionally delete *fgfr1* (Trokovic et al. [250]) and *fgfr2* (Yu et al. [279]) in the metanephric mesenchyme and not the ureteric bud at later stages of development.	*Fgfr1Mes$^{-/-}$* and *fgfr2Mes$^{-/-}$* mice develop normal-appearing kidneys. *Fgfr1Mes$^{-/-}$* display renal aplasia. *Fgfr2Mes$^{-/-}$* mice develop a ureteric bud that does not elongate or branch, and the mice do not develop an obvious metanephric mesenchyme. Mutant buds are undergoing high rates of apoptosis at E10.5 and 11.5 [333]. Postnatal *fgfr2Mes$^{-/-}$* mutants had high rates of VUR due to abnormal bladder insertion [334].
fgfr2^{fl} hGFAP-Cre; fgfr1/2^{fl} hGFAP-Cre	The *hGFAP-Cre* transgene targets Cre recombination of floxed *fgfr1* (Pirvola et al. [252]) and/or *fgfr2* (Yu et al. [279]) to radial glia progenitors of the dorsal telencephalon starting at E13.5.	The proper formation of the medial prefrontal cortex depends upon the function of FGFR2, whereas FGFR1 together with FGFR2 control excitatory cortical neuron development within the entire cerebral cortex [335].
fgfr1/2fl hGFAP-Cre fgfr3-/-	The *hGFAP-Cre* transgene targets Cre recombination of floxed *fgfr1* (Trokovic et al. [250]) and *fgfr2* (Yu et al. [279]) to radial glia progenitors of the dorsal telencephalon starting at E13.5. Mice were further crossed to *fgfr3$^{-/-}$* mice (Deng et al. [308]).	In triple knockout mice radial glia are depleted due to an increased transition from an uncommitted state to a more differentiated one, initially causing an increase in progenitors, but ultimately resulting in a smaller cortex [336].
fgfr1/2^{fl} K5-Cre (K5-R1; K5-R2; K5-R1/2)	Bovine Keratin5-Cre [292] leads to recombines floxed *fgfr1* (Pirvola et al. [252]) and/or *fgfr2* (Yu et al. [279]) in basal cells of stratified epithelia.	*K5-R2* mice show hair abnormalities, a reduction in the number of hairs, and loss of sebaceous glands whereas *K5-R1* mice are phenotypically normal. *K5R1/R2* mice show a more severe phenotype with failure of hair cycling, progressive barrier defect, and inflammatory infiltrate [337].
fgfr1/2fl MLR10-Cre fgfr3-/-	*Mlr10-Cre* mice [100] expressing Cre in the lens from the lens pit stage. *Fgfr1^{fl}* (Trokovic et al. [250]), *Fgfr2fl* (Yu et al. [279]), *Fgfr3$^{-/-}$* (Deng et al. [308])	Microphthalmia. Lack of fiber cell elongation, abnormal proliferation in prospective lens fiber cells and increased apoptosis [338].

11.7 The Insulin/IGF Receptor Family

11.7.1 Overview About Family Members

The Insulin/IGF receptor family comprises 4 genes: Insulin receptor (IR), Insulin like growth factor (IGF) 1 receptor (IGF1R), Insulin like growth factor 2 receptor (IGF2R), and insulin receptor-related (INSRR) [339]. Multiple ligands bind these receptors, the most famous being insulin, IGF1 and IGF2 [339]. INSRR does not bind growth factors but is rather activated by pH-changes [340]. IR, IGF1R, and INSRR function as homodimers composed of two identical α/β monomers, or as heterodimers composed of two different receptor monomers (IR/IGF1R). IGF2R on the other hand is a monomeric receptor, which mainly served ligand clearing functions. This gene is also famous for its complex regulatory mechanism which mediates mono-allelic expression and makes IGF2R a model system for studies of genomic imprinting [341]. IR and IGF1R signaling generally regulates cell growth and metabolic responses. Consequently, mutations in these genes have disastrous consequences and results in a variety of diseases such as diabetes. Furthermore, components of this RTK family are often deregulated in a variety of cancers [342].

11.7.2 Insulin Receptor (IR)

11.7.2.1 Constitutive IR Knockout Strains

11.7.2.1.1 $IR^{-/-}$ (Accili); $IR^{-/-}$ (Joshi)

Generation of knockout mice: Two independent null alleles of the insulin receptor gene have been generated [343, 344]. The first allele features a premature translational termination mutation in exon 4 [343]. Selection of targeted J-1 ES-cell clones was achieved using a Neomycin resistance cassette and after verification of proper gene targeting these cells were used for the generation of chimeric mice. The second constitutive loss-of-function allele was generated by deleting exon 2 of the insulin receptor gene, which encodes the ligand binding domain of the receptor [344]. Positive and negative selection of targeted clones was performed using neomycin and ganciclovir and proper recombination was tested via Southern blotting. Of three positively targeted ES-cells clones, two were used for the generation of insulin receptor null mice.

Phenotype: Neonatal lethality. Mice are born with slight growth retardation but normal metabolic parameters. After birth massive metabolic defects, such as hyperglycemia and hyperinsulinemia, subsequent beta-cell failure and animal death. Diabetic ketoacidosis [343, 344]

11.7.2.2 Conditional IR Knockout Strains

11.7.2.2.1 IR^{lox}

Generation of knockout mice: To generate a conditional allele of the insulin receptor, a targeting vector introducing neomycin resistance and HSV-Tk genes flanked by loxP sites upstream of exon 4 and a single loxP site downstream of exon 4 was generated [345]. The construct was transfected into J1 ES-cells, and neomycin resistant clones were analyzed for homologous recombination by Southern blotting. Transient transfection of Cre was used to delete the floxed NeoR-HSV-Tk cassette, selection was performed using ganciclovir and the resulting clones were again screened by Southern blotting and subsequently injected into blastocysts. Cre-mediated deletion of exon causes a frame shift mutation resulting in a stop of translation at amino acid 308 of the mature IR protein.

Phenotype:

Strain	Organ	Phenotype/features
Mck Cre	Skeletal muscle, heart muscle	Dyslipidemia. Elevated fat mass, serum triglycerides, and free fatty acids [345].
RIP-Cre	Pancreatic beta islet cells	Selective loss of insulin secretion in response to glucose and a progressive impairment of glucose tolerance [346].
Nestin-Cre	Brain	Female mice: increased food intake. Male and female mice: diet-sensitive obesity with increases in body fat and plasma leptin levels, mild insulin resistance, elevated plasma insulin levels, and hypertriglyceridemia. In addition impaired spermatogenesis and ovarian follicle maturation [347].
UCP1-Cre	Brown adipose tissue	Age-dependent loss of interscapular brown fat. Insulin-secretion defect resulting in progressive glucose intolerance [348].
aP2-Cre	Adipose tissue	Low fat mass, loss of the relationship between plasma leptin and body weight. Protected against age-related and hypothalamic lesion-induced obesity and obesity-related glucose intolerance. In addition polarization of adipocytes into populations of large and small cells [349].
TIE2-Cre	Endothelilal cells	Altered expression of vasoactive mediators (eNOS and endothelin-1), increase atherosclerosis (in Apolipoprotein E null mice) [350].
MYH6-Cre	Heart	Reduced animal size. Deregulated postnatal cardiac size, myosin isoform expression and the switching of cardiac substrate utilization from glucose to fatty acids [351].

(continued)

(continued)

Strain	Organ	Phenotype/features
Nephrin-Cre/ Podocin-Cre	Kidney	Development of albuminuria, histological features of diabetic nephropathy [352].
LysM-Cre	Myeloid cells	Protection from obesity-associated insulin resistance upon high fat feeding and protection against artherosclerosis in apolipoproteinE-deficient mice. Decreased basal hepatic glucose production and increased insulin-stimulated glucose disposal in skeletal muscle. Deregulated cytokine production [353, 354].
K14-Cre	Epidermis, basal keratinocytes	Progressive decrease of epidermal thickness, Rac-dependent defects in the regulation of clonogenicity of epidermal progenitor cells [355].
GDF9-Cre	Oocytes	No defects [356].
TPO-Cre	Thyroid gland thyrocytes	No defects [357].
GnRH-Cre	Hypothalamic gonadotropin-releasing hormone producing neurons	Normal puberty and fertility [358].

11.7.2.3 Transgenic Mouse Lines of IR

Various transgenic mice overexpressing different forms of the insulin receptor gene have been generated over the years. Table 11.16 summarizes the generation of these animals (promoters, specificity and cDNAs used) and the most important features of these transgenic mice.

11.7.3 Igf1R

11.7.3.1 Constitutive Igfr1 Knockout Strain

11.7.3.1.1 Igf1r$^{-/-}$ Mice

Generation of knockout mice: The constitutive loss-of-function allele of *Igf1r* was generated by deleting a part of exon 3 which encodes the ligand binding domain of the receptor [365]. Positive and negative selection of targeted clones was performed using neomycin and ganciclovir and proper recombination was tested via PCR and Southern blotting. Positively targeted CCE ES-cells clones were used for the generation of Igf1r null mice.

Phenotype: Death at birth due to respiratory failure. Severe growth deficiency (45 % normal size) with organ hypoplasia and developmental delays in ossification. Defective development of the central nervous system and epidermis [365].

Table 11.16 Summary of the features of insulin receptor transgenic mice

Strain	Organ	Phenotype/features
TMPE/HIR	Rat MLC1/3 enhancer/muscle/ human cDNA	Increased glucose sensitivity [359].
MLC-IR	Rat myosin light chain 1/3 promoter and enhancer coupled to a mutant human IR cDNA results in a dominant negative transgene.	Reduced receptor activity, reduced glycogen content in the muscle. Higher insulin and glucose levels. Reduced insulin sensitivity. Impaired Insulin Signaling. Older mice show decreased insulin induced glucose uptake, increased adiposity, impaired glucose tolerance and dyslipidemia [360, 361].
Ttr-Insr	Transthyretin-promoter used to express human insulin receptor cDNA in liver, brain, and pancreatic beta-cells.	Rescue of insulin receptor knockout mice from neonatal death. Prevention of diabetes, normalization of adipose tissue content, lifespan, and reproductive defects [362].
IR hIR^{K1030M}	Under the IR promoter cDNA of the human IR with a mutated tyrosine kinase domain is expressed.	Hyperinsulinemia, impaired glucose tolerance [363].
PAH-IRΔ	Under the control of the human phenylalanine hydroxylase promoter a cDNA for a constitutively active human IR is expressed.	Partial rescue of Insulin receptor deficient mice [364].

11.7.3.2 Conditional Igfr1 Knockout Strains

11.7.3.2.1 $Igfr1^{fl/fl}$ [58], $Igfr1^{fl/fl}$ [366]

Generation of knockout strains: Two independent conditional *Igf1r* alleles have been generated. Both alleles were generated by floxing exon 3 of the *Igf1r* gene. The first allele was generated by inserting a floxed NeoR cassette upstream of exon 3 and a single loxP site downstream of exon 3 [58]. Transient expression of Cre was used to delete the floxed NeoR cassette so that only two loxP sites flanking exon 3 remained. The second allele was also constructed by inserting loxP sites flanking exon 3 of the gene [366]. An FRT-flanked neomycin resistance gene was used as the selection marker. Cre-mediated recombination results in a frame shift after 213 codons which is followed by a stop codon in exon 4. Gene targeting was performed in Bruce4 ES-cells and resistant clones were analyzed by Southern blot and PCR analysis. A summary of the features of mice lacking Igf1R in specific cell types and organs can be found in Table 11.17.

Phenotype:

Table 11.17 Summary of the features of *Igfr1*$^{fl/fl}$ mice

Strain	Organ	Phenotype/features
Heat Shock Inducible Cre	General	Test system for inducible gene deletion [58].
EIIa Cre	Ubiquitous	Test system for the generation of germline mosaicism [367].
RIP-Cre	Pancreatic beta cells	Age-dependent impairment of glucose tolerance, decreased insulin release [368].
Tyr-Cre	Melanocytes	No phenotype [369].
Alfp-Cre	Hepatocytes	Decreased hepatocyte proliferation [370].
ARR2PBi-Cre	Prostate epithelium	Cell autonomous proliferation and hyperplasia, induction of p53-regulated apoptosis and cellular senescence [371].
Nestin-Cre	Brain	Microcephaly, growth retardation, infertility, increased lifespan, and abnormal behavior [372].
Wap-Cre	Mammary epithelium	Significantly delayed tumorigenesis in Kras dependent mammary carcinomas [373].
GDF9-Cre	Oocytes	No defects [356].
Villin-Cre	Intestinal epithelial cells	Impaired proliferation of crypt cells upon fasting. Impaired response to re-feeding [374].
Tie2-Cre	Endothelial cells	Hypocontractile aortas with increased basal NO generation [375].
aP2-Cre	Adipocytes	Increased adipose tissue mass, increased lipid accumulation in epigonadal fat pads. Increased somatic growth in the presence of elevated IGF-1 levels [366].
Synpasin-Cre	Neurons	Reduced cerebral amyloid accumulation and protection against premature death in a mouse model of Alzheimer's disease. Mono-allelic deletion in male mice decreases body fat accumulation and increases energy expenditure during aging [376].
MLC2v-Cre	Heart	Combined deletion with IR causes thinning of the ventricular wall [377].
Nkx2.5-Cre	Heart	Combined deletion with IR causes thinning of the ventricular wall, edema, and decreased proliferation in the compact zone [377].
rGHpCre	Somatotroph	Increased growth hormone expression and secretion as well as increased serum Igf1, however, the mice weigh less than controls after 3 months and have less fat mass and body percent fat. Igfr1/IR double deficiency causes metabolic defects which are dependent on caloric intake [378].
K14-Cre	Epidermis	Progressive decrease of epidermal thickness, Rac-dependent defects in the regulation of clonogenicity of epidermal progenitor cells [355].
GnRH-Cre	Gonadotropin-releasing hormone producing neurons	Delayed puberty but normal fertility [358].
α-MHC-CreER	Inducible in cardiomyocytes	Depressed diastolic cardiac function [379].

Table 11.18 Summary of the features of *Igf1r* transgenic mice

Strain	Organ	Phenotype/features
αMHC hIgf1r	Expression of the human cDNA encoding the IGF1R in heart muscle cells.	Cardiac hypertrophy resulting from an increase in myocyte size. No signs of histopathology [380].
MKR	Dominant negative cDNA of Igf1r cloned downstream of the muscle kreatinin promoter	Insulin resistance in fat and liver, β-cell dysfunction and type 2 diabetes, muscle hypoplasia, perturbed gene expression in myocytes. Impaired muscle regeneration [381, 382].
Nephrin IGF-1r^{486FLAGstop}	Nephrin promoter drives expression of a dominant negative cDNA in the kidney.	Small glomeruli with dilated glomerular capillaries and condensed podocyte nuclei [383].
Tg hIgf1; Tg-hIgf1R	The thyroglobulin promoter drives expression of the human cDNA in the thyroid gland.	No defects, however, increased weight of the thyroid gland in double transgenic (IGF1R/IGF-1) [384].
Mmtv-cd8-igf1r	Expression of a constitutively active fusion protein (fragment of human cDNA fused to a part of CD8α) in the mouse mammary epithelium	Aberrant development of the mammary glands. Salivary and mammary adenocarcinomas [385].
TRE-IGFIR MTb-rtTA	Human type I insulin-like growth factor receptor cDNA cloned downstream of a tetracycline responsive element. The rtTA is expressed under the control of the *mmtv-Ltr* resulting in expression in mammary epithelium.	Impaired ductal elongation. Hyperplasic lesions resulting in mammary tumors [386].

11.7.3.3 Transgenic Mouse Lines of Igfr1

Various transgenic mice overexpressing different forms of the *Igf1r* gene have been generated over the years. Table 11.18 summarizes the generation of these animals (promoters and cDNAs used) and the most important features of these transgenic mice.

11.7.4 Igf2r

11.7.4.1 Constitutive Igfr2 Knockout Strain

11.7.4.1.1 Igf2r$^{-/-}$ (Wang)

Generation of knockout strain: Two independent null alleles of the *Igf2r* have been generated. The first allele was generated by inserting a LacZ/neoR cassette into the first exon of the gene [387]. This renders the gene nonfunctional while at the same time generating a LacZ reporter. After positive selection of targeted D3 ES-cell clones Southern blotting was performed to identify clones with proper homologous recombination.

11.7.4.1.2 Igf2r$^{-/-}$ (Lau)

Generation of knockout strain: The loss-of-function allele was generated by deleting exons 13-18 of *Igf2r* by insertion of a neoR cassette [388]. Positive selection of targeted W9.5 ES-cell clones was performed using Neomycin and proper recombination was tested via Southern blotting.

11.7.4.1.3 Igf2r$^{-/-}$ (Ludwig)

Generation of knockout strain: Exon 1 and the adjacent 5′ region of *Igf2r* were replaced by a *neo* cassette (V1). Alternatively only part of exon 1 was deleted encoding for codons 28-38 (V2), targeting constructs were electroporated into CCE33 cells and positive clones were injected into C57BL/6 blastocysts [389].

11.7.4.1.4 Igf2$^{\Delta NT}$

Generation of knockout strain: A targeting construct was designed to delete region 2 located upstream of exon 3 of *Igf2r* by using a floxed *neo*/TK cassette but leaves potential splice sites intact [390]. The linearized targeting construct was electroporated into E14 ES-cells.

Phenotype: Knockout of *Igf2r* is lethal at birth. Mutants are larger and display organ and skeletal abnormalities and kinky tail. Lethality is rescued in *Igf2*$^{-/-}$ and *Igf1r*$^{-/-}$ backgrounds [387–389]. *Igf2r* is an imprinted gene and considerable effort has been spent on identifying genetic elements involved in its regulation. A major finding was the identification of an imprinting control region in the *Igf2r* gene which is crucially required for its proper expression [341]. When paternally inherited, deletion of this DNA element leads to 20 % reduction in the weight of mice [390]. Furthermore, a humanized allele of mouse *Igf2r* was generated. This humanized knockin allele caused over-growth when transmitted in a maternal manner, which is rescued by paternal transmission of a wild-type or the knockin allele or combined Igf2-deficiency [391].

11.7.4.2 Conditional Igfr2 Knockout Strain

11.7.4.2.1 Igf2Rr$^{fl/fl}$

Generation of knockout strain: A conditional allele for *Igf2r* was constructed by floxing exon 10 [392]. Deletion of exon 10 shifts the coding sequence out of frame, resulting in a stop codon and truncation of the receptor. A single loxP site was inserted 5′ of exon 10 and a flox/neoR-TK element was placed 3′ of exon 10. After successful targeting of Sv129 ES-cells positive selection and verification of homologous recombination, Cre was transiently expressed which led to the deletion of the

Table 11.19 Summary of the features of conditional *Igfr2* knockout mice

Strain	Organ	Phenotype/features
CMV-Cre	Whole animal	Embryonic lethal, extra digits, lungs underdeveloped [392].
Alb-Cre	Hepatocytes in liver	No defects [392].
Ckmm-Cre	Cre driven by the muscle creatine kinase induces recombination in heart and skeletal muscle	No defects [392].

Table 11.20 Summary of the features of *Igf1r* transgenic mice

H19eIgfMlu	YAC transgene containing the full *Igf2r* locus under the control of the H19 enhancers.	Delayed tumor formation in combination with Igf2 transgenic mice, improved survival of Igf2 transgenic mice [393, 394].
K10p-sΔIgf2r	Under control of the K10 promoter which drives transgene expression in differentiated layers of the epidermis and the alimentary canal a soluble form of Igf2r is expressed.	Reduced size of the alimentary canal. Rescues $Apc^{Min/+}$ intestinal adenoma progression induced by *Igf2* loss of imprinting [395, 396].

neoR-TK element. Subsequently, ES-cells were injected into blastocycst for the generation of chimeric mice. A summary of the features of mice lacking the insulin receptor in specific cell types and organs can be found in Table 11.19.

11.7.4.3 Transgenic Mouse Lines of Igfr2

Two transgenic mouse strains which overexpress *Igf2r* gene have been generated. Table 11.20 summarizes the generation of these animals (promoters and cDNAs used) and the most important features of these transgenic mice.

11.7.5 Insrr

11.7.5.1 Constitutive Insrr Knockout Strain

11.7.5.1.1 Insrr$^{-/-}$

Generation of knockout strain: To generate a functional null mutation of the *Insrr* gene, a neoR cassette was used to disrupt exon 3. After positive selection of Cj7 ES-cell clones, homologous recombination was tested via Southern blotting and properly recombined clones were used for blastocyst injections.

Phenotype: No developmental defects, no phenotype, no defects in glucose metabolism, no impact on IR deficiency; impaired response to artificially induced alkalosis [397, 398].

11.8 The MET Receptor Family

11.8.1 Overview About Family Members

The MET receptor tyrosine kinase (RTK) family comprises three members: MET [399–401], RON [402], and the avian Sea [403]. MET is the receptor for its physiological ligand hepatocyte growth factor (HGF) [404]/scatter factor (SF) [405, 406]. MET is expressed in epithelial cells [405, 407], endothelial cells [408, 409], myoblasts [410], spinal motor neurons [411], and hematopoietic cells [412, 413]. Met is involved in muscle and liver formation [414–416], cell proliferation [417–419], morphogenesis [420], and motility [405, 421].

RON is the receptor for macrophage-stimulating protein (MSP), also known as HGF-like protein [422]. RON is expressed in macrophages [423], epithelial cells [424], osteoclasts [425], and hematopoietic cells [426, 427]. Ron is involved in cell adhesion, motility, growth, and survival [428, 429].

Murine receptors: MET/c-met, STK/RON
Ligands: HGF/SF (=PRGF-1), MSP (HGFL)

Box: List of all mouse strains

c-met$^{-/-}$	Bladt et al. 1995 [430]
MTM1, MTM2, MTM3	Liang et al. 1996 [433]
met^{D} *and met*^{Grb2}	Maina et al. 1996 [416]
TRK-MET	Jeffers et al. 1998 [436]
TRE-MET LAP-tTA	Wang et al. 2001 [439]
c-met$^{fl/fl}$	Huh et al. 2004 [444]
met^{flox}	Borowiak et al. 2004 [448]
met mutant mice *D1226N; Y1228C; M1248T; M1248T/L1193V*	Graveel et al. 2004 [452]
MetCKO, RIP-Cre/Met$^{lox/lox}$	Roccisana et al. 2005 [445]
Met^{mut}	Graveel et al. 2009 [453]

11.8.2 Met

11.8.2.1 Constitutive Met Knockout Strain

11.8.2.1.1 c-met$^{-/-}$

Generation of knockout strain: The genomic *c-met* locus was targeted to delete 2 kb of genomic DNA containing the invariant lysine essential for kinase activity. Homologous recombination was performed in E14-1 ES-cells. ES-cells that carried a mutant *c-met* alleles were again electroporated with a similar targeting construct containing wt met and selected with high G418 to select for double targeted clones [430].

Phenotype: Mice heterozygous for the *c-met* mutation were healthy and fertile. Homozygous mutant offspring died during embryogenesis, being viable up to E12.5. At E14.5, 16 % of mutant embryos were viable, appearing normal with reduction in size and markedly reduced liver size and damaged liver parenchyma, as well as defects of placenta development. At E15.5, myogenic precursor cells did not colonize the limb bud, diaphragm, shoulders and the tongue of *c-met*$^{-/-}$ embryos, where as a consequence skeletal muscles did not form. Axial muscles and body wall developed normally in absence of c-Met signaling [430].

11.8.2.2 Transgenic Mouse Lines of Met

11.8.2.2.1 MTM1, MTM2, MTM3

Generation of transgenic strain: The cDNA clone of the constitutively active *tpr-met* oncogene [431] was inserted into an expression construct containing a mouse metallothionein 1 promoter [432] and the SV40 small T intron and polyadenylation signal. Fertilized mouse FVB/n X FVB/n eggs were injected with a linearized fragment of the construct. Three founder pups were positive for the transgene [433].

Phenotype: Mammary and other tumors were observed in all of the MTM transgenic mouse strains. None of the nontransgenic littermates developed tumors up to 1.5 years of age [433].

11.8.2.2.2 TRK-MET

Generation of transgenic strain: The murine *met* cDNA was subcloned into an expression vector containing the murine metallothionein-1 promoter flanked by 59 and 39 control regions [434]. Site-directed mutagenesis was performed. After digestion of the constructs for the expression of activated *met*, the fragment was injected into fertilized eggs derived from C57BL/6NCR x C3H/HeNCR F2 mice [435]. Founder animals harboring each mutant *met* constructs were identified [436].

Phenotype: Founders exhibited severe breeding difficulties or did not transmit the transgene. At 10 months of age, one female *met* mutant expressing mutant M1268T and one expressing mutant Y1248H developed overt tumors diagnosed as type B mammary metastatic adenocarcinomas. Remaining founders at approximately 1 year of age were examined and did not exhibit any significant pathology, while one female from construct M1268T exhibited mammary hyperplasia [436].

11.8.2.2.3 TRE-MET LAP-tTA

Generation of transgenic mice: Transgenic mice expressing human MET in hepatocytes under the control of a tetracycline responsive element were created. The DNA fragments containing TRE, hCMV minimal promoter with heptamerized upstream tet-operators, *hMET* coding sequence, rabbit betaglobin intron and Poly (A) sequence were used for microinjection. *TRE-Met* mice were derived from the FVB/N strain [437] and were crossed with *LAP-tTA* mice [438], where expression of the tetracycline transactivator protein is under control of the promoter for the liver-activating protein gene. Double transgenic mice express hMET which is suppressed upon administration of doxycycline [439].

Phenotype: The double transgenic mice overexpress hMET in hepatocytes of the normal liver parenchyma as well as hyperplastic nodules. In two of the transgenic mouse lines, the mice obtained showed enlarged and fatty livers and died within 2 months. Feeding parents and newborn pups with doxycycline could prevent death. Withdrawal of doxycycline at 4 weeks of age led to death at 10 months of age. 85 % of deaths were accompanied by HCC. Inactivation of hMET expression in mice carrying established tumors lead to tumor regression by apoptosis [439].

11.8.2.2.4 GFP–Met Transgenic Mice

Generation of transgenic mice: A GFP–Met construct was generated by fusing EGFP to the C-terminus of murine *met* cDNA in the pEGFP-N1 vector. Transgenic mice were generated expressing GFP–Met from the CMV promoter using the pGFP–Met construct. Five founder lines were obtained [440].

Phenotype: Ubiquitous GFP–Met expression in specific epithelial and endothelial cells and enhanced GFP–Met fluorescence in sebaceous glands could be observed. Three founder lines displayed high fluorescence, two displayed moderate fluorescence. High levels of GFP–Met expression could be observed in spleen, kidneys and skin, moderate levels in liver, low levels in lung, and brain. GFP–Met transgenic males, but not females of all founder lines developed spontaneous tumors in sebaceous glands such as adenomas, adenocarcinomas, and angiosarcomas with a correlation between high GFP-Met levels and early aggressive tumor onset [440].

11.8.2.2.5 MMTV/Met mice (Met^{mt})

Generation of transgenic mice: Stephanie Petkiewicz generated *MMTV/Met* mice in 2007 (PhD Thesis at McGill University Montreal "The Met receptor tyrosine kinase in mammary gland tumorigenesis and development"), a transgenic mouse model that expresses Met receptor driven by mouse mammary tumor virus (MMTV) promoter in mammary epithelium. The *met* cDNA containing fragment, MMTV promoter and SV40 polyA were injected into FVB/N zygotes and implanted into FVB/N hosts [441].

Phenotype: Oncogenic variants of met (M1248T, Y1003F/M1248T, *met*^{MT}) develop tumors with moderate penetrance and long latency. 50 % of mice developed carcinomas with solid nodular histopathology and 50 % tumors with papillary, scirrhous, adenosquamous, or spindle-cell phenotypes. A minority of animals produced multiple tumors with distinct pathology and a gene expression profile similar to breast cancer [441].

11.8.2.3 Conditional Met Knockout Strains

11.8.2.3.1 c-met$^{fl/fl}$

Generation of knockout strain: Conditional gene targeting modified exon 16 of the mouse c-met gene in HM-1 embryonic stem cells. An upstream loxP site in the intron preceding exon 16 and a floxed neomycin resistance cassette in the intron immediately downstream of exon 16 was introduced in the targeting construct. The neo cassette was deleted from the target allele by crossing with *EIIa-Cre* mice [442] to generate *c-met*$^{+/fl}$ or *c-met*$^{+/\Delta 16}$ mice.

Phenotype: Germ-line depleted *c-met*$^{\Delta 16/\Delta 16}$ mice were embryonically lethal with abnormal liver development.

Strain	Organ	Phenotype
c-met$^{fl/fl}$; *AlbCre*$^{+/-}$	Albumin-Cre (*AlbCre*) [443] express Cre exclusively in postnatal liver.	*C-met*$^{fl/fl}$;*AlbCre* mice were fertile, showed normal growth and had no apparent histological or physiological abnormalities for more than 1 year [444]. Liver regeneration was impaired after CCl4 induced necrosis and hepatocytes were hypersensitive to Fas-induced necrosis.
c-met$^{fl/fl}$ *RIP-Cre*	Cre is expressed in β-islet cells of the pancreas	Mice showed normal body weight, blood glucose and plasma insulin. After in vivo glucose challenge they displayed reduced glucose tolerance and plasma insulin, which was not caused by insulin resistance. *C-met*$^{fl/fl}$ *RIP-Cre* mice show an increased number of small islets, mainly single and doublet β-cells [445]
c-met$^{fl/fl}$; *HoxB7-Cre*	HoxB7-Cre mice express the Cre recombinase in the ureteric bud (UB) [446]	Mice have a reduction in the final nephron number. Partially rescued by Egf treatment, whereas hypomorphic Egfr$^{wa2/wa2}$ results in smaller kidneys and decreased UB branching [447].

11.8.2.3.2 Met^{flox} MxCre; Met^{flox} K14Cre

Generation of knockout strain: LoxP sites were inserted into the *met* locus upstream of exon 15 encoding the ATP-binding site of Met and downstream of exon 15 a floxed neomycin resistance cassette was inserted. By homologous recombination in E14.1

ES-cells a floxed *met* allele (*met*^{flox}) was generated [430] and the neomycin cassette removed by transient expression of cre [72]. *Met*^{flox} cells were injected into blastocysts of C57BL/6 mice resulting in chimeras used to establish the *met*^{flox} strain [448].

Phenotype: *Met*$^{flox/flox}$ mice were viable, fertile, and appeared normal. To obtain *met*$^{\Delta}$ mice, *met*^{flox} and Cre-deleter mice [146] were crossed. *met*$^{\Delta/\Delta}$ offspring die during embryogenesis being phenotypically indistinguishable from *met* $^{-/-}$ or HGF/SF$^{-/-}$ mice [415, 430, 449]. *Met*^{flox} mice were crossed to MxCre mice, which express Cre under control of the IFN-α-inducible Mx promoter [72]. Cre expression is induced by injection of pIpC resulting in complete recombination in the liver. *Met*$^{flox/flox}$ *MxCre* mice were viable, fertile and did not show overt anomalies with normal liver size and histology after recombination. Long-term loss of Met resulted in microvesicular steatosis. Regeneration was impaired in *Met*$^{flox/flox}$ *MxCre* mice after partial hepatectomy [448].

K14-cre mice expressing Cre recombinase in the epidermis starting on E15 [450] were crossed with *met*^{flox} mice to generate *c-met*$^{flox/null}$ *K14-cre* mice. No phenotypically abnormalities were detected in the epidermis of conditional c-Met mutant mice compared to control mice concerning epidermal thickness and hair cycle progression. Conditional c-Met mutant mice showed impaired wound healing. 2-year-old conditional c-Met mutant mice did not show unusual hair loss or other changes in skin appearance [451]

11.8.2.4 Met Knock in Lines

11.8.2.4.1 Met Mutant Mice (D1226N; Y1228C; M1248T; M1248T/L1193V)

Generation of mice with targeted met mutations: Mice with activating missense mutations of met were generated by using a murine *met* construct containing the 3′ portion of exons 14 through 20 and a simian virus 40 poly(A) tail. Mutations M1248T, Y1228C, D1226N, and M1248T/L1193V were introduced into individual constructs. Mice were bred on a C57BL/6J-129/SV background. Germ-line transmission was confirmed. A *wt met* line was generated to control for locus modifications [452].

Phenotype: Heterozygous mice were intercrossed to generate homozygous mutant mice, which could be only obtained for wt and the D1226N mutation. Viable homozygous embryos were not observed in Y1228C and M1248T mutant mice or M1248T/L1193V mutant mice after embryonic day 6.5 or 13.5, respectively. D1226N, Y1228C, and M1248T/L1193V mutant mice showed sarcomas and lymphomas with a high frequency, the M1248T mice developed carcinomas and lymphomas [452]. The majority of the tumors contained trisomy of chromosome 6 and duplications of *met*.

When M1248T/L1193V mutant mice were bred congenically onto the FVB/N background and maintained heterozygous for the met$^{M1248T/L1193V}$ allele (Met^{mut}) Met^{mut} females developed aggressive mammary tumors with high penetrance [453].

Tumors showed absence of progesterone receptor and ErbB2 expression that correlated to the intensity of MET expression.

11.8.2.4.2 Met^{D} and Met^{Grb2}

Generation of knockin point mutation in the met locus: To construct the targeting vector, a fragment of *hMET* cDNA was inserted into the third exon of a mouse genomic clone by homologous recombination. The resulting hybrid receptor consists of a murine extracellular domain fused to a human transmembrane and cytoplasmic domain. Three different constructs were made: A met^{wt} served as control; a double Y→F mutation of Y1349 and Y1356 was inserted into the met locus, to obtain a severe loss-of-function phenotype (met^{D}) and an N→H mutation was introduced, to obtain a partial loss-of-function mutant (met^{Grb2}), disrupting binding of Grb2. Recombinant embryonic stem cells were injected into C57BL/6J blastocysts to produce chimeras [416].

Phenotype: Heterozygous mice and homozygous met^{wt} mice did not show any obvious phenotype and were fertile. $Met^{D/D}$ mice were lethal after E13.5 and mimicked the phenotype of the met null mutants [415, 430, 449] with paler and smaller placentas, reduction in liver size, lack of muscles in limb, tip of the tongue and diaphragm. Homozygous $met^{Grb2/Grb2}$ mice died several hours after birth, showing normal placenta and liver development, but a generalized deficit of secondary fibers and reduction in limb muscle. $Met^{Grb2/D}$ mice were generated to obtain an intermediate loss-of-function mutant, being lethal within a few minutes after birth due to respiratory failure [416].

11.8.3 Ron Kinase

Overview about family members

Murine receptors: STK = murine homologue of the human RON receptor
Ligands: macrophage stimulating protein (MSP) [454]

Box: List of all mouse strains

$STK^{-/-}$	Correll et al. 1997 [455]
$Ron^{-/-}$	Muraoka et al. 1999 [456]
$Ron\ TK^{-/-}$, $Ron\ TK^{fl/fl}HNF3\alpha^{+/-}$	Waltz et al. 2001 [460]
SPCp-RON transgenic mice	Chen et al. 2002 [465]
MMTV-pMT/Ron $TK^{-/-}$ *mice*	Peace et al. 2005 [461]
$Tg.AC^{+/-}/Ron\ TK^{-/-}$ *mice*	Chan et al. 2005 [462]
WT-Ron and MT-Ron transgenic lines	Zinser et al. 2006 [466]
$TK^{-/-}$*TRAMP*	Thobe et al. 2005 [464]
Pdx-1/RON mice	Babicky et al. 2011 [467]
Pb-Ron and ARR2Pb-Ron transgenic mice	Gray et al. 2012 [468]

11.8.3.1 Constitutive Ron Knockout Strain

11.8.3.1.1 STK$^{-/-}$

Generation of knockout mice: An 850-bp region within the first translated exon of the *stk* gene (including the translational start site) was replaced with a *lacZ* reporter gene under the transcriptional control of the Ron promoter in the constructed STK targeting construct. Translation was from the *lacZ* ATG. Homologous recombination led to abrogation of translation of STK. Targeting vector was introduced into R1 embryonic cells. The *stk*$^{+/-}$ embryonic stem cells were aggregated with blastocysts from CD-1 mice, the chimeras, which resulted there from, were bred with CD-1 female mice and offspring was interbred to generate *stk*$^{-/-}$ mice. The experimental group was *stk*$^{-/-}$ mice, which appeared overtly normal, were fertile and indistinguishable from their littermates, which were the control group. Macrophages from *stk*$^{-/-}$ mice did not display STK on their cell surface and failed to respond to MSP [455].

Phenotype: Elevated levels of nitric oxide were produced in a dose dependent response to IFNγ by peritoneal macrophages from *stk*$^{-/-}$ mice, but cytokine production due to inflammation was unaltered in macrophages of those mice. *stk*$^{-/-}$ mice displayed increased inflammation and susceptibility to endotoxic shock in an IFNγ-mediated hypersensitivity reaction. IFNγ and other cytokine levels were not significantly altered after an inflammatory response. LPS injected *stk*$^{-/-}$ mice showed elevated nitric oxide serum levels compared to their littermate controls [455].

11.8.3.1.2 Ron$^{-/-}$

Generation of knockout mice: The targeting vector was constructed by replacing exons 1-15 of the mouse Ron gene by a hypoxanthine phosphoribosyltransferase (HPRT) expression cassette. An *Xba*I-*Bam*HI fragment from the 5′ end of the Ron gene, a *Sma*I fragment from the 3′ end of the Ron gene, the HPRT expression cassette and a cassette coding for the herpes simplex virus thymidine kinase gene were cloned into the Bluescript plasmid. The linearized targeting vector was introduced into mouse embryonic stem cells, which were injected into C57BL/6 blastocysts and transferred into females to generate chimeric founders. Background for F3 offspring was 129/Black Swiss mixed background [456].

Phenotype: Hemizygous Ron$^{+/-}$ mice are viable and grow to adulthood. However, they were highly susceptible to endotoxic shock and displayed an impaired ability to regulate nitric oxide and showed an enhanced response to inflammation. Homozygous Ron$^{-/-}$ embryos are viable through the blastocysts stage of development, but fail to survive past the peri-implantation period, indicating a pivotal role for Ron in embryogenesis, due to its expression in the trophectoderm at E3.5 and in extra-embryonic tissue through E7.5 [456].

11.8.3.1.3 Ron TK$^{-/-}$ Mice

Generation of mice: A targeting vector was constructed to delete the tyrosine kinase [441] domain of Ron by incorporating loxP recombinase recognition sites between exon 13 and 18. Embryonic stem cells were transfected and homologous recombination was confirmed. Clones were injected into mouse blastocysts. Germ-line transmission was obtained. Ron TK$^{-/-}$ mice contain a homozygous deletion of Ron tyrosine kinase. Ron TK$^{+/+}$ mice are referred to as wild-type mice or mice with incorporated loxP sites. Ron TK$^{fl/fl}$ mice were crossed with mice expressing Cre recombinase driven by endogenous sequences of the HNF3α locus to conditionally delete the floxed allele. The targeting vector was similar to a previously used vector [457–460].

Phenotype: Mice with a germ-line ablation of the TK domain to impair cytoplasmic signaling were generated. Ron TK$^{-/-}$ mice are viable and overtly normal. Adult mice are phenotypically normal with inflammatory defects appearing upon challenge. Alterations such as reduced ovarian size, mild epithelial hyperplasia, and reduced number of corpora lutea were observed. Ron TK$^{fl/fl}$HNF3α$^{+/-}$ mice are viable [460].

11.8.3.1.4 MMTV-pMT/Ron TK$^{-/-}$ Mice

Generation of mice: Mice with a targeted deletion of the Ron tyrosine kinase domain (TK$^{-/-}$) (Waltz et. al. [460]) were crossed with hemizygous male mice expressing pMT under control of the mouse mammary tumor virus promoter. Mice were maintained in an FVB/NJ background. The pMT$^{+/-}$ TK$^{-/-}$ mice displayed the experimental group, pMT$^{+/-}$ TK$^{+/+}$ the control group. Female mice were used for tumor examination [461].

Phenotype: Mammary tumors and lung metastases could be observed in the experimental (pMT$^{+/-}$ TK$^{-/-}$) and the control group (pMT$^{+/-}$ TK$^{+/+}$), whereas mammary tumor initiation and growth are significantly decreased in pMT$^{+/-}$ TK$^{-/-}$ mice. Microvessel density and cellular proliferation were significantly decreased. TK$^{+/+}$ and TK$^{-/-}$ pMT-expressing tumors have both increased Ron expression compared to normal mammary glands. The experiments demonstrated that Ron impacted tumorigenesis through MAPK and Akt signaling [461].

11.8.3.1.5 Tg.AC$^{+/-}$/Ron TK$^{-/-}$ Mice

Generation of mice: Mice with a targeted deletion of the Ron tyrosine kinase domain (TK$^{-/-}$) of Ron [460] were crossed with male Ras transgenic Tg.AC$^{+/-}$mice in an FVB/N background. The Tg.AC$^{+/-}$TK$^{-/-}$ mice displayed the experimental group, Tg.AC$^{+/-}$TK$^{+/+}$ the control group. Also animals without the Ras transgene (TK$^{-/-}$Tg.AC$^{-/-}$ and TK$^{+/+}$Tg.AC$^{-/-}$) were generated [462].

Phenotype: Experimental and control mice were treated with 12-O-tetradecanoylphorbol-13-acetate, also known as TPA or PMA. Resulting papillomas were analyzed, which revealed a reduction in cell proliferation and volume in Tg. $AC^{+/-}TK^{-/-}$ mice compared to control mice. Expression of Ron underwent an upregulation in papilloma formation. Impaired Ron signaling significantly reduced the number of papillomas that undergo malignant conversion and the number of other malignant tumor types normally found in these mice [462].

11.8.3.1.6 $TK^{-/-}$TRAMP

Generation of mice: Mice with a targeted deletion of the Ron tyrosine kinase domain ($TK^{-/-}$) [460] were crossed with TRAMP mice (=transgenic adenocarcinomas of the mouse prostate), a genetically engineered mouse model, that is predisposed to develop prostate tumors [463, 464].

Phenotype: $TK^{-/-}$TRAMP mice show a significant decrease in prostate tumor mass, exhibiting decreased tumor vascularization relative to $TK^{+/+}$TRAMP mice containing functional Ron. Although Ron loss did not result in altered proliferation of tumor cells, a significant decrease in cell survival was observed [464].

11.8.3.2 Transgenic Mouse Models of Ron

11.8.3.2.1 SPCp-RON Transgenic Mice

Generation of transgenic mice: Into the vector 3.7SPC/SV40 hRON cDNA was inserted between an SPC promoter and an SV40 sT intro-polyA sequence. After purification, the transgenic fragment was injected into fertilized eggs from B6C3/F1 hybrid mice (C57BL/6XC3H), which were implanted into pseudopregnant foster mice. Founder mice were bred with B6C3/F1 hybrid mice; germ-line transmission was identified. Transgenic mice specifically express human wtRON in the lung in more than 90 % of type II cells [465].

Phenotype: Newborns did not show any lung tumors. At around 4–6 months, tumor nodules could be observed. At 8–14 months, numerous tumor masses were observed and some mice showed respiratory distress and weight loss. Lung tumors appearing as solid-alveolar adenomas showed uniform morphology and were mostly located in peripheral areas of the lung, whereas overall lung structure was still intact. Other organs did not show any tumor formation [465].

11.8.3.2.2 WT-Ron and MT-Ron Transgenic Lines

Generation of transgenic mice: A construct that contains a cassette expressing wt Ron has been generated. In the constitutively active Ron transgene, a point mutation was introduced (DM1231T (MT-RON)) causing constitutive receptor activation and leading to tumorigenicity in vivo. Each construct was placed under control of the MMTV promoter. Excision of the expression cassettes from the vector was followed by injection into fertilized eggs from FVB/N mice. Founders positive for the transgene were crossed with wt FVB/N mice to generate offspring, which were bred with wt littermates [466].

Phenotype: Transgene expression occurred only in mammary glands and to a limited extent in lacrimal glands. At 12 weeks of age, transgenic mice (WT-Ron and MT-Ron) developed hyperplastic mammary glands. Nontransgenic mice showed normal mammary gland development. At 4 months of age, mammary glands from transgenic showed a large number of hyperplastic alveolar nodules. Overexpression of Ron led to a high incidence of mammary tumor development in female mice with tumors overexpressing Ron having aggressive metastatic potential [466].

11.8.3.2.3 Pdx-1/RON Mice

Generation of transgenic mice: Transgenic mice overexpressing wt-RON in a pancreas-specific manner were generated. As in Kras-LSLGD12/Pdx-1-cre(KC) mice that develop pancreatic duct neoplasia, the Pdx-1-cre promoter was used also here. Pdx-1/RON mice were bred with LSL-KRASG12D mice; offspring were bred to Pdx-1-Cre mice. This was done to combine RON overexpression when oncogenic KRAS is present (RCK mice) [467].

Phenotype: Until 12 months of age, Pdx1/RON mice showed no pancreatic phenotype. At 18 months, one quarter developed primary pancreatic adenocarcinomas with lung metastasis. RCK mice in the presence of oncogenic KRAS led to visibly accelerated pancreatic intraepithelial neoplasia progression at 6 weeks compared to KC mice. At 6 months, invasive carcinoma was detectable compared to control mice (KC) and at 9 months, metastatic disease was visible in RCK mice [467].

11.8.3.2.4 Pb-Ron and ARR_2Pb-Ron Transgenic Mice

Generation of transgenic mice: Transgenic mice that selectively overexpress Ron in the prostate epithelium were generated. Two independent transgene constructs were generated, one being the Pb-Ron and the other one, the ARR_2Pb-Ron construct. Both constructs were injected into blastocysts of FVB/N mice. Founder mice were crossed with wild type (WT) FVB/N mice to generate offspring. Only male mice were used for experiments. The Pb-Ron construct resulted in four founder lines, the ARR_2Pb-Ron construct in two founder lines [468].

Phenotype: In mice positive for the Pb-Ron transgene, which overexpress Ron, microinvasive mouse prostate intraepithelial neoplasia could be observed, together with local invasion and association with increase in prostate cell proliferation and decrease in cell death. Mice positive for the ARR_2Pb-Ron transgene developed prostate pathology over a long time frame, but the development of adenocarcinomas was further observed. Ron overexpression led to the development of prostate intraepithelial neoplasia with local invasion and was associated with increases in prostate cell proliferation and decreases in cell death [468].

11.9 The MuSK Receptor

11.9.1 Overview About Family Members

The MuSK receptor tyrosine kinase (RTK) is required for the formation of the neuromuscular junction [469]. MuSK is activated by its ligand Agrin, a nerve-derived proteoglycan [470]. Upon activation by its ligand Agrin, MuSK induces clustering of acetylcholine receptors [469].

Murine receptor: MUSK
Ligand: Agrin

Box: list of all mouse strains

MuSK$^{-/-}$	De Chiara et al. 1996 [469]
MuSK$^{loxP/loxP}$	Sander et al. 2001 [473]
MCK-Musk and MCK-MMT transgenic mice	Herbst et al. 2002 [475]
MuSK$^{loxP/loxP}$ and MuSK/Cre mice	Hesser et al. 2005 [474]
HSA::MuSK (MuSK-H or MuSK-L mice)	Kim et al. 2008 [479]
musk$^{V789M/V789M}$ mice and musk$^{V789M/-}$ mice	Chevessier et al. 2008 [480]
MuSK-tg mice	Punga et al. 2011 [482]
MV-MuSK mice	Ammar et al. 2013 [483]

11.9.2 Constitutive Musk Knockout Strain

11.9.2.1 MuSK$^{-/-}$

Generation of knockout mice: The targeting vector was designed to delete most of the third kinase exon by ligating mouse genomic DNA fragments into compatible ends of XbaI site upstream of the PGK-neo cassette. Linearization of the vector was followed by introduction into embryonic stem cells derived from 129 mice strain. Male chimeras derived from all four of the targeted clones were bred with C57BL/6 females. Two clones transmitted the mutant allele to the offspring [469].

Phenotype: *Musk$^{+/-}$* mice are viable and fertile, showing no phenotypical abnormalities. *Musk$^{-/-}$* mice did not survive the perinatal period. *musk$^{-/-}$* mice are normal in their gross anatomy, being cyanotic and immobile at birth due to a lack of neuromuscular synapses, indicating that MUSK is involved in synapse formation [469].

11.9.3 Conditional Musk Knockout Strain

11.9.3.1 Musk$^{loxP/loxP}$

Generation of transgenic mice: For targeting vector construction, a genomic fragment that contained the *musk* kinase exon domain was cloned into pSP72. LoxP sites were introduced into the introns, which flank the kinase domain. After introduction of the targeting vector into embryonic stem cells by homologous recombination, the resulting cell clones were injected into blastocysts derived from C57BL/6J mice. Kinase domain of the *musk* gene could be deleted upon expression of transgenic Cre recombinase upon DNA injection into individual muscle fibers [471, 472], which allowed conditional inactivation of endogenous MuSK in single muscle fibers [473].

Phenotype: Deletion of the kinase domain in vivo showed that kinase mediated signals are sufficient to induce the formation of acetylcholine receptor clusters [473].

11.9.3.2 MuSK$^{loxP/loxP}$ and MuSK/Cre mice

Generation of transgenic mice: A neomycin resistance cassette and *loxP* sites flanking the exons encoding the MuSK kinase domain were introduced into musk by homologous recombination. *MuSK*$^{loxP/+}$ embryonic stem cells were transfected with the Cre-recombinase expression vector to produce *musk* $^{+/-}$ embryonic stem cells, which were used to generate heterozygous *musk* $^{+/-}$ mice. *Musk* $^{+/-}$ mice were bred with *musk* $^{loxP/loxP}$ mice resulting in *musk* $^{loxP/-}$ mice. These were bred to *Mck-Cre* mice expressing Cre under the muscle creatine kinase promoter abolishing MUSK expression in skeletal muscles (*MuSK/Cre* mice) [345, 474].

Phenotype: Heterozygous mice did not show any overt phenotype. *musk* $^{loxP/loxP}$*:Cre* and *musk* $^{loxP/-}$*:Cre* mice (*MuSK/Cre* mice) were viable at birth due to the timepoint of Cre expression and showed no phenotypical abnormalities during the first 2 postnatal weeks. Thereafter *MuSK/Cre* mice developed severe muscle weakness, were impaired in their motility, did not grow any more, showed weight loss and developed a back deformation, leading to death before postnatal day 30 [474].

11.9.4 Transgenic Mouse Strains of Musk

11.9.4.1 MCK-Musk and MCK-MMT Transgenic Mice

Generation of transgenic mice: Transgenic mice for the expression of wt MUSK fused to GFP (*Mck-musk)* or a chimeric MUSK/TrkA (*Mck-Mmt*) receptor were generated. DNA was injected into male pronucleus of B6/D2 fertilized eggs, which were transferred into pseudo-pregnant foster mice. *MCK-musk* and *Mck-Mmt* transgenes are expressed specifically and strongly in skeletal muscle. Both transgenic strains were kept on a *musk*$^{-/-}$ background [475].

Phenotype: Several intracellular tyrosine domains were identified that are responsible for agrin stimulated phosphorylation of MUSK [476]. *Musk*$^{-/-}$*;Mck-Mmt* and *musk*$^{-/-}$*;MCK-musk* were smaller than wild-type littermates before weaning due to poor competition with wild-type littermates for nourishment, but gain the same size as their littermates at 1 month of age. *musk*$^{-/-}$*;Mck-Mmt* and *Musk*$^{-/-}$*;MCK-Musk* adult mice show normal motility, no signs of a shortened longevity as they have been maintained for 1 year [475].

11.9.4.2 HSA::MuSK (MuSK-H or MuSK-L mice)

Generation of transgenic mice: Two *HSA::musk* transgenic lines were generated, a low-expression line (*musk-L*) and a high-expression line (*musk-H*) expressing 3-fold more and 20-fold more Musk than wild-type mice at E15.5, respectively. Human alpha skeletal actin (HAS) induces expression of wt MuSK in skeletal muscle. The fragment was injected into zygotic male pronuclei. Three lines were generated. *HB9*^{cre}-*Isl2*^{DTA} mice lacking differentiated motor neurons have been described previously [477, 478] and *HB9*^{cre}*Isl2*^{DTA} mice carrying either the *musk-L* or *musk-H* transgene were generated [479].

Phenotype: At P0, motor axons in *musk-H* mice extended over 90 % of the muscle and did not continue to grow postnatally, compared to wt mice, where motor axons were restricted to central muscle regions and grow substantially in length postnatally. The Musk-L transgene was sufficient to initiate, but not to maintain muscle patterning [479].

11.9.4.3 MuSk$^{V789M/V789M}$ mice and MuSk$^{V789M/-}$ mice

Generation of mice: *Musk* $^{V789M/V789M}$ mice carry a missense mutation V789M in the kinase domain of musk. The targeting vector has been used before for conditional inactivation of *MuSK* during postnatal development [474]. In exon 15, encoding for the kinase domain, a nucleotide G was replaced by A. The construct was used for embryonic stem cell transfection resulting in two independent clones being used for transfer into C57BL/6 blastocysts [480].

Phenotype: Mutated *Musk* $^{V789M/+}$ and *Musk* $^{V789M/V789M}$ mice were indistinguishable from their wt littermates at birth appearing healthy during sexual maturation. *Musk* $^{V789M/V789M}$ mice were fertile and generated homozygous offspring. *Musk* $^{V789M/V789M}$ mice did not show any obvious phenotype whereas *musk* $^{V789M/-}$ mice suffered from severe muscle weakness, shrinkage of pelvic, scapular regions and hunchback; diaphragm exhibited pronounced changes in endplate architecture, distribution and innervation pattern and developed less force upon direct or nerve-induced stimulation [480].

11.9.4.4 MuSK-tg mice

Generation of transgenic mice: Mice overexpressing MuSK (*musk -tg*) in skeletal muscle were generated by using the pBS KS+ vector with a fragment of the muscle creatine kinase (mck) promoter and a polyadenylation sequence [481]. Full-length cDNA, which encodes a mouse MuSK-TAP tag fusion protein, was inserted in the polylinker. The linearized construct was injected into mouse oocytes. The transcript encoding the TAP-tagged MuSK could be detected starting at E13 and was strongly expressed in skeletal muscle [482].

Phenotype: Different muscles of 8-week-old *musk -tg* mice showed that ectopic clusters were present in the soleus and sternomastoid muscles. Compared to c-*magB8*-tg mice the formation of ectopic clusters in *musk -tg* mice was clearly lower. A high number of ectopic acetylcholine receptor clusters in both the soleus and the EDL muscles of *musk -tg* mice could be detected 10 days after sciatic nerve transsection [482].

11.9.4.5 Mv-MuSK mouse

Generation of transgenic mice: Rat-musk cDNA was cloned into the pCDNA3 vector and wt or mutant *mv-musk* expression vectors were electrotransfected into mouse tibialis anterior muscle in vivo [483].

Phenotype: The terminal axon of a remodeled neuromuscular junction showed reduced nerve ramification. The breakdown of axonal sprouts was similar to the patient situation. The first stage of ectopic NMJ formation, consisting of small clusters of acetylcholine receptor clusters, could be observed. A 50 % decrease in nicotinic acetylcholine receptor per neuromuscular junction with Mv MuSK versus WT MuSK could be noted [483].

11.10 The PDGF Receptor Family

11.10.1 Overview About Family Members

Two types of PDGF receptors, namely PDGFRα and PDGFRβ exist, which undergo distinctive but overlapping signal transduction in various cell types. Both have similar structures, consisting of five extracellular immunoglobulin (Ig)-like domains, a transmembrane domain, a juxtamembrane domain, a kinase domain split by an ~100 aa insert, and a carboxy-terminal tail. The binding of PDGFR ligands (PDGF-A, PDGF-B, PDGF-C, and PDGF-D), leads to homo- or heterodimeric receptors, which in turn induces to transphosphorylation of the intracellular domains and receptor activation. PDGFR signaling promotes proliferation, survival and migration in diverse cell types and enhanced signaling of PDGF has been implicated in the pathogenesis of atherosclerosis, balloon injury induced restenosis, pulmonary fibrosis, angiogenesis, and tumorigenesis [484–486].

11.10.2 Constitutive Pdgfr A and B Knockout Strains

11.10.2.1 Pdgfrα$^{-/-}$

Generation of knockout strain: To target PDGFRα, two different mutations were introduced into the Pdgfrα gene. The first targeting vector (α6) generates a small (~0.5 kb) deletion corresponding to the first immunoglobulin (Ig) domain. A second targeting construct (α16) carries a larger deletion (~6.5 kb) corresponding to the signal peptide, as well as the first and second Ig domains [487].

Phenotype: Homozygous mice die during embryonic development (between E8 and E16) and exhibit incomplete cephalic closure. Additionally, increased apoptosis was observed on pathways followed by migrating neural crest cells as well as alterations like abnormally patterned somites, skeletal abnormalities, subepidermal blebbing, and hemorrhaging [487]. In later studies it was found that *Pdgfrα*$^{-/-}$ males have reduced or absent fetal Leydig cells [488] and that the lungs of the knockout mice showed an alveolar septation [489].

11.10.2.2 Patch (Ph) mutation

Generation of knockout strain: Homozygous *Ph* mutant mouse embryos lack the receptor gene (*Pdgfrα*) due to an extensive spontaneous deletion at that locus [490–492].

Phenotype: Heterozygotes have a large white patch largely on the ventral side. Presumptive homozygotes, produced in crosses between heterozygotes, display gross anatomical abnormalities and die midway through gestation.

11.10.2.3 PDGFRβ$^{-/-}$

Generation of knockout strain: To generate PDGFRβ-deficient mice, ES-cells were targeted by two different targeting vectors. One of the vectors includes a promoterless beta-geo (beta-gal/neo fusion) gene inserted in-frame between the exon coding for the signal peptide and an EcoRV site in an exon coding for the second immunoglobulin domain and the other vector includes a neomycin phosphotransferase (neo) expression cassette, which was used to replace the 1.8 kb of *Pdgfrβ* genomic sequences [493].

Phenotype: Mutant mice die just prior to birth, however, no overt phenotype could be detected in mutant embryos younger than E16. *Pdgfrβ* -deficient mice exhibit cardiovascular, renal, and hematological defects. In detail vascular developmental defects attributed to the phenotype because of failed recruitment of specialized vascular smooth muscle cells (VSMCs) to develop capillaries in the brain (pericytes) and in the kidney (mesangial cells) [493, 495]. Moreover *Pdgfrβ*$^{-/-}$ hearts showed an impairment to form dominant coronary vessels on the ventral heart surface, had a thinned myocardium, and completely lacked coronary VSMCs (cVSMCs) [494, 496].

11.10.3 Conditional Pdgfr Knockout Strains

11.10.3.1 Pdgfrα$^{fl/fl}$

Generation of knockout strain: To generate a conditional mutant a loxP site was introduced upstream of exon 1 and a PGK-neo-loxP cassette was introduced downstream of exon 4 via homologous recombination, In the presence of Cre recombinase, the second and third exon of the *Pdgfrα* gene as well as the neo cassette will be removed. The *Pdgfrα* floxed allele is hypomorphic [497].

Phenotype:

Strain	Organ	Phenotype/features
Pdgfrα$^{fl/fl}$ Z/EG GFP-Cre		Permanent GFP labeling of cells that lose *Pdgfrα* [498]
Pdgfrα$^{fl/fl}$ HoxA3-Cre	Dorsal mesocardium	Anomalous pulmonary venous connection [499]
Pdgfrα$^{fl/fl}$ Lefty2-Cre	Dorsal mesocardium	Anomalous pulmonary venous connection [499]
Pdgfrα$^{fl/fl}$ RIP-Cre	Pancreatic beta cells	Defects in beta-cell proliferation [500]
Pdgfr$^{fl/fl}$Col2a1-Cre	Chondrocytes, sclerotome	No obvious defects, survived postnatally and reached adulthood [501]
Pdgfrα$^{fl/fl}$ Wnt1- Cre	Dorsal neural tube, cardiac and cephalic neural crest cells	Neonatal lethality, due to aortic arch defects and a severely cleft palate [497]
Pdgfrα$^{fl/fl}$ SM22α-Cre	Smooth muscle cells	No obvious defects [502]
Pdgfrα$^{fl/fl}$ Sf1-Cre	Neurons in the ventromedial hypothalamic nucleus, pituitary, gonad, adrenal tissue, cerebral cortex	Defects in development of steroidogenic cells, testis and ovary, delay of fertility and hormone production [503]
Pdgfrα$^{fl/fl}$ Twist2-Cre	Sclerotome, dermatome, osteoblasts	Development of spina bifida [504]

11.10.3.2 Pdgfrβ $^{fl/fl}$

Three different strains of PDGFRβ conditional mutant have been generated.

1. *Pdgfrβ*^{fl007}: In this mutant strain exons 4-7 of *Pdgfrβ*, encoding the extracellular domain have been flanked by two loxP-sequences positioned in introns 3 and 7. After linearization, the vector was electroporated into mouse embryonic stem (ES) cells. The resulting ES were injected into blastocysts to obtain chimeric mice [505].
2. *Pdgfrβ*^{fl2000}: In this mutant strain exons 3-7 of *Pdgfrβ* have been flanked by two loxP-sequences. After linearization, the vector was electroporated into mouse embryonic stem cells. Positive ES-cells were injected into blastocysts to obtain chimeric mice. After Cre-mediated recombination, deletion of the loxP-flanking region results in frame shift mutation in the following 3′ region [506].
3. *Pdgfrβ*$^{tm11Sor/J}$: The conditional *Pdgfrβ* floxed allele harbor one loxP site 500 bp upstream of the exon that encodes the first immunoglobulin domain and a second 100 bp downstream from the exon encoding the second immunoglobulin domain. Following homologous recombination in AK7 (129S4) embryonic stem cells, germline chimeras were derived by blastocyst injections [503].

Strain	Organ	Phenotype/features
Pdgfrβ$^{fl/fl}$ *Sf1-Cre*	Neurons in the ventromedial hypothalamic nucleus, pituitary, gonad, adrenal tissue, cerebral cortex	Viable and fertile, no obvious defects [503]
Pdgfrβ$^{fl/fl}$ Nkx2.5-Cre	Tamoxifen inducible α-MHC-MerCreMer/ cardiomyocytes	No obvious defects Cardiac dysfunction and heart failure after load-induced stress [507]
Pdgfrβ$^{fl/fl}$ Arf-Cre	Male germ cell and eye development	Restored vision [508]
Pdgfrβ$^{fl/fl}$ Nestin cre	Neurons, astrocytes, preserved in vascular PC/vSMCs	Increased vascular leakage and infarction volume after photothrombolic middle cerebral artery occlusion (MACO) [509] Increased cerebral damage after cryogenic injury and NMDA induced excitotoxicity) [506] Defects in cognitive and socioemotional functions [510]
Pdgfrβ$^{fl/fl}$ *CAGGCre-ER*	Systemically expressed inducible Cre.	Delayed recovery of body weight and behavior, larger infarction volume, decreased PC/vSMC coverage [505] After subtotal nephrectomy: dilated capillaries, decreased cell number and reduced mesangial matrix area with less extended mesangial cell processes [511]
Pdgfrβ$^{fl/fl}$ *Gata5Cre*	Epicardial cells	Coronary vessel disruption in hearts, vessel remodeling is later recovered, reduced cVSMCs distal to the aorta [496]
Pdgfrβ$^{fl/fl}$ *SM22α-Cre*	Smooth muscle cells	Viable and fertile, no obvious defects [502]

Pdgfrα /Pdgfrβ DKO:

Strain	Organ	Phenotype/features
$Pdgfr\alpha^{fl/fl}$ /$Pdgfr\beta^{fl/fl}$ Meox2 -Cre	All embryonic tissues and the extraembryonic mesoderm	Few embryos recovered, a complete failure in yolk sac remodeling [502]
$Pdgfr\alpha^{fl/fl}$ /$Pdgfr\beta^{fl/fl}$ Myocardin-Cre	Cardiomyocytes/VSMC precursors	Survived until birth [502]
$Pdgfr\alpha^{fl/fl}$ /$Pdgfr\beta^{fl/fl}$ SM22α-Cre	Smooth muscle cells	Pups were recovered at birth, disruption in yolk sac blood vessel development [502]
$Pdgfr\alpha^{fl/fl}$ /$Pdgfr\beta^{fl/fl}$ Tie2-Cre	Endothelial cells	No obvious defects in yolk sac development [502]
$Pdgfr\alpha^{fl/fl}$ /$Pdgfr\beta^{fl/fl}$ Sf1-Cre	Neurons in the ventromedial hypothalamic nucleus, pituitary, gonad, adrenal tissue, cerebral cortex	Can be prenatally lethal, development defects of steroidogenic cells in the ovary, testis and adrenal gland [503]
$Pdgfr\alpha^{fl/fl}$ /$Pdgfr\beta^{fl/fl}$ Meox2 -Cre	All embryonic tissues and the extraembryonic mesoderm	Few embryos recovered, a complete failure in yolk sac remodeling [502]
$Pdgfr\alpha^{fl/fl}$ /$Pdgfr\beta^{fl/fl}$ (as well as $Pdgfr\alpha^{fl/fl}$ or $Pdgfr\beta^{fl/fl}$ MesP1-Cre	Early mesoderm	No ventricular or atrial abnormalities; deficiency in the compact zone of the ventricular walls, changed morphology of vascular structures in the ventricular wall [512]
$Pdgfr\alpha^{fl/fl}$ /$Pdgfr\beta^{fl/fl}$ Nebulette-Cre	Myocardium	No morphological abnormalities [512]
$Pdgfr\alpha^{fl/fl}$ /$Pdgfr\beta^{fl/fl}$ Gata5-Cre	Epicardial cells	Temporally epicardial detachment and hemorrhaging, but resolved by birth, epicardial formation not affected [513]
$Pdgfr\alpha^{fl/fl}$ /$Pdgfr\beta^{fl/fl}$ *WT1-CreER*	Epicardial cells just prior to EMT (E12.5)	Defects in epicardial cell migration, failure of epicardial EMT and EPDCs formation [513]
$Pdgfr\alpha^{fl/fl}$ /$Pdgfr\beta^{fl/fl}$ Wnt1-Cre	Neural crest cell population	Defects in neural crest cell development [514]

11.10.4 *Pdgfr* Knockin Mouse Lines

Strain	Generation	Features/phenotype
$Pdgfr\alpha^{Src}$	*Pdgfrα* α cDNA inserts encoded specific tyrosine to phenylalanine mutations introduced by site-directed mutagenesis	Mutations eliminate the ability to bind Src family kinases and results in defects in oligodendrocyte development [515]
$Pdgfr\alpha^{PI3K}$	PDGFR-α cDNA inserts encoded specific tyrosine to phenylalanine mutations introduced by site-directed mutagenesis	Mutated PI3K binding site. neonatal lethal, few survivers until P16, abnormal skeletal development, failed alveolar septation, lungs were emphysematous, spina bifida [515]
$Pdgfr\alpha^{F7}$	Tyrosine to phenylalanine, mutations eliminate the ability to bind PI3 family kinases, SHP-2 and PLCγ	Similar phenotype as $Pdgfr\alpha^{PI3K}$, neonatal lethal, abnormal skeletal development [515]
Pdgfr(αβ)	*Pdgfrα* fused to the intracellular domain of the *Pdgfrβ* was knocked into *Pdgfrα* locus and reverse. Junction within the transmembrane domain.	Normal development, no obvious defects. The targeted *Pdgfrβ showed varying degrees of vascular disease* [516].
$Pdgfr\alpha^{K}$	The endogenous *Pdgfrα* sequence was replaced with a lox-stop-lox cassette before the mutation (D842V) generating a constitutively active kinase domain	Heterozygous mice are viable, no reported abnormalities [517]
$Pdgfr\alpha^{J}$	The endogenous *Pdgfrα* sequence was replaced with a lox-stop-lox cassette before the mutation (V561D) generating a constitutively active kinase domain	Embryonic lethal [517]
$Pdgfr\alpha^{EGFP}$	Express the H2B-eGFP fusion gene from the endogenous *Pdgfrα* locus	Fluorescence pattern mimics the expression pattern of the endogenous gene; Homozygotes embryonic lethal (E12.5–15.5) because of abnormal placenta development and vasculature [518]
$R26^{(hPDGFR\alpha^{PM})}$ *Wnt1Cre* $R26^{(hPDGFR\alpha^{PM})}$ *Sox2Cre*	Human *PDGFRα* cDNA containing D842V mutation was inserted into a targeting vector for the ubiquitously expressed ROSA26 (R26) gene locus in such a manner that transcription would be prevented by a *loxP*-flanked transcriptional stop sequence in the resulting modified allele.	*Wnt1Cre:* mice are viable; develop craniosynostosis; increased proliferation of osteoprogenitors and accelerated ossification of osteoblasts [519] *Sox2Cre:* embryonic lethal, growth retardation, massive hemorrhage in the trunk and the pericardial cavity [520]

Pdgfrβ ^{D849N}	Exchange of aspartic acid at amino acid position 849 into asparagine by homologous recombination into *Pdgfrβ* codon in the activation loop.	No obvious phenotype and no signs of tumors or fibrotic developments [521] impairments in injury models (cutaneous wound healing or liver injury) [522]
Pdgfrβ ^{D849V}	Exchange of aspartic acid at amino acid position 849 into valine	Dominantly lethal also in chimeras, due to hyperactivity of the mutant receptor, defects in chimeric fetal blood vessels and placentas [523]
Pdgfrβ ^{F1}	Tyrosine-to-phenylalanine inactivating point mutations at amino acid 770 introduces a mutation in the RasGAP-binding site	Mutant mice were viable [524] reduction of vascular smooth muscle cells/pericytes,
Pdgfrβ ^{F2}	Tyrosine-to-phenylalanine inactivating mutations at amino acid 739 and 750 introduces a mutation of the PI3K-binding site	Normal development, defects in interstitial fluid homeostasis [525]; reduction of vascular smooth muscle cells/pericytes [39]
Pdgfrβ ^{F3}	Tyrosine-to-phenylalanine inactivating mutations at amino acid 739 and 750 (PI3K binding) and at residue 1020 from tyrosine to isoleucine (PLCγ binding)	Normal development [526] reduction of vascular smooth muscle cells/pericytes [524]
Pdgfrβ ^{F5}	Tyrosine-to-phenylalanine inactivating mutations at amino acid 578 and 715 results in mutation of the Src and Grb2 binding sites	Mutant mice were viable vascular defects in the tissues of the eyes, hearts, and brains; reduction of vascular smooth muscle cells/pericytes [524];
Pdgfrβ ^{F7}	Tyrosine-to-phenylalanine inactivating mutations at amino acid 578, 715, 739 and 750, 770, 1008. The tyrosine at position 1020 was replaced with isoleucine. Disrupting Src-, Grb2-, PI3K-, RasGAP-, SHP-2 and PLCγ-dependent signal transduction	Mutant mice were viable; vascular defects in the tissues of the eyes, hearts, and brains; reduction of vascular smooth muscle cells/pericytes [524];
Pdgfrβ $^{(\beta T)}$	Introduction of a frameshift and subsequent premature stop codon downstream of the RasGAP-binding site	Homozygous: die perinatally; phenotype identical to that of the PDGFRβ null embryos; E18 embryos exhibit edema and hemorrhaging in multiple tissues, including the kidney, brain and skin [524]
Pdgfrβ $^{(S)J}$ *Sox2-Cre* *Pdgfrβ* $^{(S)J}$ *Meox-Cre*	The endogenous *Pdgfrβ* sequence was replaced with a lox-stop-lox cassette before the mutation (V536A) generating a constitutively active kinase domain.	epiblast-specific Cre deletion (Sox2Cre, Meox2Cre) mutants were growth deficient, died by P14. Differentiation defects of white adipocytes; increased cell proliferation and downregulated differentiation genes in aortic vascular smooth muscle [527]
Pdgfrβ $^{(S)K}$ *Sox2-Cre* *Pdgfrβ* $^{(S)K}$ *Meox-Cre*	The endogenous *Pdgfrβ* sequence was replaced with a lox-stop-lox cassette before the mutation (D842V) generating a constitutively active kinase domain	Homozygous: neonatal lethal. heterozygous :viable and fertile with no reported abnormalities, epiblast-specific Cre deletion (Sox2Cre, Meox2Cre) mutants were growth deficient, died by P14. Differentiation defects of white adipocytes; increased cell proliferation and downregulated differentiation genes in aortic vascular smooth muscle [527]

11.10.5 Transgenic Mouse Lines of Pdgfr

Strain	Organ	Phenotype/features
Pdgfrα$^{-/-}$/*YAC*	Transgenic mice harbor artificial chromosome (YAC) containing the human *Pdgfrα* gene	Survival until birth, rescues craniofacial and skeletal defects and spina bifida observed in *Pdgfrα*$^{-/-}$ mice, defect in lung growth [528]
PDGFRα-LacZ	lacZ expressed under the control of a 6-kb promoter sequence	monitoring of *Pdgfrα* gene expression during mouse development [529]
Alb-PDGFRβ	Expression of the soluble extracellular domain of under the control of the albumin promoter active late during embryogenesis	No obvious phenotype [530]
CD11a-Tel-Pdgfrβ	*Tel-Pdgfrβ* cDNA under the control of the human CD11a promoter induces expression of a TEL-PDGFR*β* fusion protein in myeloid cells.	Myeloproliferative syndrome [531]
MCK-PDGFRβ	A mouse muscle creatine kinase 6.5-kb enhancer/promoter/first intron segment drives expression of the human PDGFR*β* in skeletal muscle cells.	Stimulated glucose transport into skeletal muscle in vitro and in vivo [532]

Pdgfrα-cre line
A *cre* recombinase gene, SV40 late polyA site, and an-*frt*-flanked kanamycin cassette were introduced 3′ of the *Pdgfrα* promoter in mouse BAC RP23-116F7. The kanamycin cassette was excised from the vector and the linearized BAC was injected into C57BL/6 blastocysts. Hemizygous *Pdgfrα* -cre mice are viable and fertile [533–535].

Pdgfrα -cre^{ERT} line
The Bacterial artificial chromosome (BAC) library (CHORI) was used to obtain a 150 kb BAC containing the entire mouse *Pdgfrα* gene. This BAC was modified to contain a *creER*^{TM} fusion gene and a polyadenylation signal (polyA) downstream of the initiation codon of *Pdgfrα*. Hemizygous mice for the BAC *Pdgfrα -creER*^{TM} transgene are viable and fertile [536, 537].

Pdgfrβ-cre line
A −4.7/+0.1 kb fragment of the *Pdgfrβ* promoter including the first exon was amplified from mouse genomic DNA and used to express a nuclear targeted Cre recombinase [538].

11.11 The Ptk7 Receptor

11.11.1 Overview About Family Members

Ptk7 is similar to proteins of the receptor-tyrosine kinase family, however, the kinase homology domain lacks the DFG triplet that is necessary for the chelation of Mg2+ during phosphotransfer and required for kinase activity. Therefore *Ptk7* is thought to encode a pseudokinase, lacking catalytic function. So far, no ligand has been identified. Ptk7 is a versatile co-receptor for Wnt, Semaphorin and Vegf signaling, supporting a role of Ptk7 as a molecular switch between signaling pathways.

Box: List of all mouse strains

Ptk7^{Betageo}	Lu 2004 [539]
Ptk7^{chz}	Paudyal 2010 [541]

11.11.2 Constitutive Ptk7 Knockout Strain

11.11.2.1 Ptk7^{Betageo}

Generation of knockout strain: A gene trap vector including a β-galactosidase and neo was inserted into intron 2 of the *ptk7* gene in murine ES-cells, resulting in a null or a severely hypomorphic mutation, with the in-frame fusion of the first 114 amino acids to the transmembrane β-geo and neo construct. A transcript with the expected size for the fused RNA was detected by Northern blot of mutant brain lysates, and Western blot showed a peptide of the size of the predicted fusion product instead of the two peptides detected in wild-type brain lysates [539].

Phenotype: Heterozygous *Ptk7$^{betageo/+}$* embryos demonstrate that Ptk7 is dynamically expressed during gastrulation and neurulation, with higher levels in the posterior region of the embryo. However, homozygous *Ptk7$^{betageo/betageo}$* mutants die perinatally and show disruption of neural tube closure 1 accompanied by defects of the floor plate and somites. Ptk7 mutant cochleae display disrupted stereociliary bundle orientation of outer and inner hair cells at E18.5. Moreover, the mutation in Ptk7 shows genetic interactions with a mutation in the mouse *Van Gogh* homologue *vangl2* and identifies Ptk7 as a novel regulator of planar cell polarity (PCP) in vertebrates [539]. In addition *Ptk7$^{betageo/betageo}$* mice exhibited failures in epidermal wound healing, an effect similar to mice mutated in the PCP genes Vangl2, Celsr1, and Scrb1 [540].

11.11.3 ENU Mutated Strain of Ptk7

11.11.3.1 Ptk7^{chz}

Generation of knockout strain: The *chuzhoi* mutation was identified during a three-generation (G3) recessive mutagenesis screen, in which BALB/c males were injected with ENU and further outbred to C3H/HeH mice. G3 embryos were examined at E13.5 for developmental abnormalities. Affected fetuses were used for genetic mapping with a SNP panel and microsatellite markers. The mutation was identified to be present in the *Ptk7* gene by pyrosequencing and results in the insertion of three amino acids (Ala-Asp-Pro) after the 5th IG-like domain, displaying a splice acceptor site with closer consensus sequence similarity than the WT, which causes disruption of Ptk7 protein expression in *chuzhoi* mutants [541].

Phenotype: Embryos homozygous for the *Ptk7^{chz}* mutation exhibit gross developmental defects and mice die soon after birth from severe neural tube closure defects [541]. Half of the embryos display ventral body wall closure defects (probably omphalocele), with protrusion of the guts and liver and a failure of eyelid closure at E16.5. Splayed vertebrae are observed as a consequence of the open neural tube defect, as well as a shortening and skewing of the spinal cord. Fusions and bifurcations of the ribs are common and in approximately 7 % of embryos postaxial or preaxial polydactyly can be observed [541]. As for *Ptk7^{Betageo}* mutants, disruption in the orientation of inner ear hair is reported. Additionally, lung lobes are reduced in size and highly misshapened, due to thickened interstitial mesenchyme with infrequent septation. Malformations affecting the outflow region of the heart are noted. Moreover, like for the *Ptk7^{Betageo}* mutant, genetic interaction between *chuzhoi* mutants and both Vangl2^{Lp} and Celsr1^{Crsh} mutants is demonstrated [541]. Embryos carrying the *Ptk7^{chz}* mutation also have severe defects in caudal migration of facial branchimotor neurons [542].

11.12 RET Receptor

11.12.1 Overview About Family Members

The RET tyrosine kinase a 170-kDa glycoprotein, is a proto-oncogene. Gain of function mutation by DNA rearrangement are found in transformed cells [543]. It is named *ret* after this property REarranged during Transfection. RET is expressed in three isoforms in humans, designated RET51, RET43, and RET9 which arise due to alternative splicing at the C-terminus. It is a functional RTK, mainly activated by the glial cell-line derived neurotrophic factor (GDNF) family ligands, which include glial cell line-derived neurotrophic factor (GDNF), artemin (ARTN), neurturin (NRTN), and persephin (PSPN). RET signaling is essential in the development of neuronal crest, renal cells, and normal respiratory function.

11.12.2 Constitutive Ret Knockout Strain

11.12.2.1 Ret$^{-/-}$

Generation of knockout mice: *Ret*$^{-/-}$ mice were generated by targeted insertion of a neomycin cassette into the c-*ret* gene, which resulted in deletion of amino acids required for kinase activity. The targeting vector was linearized and injected into CCE ES-cells [544].

Phenotype: *Ret*$^{-/-}$ mice, though born at mendelian ratios, die within 24 h of birth. These mutants were used to characterize the essential functions of RET receptor kinases which are summarized below (Table 11.21):

11.12.3 Constitutive Ret Knock in Strain

11.12.3.1 Ret ^{Y1062F}

Generation of knockout mice: A genomic DNA clone was modified by PCR mutagenesis to introduce a point mutation in the sequence of exon 19 of *ret*. The targeting construct was electroporated into ES-cells derived from 129svj mice. Positive clones were injected into C57BL/6 blastocysts to generate chimeric mice.

Table 11.21 Summary of the features of *Ret*$^{-/-}$ mice

Organ	Phenotype
Urinary system	Kidneys absent or rudimentary with malformed nephrons, and ureters [544].
Gastrointestinal tract	Absence of neurons in the enteric nervous system, leading to inability in peristalsis [544].
Respiratory system	Reduced ventilation may lead to hypercapnia [545], possible failure to develop neurons involved in central respiratory chemosensitivity.

Phenotype: Homozygous mice show severe growth retardation at day 13.5 and most mice die after 4 weeks of age. In embryos, decrease in ureteric bud branching could be observed, whereas at later timepoints dilation of the small intestine, colon stenosis, absence of ganglion cells in both the small intestine and colon was found [546].

11.12.3.2 ret^{MEN2B}

Generation of knockout mice: A codon transition (T to C) in codon 919, was generated by site directed mutagenesis to result in threonine substitution for methionine. The targeting construct was electroporated into W9.5 ES-cells. The floxed *neo* cassette in the adjacent intron was excised by Cre mediated recombination.

Phenotype: Whereas the *ret*$^{MEN2B\text{-}neo}$ allele acts as a loss of function allele, homozygous *ret*^{MEN2B} mice are viable and overall normal. *Ret*$^{MEN2B/MEN2B}$ mice display C-cell hyperplasia (both diffused and nodular) and homozygous males fail to produce copulation plugs, despite normal functioning gonads and accessory organs and normal spermatogenesis [547].

11.12.3.3 Ret^{C620R}

Generation of knockout mice: Serine 696, a putative PKA phosphorylation site encoded by exon 11 of *ret,* was replaced with alanine (S697A) by site directed mutagenesis. The linearized targeting vector was electroporated into W9.5 ES-cells.

Phenotype: Homozygous *Ret*^{C620R} mice show reduction of enteric nervous system cells occurs in the middle and distal colon that may be caused by impaired migration of enteric neural crest cells in the developing gut [548]. RET S697A mutation causes slight suppression of kidney development, but no abnormality in the sympathetic and parasympathetic neural system or in spermatogenesis.

11.12.3.4 Ret^{C620R}

Generation of knockout mice: Targeted insertion of mutant *ret* DNA containing a point mutation in exon 10 resulting in a shift from cysteine to arginine. CCB ES-cells were electroporated and positive clones were injected into C57BL/6J blastocysts.

Phenotype: This mouse model mimics Hirschsprung's disease in humans caused by mutations in exon 10 by displaying defects in kidnay organogenesis and the enteric nervous system development and therefore causes neonatal lethality [549, 550].

11.12.3.5 Ret^{TLZ}

Generation of knockout mice: The tau-targeted lacZ (TLZ) cDNA encoding a fusion protein composed of the N-terminal region of bovine tau and full-length β-galactosidase followed by a floxed neo cassette was inserted into the first coding exon of ret thereby producing *Ret^{TLZ}* mice. The neo cassette was removed by crossing to *β-actin Cre* mice [551].

Phenotype: The *Ret^{TLZ}* allele is a reporter allele. In homozygous mice at p0 intrafusal muscle spindles displayed a significant loss of innervation in most but not all muscles [551].

11.12.3.6 Ret^{CreERT2}

Generation of knockout mice: A gene cassette composed of the CreERT2 cDNA intron polyA, and a floxed cassette consisting of AC-Cre with Tn5-neo was inserted into the endogenous promoter of the mouse Ret locus by gene targeting as described previously [552]. AC-Cre drives expression of Cre recombinase under the control of the Angiotensin Converting-enzyme promoter, allowing self-excision of the floxed gene in the germ line. Recombination was induced by gavage of 4-hydroxytamoxifen into pregnant females, e.g., at E10.5–12.5 [553].

Phenotype: In the reporter strain *Tau$^{f(mGFP)}$* [554] Cre expression recapitulates special and temporal expression of ret [553].

11.12.3.7 RET^9, RET^{51} and Docking Mutants

Generation of knockin strain and phenotype:

Strain	Organ	Phenotype/features
RET^9, RET^{51}	Human cDNA of the respective isoform (9;51) or WT *RET* was knocked into exon 1 of *ret* by homologous recombination in R1 ES-cells [555].	Both, RET$^{9/9}$ and RET$^{51/51}$ are viable and show no defects in the gonads or the enteric nervous system. The only mild defect is dilatation of the proximal ureter occurring with low penetrance [555].
$RET^{51(Y981F)}$; $RET^{51(Y1062F)}$; $RET^{51(Y1015F)}$	Y°→F mutants of RET^{51} cDNA were generated by PCR mutagenesis and knock in mice were generated as described above [555]. The Y981F mutation inhibits Src binding, Y1015F inhibits PLCγ binding whereas the Y1062F mutation inhibits binding of shc to RET [556].	Y981F and Y1062F mutations were redundant. Mice carrying the Y1015F mutation were viable only up to 5 weeks due to severe genitourinary defects like multiple cysts, megaureters with distal stenosis, reduced number of glomeruli and inability of the vas deferens/wolffian duct to separate from the ureters [555].
$Ret^{9(Y1062F)}$ *(Jain)*	Y→F mutants of RET^9 cDNA were generated by PCR mutagenesis and knock in mice were generated as described above [555].	Renal agenesis.
$miRet^9$ $miRet^{51}$	The endogenous ret gene was disrupted and replaced with alleles encoding single chimeric receptors composed of the extracellular domain of mouse RET fused in-frame to the intracellular segment of human RET9 or RET51 [557].	$miRet^{51/51}$ mice which lack RET9 and only express isoform 51, have kidney hypodysplasia and lack enteric ganglia from the colon, whereas mice did not show any symptoms when only isoform 9 was expressed [557].
$Ret^{9(Y1062F)}$ *(Wong)*	Y→F mutants of RET^9 cDNA were generated by PCR mutagenesis and knock in mice were generated as described above [557]. The linearized construct was electroporated into E14 ES-cells.	Hypoplastic and dysplastic kidneys, with no recognizable medulla, cortex, or nephrogenic zone and large regions of undifferentiated mesenchyme, decreased nephric elements, intestinal aganglionosis, inability of embryonic crest cell migration [558].
$Ret^{51\text{-}RI}$	To try to rescue the defect in $miRet^{51/51}$ mice a targeted insertion of the mutant human RET^{51} isoform in which M1064 and S1065 residues of Ret^{51} were replaced by the corresponding amino acids of Ret^9 (R1064 and I1065) [558]. The linearized construct was electroporated into W9.5 ES-cells.	Phenotype only marginally improved [558].
$Ret^{9\text{-}F1072A}$	Targeted insertion of mutant human RET9 isoform harboring base substitutions which result in amino acid substitution F1072A [558]. The linearized construct was electroporated into W9.5 ES-cells.	No overt phenotype [558].

11.12.4 Conditional Ret Knockout Strain

Generation of knockout strain and phenotype:

Strain	Organ	Phenotype/features
$Ret^{fl/CFP}$ (Uesaka)	A floxed human WT or *RET9* cDNA followed by CFP and the FRT site-enclosed Tn5 *neo* sequence was introduced into the first coding exon. Deletion disrupts expression of the endogenous *ret* gene but allows the expression of the inserted gene [552].	Viable and apparently healthy. Homozygous floxed $RET^{9/9}$ mice behave like a hypomorph mimicking Hirschsprung's disease. Homozygous deletion of the floxed cDNA cassette ($Ret^{CFP/CFP}$) results in a null allele.
CAGGCre-ER $Ret^{fl/fl}$	Treatment with 4-hydroxytamoxifen results in ubiquitous deletion of the *RET* cDNA.	Mice in which the function of Ret was inactivated after P5 failed to exhibit denervation of muscle spindles or motorneuron loss [551]. Conditional ablation of Ret in postmigratory enteric neurons caused widespread neuronal death in the colon, which led to colonic aganglionosis [552].
HB9-Cre/$Ret^{fl/-}$	HB9-Cre/ motor neurons	Loss of small myelinated axons in the lumbar ventral roots of adult mice in which the function of either Ret was inactivated in motor neurons early in development [551].
Ret^{fl} (Kramer)	Exon 12 of *ret* was flanked by 2 loxPsites and a flirted neo cassette was inserted into the 5′intron. Deletion in neurons and glial cells by Nestin –Cre (Nes) [559, 560].	Aberrant projection of lateral motor column axons in ret^{nes} mice. GDNF/Ret system functions as an instructive guidance signal for motor axons in coordination with ephrin4 signaling.
Ret^{fl} Dat-Cre *Ret^{fl} Nes-Cre*	*Dat-Cre* mice were generated to induce resombination in tyrosine hydroxylase positive cells in substantia nigra compacta and ventral tegmental areas [561].	Mice are viable and fertile. In 1 year old mice loss of neurons in Substantia Nigra pars compacta, reduction in dopaminergic nerve endings in the Striatum, reduced dopamine release and reuptake, postsynaptic dysfunction, neuroinflammation [562] .

Strain	Organ	Phenotype/features
Ret^{f/f};Wnt1-Cre (Luo)	A 129J BAC clone was used to flank exons 14 and 15 by loxP sites and to introduce a point mutation to change valin 805 to alanine. The targeting construct was injected into 129.1 ES-cells. *Wnt1-Cre* results in deletion of *ret* in premigratory neural crest cells [553].	Enlarged small and large bowel, complete loss of enteric neurons the intestines (intestinal aganglionosis), hypotrophy of nonpeptidergic nociceptors in dorsal root ganglion.
Ret^{fl/GFP} (Jain) *Ret^{fl/fl Dat-Cre}*	A floxed human WT or *RET9* cDNA followed by EGFP and the flirted *neo* sequence was introduced into the first coding exon. homologous recombination was performed in R1 ES-cells [563].	Cre-mediated recombination results in ablation of endogenous *Ret* and expression of EGFP. In this model no effect of ret deletion on midbrain dopaminergic neurons was observed.
Ret^{fl/fl} Nav1.8 Cre	*Nav1.8 Cre* mice express Cre recombinase in nociceptors [564] resulting in deletion of ret in *Ret^{fl/fl}*mice [551].	Decrease in peripheral nonpeptidergic nociceptive neurons of dorsal root ganglia and epidermal innervations. Increased sensitivity to cold and formalin induced pain [565].
Ret^{TGM} (tau-EGFP-myc)	Targeted insertion into the first coding exon of *ret* by homologous recombination in RW4 ES-cells results in expression of a fusion protein of bovine tau, EGFR and 3 repeats of human MYC tag [566] .	Neonatal lethality, increased neural apoptosis and proliferation and defect migration in sympathetic neuron precursors, abnormal morphology in sympathetic neurons and ganglions, intestinal agangliosis, absence of kidneys [566]. Strongly decreased density of neurons in esophagus [567].
Ret^{fl} (Martins Almeida) *Ret^{fl/fl} CD2-Cre*	Exon 1 and the *ret* promoter region was flanked by 2 loxP sites. The linearized targeting construct was electroporated into R1 ES-cells [568]. *CD2 Cre* expresses Cre recombinase in T-cells [569].	Thymic T-cell development is not affected by the lack of ret in T-cells [568].
Ret^{fl} (Baudet) *Ret^{fl/fl} Nkx6.2 Cre*	A 129/Sv BAC clone was modified to contain floxed exons 12 and 13 of ret. The linearized construct was electroporated into E14 ES-cells for homologous resombination [570]. Nkx-Cre is expressed in cranial motor neurons [571] .	*Ret^{fl/fl}* are normal. Ubiquitous deletion by crossing to PGL-Cre results in a null allele. *Ret^{fl/fl} Nkx6.2 Cre* have a deficit of maturation and specialization of presynaptic neuromuscular terminals and presynaptic axon terminal maturation is severely compromised [570].
Ret^{DN/+}	Human *RET* cDNA constructs encoding a mutated form of RET9 that contained a L985P and Y1062F mutation were inserted by homologous recombination into the first coding exon of the mouse *Ret* gene [572].	The DN allele results in diminished kinase activity of ret. *Ret^{DN/+}* mice died by 1 month of age and had distal intestinal aganglionosis reminiscent of Hirschsprung disease (HSCR) in humans. Unlike *Ret^{Null}* mice *Ret^{DN/+}* mice only had defects in the parasympathetic nervous system [572].

11.12.5 Transgenic Mouse Lines of Ret

11.12.5.1 MT/ret

Generation of transgenic mice: Transgenic mice expressing ret under the control of the metallothionein-I promoter-enhancer, which is known to function in almost all tissues, were developed by injecting a construct containing the ret oncogene cDNA fused to the mouse MT-I promoter-enhancer fragment into the fertilized eggs of (balb/c × C57BL/6) × balb/c mice [573, 574]. A number of different lines were generated differing in transgene copy number and phenotype.

Phenotype: Depending on the transgenic line generated *MT/ret* mice develop tumors and show aberrant melanogenesis and a variety of other phenotypes that are influenced by genetic and environmental cofactors.

Organ	Phenotype/features
Salivary gland	Salivary gland adenocarcinoma [574].
Liver	Hepatocellular carcinoma [574].
Abdomen	Retroperitoneal tumors [574].
Skin	Hyper pigmentation, increased melanocytes in dermis [573], Tumor formation in dermis of the face and neck [574], metastatic melanoma [575].
Skeletal muscle	Abundant melanocytes in skeletal muscle, Tumor formation in the leg muscles [574].
Eyes	Exophthalmos due to tumor formation in the choroid of eyes, abundance of melanocytes in choroid and Harderian glands of lines [574].
Ears	Cochleae contained double amount of eumelanin and richer in interdigitations [10].
Hair	Enhanced hair growth [576]. *Ret* seems to partially complement the hairless defect. After crossing to *MT/ret*, hairless mice display hair follicles that progress to anagen, however, no hair are macroscopically visible [577] .

11.12.5.2 Ret/ptc1

Generation of transgenic mouse: As the *RET/PTC* proto-oncogenes are frequently associated with human papillary thyroid carcinomas and are caused by rearrangements of *RET*. Independent studies were made to understand the role of *RET* proto-oncogene in thyroid carcinomas by using transgenic mice lines expressing the *ret/ptc1* oncogene under the thyroid-specific rat [578] or bovine thyroglobulin promoter [579, 580].

Phenotype: Development of thyroid hyperplasia and thyroid tumors displaying the histological aspect of papillary carcinomas [578, 580]. Bilateral thyroid tumors in 4-day-old high copy number mice, while similar tumors emerged in 3 weeks in low copy number expressing mice, early embryonic onset of thyroid follicle development, progressing to distorted thyroid follicles, congenital hypothyroidism in high copy number mice, defect in iodine uptake [579].

11.12.5.3 Mmtv/ret

Generation of transgenic mouse: Transgenic mice expressing *ret* oncogene under the control of mouse mammary tumor virus promoter/enhancer were generated [581].

Phenotype:

Organ	Phenotype/features
Mammary glands	Adenocarcinomas [581].
Salivary gland	Adenocarcinomas [581].
Eyes	Premalignant tumors with hyperplasic and dysplastic lesions of Harderian glands [581].
Reproductive tracts	Premalignant tumors in males [581].

11.12.5.4 D^{β}H-RET^{MEN2B} (Sweetster)

Generation of transgenic mouse: The *RET^{MEN2B}* mutation (Met918Thr) was introduced by PCR into a cDNA encoding the short isoform (1072 amino acids) of human *RET*. This cDNA is expressed under the control of human dopamine *ß*-hydroxylase (*$D^{\beta}H$*) promoter. The transgenic construct was injected into fertilized DBA × C57BL/6F1 eggs and these were transferred into blastocysts. Transgene copy numbers and number of insertions were identified by dot and Southern blot [582].

Phenotype:

Organ	Phenotype/features [582]
viability	Low copy number mice were viable and bred to homozygosity, while intermediate copy number mice survived only until weaning age and high copy number mice died with 24 h after birth.
Nervous system	Increase in neurons, Schwann cells and nerve fibers, hyperplasia of sympathetic ganglia, Neuroglial hyperplasia in preaortic sympathetic complex near kidneys, Hyperplasia of the paravertebral sympathetic chain.
Adrenal glands	Enlarged adrenal glands with disrupted cortex and abundant neuroglial tissue in adrenal medulla.
Enteric nervous system	Not affected.
Kidneys	Renal agenesis and small dysplastic kidneys with microcysts, tubular atrophy, and few or absent glomeruli. Failure of ureter development.

11.12.5.5 MoMuLV/RET^{MEN2A}

Generation of transgenic mouse: A multiple endocrine neoplasia type IIa (MEN2A) mutation (Cys634 → Arg) was introduced by PCR into a full length cDNA of the human *RET*. *RET^{MEN2A}* is expressed from the moloney murine leukemia virus long terminal repeat. The linearized fragment was injected into fertilized eggs from C57BL/6 × BALB/cA mice. Expression of the transgene and its product was detected at variable levels in a variety of tissues including thyroid, heart, liver, colon, parotid gland, and brain [583].

Phenotype: Similar to human patients carrying the mutation transgenic mice developed thyroid C-cell hyperplasia or medullary carcinoma, accompanied by high levels of serum calcitonin. Additionally 50 % of the mice developed adenocarcinoma of mammary or parotid gland [583].

11.12.5.6 CALC-$^{MEN2B\text{-}RET}$

Generation of transgenic mice: The *RET*^{MEN2B} mutation (Met918Thr) was introduced by PCR into a cDNA encoding the short isoform (1072 amino acids) of human *RET*. This cDNA is expressed under the control of human calcitonin (CALC) promoter. The transgenic construct was injected into fertilized DBA2×C57BL/6F1 eggs and these were transferred into blastocysts. Transgenic mice expressing WT *RET* were generated as control [584].

Phenotype: *CALC-*$^{MEN2B\text{-}RET}$ develop diffuse C-cell hyperplasia accompanied by high plasma calcitonin levels, large bilateral malignant tumors of thyroid and multiple microscopic foci at 20 months of age. Additional pathologies were lung carcinoma and lymphoma. Ectopic expression of WT RET in the kidney caused renal cystic disease [584].

11.12.5.7 CALC-$^{MEN2A\text{-}RET}$

Generation of transgenic mice: A point mutation in human *RET51* cDNA was generated resulting in a Cys-634-Arg shift and cloned after the human calcitonin (CALC) promoter. The linearized DNA was injected into the pronuclei of fertilized oocytes from either FVB or C57BL/6/CBA mice. Embryos surviving to two-cell stage were transferred into the oviducts of pseudopregnant MF-1 females. Mice expressing WT RET were generated as controls [585].

Phenotype: CALC$^{WT\text{-}RET}$ mice displayed an altered development of the thyroid gland with cystic follicles, presence of ciliated follicular epithelial cells and non metastatic tumors. In some mice, metastatic exocrine cystadenoma or cystadenocarcinoma was detected [585].

11.12.5.8 CGRP/CT RET

Generation of transgenic mice: A point mutation in human *RET51* cDNA was generated resulting in a Cys-634-Arg shift and cloned after the rat CGRP/CT promoter. The linearized DNA was injected into the pronuclei of fertilized oocytes from either FVB or C57BL/6/CBA mice [586].

Phenotype: Thyroid developmental defect with bilateral C cell hyperplasia and medullary thyroid carcinoma and follicular lesions resembling human papillary thyroid carcinoma [586]. Cystic pancreatic cancer was detected in some mice.

11.12.5.9 Hoxb7-RET^{9}; Hoxb7-RET^{PTC2} (Srinivas)

Generation of transgenic mice: cDNA for the *RET*^{9} isoform or a ligand independent mutant (PTC2) was cloned behind a *hoxb7* promoter fragment to express RET throughout the ureteric bud branches, in contrast to its normal expression only at the bud tips. The linearized transgenic construct was injected into (B6×CBA) F1 zygotes [587].

Phenotype: Inhibition of bud growth and branching weaker compared to *Ret*$^{-/-}$ mice, however, in a Ret$^{-/-}$ background transgene expression rescued kidney development in some mice [587]. *Hoxb7-RET*^{PTC2} causes ectopic nodule formation.

11.12.5.10 Hoxb7-RET^{9} (Batourina)

Generation of transgenic mice: Not described. Uses of a heterologous promoter.

Phenotype: Does not induce dominant renal defects [588].

11.12.5.11 Hoxb7-RET^{51}; Hoxb7-RET^{Man2A}; Hoxb7-RET^{Men2B}

Generation of transgenic mice: cDNA for the WT or mutated *RET*^{51} isoform or mutated *RET*^{9} was cloned behind a *hoxb7* promoter fragment to express RET throughout the ureteric bud branches, in contrast to its normal expression only at the bud tips. Mutations were introduced by PCR mutagenesis and resulted in constitutively active MEN2A (Cys634Lys) and MEN2B (Met918Thr) alleles of the respective isoform. The linearized transgenic construct was injected into (B6 x CBA) F1 zygotes [557].

Phenotype: Only constitutively active isoforms of *RET*^{9} but not of *RET*^{51} induce cystic kidneys [557].

11.13 The ROR Receptor Family

11.13.1 Overview About Family Members

Receptor tyrosine kinase-like orphan receptor (ROR) proteins are a conserved family of receptors that function in developmental processes including skeletal and neuronal development, cell movement and cell polarity [589]. Recently, the term orphan became obsolete when several studies established ROR2 as WNT-receptor, which predominantly binds WNT5A and plays a role in non-canonical Wnt signaling [590, 591]. Wnt5a—ROR2 signaling also plays a role in cancer development and metastasis [592, 593].

Murine receptors: *mRor1*, *mRor2* (former NTRKrelated-1,2)
Ligands: *Wnt5a, Gdf5*

Box: list of all mouse strains

mRor1$^{-/-}$	Nomi et al. 2001 [594]
mRor1$^{fl/fl}$	Ho et al. 2012 [595]
mRor2$^{-/-}$	Takeuchi et al. 2000 [599]
mRor2$^{TMlacZ/TMlacZ}$	DeChiara et al. 2000 [598]
Ror2$^{fl/fl}$	Ho et al. 2012 [595]
ROR2^{W749FLAG}	Raz et al. 2008 [601]

11.13.2 mRor1

11.13.2.1 Constitutive Ror Knockout Strain

11.13.2.1.1 mRor1$^{-/-}$

Generation of knockout strain: The wild-type *mRor1* locus containing the exon including the Ig-like domain was targeted in E14 ES-cells by homologous recombination and replaced by a *PGK-neo* resistance cassette using SV129 genomic DNA. Integration was confirmed by Southern blot [594]. However, Western blot analysis by a different group suggests that this line might not represent a true null mutant [595].

Phenotype: mRor1$^{-/-}$ mice are born at mendelian ratio and have no obvious phenotype at birth [594]. However, after birth they exhibit forced respiration and cyanosis and die within 24 h. For currently unknown reasons alveoli are not fully inflated. *mRor1*$^{-/-}$ mice did not show any abnormalities of skeleton or heart. In a later study mRor1$^{-/-}$ mice were viable showing postnatal growth retardation and age-related skeletal changes. Furthermore mice displayed kidney defects and female infertility [596]. *mRor1*$^{-/-}$*mRor2*$^{-/-}$ double knockout mice show more severe limb and face malformations compared to *mRor2*$^{-/-}$mice indicating redundant expression [597].

11.13.2.2 Conditional Ror Knockout Strain

11.13.2.2.1 mRor1$^{fl/fl}$

Generation of knockout strain: The *mRor1*conditional allele (*mRor1*^{fl}) was generated by flanking exons 3 and 4 of the mRor1 genomic locus with loxP sequences [595]. The mRor1$^{-/-}$allele was generated by crossing the *mRor 1*$^{fl/fl}$mice to the EIIA-Credeleter line [88].

Phenotype: *mRor 1*$^{fl/fl}$mice showed normal protein expression. Most *mRor1*$^{-/-}$mice were viable at birth and showed no obvious phenotype [595]. Occasionally, embryonic exencephaly is observed. Detailed analysis is currently ongoing.

11.13.3 mRor2

11.13.3.1 Constitutive Ror2 Knockout Strain and Phenotype

11.13.3.1.1 mRor2$^{TMlacZ/TMlacZ}$ (DeChiara)

Generation of knockout strain: The wild-type *mRor2* was targeted at the sequence encoding the kinase domain in E14.1 ES-cells by homologous recombination and replaced by a *lacZ/neo* cassette using SV129 genomic DNA [598]. Cells expressing ROR2 can be identified by staining for β-gal expression.

11.13.3.1.2 mRor2$^{-/-}$ (Takeuchi)

Generation of knockout strain: The wild-type *mRor2* locus containing the exon including the Ig-like domain was targeted in E14 ES-cells by homologous recombination and replaced by a *Pgk-neo* resistance cassette using SV129 genomic DNA [599]. Integration was confirmed by Southern blot.

Phenotype: All *Ror2* knockout and knockin mouse lines may serve as intermediary models for Recessive Robinow Syndrome and dominant Brachydactyly Type B1.

Both mouse models, DeChiara as well as Takeuchi, display overlapping phenotypes. Mice exhibited perinatal lethality. Alive mice display forced respiration and cyanosis with less inflated alveolar sacks [599]. Embryos are smaller and have less weight after E12.5 and show skeletal and heart defects [599]. At E15.5 short limbs and tails and malformation of facial structures can be observed. Bones formed by endochondral ossification are foreshortened, with a tendency toward greater abnormalities more distally [598]. *mRor2*$^{TMlacZ/\ TMlacZ}$ lack middle phalanges [598]. The heart of *mRor2*$^{-/-}$ mice shows a ventricular septum defect.

mRor1$^{-/-}$*mRor 2*$^{-/-}$*DKO:* The skeletal phenotype is enhanced in *mRor1*$^{-/-}$*mRor 2*$^{-/-}$ double knockout mice accompanied by abnormal chondrocytes [594]. In addition to ventricular septum defect, *mRor1*$^{-/-}$*mRor 2*$^{-/-}$ DKO embryos exhibited complete transposition of the great arteries [594].

11.13.3.2 Conditional Ror2 Knockout Strain

11.13.3.2.1 Ror2$^{fl/fl}$

Generation of knockout strain: The *mRor2* conditional allele (*mRor 2*^{fl}) was generated by flanking exons 3 and 4 of the *mRor2* genomic locus with lox2722 sequences [595]. The *mRor2*$^{-/-}$ allele was generated by crossing the *Ror 2*$^{fl/fl}$ mice to the ubiquitous *EIIA-Cre* deleter line [88].

Phenotype:

Strain name	Organ	Phenotype
mRor2$^{fl/fl}$ *EIIACre*	Early embryo/ most tissues	Facial malformations and truncation of the limbs and posterior region of the embryo consistent with previously published *mRor2* mutant [595]
mRor1$^{-/-}$*mRor 2*$^{-/-}$ *EIIA DKO*	Early embryo/ most tissues	Die in utero, most around E15,5.; severe facial malformations, truncation of the caudal axis and limbs, edema in the trunk. Occasionally, exencephaly is observed [595].
mRor1$^{fl/fl}$*mRor 2*$^{fl/fl}$ *Wnt1-Cre*	Embryo	Defect in axonal branching in organs like kidney and bladder at E17.5, similar to *Wnt5*-null mutants. Indeed Ror RTKs function as receptors for Wnt5a-dependent dishevelled phosphorylation [595]

11.13.3.3 Ror2 Knockin Mouse Lines

11.13.3.3.1 ROR2^{W749FLAG}

Generation of knockin strain: *mRor2*^{W749FLAG} mice were generated using VelociGene technology [600]. A bacterial artificial chromosome-based targeting vector was used to replace the sequence coding for domain G750-D930 of *mRor2* with the FLAG epitope, followed by a neomycin resistance cassette [601]. Knockin mice are expressing ROR2 truncated at W749 and tagged at the carboxyl terminus with the FLAG epitope.

Phenotype: *mRor2*$^{W749FLAG/W749FLAG}$ mice are born at mendelian ratio and viable. Homozygous mice display a decrease in body weight before weaning age and multiple craniofacial defects [601]. At E14.5, *mRor2*$^{W749FLAG/W749FLAG}$ embryos displayed hypoplastic skeletons, with a reduction in the size of all the anlagen. Adult mice exhibited a reduced body length, but not width, and brachydactyly in fore and hind paws. Bone remodeling appears normal, however, bone area is reduced. Male *mRor2*$^{W749FLAG/W749FLAG}$ mice show reduced fertility [601].

11.14 The-ROS Receptor

11.14.1 Overview About Family Members

The c-ROS gene was cloned in 1986 due to the similarity to the transforming gene v-ROS [602, 603]. The c-ROS gene is highly conserved in vertebrates. Structurally, c-ROS resembles the proteins from the ERBB family, which strongly suggested that it functions as a receptor [604, 605]. Importantly, a ligand for c-ROS is still elusive. In humans, c-ROS has important roles as an oncogene in various cancers such as brain tumors, NSCLC and others [606–608]. Biochemically, it plays a role in various signaling networks and functions via molecules such as STAT3 or the PI3K-AKT-mTOR axis. Consequently, c-ROS is potential target for anticancer drugs [606].

11.14.2 Constitutive Ros Knockout Strain

11.14.2.1 c-Ros$^{-/-}$ (ros^{ex};ros^{kin})

Generation of knockout mice: To inactivate the c-Ros allele in mice, Birchmaier and colleagues generated 2 targeting vectors [609]. One was used to delete 2 exons which encode the extracellular domain of the receptor (*ros^{ex}*), the second targeted 2 exons in the kinase domain (*ros^{kin}*). NeomycinR and HSV-TK were used as selection markers. Both constructs were successfully used to perform gene targeting in E14-1 ES-cells and the subsequent generation of germ-line competent chimeric mice. Both, the *ros^{ex}* and the *ros^{kin}* alleles, when homozygous, caused male infertility.

Phenotype:

Organ/process analyzed	Phenotype/features
Development/General	No developmental phenotype, sterile [609]
Epididymis	Defective regionalization and differentiation of epithelial cells in the epididymis, perturbed gene expression [609, 610]
Spermatozoa	Flagellar defects, impaired regulation of cell volume [611]

11.15 RYK Receptor

11.15.1 Overview About Family Members

The receptor tyrosine kinase-related (RYK) molecule got its name for containing a hydrophobic transmembrane domain and a conserved kinase domain similar to other members of the RTK family, however, it has an unsually short extracellular domain and is catalytically inactive. The protein is involved in activating Wnt signaling pathways. Few mouse models of RYK have been generated so far.

11.15.2 Constitutive Ryk Knockout Strain

11.15.2.1 Ryk$^{-/-}$

Generation of knockout mouse: Homologous recombination deleted 14.5 kb of genomic DNA including exons (numbered boxes) encoding >95 % of the Ryk extracellular domain and the entire transmembrane (TM) domain. This was replaced with a promoterless IRES βgeo.pA cassette which, using the endogenous exon 2 splice acceptor site, was expected to report the activity of the *Ryk* promoter [612]. The linearized construct was electroporated into W9.5 ES-cells.

Phenotype: Most *Ryk*$^{-/-}$ mice died on the day of birth, few survived up to 1 week but were cachectic. Neonates were not able to breathe and suckle and therefore cyanotic. The displayed defects in craniofacial development and morphogenesis like shortened snout and a completely cleft secondary palate. Other skeletal defects were, shorter limbs, with the hind limbs splayed laterally, shorter and thicker long bones, and absence of a tail. In the few mice that survived bilateral microphthalmia and abnormal gait with dragging of the hind limbs while walking was observed [612].

Ryk$^{-/-}$ mice display a novel callosal guidance phenotype where cortical axons project aberrantly across the major forebrain commissure, the corpus callosum. Wnt5a acts as a chemorepulsive ligand for Ryk, driving callosal axons toward the contralateral hemisphere after crossing the midline [613] .

11.15.3 Transgenic Mouse Line of Ryk

11.15.3.1 Ryk siRNA

Generation of transgenic line: Mouse one-cell-stage embryos were infected with lentivirus containing siRNA against conserved sequences of mouse and human *syk*. Around 40 embryos were transferred to pseudopregnant recipient mice and brought to term. Genomic DNA from transgenic mice was analyzed for GFP expression and single-copy transgenic mice were used for analysis [614].

Phenotype: The study identifies Ryk as a coreceptor for Wnt, that via Dishevelled activates the canonical Wnt pathway.

Many mice died on the day of birth, few survived. They had less weight and abnormal gait. These mice displayed defects in axon guidance and neurite outgrowth [614].

11.16 The Tie Receptor Family

11.16.1 Overview About Family Members

Tie-1 and Tie-2 (also known as Tek) comprise the two members of the Tie receptors [615]. The acronym Tie stands for tyrosine kinase with Ig and EGF homology domains and implies structural properties of the extracellular domain of the Tie receptor family [615]. Tie receptors have been identified as the second vascular tissue specific tyrosine kinase receptor system besides the VEGF-receptor system [615].

Tie-1 is expressed in endothelial cells and immature hematopoietic cells [616] whereas Tie-2 is expressed in endothelial cells within blood vessels and the lymphatic vasculature [617, 618]. Moreover, Tie-2 is reportedly expressed in a subpopulation of hematopoietic stem cells and Tie-2 null mice display impaired hematopoiesis [619, 620].

Angiopoietins (Ang1, Ang2, Ang3, Ang4) are the known ligands for TIE receptors and the Ang-Tie system regulates vascular development and remodeling, vessel maturation, basal endothelial barrier function, vascular inflammation, and controls vessel quiescence in adults [617, 618, 621].

The Ang-Tie system might be involved in lymph-angiogenesis, but the exact mechanism of function remains elusive [617]. In addition, Ang-Tie signaling affects tumor angiogenesis [618]. Tumor growth activates high Ang2 expression in endothelial cells resulting in disturbed Ang1/Ang2 balance resulting in vessel regression, hypoxia as well as VEGF production and tumor vascularization [615, 622].

Tie-1 is activated by Ang1 and Ang4, however, Tie-1 does not directly bind these ligands and a specific Tie-1 ligand has not been identified to date [615, 618, 623]. Interestingly, Tie-1activation seems to require the presence of Tie-2 [624, 625]. Importantly, the physiologic role of Tie-1 has not been clarified yet [618].

Tie-2 activation in endothelial cells is controlled by the balance of expression between agonist Ang1 and the competitive inhibitor Ang2, which exerts antagonistic effects on Tie-2 in the blood vascular system [621]. Interestingly, Ang2 can also function as an agonist in a context dependent manner [626]. Ang1-Tie-2 signaling supposedly maintains endothelial cells in a quiescent state [627]. Furthermore, Ang1-mediated Tie-2 activation in endothelium is associated with anti-inflammatory, anti-apoptotic, and anti-permeable effects [621]. Moreover, Tie-2 activation is important for endothelial cell migration [615]. Ang3 and Ang4 can bind to the Tie-2 receptor, however, Ang1 and Ang2 have been studied more comprehensively and the roles of Ang3/4 in the context of Ang-Tie signaling have to be further clarified [626].

Murine receptors: Tie1, Tie2 (Tek)
Ligands: Tie1: Ang1, Ang4
Tie2: Ang1-4

11.16.2 Tie 1

11.16.2.1 Constitutive Tie1 Knockout Strains

11.16.2.1.1 Tie1^{lacZ}

Generation of knockout strain: A targeting vector was designed that, after homologous recombination in R1 ES-cells, replaced the coding sequences in the first exon of *Tie* by a *lacZ* gene followed by a loxP-flanked neomycin selection cassette under the control of the endogenous promoter [628]. The neo cassette was removed in ES-cells by transient Cre expression.

Phenotype: $Tie1^{lacZ}$ is expressed in vascular structures from E8.0 on. Homozygous *Tie-1*^{lacZ} is embryonic lethal. *Tie-1*$^{-/-}$ embryos die between E13,5 and E14,5 due to impaired endothelial cell integrity and hemorrhage, development of severe edema and rupture of the microvasculature. Tie-1 is required for endothelium formation in adult chimeric mice [628]. To be able to create chimeric mice, homozygous $Tie1^{lacZ/lacZn-}$ ES-cells were generated. Diploid chimera between CD1 morula and $Tie1^{lacZ/lacZn-}$ ES-cells were generated. No lacZ staining in endothelial cells could be detected indicating that homozygous Tie1lacZ cells did not contribute to adult vessels [628]. $Tie1^{lacZ/+}$ mice show reduced expression of Tie1 RNA and protein [629] .

11.16.2.1.2 $Tie1^{-/-}$

Generation of knockout strain: Tie-1 was deleted via a neomycin selection cassette which was inserted into the first coding exon of *Tie-1* interrupting *Tie-1* at the second amino acid within the signal peptide sequence. Linearized DNA was injected into CJ7 ES-cells [630].

Phenotype: Tie-1 mutants die before or immediately after birth due to breathing difficulties. *Tie-1*$^{-/-}$ embryos develop edema and hemorrhage due to vascular endothelial cell defects and develop smaller hearts [630]. Tie-$1^{-/-}$ embryos show increased vascular density in numerous organs [631]. Tie-1 deficiency results in "hyperactive" endothelial cells [631].

11.16.2.2 Conditional Tie1 Knockout Strains

11.16.2.2.1 mTie-$1^{neo/neo}$; mTie-$1^{fl/fl}$E2A-Cre (Qu)

Generation of knockout strain: The targeting vector was constructed so that *loxP* sites flank the *Tie-1* minimal promoter and exon 1 and a flirted *neo* cassette downstream thereof. Southern blot analysis was used to confirm homologous recombination of targeted R1 ES-cell clones (*Tie-1*^{neo}) [632].

Phenotype: The *Tie-1*^{neo} allele is a hypomorph. *Tie-1*$^{neo/neo}$ hypomorphic embryos displayed abnormal lymphatic pattering and disorganized lymphatic vessels as well as abnormally dilated jugular lymphatic vessels and increased Prox1-positive lymphatic endothelial cells at early lymph angiogenesis accompanied by mild edema and no hemorrhage.. Tie-1 attenuation leads to abnormal regression of the lymphatic vasculature [632]. Ubiquitous Cre-mediated deletion in embryos (*mTie-1*$^{fl/fl}$*E2A-Cre)* resulted in death before birth due to hemorrhage and severe edema.

11.16.2.2.2 mTie-$1^{fl/fl}$SCL-ER^{T} Cre

Tie-1$^{fl/fl}$ (Qu et al) were crossed to endothelial cell specific, tamoxifen inducible stem cell leukemia (SCL)-Cre ER^{T} mice [629].

Phenotype: In an *ApoE*$^{-/-}$ mouse model, *mTie-1*$^{fl/fl}$*SCL-ER*^{T} Cre mice showed reduced atherosclerosis progression along with reduced expression of cell adhesion molecules and Rho-associated, coiled-coil containing protein kinases (ROCK1 and ROCK2) and increased eNOS expression and Tie2 phosphorylation in the aorta [629].

11.16.2.2.3 mTie-1$^{neo/neo}$; mTie-1$^{fl/fl}$ (D´Amico)

Generation of knockout strain: The conditional allele was generated by flanking the *Tie1* minimal promoter and exon 1 of the *Tie-1* gene by loxP sites and insertion of a flirted neomycin resistance cassette into the first intron of *Tie-1* by homologous recombination [633]. The linearized construct was electroporated into R1 ES-cells and positive clones were identified via Southern blot analysis and injected into C57BL/6 blasocysts [633].

Phenotype: Homozygous and hypomorphic (*Tie-1*$^{neo/neo}$) embryos displayed lymphatic malformations due to abnormal pattering of lymph sacs. Hypomorphic embryos were swollen and survived until E15,5 with lymphatic malformations resembling those observed in *Tie-1*$^{-/-}$embryos. Excision of *neo* by Flp recombinase (Tie$^{fl/fl}$) abolished the hypomorphic phenotype of *Tie-1*$^{neo/neo}$ embryos [633].

11.16.3 Tie2

11.16.3.1 Constitutive Tie2 Knockout Strain

11.16.3.1.1 Tie2$^{-/-}$ (Dumont)

Generation of knockout strain: Tie-2 was targeted via a PGK-*neo* selection cassette that replaced the translation start site and the first 17 amino acids in exon 1 of *Tie-2* by homologous recombination in R1 ES-cells [634].

Phenotype: Tie-2$^{-/-}$ mice died during midgestation due to defects in endothelial integrity. *Tie-2*$^{-/-}$ mice have reduced numbers of endothelial cells at E8.5 and displayed impaired heart development and abnormal yolk sac morphology [19].

11.16.3.1.2 Tie2$^{-/-}$ (Sato)

Generation of knockout strain: The Tie-2 knockout strain was generated by a neomycin selection cassette that interrupted the *Tie-2* gene in the second exon. Linearized DNA was injected into CJ7 ES-cells [630].

Phenotype: Tie-2 knockout is embryonic lethal (E10.5) and leads to growth retardation of head and heart. Embryonic Tie-2 deficiency results in abnormal vascular networks also in extraembryonic yolk sack vessels [630]. Defect in cardiac vessel branching and cardiac malformations. Vessels in the perineural plexus lack periendothelial cells and display malformed tissue folds [631].

11.16.3.2 Conditional Tie2 Knockin Strain

11.16.3.2.1 Tie-2$^{F1100/F1100}$

Generation of knockin mice: A tyrosine 1100 to phenylalanine point mutation was performed with a targeting construct that contained *Tie-2* cDNA, a simian virus 40 poly(A) sequence, a neomycin resistance gene flanked by loxP sites and the thymidine kinase gene. Homologous recombination was performed in R1 ES-cells and identified by Southern blot [635].

Phenotype: Homozygous *Tie-2*$^{F1100/F1100}$ is embryonic lethal. Embryonic lethality is caused due to defects in cardiac development and angiogenesis, impaired hematopoietic development and impaired endothelial cell development at E10.5. In contrast toTie-2$^{-/-}$ mice normal association of perivascular cells with nascent blood vessels and normal vascular integrity [635].

11.16.3.3 Tie1;Tie2 Double Knockout Strains

11.16.3.3.1 Tie1$^{lacZ/lacZ}$;Tie2$^{-/-}$

Phenotype: Double knockout of Tie1 and Tie2 is embryonic lethal. Double mutant embryos are smaller and display impaired heart development and deficiency in forebrain structures. Moreover, double mutants show an underdeveloped yolk sac vasculature and edema. Tie2 is essential for endocardial development whereas both, Tie1 and 2, are necessary for the microvasculature during organogenesis [636].

11.16.3.4 Transgenic Mouse Lines of Tie2

Generation of transgenic strain and phenotype:

Strain	Organ	Phenotype/features
β-actin-Tek^{A853} *tek-Tek*^{A853} *polyoma-Tek*^{A853}	The codon for lysine-853 of Tie2 was altered by oligonucleotide-directed mutagenesis to the codon encoding an alanine residue. The cDNA was cloned behind different promoters to express dominant active Tie2 in endothelial cells [634].	*β-actin-Tek*^{A853} show no delay in development. *tek-Tek*^{A853} and *polyoma-TekA853* show developmental delay in 30–50 % of embryos.
Tek-tTA/tetO-Tek *Tie1*$^{lacZ/lacZ}$	In double transgenic mice the *Tek* promoter drives tTA expression inducing mouse *Tek* cDNA expression in embryos in endothelial cells after DOX treatment [637]. Expression level correlates to wt levels. Mice were bred to *Tie1*$^{lacZ/lacZ}$ mice [628, 638].	Dox treatment of *Tie1*$^{lacZ/lacZ}$ mothers from E8.5 on could not rescue embryonic lethality but delayed heart phenotype. Loss of Tek expression after single DOX treatment results in EC apoptosis in hemorrhagic regions of the embryo [638].

(continued)

(continued)

Strain	Organ	Phenotype/features
Tek-tTA/tetO-Tek	Similar to above however, in adult mice in a wt *Tie2* background transgene expression was also found in keratinocytes and epithelial cells of the hair follicle.	Conditional overexpression of Tek results in a reversible psoriasis like skin disorder. Increased number of large and tortuous dermal capillaries, inflammatory infiltrate and hyperplastic epidermis [639].
K5-tTA/tetO-Tek *Tie1-tTA/tetO-Tek*	Tet-inducible *Tek* transgenic mice [638] were crossed to keratinocyte (*K5*-tTA [640]) or endothelial cell specific (*Tie-1*-tTA) *Tet*-transgenic mice [637].	Increased dermal vasculature in both $Tet^{OS}Tek$ responder lines. Only KC-specific overexpression of Tek results in a reversible psoriasis like phenotype with increased dermal CD4+ cells, infiltration of epidermal CD8+ cells, development of acanthosis, increased proinflammatory cytokines and hyperproliferative skin [641].

11.17 The NTRK Receptor Family

11.17.1 Overview About Family Members

The Neurotrophic receptor kinase receptor family is comprised of three receptors that play important roles in the development of the nervous system, specifically the sympathetic nervous system. The NTRK family consists of the tyrosine kinase receptors TrkA (NTRK1), TrkB (NTRK2) and TrkC (NTRK3), which are activated by the ligands nerve growth factor (NGF), brain-derived neurotrophic factor (BDNF) and neurotrophin-3 (NT3), respectively. Other ligands include Neurotrophin-4 (NT4) which can bind to TrkB with a lower affinity than BDNF and P75, a transmembrane receptor, which can bind all the neurotrophin receptors albeit with lower affinity than the ligands mentioned above [642]. The three receptors share approximately 50 % homology in their cDNA sequences, with the tyrosine kinase being the most conserved and the ligand binding domains being the most different [643]. The Trk transduction signal is mediated by ligand binding upon which the receptors homodimerize, activate tyrosine kinase and transphosphorylate tyrosine residues which leads to the phosphorylation and activation of SHC, PI3K and PLCγ1 which eventually activate Ras and Mek signaling pathways [644]. Collectively, the Trk receptors and their ligands regulate synaptic strength and plasticity and play a key role in neuronal survival and differentiation. In sympathetic neurons it has been observed that TrkC receptor is expressed early during their development and the TrkA receptor in later stages with the TrkB receptor rarely seen illustrating that there is a gradient of Trk receptor activity that mediates sympathetic neuron development [645, 646]. In addition to neuron survival and differentiation the Trk receptor family has been linked to axon and dendrite growth, retrograde signaling, assembling of the cytoskeleton and receptor cross talking [647]. These observations have been identified through the use of Trk animal models including receptor null knockout mice, conditional knockout mice, and transgenic mice. Knockout animals have been generated for all three receptors (detailed below) and all models present similar phenotypes of central and peripheral nervous system abnormalities but ablation of *TrkB* expression is the most severe [648–650].

11.17.2 TrkA

NGF-induced TrkA activation has shown to be critical for the survival and development of sympathetic neurons [651, 652].

11.17.2.1 Constitutive TrkA Knockout Strains

The first $TrkA^{-/-}$ animal was generated in 1994 via homologous recombination of the kinase catalytic domain giving rise to mice with an average lifespan of 20 days [650]. Another constitutive knockout model was generated in 2000 via homologous recombination of the translation initiation codon [653] which show the same phenotype as the 1994 knockout model, that is, a reduction in the size of the dorsal root ganglion and mottled fur (Table 11.22) [653].

11.17.2.2 Conditional TrkA Knockout Strains

Because of the rapid mortality of $TrkA^{-/-}$ mice conditional *TrkA* mouse models were generated to better be able to study loss of TrkA during development. Two conditional *TrkA* models have been generated via homologous recombination or point mutations. In both models, cre-mediated or NMPP1-chemical mediated *TrkA* deletion gave rise to viable, fertile mice that display no abnormal phenotypes [654, 655]. Site-specific deletion of *TrkA* identified a role for TrkA in B-cell proliferation and neurotrophin signaling which could not be identified in null models (Table 11.23).

11.17.3 TrkB

TrkB signaling is one of the most studied in Trk receptor family members and ablation of its activity has severe consequences for mouse survival as illustrated but the numerous mouse models generated. The first *TrkB* null model was generated in 1993 via homologous recombination of a neomycin cassette into exon 2 which gives rise to a protein that lacks tyrosine kinase activity [649]. Subsequent knockout

Table 11.22 *TrkA* constitutive knockout models

Strain name	Organ	Phenotype
TrkA$^{-/-}$ (Smeyne)	Constitutive loss of TrkA signaling. Sequences encoding 51 amino acids of the kinase catalytic domain were replaced with a neomycin cassette by homologous recombination [650].	Normal at birth but approximately 50 % are dead by P20, 100 % dead by P55. Reduction in the size of some sections of the nervous system including the dorsal root ganglion. Mottled fur and abnormal digit morphology.
TrkA$^{-/-}$ [653]	Constitutive loss of TrkA signaling. A PGK-neomycin resistance cassette replaced all coding sequences except the first 53 nucleotides of the exon that contains the translation initiation codon [653]. Homologous recombination was performed in CJ7 ES-cells.	Normal at birth. Small dorsal root ganglion. Exhibit reduction in sensory neuron number by E13.

Table 11.23 *TrkA* conditional knockout models

Strain name	Organ	Phenotype
TrkA$^{lox/lox}$	Central nervous system. A floxed PGK-neo pA cassette was inserted into intron 11 that serves as a STOP signal. Homologous recombination was performed in CJ7 ES-cells [655]. Cre mediated recombination results in reactivation of gene expression of wt TrkA.	*Tα1-Cre* [656] mediated *TrkA* reexpression only in neuronal tissues results in mice that are viable and display no gross abnormal phenotype but show evidence of increased B1-B cells and IgG abnormalities.
TrkA^{F592A}	Central nervous system. Point mutation: phenylalanine to alanine substitution mutation at residue 592 [654].	Mice are viable and fertile, with no obvious behavioral abnormalities. Addition of NMPP1, a derivative of the general kinase inhibitor PP1, inhibited neurotrophin signaling in conditional neurons but not in wild-type neurons.

models were generated via point mutations or homologous recombination of selection cassettes into regions that remove the start codon, effect protein binding, remove the extracellular domain or intracellular domain giving a plethora of models which differ in ways to allow for better studies of TrkB activity (Table 11.24) [234, 657–663]. In most of these null models the mice have a short lifespan (48 h to 2 weeks), particularly where mutations are made in the tyrosine kinase domain, although some *TrkB* mutant mice have normal life spans such as those models where point mutations were made in the *Shc* introns or *Plc* domains (Table 11.24). In nearly all cases the mice exhibit a severe defect in vestibular development, show small sensory numbers and have abnormal neural system morphology. In one case ablation of TrkB activity was associated with obesity [664] whereas another reported smaller body size and hormone levels [663] illustrating the significant differences that can result from modifications of alternate domains of the *TrkB* sequence.

Table 11.24 TrkB constitutive knockout models

Strain name	Organ	Phenotype
TrkB$^{-/-}$ *(Minichiello)*	Constitutive loss of TrkB signaling. Homologous recombination of a *PGK-neomycin* cassette into exon 2. Resulting protein lacks subdomains 3–5 of the tyrosine kinase domain [649].	Homozygous knock out results in death latest by 2–3 weeks of age. Mice lack a number of neuronal cell types including vestibular, nodose and cochlear neurons and display abnormal cerebellum development [662].
TrkB^{shc} *(Minichiello)*	Central nervous system; sensory neurons. Point mutation of tyrosine 515 to phenylalanine and insertion of selective *neomycin* cassette into the adjacent intron [662].	Normal lifespan but mutant mice show disrupted vestibular system; smaller sensory neurons and failure of many synapses to form in the vestibular canal epithelia.
TrkB$^{-/-}$ *(Rohrer)*	Constitutive loss of TrkB signaling. Homologous recombination of the first coding exon, specifically 31 codons which encode for the signal peptide and the start codon [663].	Average lifespan 48 h, some survival until 3 weeks. Small body size, abnormal breathing patterns, abnormal visual development [665], decreased follicle stimulating hormone levels [666].
TrkB^{plc} *(Minichiello)*	Central nervous system. Point mutation at the PLCγ site in the mouse *TrkB* Locus (Tyrosine 816 (Y816) to phenylalanine (F) change). Resulting protein fails to bind adaptor proteins [661].	Normal lifespan. Hippocampal long-term potentiation was impaired in these mice (unlike *TrkB*^{shc} mice). Hippocampal morphology normal.
TrkB^{lacZ}	Constitutive loss of TrkB signaling. Insertion of a*EphB2*^{lacZ} reporter cassette into the endogenous *TrkB* locus [664].	Mice survive but show a reduction in neuron regeneration rate and show early onset obesity.
TrkB$^{-/-}$ *(Luikart)*	Constitutive loss of TrkB signaling. Homologous recombination of the two start codons within the first exon [659].	Average lifespan 3–4 days, all dead by P15. Reduction in gross sensory neuron numbers
TrkBD	Central nervous system. Point mutation of phenylalanine for tyrosine at positions 515 and 816 (Y515F and Y816F) [660]	Mice die between 2 and 4 weeks. Show growth retardation. Abnormal neuron differentiation and migration. Abnormal cerebral cortex morphology.
TrkB$^{S478A/S478A}$	Central nervous system. Point mutation in Exon 12 with a modified exon in where alanine was substituted for serine at position 478 (S478A) [658].	Mice are viable and fertile with no gross morphological abnormalities but mutant mice show abnormal axon outgrowth and memory impairments.
TrkB.T1$^{-/-}$	Central nervous system. Homologous recombination of a *loxP* site 300pb upstream of exon 11 and a neo cassette downstream in CJ7 ES-cells. Neo and exon 11 were deleted in the germ line. By this strategy, the T1 truncated receptor isoform of TrkB is deleted [657].	T1 mutation rescues neurons from trisomy 16 mice from premature death, normalizes Ca2+ levels and restores BDNF-induced intracellular signaling.

11.17.3.1 Constitutive TrkB Knockout Strains

See Table 11.24.

11.17.3.2 Conditional TrkB Knockout Strains

Generation of knockout strain and phenotype:

Strain name	Organ	Phenotype
TrkB$^{lox/lox}$ *(He)*	Central nervous system. Insertion of *loxP* sites in between the signal peptide and the first 40 amino acids of the N terminus in exon 1 [667].	*Synapsin-Cre* (neuron specific) mediated deletion of *TrkB* in the hippocampus showed reduced kindling response (epileptogenesis)
TrkB^{F616A}	Central nervous system. Point mutation at amino acid coding site 616 [654]. The point mutation allows selective inhibition induced by kinase inhibitors.	Viable and fertile, with no obvious behavioral abnormality. 60 % loss in nodose ganglion neurons in TrkBF616A homozygous offspring whose mothers were treated with NMPP1 [654].
TrkB$^{lox/lox}$ *(Liu)*	Eye. *LoxP* sites were inserted into the first exon to generate a truncated protein following cre-mediated recombination [668].	*Six3-Cre* mediated TrkB deleted mice are healthy and fertile but show abnormal retinal ganglion cell morphology
TrkB$^{lox/lox}$ *(Minichiello)*	Exon 2 of TrkB was flanked by 2 loxP sites. Recombination results in deletion of the catalytically active pg145^{TrkB} receptors but expression of truncated isoforms of TrkB receptors [669].	Recombination after crossing to *CaMKII–Cre* active in hippocampus, cortex and olfactory bulb resulted in impaired learning behavior and reduced long term potentiation at synapses [669]. Crossing to *Glast-CreERT2* mice deleted TrkB in astrocates and neural progenitors after Tamoxifen application. Deletion resulted in reduced growth of dendrites and spines [670].

11.17.3.3 TrkB Conditional Knock in Strains

Generation of Knockin strain and phenotype:

Strain name	Organ	Phenotype
TrkB^{eGFP}	Central nervous system. Transgenic mouse contains an insertion into the *TrkB* locus that contains the Thy1 promoter followed by a loxP flanked stop signal and a truncated form of the murine *Ntrk2* gene fused to *eGFP*. Transcribed protein is a truncated form of TrkB fused to eGFP[671].	Mice are viable and show GFP fluorescence from 6 weeks of age in pyramidal neurons within layers II/III and V of the neocortex as well as pyramidal and granule cells of the hippocampus. Some abnormal pyramidal cell morphology was observed.

11.17.4 TrkC

TrkC receptor activity has been documented as an important Trk receptor in the early development of sympathetic neurons [645, 646] but also in sensory neurons and the vestibular system [648, 672]. Together *TrkC* knockout models illustrate the importance of the tyrosine domain in mouse viability whereas the Shc domain is important for neuronal survival.

11.17.4.1 Constitutive TrkC Knockout Strains

Generation of knockout strain and phenotype:

Strain name	Organ	Phenotype
TrkC$^{-/-}$ (Klein)	Constitutive loss of TrkC signaling. Homologous recombination of a neomycin cassette into second kinase exon K2 was performed in D3 ES-cells [648].	Most mice die by P21; some live up to 4 months of age. All mutant mice are smaller is size compared to equivalent aged wild-type littermates. Homozygous mice show abnormal movements and posture. Vestibular system abnormalities. They lack proprioceptive and cochlear innervation. Vestibular canal epithelium is smaller.
TrkC$^{-/-}$ *(Tessarollo)*	Constitutive loss of TrkC signaling. Homologous recombination of a neomycin cassette into the first exon resulting in lack of all isoforms of TrkC [672].	90 % animal mortality by day P6. All dead by P25. Decreased numbers of sensory neurons. Abnormal vein morphology. Severe heart defects and lung hemorrhaging
TrkC^{shc}	Vestibular system. Tyrosine to phenylalanine substitution via point mutation in the *Shc* adaptor binding site at residue 516 and homologous recombination of a neomycin resistance cassette in the adjacent upstream intron [673].	Normal lifespan but mice show a reduction in cochlear neuron number (25–50 %) throughout aging. But, unlike *TrkB*^{shc}mice [662] there is no loss of target innervation by cochlear neurons.

11.17.4.2 Conditional TrkC Knockout Strains

Generation of knockout strain and phenotype:

Strain name	Organ	Phenotype
TrkC^{F617A}	Central nervous system. Point mutation creating an amino acid substitution at 617 [654]. | The point mutation allows selective inhibition induced by kinase inhibitors.	Mice are viable and fertile, with no obvious behavioral abnormality. Dramatic decrease of parvalbumin-positive fibers in the lumbar spinal cord and PV-positive DRG neurons in 1NMPP1-treated TrkCF617A mice [654].

(continued)

(continued)

Strain name	Organ	Phenotype
TrkCT1$^{lox/lox}$	Homologous recombination in CJ7 ES-cells results in the exons encoding the T1 specific isoform flanked by loxP sites [674].	Mice are indistinguishable from wild-type littermates but have a reduced retinal ganglion cell loss in a model for glaucoma induction.

11.17.4.3 Conditional TrkC Knock in Strains

11.17.4.3.1 TrkA$^{TrkC/TrkC}$

Generation of knockout line: In this model a construct containing the *TrkC*-cDNA, preceded by an internal ribosomal entry site (*IRES*), was inserted into the second exon of *TrkA* genomic locus. *TrkA*$^{TrkC/TrkC}$ mice express both the endogenout*TrkA* and transgenic *TrkC* mRNA.

Phenotype: TrkC expression rescued the loss of sensory neuron phenotype observed in *TrkA*$^{-/-}$ mice and thus that TrkA/TrkC switch is important for the development of proprioceptive neurons [675].

11.17.4.3.2 TrkA$^{lacZ/lacZ}$

Generation of knockout line: A τ-lacZ–SV40pA selection cassette was inserted at the endogenous start site of TrkA. The removal of the selection cassette ensured proper transcription of the transgene.

Phenotype: At P2, the transgene was expressed in dorsal root ganglion (DRG) cell bodies and TrkA-positive DRG projections in dorsal spinal cord and skin (bottom). Robust expression of transgene was detectable at E13.5 [675].

11.18 The VEGF Receptor Family

11.18.1 Overview About Family Members

Vascular endothelial growth factor receptors (VEGFRs) are crucial regulators of vascular development during embryogenesis (vasculogenesis) as well as blood-vessel formation (angiogenesis) in the adult. The VEGFR family encompasses the three conserved receptors VEGFR-1, -2, and -3, as well as the coreceptors Neuropilin (NRP)-1 and -2, as well as heparan sulphate proteoglycans (HSPGs). Their ligands are VEGF-A, -B, -C, -D and the placental growth Factor (PlGF), structurally related dimeric glycoproteins which they bind in an overlapping pattern (Table 11.25).

Table 11.25 List of receptors and their respective ligands

Receptor	Alternative names	Ligands	Expression pattern	Function
VEGFR-1	Flt-1	VEGF-A, VEGF-B, PlGF	Widespread	Endoth.c. differentiation during vasculogenesis, modulation of VEGFR-2 signaling, monocyte migration
VEGFR-2	Flk-1, KDC	VEGF-A, VEGF-C, VEGF-D	Endoth. Cells	Master regulator of vasculogenesis and angiogenesis, inhibition of VEGFR-3
VEGFR-3	Flt-4	VEGF-C, VEGF-D	Embryonic: vascular endoth. Cells. Adult: lymphatic endoth. Cells, osteoblasts, neuronal progenitors	vasculogenesis, lymphangiogenesis
NRP-1		VEGF-A_{165}, VEGF-A_{121}	Endoth. c., developing neurons	Coreceptor for VEGFR-1/2, embryonic vessel formation
NRP-2		PlGF	Embryonic veins, lymphatic vessels	Coreceptor for VEGFR-3
HSPGs			Widespread	Concentration of VEGF-A isoforms, modulation of VEGFR-2 signal duration and intensity

VEGFR-1 and -2 may occur in a full length transmembrane form, or as truncated, alternatively spliced isoforms, that bind to and sequester their cognate ligands. VEGFR-1 binds VEGF-A with high affinity, whereas its kinase activity is poor and redundant for endothelial cell development [676]. In endothelial cells, its main function seems to be the control of VEGFR-2 activity by spatially regulating VEGF-A availability. Alternative splicing of VEGF-A gives rise to isoforms with diverse biological functions. In mice, they are denoted VEGF-$A_{120, 144, 164, 187,}$ and $_{205}$. Their functions are determined by their differential ability to bind to the NRP coreceptors [677]. Furthermore, VEGFR-1 is involved in angiogenesis during inflammation and cancer and has a role in monocyte and macrophage migration [677]. VEGFR-2 orchestrates many aspects of endothelial cell biology, controlling embryonic and adult vasculogenesis, and angiogenesis by regulating endothelial cell proliferation, survival, migration, and vessel permeability [676]. Most of its actions are mediated by VEGF-A, which it binds however with a tenfold lower affinity compared to VEGFR-1 [678, 679]. Additionally, VEGFR-2 binds to VEGF-C, thereby inhibiting VEGFR-3 signaling [680]. VEGFR-3 and its ligand VEGF-C are the master regulators of lymphatic vessel formation [676]. Whereas in most situations, VEGFR proteins act as homodimers, some VEGFR family proteins are able to form heterodimers. The NRP homologs NRP-1 and NRP-2 are single spanning transmembrane proteins with a short cytoplasmic domain that lack intrinsic catalytic function [681]. They are receptors for class 3 semaphorins, with a crucial role in neural development [682–685]. NRP-1 and -2 have been shown to modulate VEGFR-1/2 and VEGFR-3 function, playing important roles during vasculogenesis [686, 687]. NRP-1 binds to the VEGF-A isoforms VEGF-A_{165} and VEGF-A_{121}. This interaction has been shown to be relevant in tumor vessel formation [688].

Murine receptors: VEGFR-1,-2,-3
Coreceptors: Nrp-1, Nrp-2, HSPGs
Ligands: VEGF-A, -B, -C, -D, PlGF

Box: list of all mouse strains

Vegfr1$^{-/-}$	Fong et al. 1995 [689]
Vegfr1TK$^{-/-}$	Hiratsuka et al. 1998 [690], Hiratsuka et al. 2001 [691]
Vegfr1(TM-TK)$^{-/-}$	Hiratsuka et al. 2005 [692]
Pod/sVEGFR-1 mice	Ku et al. 2008 [693]
Vegfr1$^{loxP/loxP}$	Ambati et al. 2006 [694]
Vegfr2$^{-/-}$	Shalaby et al. 1995 [695]
Vegfr2^{I173F}	Sakurai et al. 2005 [696]
Vegfr2^{GFP}	Ema et al. 2006 [697]
Vegfr2$^{loxP/loxP}$	Albuquerque et al. 2009 [680]
Vegfr2^{Cre}	Licht et al. 2004 [698]
Vegfr3$^{-/-}$	Dumont et al. 1998 [699]
Vegfr3^{I1053F}	Karkkainen et al. 2001 [700]
Vegfr3^{Cre}	Ichise et al. 2010 [701]
Vegfr3^{EGFP}	Ichise et al. 2010 [701]
Vegfr3$^{loxP/loxP}$	Haiko et al. 2008 [702]
Vegfr3$^{\Delta LBD}$	Zhang et al. 2010 [703]
Vegfr3^{EGFPLuc}	Martinez-Corral et al. 2012 [704]
K14-VEGFR3-Ig transgenic	Mäkinen et al. 2001 [705]
Nrp1$^{-/-}$	Kitsukawa et al. 1997 [684] , Kawasaki et al. 1999 [706]
Nrp2$^{-/-}$	Giger et al. 2000 [707], Yuan et al. 2002 [708]
Nrp1$^{-/-}$ *Nrp2*$^{-/-}$	Takashima et al. 2002 [709]
Nrp-1 overexpression	Kitsukawa et al. 1995 [686]
Nrp-1^{sema}	Gu et al. 2003 [710]
Nrp-1$^{loxP/loxP}$	Goel et al. 2011 [711]
Nrp-2$^{loxP/loxP}$	Walz et al. 2002 [712]
Nrp-2^{mut}	Chen et al. 2000 [713]

11.18.2 Vegfr1

11.18.2.1 Constitutive Vegfr1 Knockout Strain

11.18.2.1.1 Vegfr1$^{-/-}$

Generation of knockout strain: Null mutation of the *Vegfr1* was achieved by targeting of the exon containing the translation initiation sequence of the *Vegfr1* gene using a construct derived from SV129 genomic DNA, replacing the signal peptide and the splicing donor site with a bacterial *β-galactosidase* gene (*LacZ*) and a neo^R gene. Due to expression of LacZ, Vegfr-1 expression can be located [689].

Phenotype: Homozygous embryos died around E8.5 due to abnormal development of major embryonic and extraembryonic vessels, affecting brain, heart, and yolk sac development [689]. This was due to increased hemangioblast commitment in *Vegfr1*$^{-/-}$ embryos during the primitive streak stage, leading to higher abundance of endothelial cells and dysfunctional vessel formation [714].

11.18.2.2 Conditional Vegfr1 Knockout Strain

11.18.2.2.1 *Vegfr1*$^{loxP/loxP}$ mice

Generation of knockout strain: Mice harboring a loxP flanked *Vegfr1* allele have been generated by Genentech and been mentioned for the first time by Ambati et al. 2006 [694]. The generation of the strain will be described elsewhere.

Phenotype: Intracorneal injection of a plasmid encoding Cre recombinase induced corneal neovascularization in *Vegfr1*$^{loxP/loxP}$ mice, accompanied by significantly reduced sVEGFR1 levels, demonstrating that sVEGFR1 mediates avascularity of the cornea [694]. Deletion of VEGFR1 in podocytes by intercrossing *Vegfr1*$^{loxP/loxP}$ mice with *Nphs-Cre* mice [715] results in kidney glomerulus injury [716].

11.18.2.3 Vegfr1 Knockin Strains

11.18.2.3.1 Vegfr1 TK$^{-/-}$

Generation of knockin strain: To generate mice lacking the tyrosine kinase domain of VEGFR1, the exon 17 encoding the 3′ juxtamembrane region and the amino terminal part of the tyrosine kinase domain were targeted in J1 ES-cells via homologous recombination using a 129Sv genomic library, replacing them with a neo^R cassette carrying its own promoter and polyadenylation signal. Alternative splicing without the neo^R gene to exon 18 caused an out-of-frame mutation. The exon encoding the transmembrane domain remained without mutation. Southern blot confirmed correct integration and Western blot confirmed the presence of the truncated VEGFR-1 [690].

Phenotype: VEGFR1 tyrosine kinase deficient homozygous mice were born at the expected Mendelian frequency, developed normal vessels and survived. However, VEGF-induced macrophage migration was strongly suppressed [690].

11.18.2.3.2 Vegfr1(TM-TK)$^{-/-}$

Generation of knockin strain: To generate *Vegfr1(TM)*$^{-/-}$ mice lacking the transmembrane domain of VEGFR1, exon 16 of the *Vegfr* gene encoding the transmembrane domain was targeted in CCE ES-cells via homologous recombination using a 129Sv library, replacing it with a neo^R gene [692]. To Vegfr(TM-TK) double mutant mice leaving only the extracellular domain, which can absorb VEGF but cannot induce the recruitment of VEGF to the cell membrane, were generated by intercrossing heterozygous Vegr-TM and Vegfr-TK mice [692]. Expression of the truncated VEGFR1 protein was confirmed by Western blot with an antibody raised against the polypeptide sequence of exon 15 [692].

Phenotype: 50 % of *Vegfr1(TM-TK)*$^{-/-}$ embryos died in utero at around E8.5 as a consequence of abnormal blood vessel formation with insufficient development of large vessels and fragmentation of endothelial cells. The other half developed normally and was healthy. *Vegfr1(TM-TK)*$^{-/-}$ with a growth arrest displayed diminished membrane-targeted VEGF, resulting in the suppression of VEGFR2 phosphorylation. Embryonic lethality in *Vegfr1(TM-TK)*$^{-/-}$ mice was increased to 80–90 % when they were bred into a *Vegfr2*$^{+/-}$ background [692].

11.18.2.4 Transgenic mouse lines of Vegfr1

11.18.2.4.1 Pod/sVegfr1

Generation of overexpressing strain: Mice overexpressing soluble VEGFR-1 (sVEGFR-1) specifically in podocytes in a doxycycline-inducible (Tet-on) manner were obtained by microinjection of a tetracycline responsive element construct [718] containing human *sVegfr-1* cDNA and the *LacZ* gene. Genomic transgene integration was confirmed by Southern blot [693]. The so generated heterozygous *pTRE-LacZ/sVegfr1* mice were then bred to *Podocin-rtTA* mice [96]. Expression of sVEGFR-1 was induced by administration of doxycycline (dox) in the drinking water. Dox-treated *Pod/sVegfr1* mice showed inducible sVEGFR-1 overexpression in the urine and in the kidney cortex [693].

Phenotype: Pod-Vegfr1 mice showed no evidence of changes in renal morphology and function. In a model of streptozocotin-induced diabetes, overexpression of VEGFR-1 in podocytes ameliorated diabetes-induced glomerulopathy, manifesting in reduced albuminuria [693].

11.18.3 Vegfr2

11.18.3.1 Constitutive Vegfr2 Knockout Strain

11.18.3.1.1 Vegfr2$^{-/-}$

Generation of knockout strain: The translated portion of the first coding exon and the proximal part of the first intron were replaced with a promoterless *LacZ* by homologous recombination in R1 ES-cells [695].

Phenotype: Homozygotes died of circulatory failure between E8.5 and E9. *Vegfr2*$^{-/-}$ mice exhibited poor blood island formation and were completely devoid of organized vessels, emphasizing VEGFR2 as a master regulator in endothelial cell biology and vasculogenesis [695].

11.18.3.2 Conditional Vegfr2 Knockout Strain

11.18.3.2.1 Vegfr2$^{loxP/loxP}$

Generation of knockout strain: A loxP site was inserted upstream of the first exon of *Vegfr2*. Another loxP site and a PGK-neo cassette flanked by *frt* sites were inserted into the first intron of the *Vegfr2* gene. The PGK-neo cassette was removed by FLPe-mediated recombination. Successful targeting was confirmed by Southern blot, absence of Vegfr-2 protein was confirmed by immunostaining [680].

Phenotype: Deletion of VEGFR2 in the epidermis and cornea by intercrossing *Vegfr2*$^{loxP/loxP}$ mice with *K14-Cre* [719] or *Le-Cre* [720] mice, respectively, resulted in enhanced lymphatic vasculature, establishing VEGFR2 as an antagonist of lymphangiogenesis [680].

11.18.3.3 Vegfr2 Knockin Strains

11.18.3.3.1 Vegfr2^{1173F} and Vegfr2^{1212F}

Generation of knockin strain: Site-directed mutagenesis was performed by homologous recombination with a genomic fragment isolated from a 129Sv library. Thereby, tyrosine to phenylalanine substitutions were introduced either at Y1173 or Y1212 (Y1173F or Y1212F, respectively) and a loxP-flanked *neo*^{R} gene was inserted. Targeting vectors were electroporated into 129 E14 ES-cells. Mutagenesis was confirmed by sequencing analysis with mRNA from homozygous mutants [696].

Phenotype: While *Vegfr2*$^{1212F/1212F}$ mice were viable and fertile, Homozygous *Vegfr2*$^{1173F/1173F}$ mutants died between E8.5 and 9.5 due to the absence of organized blood vessels and yolk sac blood islands, as well as a reduction of hematopoietic precursors [696].

11.18.3.3.2 VEGFR2^{EGFP}

Generation of the knockin strain: EGFP cDNA followed by a loxP flanked *PGK-neo* cassette was introduced into the first exon of the *Vegfr2* gene via homologous recombination in R1 ES-cells. After successful germline transmission of the targeted allele, the floxed *PGK-neo* cassette was removed by crossing with a Cre deleter strain, verified by Southern blot [697].

Phenotype: GFP expression was seen upon excision of the PGK-neo cassette in the E10.5 embryo within embryonic as well as extraembryonic tissues colocalizing with VEGFR2 expression [697].

11.18.4 Vegfr3

11.18.4.1 Constitutive Vegfr3 Knockout Strain

11.18.4.1.1 Vegfr3$^{-/-}$

Generation of knockout strain: The *Vegfr3* wild-type gene was disrupted by targeting of a *LacZ*-containing cassette into the first exon, under the control of transcriptional regulatory sequences via homologous recombination. Hereby, the initiation codon and signal peptide of the *Vegfr3* gene were deleted. Successful integration was confirmed by Southern blot, in situ hybridization confirmed the absence of mature mRNAs [699].

Phenotype: Vegfr3$^{-/-}$ embryos die between E10 and E11 of cardiovascular remodeling defects. While blood island and small vessel formation was intact in Vegfr-3$^{-/-}$ mice, large vessels were abnormally enlarged. LacZ staining revealed localization of the VEGFR-3 at the lymphatic endothelium [699].

11.18.4.2 Conditional Vegfr3 Knockout Strain

11.18.4.2.1 *Vegfr3$^{loxP/loxP}$* and *Vegfr3*^{neo}

Generation of knockout strain: The first exon and a part of the first intron of the *Vegfr3* gene were flanked with loxP sites in R1 ES-cells. The first intron of *Vegfr3* contains a neo^{R} gene flanked by *frt* sites. The inserted neomycin cassette (*Vegfr3$^{neo/neo}$*) produced a hypomorphic allele. To remove the *neo* cassette to generate the conditional allele, mice were crossed with mice expressing FLPe recombinase under the *β-actin* promoter. The resulting *Vegfr3$^{loxP/loxP}$* mice were crossed with *PgkCre* mice [721] to achieve germline deletion, and successful VEGFR3 deletion was confirmed by Western blot [702].

Phenotype: E14.5 *Vegfr3*$^{neo/neo}$ and *Vegfr3*$^{+/neo}$ embryos were swollen, due to the absence of lymphangiogenesis, whereas the blood vasculature was normal [702]. Germline deletion in *Vegfr3*$^{loxP/loxP}$ mice recapitulated the *Vegfr3*$^{-/-}$ phenotype [702]. Deletion of VEGFR3 in the epiblast using *K19-Cre* [722], produced an identical phenotype, suggesting that the vascular defects in *Vegfr3*$^{-/-}$ embryos are not caused by defects in placental development but that cardiovascular development is dependent on VEGFR3 signaling in the embryo. A small percentage of the *Vegfr3*$^{loxP/loxP}$*K19 Cre* mice survived without an apparent phenotype; however, in these mice, only partial recombination had taken place [702]. Postnatal deletion of VEGFR3 in the endothelium by crossing *Vegfr3*$^{loxP/loxP}$ mice with *PdcfbiCreER*^{T} mice [723] that express tamoxifen-activated Cre recombinase in endothelial cells elicited excessive angiogenic sprouting and branching, with concomitantly decreased NOTCH signaling [724].

11.18.4.2.2 Vegfr3Δ^{LBD}

Generation of knockout strain: Vegfr3^{loxP} mice, targeting the ligand binding domain of the VEGFR3, including Ig-like domain 2 and part of Ig-like domain 3 encoded by exon 4-6 of *Vegfr3* gene were created via homologous recombination using the BMQ433M21 BAC library, introducing a *loxP* into intron 3, and the second *loxP* together with a neo^{R} gene flanked by *frt* sites into intron 6 and electroporated into R1 ES-cells. The neo cassette was removed using FlPeR mice. The so generated *Vegfr3*^{loxP} mice were crossed with *EIIA-Cre* mice. Germ-line deletion of the floxed region of the *Vegfr3* gene was confirmed by expression analysis of E10.5 embryos at the mRNA and protein level [703].

Phenotype: In *Vegfr3*$^{\Delta LBD/\Delta LBD}$ mice, lymphatic growth was abrogated, whereas blood vessels developed normally in both embryo and yolk sac. Therefore, VEGFR-3 activation is not required for angiogenesis, and VEGFR-3 may play a role in this process by modulating VEGFR-2 mediated signals [703].

11.18.4.3 Vegfr3 Knock in Strains

11.18.4.3.1 Vegfr3^{EGFPLuc}

Generation of knock in strain: An EGFP-Luciferase-encoding sequence, preceded by an IRES element, was introduced into the 3′UTR of the mouse *Vegfr3* locus by homologous recombination. Introduction of the IRES-EGFPLuc cassette did not significantly alter the expression of VEGFR3. Knockin was confirmed by Northern blot and Western blot analysis [704].

Phenotype: Lymphatic vessels express the EGFP-luciferase fusion protein under the endogenous transcriptional control of the *Vegfr3* gene [704].

11.18.4.3.2 Vegfr3^{EGFP}

Generation of knock in strain: A fragment encompassing the promoter region as well as Exon 1 of *Vegfr3* amplified from the RPCI23-118J11 BAC DNA fused to an EGFP cDNA with a *Pgk* pA signal sequence followed by an *frt*-flanked *Pgk-gb3-neo* cassette was inserted upstream of the first ATG of the *Vegfr3* gene and introduced into ES-cells. Heterozygous *Vegfr3-EGFP* knockin mice without the neo gene were generated by crossing to the FLP-deleter strain FLP66 [701].

Phenotype: EGFP fluorescence was observed in lymphatic vessels [701].

11.18.4.4 ENU Mutated Mice

11.18.4.4.1 Vegfr3^{I1053F} (Chy mice)

Generation of mutagenic strain: Mice harboring a heterozygous kinase domain inactivating *Vegfr3* mutation occurred among the offspring of a C3H male treated with ethylnitrosurea. Sequencing revealed a heterozygous A3157T mutation resulting in I1053F substitution in the VEGFR3 tyrosine kinase domain. Lack of tyrosine phosphorylation was confirmed by immunoprecipitation of the VEGFR3 protein and subsequent probing with a phosphotyrosin specific antibody [700].

Phenotype: The heterozygous *Vegfr3^{I1053F}* mice developed a chylous fluid accumulation in the abdomen within the 3 first postnatal weeks, which was lethal in 10 %, and was spontaneously resorbed in the rest of the mice, which were healthy and fertile. However, they developed severe limb swelling because of hypoplastic cutaneous, but not visceral lymphatic vessels. They are a proposed model for human chylous ascites (Milroy's disease) because of the similarity of the mutation and phenotype [700].

11.18.4.5 Transgenic Mice of Vegfr3

11.18.4.5.1 Vegfr3^{Cre} Mice

Generation of transgenic strain: A 7 kb fragment harboring the mouse *Vegfr3* promoter region was subcloned from the RPCI23-118J11 BAC clone. The *Vegfr3* promoter and a nlsCre cDNA followed by a polyadenylation site from the mouse Pgk gene was injected into oocytes of C57BL/6J mice [701]. Cre activity was initially assessed by crossing with ROSA26-loxO-stop-loxP-β-geo knock in mice [725].

Phenotype: *VEGFR-3^{Cre}* mice that were bred to mice overexpressing RAS upon loxP-mediated recombination (*CAG-β-geo-Hras IRES EGFP* mice), exhibited lymphatic hyperplasia in the heart and mesentery [701].

11.18.4.5.2 K14-Vegfr3-Ig Transgenic

Generation of overexpressing strain: To generate mice overexpressing a chimeric soluble Ig-VEGFR3 protein in the basal epidermal cells of the skin, a fusion construct of human *VEGFR3* Ig homology domains 1–3 and the human *IgG1 Fc* tail was transferred into the human keratin-14 promoter-expression vector and subsequently injected into fertilized FVB/NIH and DBAxBalbC hybrid oocytes. Transgene expression in the founder lines was analyzed by Northern blotting and Western blotting from skin RNA and protein extracts, respectively [705].

Phenotype: K14-VEGFR3-Ig transgenic mice appeared healthy and fertile and had a normal lifespan. Upon transgene activation at around E16.5, regression of skin lymphatic vasculature and associated endothelial cell apoptosis was found, whereas the blood vasculature remained normal. By the postnatal age of 2 weeks, only few thin and rudimentary lymphatic vessels in organs such as heart and caecum, mesenterium, esophagus, diaphragm, lungs, and pancreas were present that regrew until 3 weeks of age. At 2 weeks, occasionally, pericardial fluid accumulation was observed in *K14-VEGFR3*-Ig mice. In adult mice, swelling of the extremities, edema and dermal fibrosis as well as a complete lack of lymphatic vessels in the skin was observed and additionally, some but not all lymph nodes were reduced in size [705].

11.18.5 Nrp1

11.18.5.1 Constitutive Nrp1 Knockout Strain

11.18.5.1.1 Nrp1$^{-/-}$

Generation of knockout strain: The wild-type *Nrp-1* gene was targeted by homologous recombination in TT-2 ES-cells. A segment was inserted into the exon encoding the N-terminal half of the a1 domain of the NRP1 protein. Thereby, the 5′ half of the exon was deleted and a neo^R gene was inserted out of frame into the deleted exon. Successful targeting was confirmed by Southern blotting and PCR analysis. Immunohistochemical staining of embryos against the fusion protein encoding the b1 and the c domain of the protein confirmed the lack of mature NRP1 protein in homozygous embryos [684].

Phenotype: Nrp1$^{-/-}$ embryos died in utero at E10.5–12.5 with defects in the development of the nervous system [684], heart, and embryonic and extraembryonic vasculature (heart, large and small vessels, yolk sac vasculature) [706].

11.18.5.2 Conditional Nrp1 Knockout Strain

11.18.5.2.1 Nrp1$^{f/f}$

Generation of knockout strain: The wild-type *Nrp1* gene was targeted by homologous recombination in 129.1 ES-cells, introducing a loxP flanked Exon 2 as well as a neo^R cassette flanked by FRT sites. The neo cassette was removed by breeding with germline FlpE expressing mice [726]. After crossing with mice carrying the Cre recombinase under the control of the *Tie-2* promoter [727], exon 2 was excised to generate an endothelial-cell specific *Nrp1* null allele. Successful recombination was confirmed by Southern blot [710]

Phenotype: Nrp1$^{-/}\Delta^{Tie2}$ mice died at mid-to-late gestation. Nrp-1$\Delta^{/}\Delta^{Tie2}$ mice died perinatally due to severe vascular and heart formation defects [710].

11.18.5.3 Nrp1 Knockin Strains

11.18.5.3.1 Nrp-1^{sema}

Generation of knockin strain: A 7 amino acid substitution was introduced into Exon 2 of the Nrp1 gene by homologous recombination using a loxP flanked neo cassette in 129.1 ES-cells. The neo cassette was removed by breeding heterozygous chimeras with mice expressing cre in germ cells. Successful recombination events were confirmed by Southern blot. Alkaline phosphatase section binding experiments showed that *Nrp-1^{sema}* mice display no binding of SEMA3A or SEMA 3C, while retaining $VEGF_{165}$ binding [710].

Phenotype: The *Nrp-1^{sema}* mice express a NRP1 protein in which signaling of Semaphorin is abolished while that of VEGF is retained. *Nrp-1^{sema}* mice are born with the expected Mendelian frequency, however, 40 % die within the first 7 days after birth. The majority of surviving mice exhibited growth retardation and nervous system abnormalities, while having a normal vasculature except for a cardiac atrial enlargement [710].

11.18.5.4 Transgenic Lines of Nrp1

11.18.5.4.1 Nrp1 Overexpression

Generation of overexpressing strain: A cDNA fragment corresponding to the protein-coding region of the mouse neuropilin cDNA (348th–119th bases of the clone M171) was inserted into the eukaryotic expression vector Miw [728], containing an RSV enhancer and chick *β-actin* promoter, expected to promote insert gene expression ubiquitously, was transfected in E14 ES-cells. Ectopic expression of NRP1 was confirmed by immunohistochemistry [686].

Phenotype: Transgenic embryos died in utero between E12.5 and E17.5, which was construct-dependent. They exhibited anomalies in the vascular system with excess capillaries, dilated vessels and cranial hemorrhage, accompanied by cardial malformation. In addition, neural defects were found, including incorrect sprouting of nerve fibers, sparse fasciculation of the peripheral nerve fibers. Embryos often exhibited extra digits [686].

11.18.6 Nrp2

11.18.6.1 Constitutive Nrp2 Knockout Strain

11.18.6.1.1 $Nrp2^{-/-}$ (Giger)

Generation of knockout strain: The first exon of the *Nrp2* gene including the entire signal sequence was targeted by homologous recombination using a 129SVJ lambda Fix II library and replaced by a neo^R expression cassette. Successful recombination was confirmed by Southern blot, Northern blot and Western blot analysis revealed, that the $Nrp2^{-/-}$ strain was a null mutant [707].

Phenotype: $Nrp\text{-}2^{-/-}$ mutants were born with reduced frequency and died more frequently as neonates and during early adulthood. Adult $Nrp2^{-/-}$ mice exhibited decreased fertility, multiple nervous system defects [707], and a paucity of small lymphatic vessels [708].

11.18.6.1.2 $Nrp2^{-/-}$ and $Nrp1^{-/-}Nrp2^{-/-}$ DKO (Takashima)

Generation of knockout strain: To generate $Nrp2^{-/-}$ mice, the translated portion of the first coding exon and the proximal part of the next intron of Nrp-2 were replaced with a promoterless *LacZ-neo* cassette by homologous recombination. Successful integration was confirmed by Southern blot. $Nrp1^{+/-}$ mice described in [684] were used to generate double knockouts [709].

Phenotype: $Nrp2^{-/-}$ mice were viable and survived up to 2 weeks after birth with no apparent vascular phenotype. Mice in which both *Nrp1* and *Nrp2* were targeted died in utero at E8.5. At this timepoint, yolk sac vascularization was completely absent. $Nrp^{+/-}$ $Nrp2^{-/-}$ or $Nrp\text{-}1^{-/-}Nrp2^{+/-}$ survived until E10-E10.5, but had a more abnormal vascular phenotype than single knockouts. These embryos exhibited a growth reduction of about 50 % at E10 and had multiple hemorrhages. The vasculature was disorganized; in the yolk sac, branching arteries and veins, as well as a capillary bed were absent, and large avascular spaces were present between the blood vessels. Embryos displayed blood vessels heterogeneous in size, large avascular regions in the head and trunk, as well as unconnected blood vessel sprouts [709].

11.18.6.2 Conditional Nrp2 Knockout Strain

11.18.6.2.1 Nrp2$^{f/f}$

Generation of knockout strain: The first intron of the *Nrp2* gene was targeted via homologous recombination using a 129/SvJ mouse bacterial artificial chromosome genomic library, inserting a cassette consisting of the gene for the axonal marker tauEGFP, preceded by a loxP sequence and followed by a polyadenylation site and a Pgk-neo cassette flanked by FRT sites. One loxP site was inserted upstream of the translational start site. The *neo*-selectable marker was removed from the targeted mutations by crossing heterozygous mice to *hACTB-Flp* transgenic mice [729]. Deletion of exon 1 and the first part of intron 1 was achieved by crossing to *EIIA-Cre* transgenic mice [88]. The final recombined allele encodes an mRNA lacking half of the first exon of Nrp-2, greatly reducing the ability of the mutant NRP2 to be targeted to the plasma membrane. Loss of the coding sequence encoding the signal peptide in the Nrp-2 mRNA was verified by RT-PCR. Cre recombination brings the axonal marker tauGFP under the control of the Nrp2 promoter. Because tauGFP is followed by a polyadenylation site, the chances of the formation of a functional Nrp-2 protein are further reduced. Consistent with a null phenotype, no NRP2 protein could be detected in immunostained sections of homozygous *Nrp2*$^{loxP/loxP}$ Cre brains [712].

Phenotype: Mice that lacked NRP2 in all cells (*Nrp2*$^{\Delta/\Delta EIIa\text{-}Cre}$) were born with a decreased frequency, were often sterile, had a growth retardation until weaning, and were more inclined to develop hydrocephalus compared to their heterozygous or wild-type littermates. Adult *Nrp2*$^{\Delta/\Delta EIIa\text{-}Cre}$ mice were morphologically indiscriminable, exhibited however aberrant sensory innervation of the olfactory bulb [712].

11.18.6.3 Mutated Nrp2 Mouse Lines

11.18.6.3.1 Nrp2^{mut}

Generation of mutated strain: Mice expressing a hypomorphic NRP2 were generated by gene trap using the secretory trap vector pGT1.8TM [730]. The insertion, interrupting the *Nrp2* coding sequence at cDNA position 2,069 nt, one-third of the way into the MAM domain of NRP2, was confirmed by RT-PCR and 5′-RACE. It constitutes a loss-of function mutation, with 0.5 % of residual protein, as confirmed by RNase protection assay [713].

Phenotype: *Nrp2*^{mut} mice exhibited β-galactosidase staining in NRP2 expressing tissues. Mice homozygous for this mutant allele were viable, and exhibited only small fertility disadvantages, but had nervous system defects [713].

11.19 NOK Receptor

No publications about mice with Nok alterations were found.

References

1. Webb TR et al. Anaplastic lymphoma kinase: role in cancer pathogenesis and small-molecule inhibitor development for therapy. Expert Rev Anticancer Ther. 2009;9(3):331–56.
2. Roll JD, Reuther GW. ALK-activating homologous mutations in LTK induce cellular transformation. PLoS One. 2012;7(2):e31733.
3. Morris SW et al. Fusion of a kinase gene, ALK, to a nucleolar protein gene, NPM, in non-Hodgkin's lymphoma. Science. 1994;263(5151):1281–4.
4. Bilsland JG et al. Behavioral and neurochemical alterations in mice deficient in anaplastic lymphoma kinase suggest therapeutic potential for psychiatric indications. Neuropsychopharmacology. 2008;33(3):685–700.
5. Chiarle R et al. NPM-ALK transgenic mice spontaneously develop T-cell lymphomas and plasma cell tumors. Blood. 2003;101(5):1919–27.
6. Wellstein A. ALK receptor activation, ligands and therapeutic targeting in glioblastoma and in other cancers. Front Oncol. 2012;2:192.
7. Lasek AW et al. An evolutionary conserved role for anaplastic lymphoma kinase in behavioral responses to ethanol. PLoS One. 2011;6(7):e22636.
8. Duyster J, Bai RY, Morris SW. Translocations involving anaplastic lymphoma kinase (ALK). Oncogene. 2001;20(40):5623–37.
9. Allouche M. ALK is a novel dependence receptor: potential implications in development and cancer. Cell Cycle. 2007;6(13):1533–8.
10. Pulford K, Morris SW, Turturro F. Anaplastic lymphoma kinase proteins in growth control and cancer. J Cell Physiol. 2004;199(3):330–58.
11. Pulford K et al. The emerging normal and disease-related roles of anaplastic lymphoma kinase. Cell Mol Life Sci. 2004;61(23):2939–53.
12. Heukamp LC et al. Targeted expression of mutated ALK induces neuroblastoma in transgenic mice. Sci Transl Med. 2012;4(141):141ra91.
13. Stanke M et al. Target-dependent specification of the neurotransmitter phenotype: cholinergic differentiation of sympathetic neurons is mediated in vivo by gp 130 signaling. Development. 2006;133(1):141–50.
14. Lindeberg J et al. Transgenic expression of Cre recombinase from the tyrosine hydroxylase locus. Genesis. 2004;40(2):67–73.
15. Giuriato S et al. Conditional TPM3-ALK and NPM-ALK transgenic mice develop reversible ALK-positive early B-cell lymphoma/leukemia. Blood. 2010;115(20):4061–70.
16. Turner SD et al. Vav-promoter regulated oncogenic fusion protein NPM-ALK in transgenic mice causes B-cell lymphomas with hyperactive Jun kinase. Oncogene. 2003;22(49):7750–61.
17. Turner SD et al. CD2 promoter regulated nucleophosmin-anaplastic lymphoma kinase in transgenic mice causes B lymphoid malignancy. Anticancer Res. 2006;26(5A):3275–9.
18. Jager R et al. Mice transgenic for NPM-ALK develop non-Hodgkin lymphomas. Anticancer Res. 2005;25(5):3191–6.
19. Soda M et al. A mouse model for EML4-ALK-positive lung cancer. Proc Natl Acad Sci U S A. 2008;105(50):19893–7.
20. Honda H et al. Heart-specific activation of LTK results in cardiac hypertrophy, cardiomyocyte degeneration and gene reprogramming in transgenic mice. Oncogene. 1999;18(26):3821–30.

21. Linger RM et al. TAM receptor tyrosine kinases: biologic functions, signaling, and potential therapeutic targeting in human cancer. Adv Cancer Res. 2008;100:35–83.
22. Lemke G, Rothlin CV. Immunobiology of the TAM receptors. Nat Rev Immunol. 2008;8(5):327–36.
23. Lemke G, Burstyn-Cohen T. TAM receptors and the clearance of apoptotic cells. Ann N Y Acad Sci. 2010;1209:23–9.
24. Lu Q et al. Tyro-3 family receptors are essential regulators of mammalian spermatogenesis. Nature. 1999;398(6729):723–8.
25. Angelillo-Scherrer A et al. Role of Gas6 receptors in platelet signaling during thrombus stabilization and implications for antithrombotic therapy. J Clin Invest. 2005;115(2):237–46.
26. Zhu D et al. Protein S controls hypoxic/ischemic blood-brain barrier disruption through the TAM receptor Tyro3 and sphingosine 1-phosphate receptor. Blood. 2010;115(23):4963–72.
27. Hoehn HJ et al. Axl-/- mice have delayed recovery and prolonged axonal damage following cuprizone toxicity. Brain Res. 2008;1240:1–11.
28. Ruan GX, Kazlauskas A. Axl is essential for VEGF-A-dependent activation of PI3K/Akt. EMBO J. 2012;31(7):1692–703.
29. Augustine KA et al. Noninsulin-dependent diabetes mellitus occurs in mice ectopically expressing the human Axl tyrosine kinase receptor. J Cell Physiol. 1999;181(3):433–47.
30. Burstyn-Cohen T et al. Genetic dissection of TAM receptor-ligand interaction in retinal pigment epithelial cell phagocytosis. Neuron. 2012;76(6):1123–32.
31. Camenisch TD et al. A novel receptor tyrosine kinase, Mer, inhibits TNF-alpha production and lipopolysaccharide-induced endotoxic shock. J Immunol. 1999;162(6):3498–503.
32. Scott RS et al. Phagocytosis and clearance of apoptotic cells is mediated by MER. Nature. 2001;411(6834):207–11.
33. Adams JM et al. Transgenic models of lymphoid neoplasia and development of a pan-hematopoietic vector. Oncogene. 1999;18(38):5268–77.
34. Keating AK et al. Lymphoblastic leukemia/lymphoma in mice overexpressing the Mer (MerTK) receptor tyrosine kinase. Oncogene. 2006;25(45):6092–100.
35. Maddox DM et al. An ENU-induced mutation in the Mertk gene (Mertknmf12) leads to a slow form of retinal degeneration. Invest Ophthalmol Vis Sci. 2011;52(7):4703–9.
36. Bauer T et al. Identification of Axl as a downstream effector of TGF-beta1 during Langerhans cell differentiation and epidermal homeostasis. J Exp Med. 2012;209(11):2033–47.
37. Caraux A et al. Natural killer cell differentiation driven by Tyro3 receptor tyrosine kinases. Nat Immunol. 2006;7(7):747–54.
38. Valiathan RR et al. Discoidin domain receptor tyrosine kinases: new players in cancer progression. Cancer Metastasis Rev. 2012;31(1–2):295–321.
39. Bargal R et al. Mutations in DDR2 gene cause SMED with short limbs and abnormal calcifications. Am J Hum Genet. 2009;84(1):80–4.
40. Ali BR et al. Trafficking defects and loss of ligand binding are the underlying causes of all reported DDR2 missense mutations found in SMED-SL patients. Hum Mol Genet. 2010;19(11):2239–50.
41. Vogel WF et al. Discoidin domain receptor 1 tyrosine kinase has an essential role in mammary gland development. Mol Cell Biol. 2001;21(8):2906–17.
42. Labrador JP et al. The collagen receptor DDR2 regulates proliferation and its elimination leads to dwarfism. EMBO Rep. 2001;2(5):446–52.
43. Kano K et al. A novel dwarfism with gonadal dysfunction due to loss-of-function allele of the collagen receptor gene, Ddr2, in the mouse. Mol Endocrinol. 2008;22(8):1866–80.
44. Gross O et al. DDR1-deficient mice show localized subepithelial GBM thickening with focal loss of slit diaphragms and proteinuria. Kidney Int. 2004;66(1):102–11.
45. Curat CA, Vogel WF. Discoidin domain receptor 1 controls growth and adhesion of mesangial cells. J Am Soc Nephrol. 2002;13(11):2648–56.
46. Meyer zum Gottesberge AM et al. Inner ear defects and hearing loss in mice lacking the collagen receptor DDR1. Lab Invest. 2008;88(1):27–37.

47. Olaso E et al. Loss of discoidin domain receptor 2 promotes hepatic fibrosis after chronic carbon tetrachloride through altered paracrine interactions between hepatic stellate cells and liver-associated macrophages. Am J Pathol. 2011;179(6):2894–904.
48. Olaso E et al. DDR2 receptor promotes MMP-2-mediated proliferation and invasion by hepatic stellate cells. J Clin Invest. 2001;108(9):1369–78.
49. Badiola I et al. Discoidin domain receptor 2 deficiency predisposes hepatic tissue to colon carcinoma metastasis. Gut. 2012;61(10):1465–72.
50. Kano K et al. Discoidin domain receptor 2 (DDR2) is required for maintenance of spermatogenesis in male mice. Mol Reprod Dev. 2010;77(1):29–37.
51. Matsumura H et al. Transcriptome analysis reveals an unexpected role of a collagen tyrosine kinase receptor gene, Ddr2, as a regulator of ovarian function. Physiol Genomics. 2009;39(2): 120–9.
52. Black J. Low birth weight dwarfism. Arch Dis Child. 1961;36:633–44.
53. Sibilia M, Wagner EF. Strain-dependent epithelial defects in mice lacking the EGF receptor. Science. 1995;269(5221):234–8.
54. Threadgill DW et al. Targeted disruption of mouse EGF receptor: effect of genetic background on mutant phenotype. Science. 1995;269(5221):230–4.
55. Miettinen PJ et al. Epithelial immaturity and multiorgan failure in mice lacking epidermal growth factor receptor. Nature. 1995;376(6538):337–41.
56. Natarajan A, Wagner B, Sibilia M. The EGF receptor is required for efficient liver regeneration. Proc Natl Acad Sci U S A. 2007;104(43):17081–6.
57. Lee TC, Threadgill DW. Generation and validation of mice carrying a conditional allele of the epidermal growth factor receptor. Genesis. 2009;47(2):85–92.
58. Dietrich P et al. Conditional mutagenesis in mice with heat shock promoter-driven cre transgenes. Mamm Genome. 2000;11(3):196–205.
59. Sibilia M et al. Mice humanised for the EGF receptor display hypomorphic phenotypes in skin, bone and heart. Development. 2003;130(19):4515–25.
60. Luetteke NC et al. The mouse waved-2 phenotype results from a point mutation in the EGF receptor tyrosine kinase. Genes Dev. 1994;8(4):399–413.
61. Lee D et al. Wa5 is a novel ENU-induced antimorphic allele of the epidermal growth factor receptor. Mamm Genome. 2004;15(7):525–36.
62. Fitch KR et al. Genetics of dark skin in mice. Genes Dev. 2003;17(2):214–28.
63. Du X et al. Velvet, a dominant Egfr mutation that causes wavy hair and defective eyelid development in mice. Genetics. 2004;166(1):331–40.
64. Hansen LA et al. Genetically null mice reveal a central role for epidermal growth factor receptor in the differentiation of the hair follicle and normal hair development. Am J Pathol. 1997;150(6):1959–75.
65. Hom YK et al. Uterine and vaginal organ growth requires epidermal growth factor receptor signaling from stroma. Endocrinology. 1998;139(3):913–21.
66. Sibilia M et al. A strain-independent postnatal neurodegeneration in mice lacking the EGF receptor. EMBO J. 1998;17(3):719–31.
67. Wagner B et al. Neuronal survival depends on EGFR signaling in cortical but not midbrain astrocytes. EMBO J. 2006;25(4):752–62.
68. Viti J et al. Epidermal growth factor receptors control competence to interpret leukemia inhibitory factor as an astrocyte inducer in developing cortex. J Neurosci. 2003;23(8): 3385–93.
69. Maklad A et al. The EGFR is required for proper innervation to the skin. J Invest Dermatol. 2009;129(3):690–8.
70. Tallquist MD, Soriano P. Epiblast-restricted Cre expression in MORE mice: a tool to distinguish embryonic vs. extra-embryonic gene function. Genesis. 2000;26(2):113–5.
71. Kellendonk C et al. Hepatocyte-specific expression of Cre recombinase. Genesis. 2000;26(2):151–3.
72. Kuhn R et al. Inducible gene targeting in mice. Science. 1995;269(5229):1427–9.

73. Murthy A et al. Ectodomain shedding of EGFR ligands and TNFR1 dictates hepatocyte apoptosis during fulminant hepatitis in mice. J Clin Invest. 2010;120(8):2731–44.
74. Indra AK et al. Temporally-controlled site-specific mutagenesis in the basal layer of the epidermis: comparison of the recombinase activity of the tamoxifen-inducible Cre-ER(T) and Cre-ER(T2) recombinases. Nucleic Acids Res. 1999;27(22):4324–7.
75. Lichtenberger BM et al. Autocrine VEGF signaling synergizes with EGFR in tumor cells to promote epithelial cancer development. Cell. 2010;140(2):268–79.
76. Eberl M et al. Hedgehog-EGFR cooperation response genes determine the oncogenic phenotype of basal cell carcinoma and tumour-initiating pancreatic cancer cells. EMBO Mol Med. 2012;4(3):218–33.
77. Tarutani M et al. Tissue-specific knockout of the mouse Pig-a gene reveals important roles for GPI-anchored proteins in skin development. Proc Natl Acad Sci U S A. 1997;94(14): 7400–5.
78. Lichtenberger BM et al. Epidermal EGFR controls cutaneous host defense and prevents inflammation. Sci Transl Med. 2013;5(199):199ra111.
79. Guerra C et al. Chronic pancreatitis is essential for induction of pancreatic ductal adenocarcinoma by K-Ras oncogenes in adult mice. Cancer Cell. 2007;11(3):291–302.
80. Navas C et al. EGF receptor signaling is essential for k-ras oncogene-driven pancreatic ductal adenocarcinoma. Cancer Cell. 2012;22(3):318–30.
81. Kawaguchi Y et al. The role of the transcriptional regulator Ptf1a in converting intestinal to pancreatic progenitors. Nat Genet. 2002;32(1):128–34.
82. Hingorani SR et al. Preinvasive and invasive ductal pancreatic cancer and its early detection in the mouse. Cancer Cell. 2003;4(6):437–50.
83. Ardito CM et al. EGF receptor is required for KRAS-induced pancreatic tumorigenesis. Cancer Cell. 2012;22(3):304–17.
84. Holtwick R et al. Smooth muscle-selective deletion of guanylyl cyclase-A prevents the acute but not chronic effects of ANP on blood pressure. Proc Natl Acad Sci U S A. 2002;99(10): 7142–7.
85. Schreier B et al. Consequences of epidermal growth factor receptor (ErbB1) loss for vascular smooth muscle cells from mice with targeted deletion of ErbB1. Arterioscler Thromb Vasc Biol. 2011;31(7):1643–52.
86. Schreier B et al. Loss of epidermal growth factor receptor in vascular smooth muscle cells and cardiomyocytes causes arterial hypotension and cardiac hypertrophy. Hypertension. 2013;61(2):333–40.
87. Zaiss DM et al. Amphiregulin enhances regulatory T cell-suppressive function via the epidermal growth factor receptor. Immunity. 2013;38(2):275–84.
88. Lakso M et al. Efficient in vivo manipulation of mouse genomic sequences at the zygote stage. Proc Natl Acad Sci U S A. 1996;93(12):5860–5.
89. Dassule HR et al. Sonic hedgehog regulates growth and morphogenesis of the tooth. Development. 2000;127(22):4775–85.
90. Bichsel KJ et al. Role for the epidermal growth factor receptor in chemotherapy-induced alopecia. PLoS One. 2013;8(7):e69368.
91. Franzke CW et al. Epidermal ADAM17 maintains the skin barrier by regulating EGFR ligand-dependent terminal keratinocyte differentiation. J Exp Med. 2012;209(6):1105–19.
92. Ramirez A et al. A keratin K5Cre transgenic line appropriate for tissue-specific or generalized Cre-mediated recombination. Genesis. 2004;39(1):52–7.
93. Mascia F et al. EGFR regulates the expression of keratinocyte-derived granulocyte/macrophage colony-stimulating factor in vitro and in vivo. J Invest Dermatol. 2010;130(3): 682–93.
94. Mascia F et al. Genetic ablation of epidermal EGFR reveals the dynamic origin of adverse effects of anti-EGFR therapy. Sci Transl Med. 2013;5(199):199ra110.
95. Eremina V et al. VEGF inhibition and renal thrombotic microangiopathy. N Engl J Med. 2008;358(11):1129–36.
96. Shigehara T et al. Inducible podocyte-specific gene expression in transgenic mice. J Am Soc Nephrol. 2003;14(8):1998–2003.

97. Bollee G et al. Epidermal growth factor receptor promotes glomerular injury and renal failure in rapidly progressive crescentic glomerulonephritis. Nat Med. 2011;17(10):1242–50.
98. Iwano M et al. Evidence that fibroblasts derive from epithelium during tissue fibrosis. J Clin Invest. 2002;110(3):341–50.
99. Chen J, Chen JK, Harris RC. Deletion of the epidermal growth factor receptor in renal proximal tubule epithelial cells delays recovery from acute kidney injury. Kidney Int. 2012;82(1):45–52.
100. Zhao H et al. Role of fibroblast growth factor receptors 1 and 2 in the ureteric bud. Dev Biol. 2004;276(2):403–15.
101. Zhang Z et al. Targeted inactivation of EGF receptor inhibits renal collecting duct development and function. J Am Soc Nephrol. 2010;21(4):573–8.
102. Liu F et al. Expression and activity of osteoblast-targeted Cre recombinase transgenes in murine skeletal tissues. Int J Dev Biol. 2004;48(7):645–53.
103. Zhang X et al. Epidermal growth factor receptor plays an anabolic role in bone metabolism in vivo. J Bone Miner Res. 2011;26(5):1022–34.
104. Ovchinnikov DA et al. Col2a1-directed expression of Cre recombinase in differentiating chondrocytes in transgenic mice. Genesis. 2000;26(2):145–6.
105. Zhang X et al. The critical role of the epidermal growth factor receptor in endochondral ossification. J Bone Miner Res. 2011;26(11):2622–33.
106. Zhang X et al. Epidermal growth factor receptor (EGFR) signaling regulates epiphyseal cartilage development through beta-catenin-dependent and -independent pathways. J Biol Chem. 2013;288(45):32229–40.
107. Clausen BE et al. Conditional gene targeting in macrophages and granulocytes using LysMcre mice. Transgenic Res. 1999;8(4):265–77.
108. Lu N et al. Activation of the epidermal growth factor receptor in macrophages regulates cytokine production and experimental colitis. J Immunol. 2014;192(3):1013–23.
109. Chen B et al. Mice mutant for Egfr and Shp2 have defective cardiac semilunar valvulogenesis. Nat Genet. 2000;24(3):296–9.
110. Barrick CJ et al. Reduced EGFR causes abnormal valvular differentiation leading to calcific aortic stenosis and left ventricular hypertrophy in C57BL/6J but not 129S1/SvImJ mice. Am J Physiol Heart Circ Physiol. 2009;297(1):H65–75.
111. Sibilia M et al. The EGF receptor provides an essential survival signal for SOS-dependent skin tumor development. Cell. 2000;102(2):211–20.
112. Egger B et al. Mice harboring a defective epidermal growth factor receptor (waved-2) have an increased susceptibility to acute dextran sulfate-induced colitis. Scand J Gastroenterol. 2000;35(11):1181–7.
113. Roberts RB et al. Importance of epidermal growth factor receptor signaling in establishment of adenomas and maintenance of carcinomas during intestinal tumorigenesis. Proc Natl Acad Sci U S A. 2002;99(3):1521–6.
114. Dahlhoff M et al. Decreased incidence of papillomas in mice with impaired EGFR function during multi-stage skin carcinogenesis. Exp Dermatol. 2011;20(3):290–3.
115. Brandl K et al. MyD88 signaling in nonhematopoietic cells protects mice against induced colitis by regulating specific EGF receptor ligands. Proc Natl Acad Sci U S A. 2010; 107(46):19967–72.
116. Murillas R et al. Expression of a dominant negative mutant of epidermal growth factor receptor in the epidermis of transgenic mice elicits striking alterations in hair follicle development and skin structure. EMBO J. 1995;14(21):5216–23.
117. Xie W et al. Targeted expression of a dominant negative epidermal growth factor receptor in the mammary gland of transgenic mice inhibits pubertal mammary duct development. Mol Endocrinol. 1997;11(12):1766–81.
118. Roh M et al. Stage-sensitive blockade of pituitary somatomammotrope development by targeted expression of a dominant negative epidermal growth factor receptor in transgenic mice. Mol Endocrinol. 2001;15(4):600–13.

119. Ling BC et al. Role for the epidermal growth factor receptor in neurofibromatosis-related peripheral nerve tumorigenesis. Cancer Cell. 2005;7(1):65–75.
120. Nir T, Melton DA, Dor Y. Recovery from diabetes in mice by beta cell regeneration. J Clin Invest. 2007;117(9):2553–61.
121. Politi K et al. Lung adenocarcinomas induced in mice by mutant EGF receptors found in human lung cancers respond to a tyrosine kinase inhibitor or to down-regulation of the receptors. Genes Dev. 2006;20(11):1496–510.
122. Hakonen E et al. In vivo activation of the PI3K-Akt pathway in mouse beta cells by the EGFR mutation L858R protects against diabetes. Diabetologia. 2014;57(5):970–9.
123. Tichelaar JW, Lu W, Whitsett JA. Conditional expression of fibroblast growth factor-7 in the developing and mature lung. J Biol Chem. 2000;275(16):11858–64.
124. Korfhagen TR et al. Cis-acting sequences from a human surfactant protein gene confer pulmonary-specific gene expression in transgenic mice. Proc Natl Acad Sci U S A. 1990; 87(16):6122–6.
125. Ohashi K et al. Induction of lung adenocarcinoma in transgenic mice expressing activated EGFR driven by the SP-C promoter. Cancer Sci. 2008;99(9):1747–53.
126. Ohashi K et al. Chemopreventive effects of gefitinib on nonsmoking-related lung tumorigenesis in activating epidermal growth factor receptor transgenic mice. Cancer Res. 2009;69(17):7088–95.
127. Li D et al. Bronchial and peripheral murine lung carcinomas induced by T790M-L858R mutant EGFR respond to HKI-272 and rapamycin combination therapy. Cancer Cell. 2007;12(1):81–93.
128. Ji H et al. Epidermal growth factor receptor variant III mutations in lung tumorigenesis and sensitivity to tyrosine kinase inhibitors. Proc Natl Acad Sci U S A. 2006;103(20):7817–22.
129. Zhou W et al. Novel mutant-selective EGFR kinase inhibitors against EGFR T790M. Nature. 2009;462(7276):1070–4.
130. Ding H et al. Oligodendrogliomas result from the expression of an activated mutant epidermal growth factor receptor in a RAS transgenic mouse astrocytoma model. Cancer Res. 2003;63(5):1106–13.
131. Zhu H et al. Oncogenic EGFR signaling cooperates with loss of tumor suppressor gene functions in gliomagenesis. Proc Natl Acad Sci U S A. 2009;106(8):2712–6.
132. Holland EC et al. A constitutively active epidermal growth factor receptor cooperates with disruption of G1 cell-cycle arrest pathways to induce glioma-like lesions in mice. Genes Dev. 1998;12(23):3675–85.
133. Meyer D, Birchmeier C. Multiple essential functions of neuregulin in development. Nature. 1995;378(6555):386–90.
134. Erickson SL et al. ErbB3 is required for normal cerebellar and cardiac development: a comparison with ErbB2-and heregulin-deficient mice. Development. 1997;124(24):4999–5011.
135. Lee KF et al. Requirement for neuregulin receptor erbB2 in neural and cardiac development. Nature. 1995;378(6555):394–8.
136. Park SK et al. The erbB2 gene is required for the development of terminally differentiated spinal cord oligodendrocytes. J Cell Biol. 2001;154(6):1245–58.
137. Britsch S et al. The ErbB2 and ErbB3 receptors and their ligand, neuregulin-1, are essential for development of the sympathetic nervous system. Genes Dev. 1998;12(12):1825–36.
138. Garratt AN et al. A dual role of erbB2 in myelination and in expansion of the schwann cell precursor pool. J Cell Biol. 2000;148(5):1035–46.
139. Crone SA et al. ErbB2 is essential in the prevention of dilated cardiomyopathy. Nat Med. 2002;8(5):459–65.
140. Leu M et al. Erbb2 regulates neuromuscular synapse formation and is essential for muscle spindle development. Development. 2003;130(11):2291–301.
141. Andrechek ER et al. ErbB2 is required for muscle spindle and myoblast cell survival. Mol Cell Biol. 2002;22(13):4714–22.
142. Chan R et al. The catalytic activity of the ErbB-2 receptor tyrosine kinase is essential for embryonic development. Mol Cell Biol. 2002;22(4):1073–8.

143. Chan R et al. Modulation of Erbb2 signaling during development: a threshold level of Erbb2 signaling is required for development. Development. 2004;131(22):5551–60.
144. Andrechek ER et al. Amplification of the neu/erbB-2 oncogene in a mouse model of mammary tumorigenesis. Proc Natl Acad Sci U S A. 2000;97(7):3444–9.
145. Lin W et al. Aberrant development of motor axons and neuromuscular synapses in erbB2-deficient mice. Proc Natl Acad Sci U S A. 2000;97(3):1299–304.
146. Schwenk F, Baron U, Rajewsky K. A cre-transgenic mouse strain for the ubiquitous deletion of loxP-flanked gene segments including deletion in germ cells. Nucleic Acids Res. 1995; 23(24):5080–1.
147. Voiculescu O, Charnay P, Schneider-Maunoury S. Expression pattern of a Krox-20/Cre knock-in allele in the developing hindbrain, bones, and peripheral nervous system. Genesis. 2000;26(2):123–6.
148. Chen J, Kubalak SW, Chien KR. Ventricular muscle-restricted targeting of the RXRalpha gene reveals a non-cell-autonomous requirement in cardiac chamber morphogenesis. Development. 1998;125(10):1943–9.
149. Wang J et al. Dilated cardiomyopathy and atrioventricular conduction blocks induced by heart-specific inactivation of mitochondrial DNA gene expression. Nat Genet. 1999; 21(1):133–7.
150. Andrechek ER et al. Germ-line expression of an oncogenic erbB2 allele confers resistance to erbB2-induced mammary tumorigenesis. Proc Natl Acad Sci U S A. 2004;101(14):4984–9.
151. Huang AL et al. Glucocorticoid regulation of the Ha-MuSV p21 gene conferred by sequences from mouse mammary tumor virus. Cell. 1981;27(2 Pt 1):245–55.
152. Muller WJ et al. Single-step induction of mammary adenocarcinoma in transgenic mice bearing the activated c-neu oncogene. Cell. 1988;54(1):105–15.
153. Muller WJ et al. Synergistic interaction of the Neu proto-oncogene product and transforming growth factor alpha in the mammary epithelium of transgenic mice. Mol Cell Biol. 1996;16(10):5726–36.
154. Guy CT et al. Expression of the neu protooncogene in the mammary epithelium of transgenic mice induces metastatic disease. Proc Natl Acad Sci U S A. 1992;89(22):10578–82.
155. Bouchard L et al. Stochastic appearance of mammary tumors in transgenic mice carrying the MMTV/c-neu oncogene. Cell. 1989;57(6):931–6.
156. Boggio K et al. Interleukin 12-mediated prevention of spontaneous mammary adenocarcinomas in two lines of Her-2/neu transgenic mice. J Exp Med. 1998;188(3):589–96.
157. Lucchini F et al. Early and multifocal tumors in breast, salivary, harderian and epididymal tissues developed in MMTY-Neu transgenic mice. Cancer Lett. 1992;64(3):203–9.
158. Dankort D et al. Grb2 and Shc adapter proteins play distinct roles in Neu (ErbB-2)-induced mammary tumorigenesis: implications for human breast cancer. Mol Cell Biol. 2001; 21(5):1540–51.
159. Oshima RG et al. Angiogenic acceleration of Neu induced mammary tumor progression and metastasis. Cancer Res. 2004;64(1):169–79.
160. Weinstein EJ, Leder P. The extracellular region of heregulin is sufficient to promote mammary gland proliferation and tumorigenesis but not apoptosis. Cancer Res. 2000;60(14): 3856–61.
161. Bol D et al. Severe follicular hyperplasia and spontaneous papilloma formation in transgenic mice expressing the neu oncogene under the control of the bovine keratin 5 promoter. Mol Carcinog. 1998;21(1):2–12.
162. Kiguchi K et al. Constitutive expression of erbB2 in epidermis of transgenic mice results in epidermal hyperproliferation and spontaneous skin tumor development. Oncogene. 2000;19(37):4243–54.
163. Xie W et al. Targeted expression of activated erbB-2 to the epidermis of transgenic mice elicits striking developmental abnormalities in the epidermis and hair follicles. Cell Growth Differ. 1998;9(4):313–25.
164. Xie W et al. Conditional expression of the ErbB2 oncogene elicits reversible hyperplasia in stratified epithelia and up-regulation of TGFalpha expression in transgenic mice. Oncogene. 1999;18(24):3593–607.

165. Piechocki MP et al. Human ErbB-2 (Her-2) transgenic mice: a model system for testing Her-2 based vaccines. J Immunol. 2003;171(11):5787–94.
166. Martin JF, Olson EN. Identification of a prx1 limb enhancer. Genesis. 2000;26(4):225–9.
167. Fisher MC et al. Requirement for ErbB2/ErbB signaling in developing cartilage and bone. Dev Growth Differ. 2007;49(6):503–13.
168. Sysa-Shah P et al. Cardiac-specific over-expression of epidermal growth factor receptor 2 (ErbB2) induces pro-survival pathways and hypertrophic cardiomyopathy in mice. PLoS One. 2012;7(8):e42805.
169. Siegel PM et al. Elevated expression of activated forms of Neu/ErbB-2 and ErbB-3 are involved in the induction of mammary tumors in transgenic mice: implications for human breast cancer. EMBO J. 1999;18(8):2149–64.
170. Riethmacher D et al. Severe neuropathies in mice with targeted mutations in the ErbB3 receptor. Nature. 1997;389(6652):725–30.
171. Qu S et al. Gene targeting of ErbB3 using a Cre-mediated unidirectional DNA inversion strategy. Genesis. 2006;44(10):477–86.
172. Lee D et al. Tumor-specific apoptosis caused by deletion of the ERBB3 pseudo-kinase in mouse intestinal epithelium. J Clin Invest. 2009;119(9):2702–13.
173. Brinkmann BG et al. Neuregulin-1/ErbB signaling serves distinct functions in myelination of the peripheral and central nervous system. Neuron. 2008;59(4):581–95.
174. Lahlou H et al. Uncoupling of PI3K from ErbB3 impairs mammary gland development but does not impact on ErbB2-induced mammary tumorigenesis. Cancer Res. 2012;72(12): 3080–90.
175. Buac K et al. A Sox10 expression screen identifies an amino acid essential for Erbb3 function. PLoS Genet. 2008;4(9):e1000177.
176. Vaught DB et al. HER3 is required for HER2-induced preneoplastic changes to the breast epithelium and tumor formation. Cancer Res. 2012;72(10):2672–82.
177. Balko JM et al. The receptor tyrosine kinase ErbB3 maintains the balance between luminal and basal breast epithelium. Proc Natl Acad Sci U S A. 2012;109(1):221–6.
178. Madison BB et al. Cis elements of the villin gene control expression in restricted domains of the vertical (crypt) and horizontal (duodenum, cecum) axes of the intestine. J Biol Chem. 2002;277(36):33275–83.
179. Lappe-Siefke C et al. Disruption of Cnp1 uncouples oligodendroglial functions in axonal support and myelination. Nat Genet. 2003;33(3):366–74.
180. Gassmann M et al. Aberrant neural and cardiac development in mice lacking the ErbB4 neuregulin receptor. Nature. 1995;378(6555):390–4.
181. Golub MS, Germann SL, Lloyd KC. Behavioral characteristics of a nervous system-specific erbB4 knock-out mouse. Behav Brain Res. 2004;153(1):159–70.
182. Jackson-Fisher AJ et al. Formation of Neu/ErbB2-induced mammary tumors is unaffected by loss of ErbB4. Oncogene. 2006;25(41):5664–72.
183. Garcia-Rivello H et al. Dilated cardiomyopathy in Erb-b4-deficient ventricular muscle. Am J Physiol Heart Circ Physiol. 2005;289(3):H1153–60.
184. Veikkolainen V et al. ErbB4 modulates tubular cell polarity and lumen diameter during kidney development. J Am Soc Nephrol. 2012;23(1):112–22.
185. Madisen L et al. A robust and high-throughput Cre reporting and characterization system for the whole mouse brain. Nat Neurosci. 2010;13(1):133–40.
186. Tidcombe H et al. Neural and mammary gland defects in ErbB4 knockout mice genetically rescued from embryonic lethality. Proc Natl Acad Sci U S A. 2003;100(14):8281–6.
187. Golding JP et al. Defects in pathfinding by cranial neural crest cells in mice lacking the neuregulin receptor ErbB4. Nat Cell Biol. 2000;2(2):103–9.
188. Zimmerman L et al. Independent regulatory elements in the nestin gene direct transgene expression to neural stem cells or muscle precursors. Neuron. 1994;12(1):11–24.
189. Long W et al. Impaired differentiation and lactational failure of Erbb4-deficient mammary glands identify ERBB4 as an obligate mediator of STAT5. Development. 2003;130(21):5257–68.
190. Bouchard M, Souabni A, Busslinger M. Tissue-specific expression of cre recombinase from the Pax8 locus. Genesis. 2004;38(3):105–9.

191. Unified nomenclature for Eph family receptors and their ligands, the ephrins. Eph Nomenclature Committee. Cell. 1997;90(3): 403–4.
192. Gale NW et al. Eph receptors and ligands comprise two major specificity subclasses and are reciprocally compartmentalized during embryogenesis. Neuron. 1996;17(1):9–19.
193. Henkemeyer M et al. Nuk controls pathfinding of commissural axons in the mammalian central nervous system. Cell. 1996;86(1):35–46.
194. Drescher U et al. In vitro guidance of retinal ganglion cell axons by RAGS, a 25 kDa tectal protein related to ligands for Eph receptor tyrosine kinases. Cell. 1995;82(3):359–70.
195. Iwasato T et al. Rac-GAP alpha-chimerin regulates motor-circuit formation as a key mediator of EphrinB3/EphA4 forward signaling. Cell. 2007;130(4):742–53.
196. Shamah SM et al. EphA receptors regulate growth cone dynamics through the novel guanine nucleotide exchange factor ephexin. Cell. 2001;105(2):233–44.
197. Foo SS et al. Ephrin-B2 controls cell motility and adhesion during blood-vessel-wall assembly. Cell. 2006;124(1):161–73.
198. Gerety SS et al. Symmetrical mutant phenotypes of the receptor EphB4 and its specific transmembrane ligand ephrin-B2 in cardiovascular development. Mol Cell. 1999;4(3):403–14.
199. Kadison SR et al. EphB receptors and ephrin-B3 regulate axon guidance at the ventral midline of the embryonic mouse spinal cord. J Neurosci. 2006;26(35):8909–14.
200. Brantley-Sieders DM et al. Impaired tumor microenvironment in EphA2-deficient mice inhibits tumor angiogenesis and metastatic progression. FASEB J. 2005;19(13):1884–6.
201. Brantley-Sieders DM et al. Ephrin-A1 facilitates mammary tumor metastasis through an angiogenesis-dependent mechanism mediated by EphA receptor and vascular endothelial growth factor in mice. Cancer Res. 2006;66(21):10315–24.
202. Pasquale EB. The Eph family of receptors. Curr Opin Cell Biol. 1997;9(5):608–15.
203. Duffy SL et al. Generation and characterization of EphA1 receptor tyrosine kinase reporter knockout mice. Genesis. 2008;46(10):553–61.
204. Chen J et al. Germ-line inactivation of the murine Eck receptor tyrosine kinase by gene trap retroviral insertion. Oncogene. 1996;12(5):979–88.
205. Brantley-Sieders DM et al. EphA2 receptor tyrosine kinase regulates endothelial cell migration and vascular assembly through phosphoinositide 3-kinase-mediated Rac1 GTPase activation. J Cell Sci. 2004;117(Pt 10):2037–49.
206. Guo H et al. Disruption of EphA2 receptor tyrosine kinase leads to increased susceptibility to carcinogenesis in mouse skin. Cancer Res. 2006;66(14):7050–8.
207. Jun G et al. EPHA2 is associated with age-related cortical cataract in mice and humans. PLoS Genet. 2009;5(7):e1000584.
208. Okazaki T et al. Capillary defects and exaggerated inflammatory response in the airways of EphA2-deficient mice. Am J Pathol. 2009;174(6):2388–99.
209. Day BW et al. EphA3 maintains tumorigenicity and is a therapeutic target in glioblastoma multiforme. Cancer Cell. 2013;23(2):238–48.
210. Vaidya A et al. EphA3 null mutants do not demonstrate motor axon guidance defects. Mol Cell Biol. 2003;23(22):8092–8.
211. Egea J et al. Regulation of EphA 4 kinase activity is required for a subset of axon guidance decisions suggesting a key role for receptor clustering in Eph function. Neuron. 2005;47(4): 515–28.
212. Orioli D et al. Sek4 and Nuk receptors cooperate in guidance of commissural axons and in palate formation. EMBO J. 1996;15(22):6035–49.
213. Dottori M et al. EphA4 (Sek1) receptor tyrosine kinase is required for the development of the corticospinal tract. Proc Natl Acad Sci U S A. 1998;95(22):13248–53.
214. Kullander K et al. Kinase-dependent and kinase-independent functions of EphA4 receptors in major axon tract formation in vivo. Neuron. 2001;29(1):73–84.
215. Helmbacher F et al. Targeting of the EphA4 tyrosine kinase receptor affects dorsal/ventral pathfinding of limb motor axons. Development. 2000;127(15):3313–24.
216. Leighton PA et al. Defining brain wiring patterns and mechanisms through gene trapping in mice. Nature. 2001;410(6825):174–9.

217. Grunwald IC et al. Hippocampal plasticity requires postsynaptic ephrinBs. Nat Neurosci. 2004;7(1):33–40.
218. Herrmann JE et al. Generation of an EphA4 conditional allele in mice. Genesis. 2010; 48(2):101–5.
219. Filosa A et al. Neuron-glia communication via EphA4/ephrin-A3 modulates LTP through glial glutamate transport. Nat Neurosci. 2009;12(10):1285–92.
220. Hanell A et al. Functional and histological outcome after focal traumatic brain injury is not improved in conditional EphA4 knockout mice. J Neurotrauma. 2012;29(17):2660–71.
221. Dudanova I et al. Genetic evidence for a contribution of EphA:ephrinA reverse signaling to motor axon guidance. J Neurosci. 2012;32(15):5209–15.
222. Kullander K et al. Ephrin-B3 is the midline barrier that prevents corticospinal tract axons from recrossing, allowing for unilateral motor control. Genes Dev. 2001;15(7):877–88.
223. Feldheim DA et al. Loss-of-function analysis of EphA receptors in retinotectal mapping. J Neurosci. 2004;24(10):2542–50.
224. Savelieva KV et al. Learning and memory impairment in Eph receptor A6 knockout mice. Neurosci Lett. 2008;438(2):205–9.
225. Rashid T et al. Opposing gradients of ephrin-As and EphA7 in the superior colliculus are essential for topographic mapping in the mammalian visual system. Neuron. 2005;47(1):57–69.
226. Park S, Frisen J, Barbacid M. Aberrant axonal projections in mice lacking EphA8 (Eek) tyrosine protein kinase receptors. EMBO J. 1997;16(11):3106–14.
227. Aasheim HC et al. Characterization of a novel Eph receptor tyrosine kinase, EphA10, expressed in testis. Biochim Biophys Acta. 2005;1723(1–3):1–7.
228. Williams SE et al. Ephrin-B2 and EphB1 mediate retinal axon divergence at the optic chiasm. Neuron. 2003;39(6):919–35.
229. Chenaux G, Henkemeyer M. Forward signaling by EphB1/EphB2 interacting with ephrin-B ligands at the optic chiasm is required to form the ipsilateral projection. Eur J Neurosci. 2011;34(10):1620–33.
230. Thakar S, Chenaux G, Henkemeyer M. Critical roles for EphB and ephrin-B bidirectional signalling in retinocollicular mapping. Nat Commun. 2011;2:431.
231. Cowan CA et al. EphB2 guides axons at the midline and is necessary for normal vestibular function. Neuron. 2000;26(2):417–30.
232. Genander M et al. Dissociation of EphB2 signaling pathways mediating progenitor cell proliferation and tumor suppression. Cell. 2009;139(4):679–92.
233. Holmberg J et al. EphB receptors coordinate migration and proliferation in the intestinal stem cell niche. Cell. 2006;125(6):1151–63.
234. Xu NJ, Henkemeyer M. Ephrin-B3 reverse signaling through Grb4 and cytoskeletal regulators mediates axon pruning. Nat Neurosci. 2009;12(3):268–76.
235. Freywald A, Sharfe N, Roifman CM. The kinase-null EphB6 receptor undergoes transphosphorylation in a complex with EphB1. J Biol Chem. 2002;277(6):3823–8.
236. Gurniak CB, Berg LJ. A new member of the Eph family of receptors that lacks protein tyrosine kinase activity. Oncogene. 1996;13(4):777–86.
237. Luo H et al. EphB6-null mutation results in compromised T cell function. J Clin Invest. 2004;114(12):1762–73.
238. Shimoyama M et al. Developmental expression of EphB6 in the thymus: lessons from EphB6 knockout mice. Biochem Biophys Res Commun. 2002;298(1):87–94.
239. Wang L et al. Anatomical coupling of sensory and motor nerve trajectory via axon tracking. Neuron. 2011;71(2):263–77.
240. Gallarda BW et al. Segregation of axial motor and sensory pathways via heterotypic trans-axonal signaling. Science. 2008;320(5873):233–6.
241. Chen L, Deng CX. Roles of FGF signaling in skeletal development and human genetic diseases. Front Biosci. 2005;10:1961–76.
242. Hou JZ et al. Fibroblast growth factor receptors from liver vary in three structural domains. Science. 1991;251(4994):665–8.

243. Du X et al. Role of FGFs/FGFRs in skeletal development and bone regeneration. J Cell Physiol. 2012;227(12):3731–43.
244. Turner N, Grose R. Fibroblast growth factor signalling: from development to cancer. Nat Rev Cancer. 2010;10(2):116–29.
245. Deng CX et al. Murine FGFR-1 is required for early postimplantation growth and axial organization. Genes Dev. 1994;8(24):3045–57.
246. Deng C et al. Fibroblast growth factor receptor-1 (FGFR-1) is essential for normal neural tube and limb development. Dev Biol. 1997;185(1):42–54.
247. Ciruna B, Rossant J. FGF signaling regulates mesoderm cell fate specification and morphogenetic movement at the primitive streak. Dev Cell. 2001;1(1):37–49.
248. Yamaguchi TP et al. fgfr-1 is required for embryonic growth and mesodermal patterning during mouse gastrulation. Genes Dev. 1994;8(24):3032–44.
249. Partanen J, Schwartz L, Rossant J. Opposite phenotypes of hypomorphic and Y766 phosphorylation site mutations reveal a function for Fgfr1 in anteroposterior patterning of mouse embryos. Genes Dev. 1998;12(15):2332–44.
250. Trokovic R et al. FGFR1 is independently required in both developing mid- and hindbrain for sustained response to isthmic signals. EMBO J. 2003;22(8):1811–23.
251. Hebert JM, McConnell SK. Targeting of cre to the Foxg1 (BF-1) locus mediates loxP recombination in the telencephalon and other developing head structures. Dev Biol. 2000;222(2): 296–306.
252. Pirvola U et al. FGFR1 is required for the development of the auditory sensory epithelium. Neuron. 2002;35(4):671–80.
253. Hebert JM et al. FGF signaling through FGFR1 is required for olfactory bulb morphogenesis. Development. 2003;130(6):1101–11.
254. Dacquin R et al. Mouse alpha1(I)-collagen promoter is the best known promoter to drive efficient Cre recombinase expression in osteoblast. Dev Dyn. 2002;224(2):245–51.
255. Jacob AL et al. Fibroblast growth factor receptor 1 signaling in the osteo-chondrogenic cell lineage regulates sequential steps of osteoblast maturation. Dev Biol. 2006;296(2):315–28.
256. Zhuo L et al. hGFAP-cre transgenic mice for manipulation of glial and neuronal function in vivo. Genesis. 2001;31(2):85–94.
257. Fagel DM et al. Fgfr1 is required for cortical regeneration and repair after perinatal hypoxia. J Neurosci. 2009;29(4):1202–11.
258. Ohkubo Y et al. Fibroblast growth factor receptor 1 is required for the proliferation of hippocampal progenitor cells and for hippocampal growth in mouse. J Neurosci. 2004; 24(27):6057–69.
259. Xu X et al. Generation of Fgfr1 conditional knockout mice. Genesis. 2002;32(2):85–6.
260. Nelson DK, Williams T. Frontonasal process-specific disruption of AP-2alpha results in postnatal midfacial hypoplasia, vascular anomalies, and nasal cavity defects. Dev Biol. 2004;267(1):72–92.
261. Li C et al. FGFR1 function at the earliest stages of mouse limb development plays an indispensable role in subsequent autopod morphogenesis. Development. 2005;132(21):4755–64.
262. Lowe LA, Yamada S, Kuehn MR. HoxB6-Cre transgenic mice express Cre recombinase in extra-embryonic mesoderm, in lateral plate and limb mesoderm and at the midbrain/hindbrain junction. Genesis. 2000;26(2):118–20.
263. Niwa Y et al. The initiation and propagation of Hes7 oscillation are cooperatively regulated by Fgf and notch signaling in the somite segmentation clock. Dev Cell. 2007;13(2):298–304.
264. Wahl MB et al. FGF signaling acts upstream of the NOTCH and WNT signaling pathways to control segmentation clock oscillations in mouse somitogenesis. Development. 2007; 134(22):4033–41.
265. Chen M et al. Generation of a transgenic mouse model with chondrocyte-specific and tamoxifen-inducible expression of Cre recombinase. Genesis. 2007;45(1):44–50.
266. Weng T et al. Genetic inhibition of fibroblast growth factor receptor 1 in knee cartilage attenuates the degeneration of articular cartilage in adult mice. Arthritis Rheum. 2012;64(12): 3982–92.

267. Werner S et al. Targeted expression of a dominant-negative FGF receptor mutant in the epidermis of transgenic mice reveals a role of FGF in keratinocyte organization and differentiation. EMBO J. 1993;12(7):2635–43.
268. Jackson D et al. Fibroblast growth factor receptor signalling has a role in lobuloalveolar development of the mammary gland. J Cell Sci. 1997;110(Pt 11):1261–8.
269. Saffell JL et al. Expression of a dominant negative FGF receptor inhibits axonal growth and FGF receptor phosphorylation stimulated by CAMs. Neuron. 1997;18(2):231–42.
270. Robinson ML et al. Expression of a truncated FGF receptor results in defective lens development in transgenic mice. Development. 1995;121(12):3959–67.
271. Peng H et al. Novel nuclear signaling pathway mediates activation of fibroblast growth factor-2 gene by type 1 and type 2 angiotensin II receptors. Mol Biol Cell. 2001;12(2):449–62.
272. Banerjee SA et al. 5' flanking sequences of the rat tyrosine hydroxylase gene target accurate tissue-specific, developmental, and transsynaptic expression in transgenic mice. J Neurosci. 1992;12(11):4460–7.
273. Klejbor I et al. Fibroblast growth factor receptor signaling affects development and function of dopamine neurons - inhibition results in a schizophrenia-like syndrome in transgenic mice. J Neurochem. 2006;97(5):1243–58.
274. Klejbor I et al. Serotonergic hyperinnervation and effective serotonin blockade in an FGF receptor developmental model of psychosis. Schizophr Res. 2009;113(2–3):308–21.
275. Arman E et al. Targeted disruption of fibroblast growth factor (FGF) receptor 2 suggests a role for FGF signaling in pregastrulation mammalian development. Proc Natl Acad Sci U S A. 1998;95(9):5082–7.
276. Nagy A et al. Derivation of completely cell culture-derived mice from early-passage embryonic stem cells. Proc Natl Acad Sci U S A. 1993;90(18):8424–8.
277. Arman E et al. Fgfr2 is required for limb outgrowth and lung-branching morphogenesis. Proc Natl Acad Sci U S A. 1999;96(21):11895–9.
278. Xu X et al. Fibroblast growth factor receptor 2 (FGFR2)-mediated reciprocal regulation loop between FGF8 and FGF10 is essential for limb induction. Development. 1998;125(4):753–65.
279. Yu K et al. Conditional inactivation of FGF receptor 2 reveals an essential role for FGF signaling in the regulation of osteoblast function and bone growth. Development. 2003;130(13):3063–74.
280. Walker KA et al. Deletion of fibroblast growth factor receptor 2 from the peri-wolffian duct stroma leads to ureteric induction abnormalities and vesicoureteral reflux. PLoS One. 2013;8(2):e56062.
281. Sun X et al. Conditional inactivation of Fgf4 reveals complexity of signalling during limb bud development. Nat Genet. 2000;25(1):83–6.
282. Mukhopadhyay A et al. Negative regulation of Shh levels by Kras and Fgfr2 during hair follicle development. Dev Biol. 2013;373(2):373–82.
283. Kaga Y et al. Mice with conditional inactivation of fibroblast growth factor receptor-2 signaling in oligodendrocytes have normal myelin but display dramatic hyperactivity when combined with Cnp1 inactivation. J Neurosci. 2006;26(47):12339–50.
284. Okubo T et al. Nmyc plays an essential role during lung development as a dosage-sensitive regulator of progenitor cell proliferation and differentiation. Development. 2005;132(6):1363–74.
285. Abler LL, Mansour SL, Sun X. Conditional gene inactivation reveals roles for Fgf10 and Fgfr2 in establishing a normal pattern of epithelial branching in the mouse lung. Dev Dyn. 2009;238(8):1999–2013.
286. Lin Y et al. Fibroblast growth factor receptor 2 tyrosine kinase is required for prostatic morphogenesis and the acquisition of strict androgen dependency for adult tissue homeostasis. Development. 2007;134(4):723–34.
287. Chi CL et al. The isthmic organizer signal FGF8 is required for cell survival in the prospective midbrain and cerebellum. Development. 2003;130(12):2633–44.
288. Blak AA et al. Fgfr2 and Fgfr3 are not required for patterning and maintenance of the midbrain and anterior hindbrain. Dev Biol. 2007;303(1):231–43.

289. Eswarakumar VP et al. Attenuation of signaling pathways stimulated by pathologically activated FGF-receptor 2 mutants prevents craniosynostosis. Proc Natl Acad Sci U S A. 2006;103(49):18603–8.
290. De Moerlooze L et al. An important role for the IIIb isoform of fibroblast growth factor receptor 2 (FGFR2) in mesenchymal-epithelial signalling during mouse organogenesis. Development. 2000;127(3):483–92.
291. Pirvola U et al. FGF/FGFR-2(IIIb) signaling is essential for inner ear morphogenesis. J Neurosci. 2000;20(16):6125–34.
292. Ramirez A et al. Sequences 5' of the bovine keratin 5 gene direct tissue- and cell-type-specific expression of a lacZ gene in the adult and during development. Differentiation. 1994;58(1):53–64.
293. Grose R et al. The role of fibroblast growth factor receptor 2b in skin homeostasis and cancer development. EMBO J. 2007;26(5):1268–78.
294. Revest JM et al. Fibroblast growth factor receptor 2-IIIb acts upstream of Shh and Fgf4 and is required for limb bud maintenance but not for the induction of Fgf8, Fgf10, Msx1, or Bmp4. Dev Biol. 2001;231(1):47–62.
295. Peters K et al. Targeted expression of a dominant negative FGF receptor blocks branching morphogenesis and epithelial differentiation of the mouse lung. EMBO J. 1994;13(14):3296–301.
296. Werner S et al. The function of KGF in morphogenesis of epithelium and reepithelialization of wounds. Science. 1994;266(5186):819–22.
297. Foster BA et al. Enforced expression of FGF-7 promotes epithelial hyperplasia whereas a dominant negative FGFR2iiib promotes the emergence of neuroendocrine phenotype in prostate glands of transgenic mice. Differentiation. 2002;70(9–10):624–32.
298. Wang Y et al. p38 Inhibition ameliorates skin and skull abnormalities in Fgfr2 Beare-Stevenson mice. J Clin Invest. 2012;122(6):2153–64.
299. Eswarakumar VP et al. The IIIc alternative of Fgfr2 is a positive regulator of bone formation. Development. 2002;129(16):3783–93.
300. Suzuki H et al. Apert syndrome mutant FGFR2 and its soluble form reciprocally alter osteogenesis of primary calvarial osteoblasts. J Cell Physiol. 2012;227(9):3267–77.
301. Wang Y et al. Activation of p38 MAPK pathway in the skull abnormalities of Apert syndrome Fgfr2(+P253R) mice. BMC Dev Biol. 2010;10:22.
302. Hajihosseini MK et al. A splicing switch and gain-of-function mutation in FgfR2-IIIc hemizygotes causes Apert/Pfeiffer-syndrome-like phenotypes. Proc Natl Acad Sci U S A. 2001;98(7):3855–60.
303. Celli G et al. Soluble dominant-negative receptor uncovers essential roles for fibroblast growth factors in multi-organ induction and patterning. EMBO J. 1998;17(6):1642–55.
304. Matsunaga S et al. Endothelium-targeted overexpression of constitutively active FGF receptor induces cardioprotection in mice myocardial infarction. J Mol Cell Cardiol. 2009;46(5):663–73.
305. Shigematsu A et al. Signaling from fibroblast growth factor receptor 2 in immature hematopoietic cells facilitates donor hematopoiesis after intra-bone marrow-bone marrow transplantation. Stem Cells Dev. 2010;19(11):1679–86.
306. Somarelli JA et al. Fluorescence-based alternative splicing reporters for the study of epithelial plasticity in vivo. RNA. 2013;19(1):116–27.
307. Colvin JS et al. Skeletal overgrowth and deafness in mice lacking fibroblast growth factor receptor 3. Nat Genet. 1996;12(4):390–7.
308. Deng C et al. Fibroblast growth factor receptor 3 is a negative regulator of bone growth. Cell. 1996;84(6):911–21.
309. Arnaud-Dabernat S et al. FGFR3 is a negative regulator of the expansion of pancreatic epithelial cells. Diabetes. 2007;56(1):96–106.
310. Carli A et al. FGF18 augments osseointegration of intra-medullary implants in osteopenic FGFR3(-/-) mice. Eur Cell Mater. 2012;24:107–16. discussion 116–7.
311. Moldrich RX et al. Fgfr3 regulates development of the caudal telencephalon. Dev Dyn. 2011;240(6):1586–99.

312. Su N et al. Generation of Fgfr3 conditional knockout mice. Int J Biol Sci. 2010; 6(4):327–32.
313. Wang JM et al. [Gly374Arg mutation in Fgfr3 causes achondroplasia in mice]. Zhonghua Yi Xue Yi Chuan Xue Za Zhi. 2004;21(6):537–41.
314. Wang Y et al. A mouse model for achondroplasia produced by targeting fibroblast growth factor receptor 3. Proc Natl Acad Sci U S A. 1999;96(8):4455–60.
315. Murakami S et al. Constitutive activation of MEK1 in chondrocytes causes Stat1-independent achondroplasia-like dwarfism and rescues the Fgfr3-deficient mouse phenotype. Genes Dev. 2004;18(3):290–305.
316. Chen L et al. Gly369Cys mutation in mouse FGFR3 causes achondroplasia by affecting both chondrogenesis and osteogenesis. J Clin Invest. 1999;104(11):1517–25.
317. Eswarakumar VP, Schlessinger J. Skeletal overgrowth is mediated by deficiency in a specific isoform of fibroblast growth factor receptor 3. Proc Natl Acad Sci U S A. 2007; 104(10):3937–42.
318. Chen J et al. Constitutively activated FGFR3 mutants signal through PLCgamma-dependent and -independent pathways for hematopoietic transformation. Blood. 2005;106(1):328–37.
319. Weinstein M et al. FGFR-3 and FGFR-4 function cooperatively to direct alveogenesis in the murine lung. Development. 1998;125(18):3615–23.
320. Yu C et al. Elevated cholesterol metabolism and bile acid synthesis in mice lacking membrane tyrosine kinase receptor FGFR4. J Biol Chem. 2000;275(20):15482–9.
321. French DM et al. Targeting FGFR4 inhibits hepatocellular carcinoma in preclinical mouse models. PLoS One. 2012;7(5):e36713.
322. Gutierrez A et al. Bile acids decrease hepatic paraoxonase 1 expression and plasma high-density lipoprotein levels via FXR-mediated signaling of FGFR4. Arterioscler Thromb Vasc Biol. 2006;26(2):301–6.
323. Yu C et al. Increased carbon tetrachloride-induced liver injury and fibrosis in FGFR4-deficient mice. Am J Pathol. 2002;161(6):2003–10.
324. Sims-Lucas S et al. Independent roles of Fgfr2 and Frs2alpha in ureteric epithelium. Development. 2011;138(7):1275–80.
325. Chen J et al. Selective requirement of myosin light chain 2v in embryonic heart function. J Biol Chem. 1998;273(2):1252–6.
326. Lavine KJ et al. Endocardial and epicardial derived FGF signals regulate myocardial proliferation and differentiation in vivo. Dev Cell. 2005;8(1):85–95.
327. Lavine KJ et al. Fibroblast growth factor signals regulate a wave of Hedgehog activation that is essential for coronary vascular development. Genes Dev. 2006;20(12):1651–66.
328. Hayashi S, McMahon AP. Efficient recombination in diverse tissues by a tamoxifen-inducible form of Cre: a tool for temporally regulated gene activation/inactivation in the mouse. Dev Biol. 2002;244(2):305–18.
329. Gutin G et al. FGF signalling generates ventral telencephalic cells independently of SHH. Development. 2006;133(15):2937–46.
330. Paek H, Gutin G, Hebert JM. FGF signaling is strictly required to maintain early telencephalic precursor cell survival. Development. 2009;136(14):2457–65.
331. Furusho M et al. Fibroblast growth factor signaling is required for the generation of oligodendrocyte progenitors from the embryonic forebrain. J Neurosci. 2011;31(13):5055–66.
332. Li W et al. Squamous cell carcinoma and mammary abscess formation through squamous metaplasia in Smad4/Dpc4 conditional knockout mice. Development. 2003;130(24): 6143–53.
333. Poladia DP et al. Role of fibroblast growth factor receptors 1 and 2 in the metanephric mesenchyme. Dev Biol. 2006;291(2):325–39.
334. Hains DS et al. High incidence of vesicoureteral reflux in mice with Fgfr2 deletion in kidney mesenchyma. J Urol. 2010;183(5):2077–84.
335. Stevens HE et al. Fgfr2 is required for the development of the medial prefrontal cortex and its connections with limbic circuits. J Neurosci. 2010;30(16):5590–602.
336. Kang W et al. The transition from radial glial to intermediate progenitor cell is inhibited by FGF signaling during corticogenesis. J Neurosci. 2009;29(46):14571–80.

337. Yang J et al. Fibroblast growth factor receptors 1 and 2 in keratinocytes control the epidermal barrier and cutaneous homeostasis. J Cell Biol. 2010;188(6):935–52.
338. Zhao H et al. Fibroblast growth factor receptor signaling is essential for lens fiber cell differentiation. Dev Biol. 2008;318(2):276–88.
339. Kitamura T, Kahn CR, Accili D. Insulin receptor knockout mice. Annu Rev Physiol. 2003;65:313–32.
340. Petrenko AG, et al. Insulin receptor-related receptor as an extracellular pH sensor involved in the regulation of acid-base balance. Biochim Biophys Acta. 2013;1834(10):2170-5. doi:10.1016/j.bbapap.2012.11.011. Epub 2012 Dec 7.
341. Wutz A et al. Imprinted expression of the Igf2r gene depends on an intronic CpG island. Nature. 1997;389(6652):745–9.
342. Pollak M. The insulin and insulin-like growth factor receptor family in neoplasia: an update. Nat Rev Cancer. 2012;12(3):159–69.
343. Accili D et al. Early neonatal death in mice homozygous for a null allele of the insulin receptor gene. Nat Genet. 1996;12(1):106–9.
344. Joshi RL et al. Targeted disruption of the insulin receptor gene in the mouse results in neonatal lethality. EMBO J. 1996;15(7):1542–7.
345. Bruning JC et al. A muscle-specific insulin receptor knockout exhibits features of the metabolic syndrome of NIDDM without altering glucose tolerance. Mol Cell. 1998;2(5):559–69.
346. Kulkarni RN et al. Tissue-specific knockout of the insulin receptor in pancreatic beta cells creates an insulin secretory defect similar to that in type 2 diabetes. Cell. 1999;96(3): 329–39.
347. Bruning JC et al. Role of brain insulin receptor in control of body weight and reproduction. Science. 2000;289(5487):2122–5.
348. Guerra C et al. Brown adipose tissue-specific insulin receptor knockout shows diabetic phenotype without insulin resistance. J Clin Invest. 2001;108(8):1205–13.
349. Bluher M et al. Adipose tissue selective insulin receptor knockout protects against obesity and obesity-related glucose intolerance. Dev Cell. 2002;3(1):25–38.
350. Rask-Madsen C et al. Loss of insulin signaling in vascular endothelial cells accelerates atherosclerosis in apolipoprotein E null mice. Cell Metab. 2010;11(5):379–89.
351. Belke DD et al. Insulin signaling coordinately regulates cardiac size, metabolism, and contractile protein isoform expression. J Clin Invest. 2002;109(5):629–39.
352. Welsh GI et al. Insulin signaling to the glomerular podocyte is critical for normal kidney function. Cell Metab. 2010;12(4):329–40.
353. Baumgartl J et al. Myeloid lineage cell-restricted insulin resistance protects apolipoproteinE-deficient mice against atherosclerosis. Cell Metab. 2006;3(4):247–56.
354. Mauer J et al. Myeloid cell-restricted insulin receptor deficiency protects against obesity-induced inflammation and systemic insulin resistance. PLoS Genet. 2010;6(5):e1000938.
355. Stachelscheid H et al. Epidermal insulin/IGF-1 signalling control interfollicular morphogenesis and proliferative potential through Rac activation. EMBO J. 2008;27(15):2091–101.
356. Pitetti JL et al. Insulin receptor and IGF1R are not required for oocyte growth, differentiation, and maturation in mice. Sex Dev. 2009;3(5):264–72.
357. Ock S et al. Conditional deletion of insulin receptor in thyrocytes does not affect thyroid structure and function. Endocr J. 2011;58(11):1013–9.
358. Divall SA et al. Divergent roles of growth factors in the GnRH regulation of puberty in mice. J Clin Invest. 2010;120(8):2900–9.
359. Benecke H et al. Muscle-specific expression of human insulin receptor in transgenic mice. Diabetes. 1993;42(1):206–12.
360. Chang PY et al. Expression of a dominant-negative mutant human insulin receptor in the muscle of transgenic mice. J Biol Chem. 1994;269(23):16034–40.
361. Moller DE et al. Transgenic mice with muscle-specific insulin resistance develop increased adiposity, impaired glucose tolerance, and dyslipidemia. Endocrinology. 1996;137(6): 2397–405.
362. Okamoto H et al. Transgenic rescue of insulin receptor-deficient mice. J Clin Invest. 2004;114(2):214–23.

363. Wang L et al. Hyperinsulinemia but no diabetes in transgenic mice homozygously expressing the tyrosine kinase-deficient human insulin receptor. Biochem Biophys Res Commun. 1997;240(2):446–51.
364. Baudry A et al. Partial rescue of insulin receptor-deficient mice by transgenic complementation with an activated insulin receptor in the liver. Gene. 2002;299(1–2):219–25.
365. Liu JP et al. Mice carrying null mutations of the genes encoding insulin-like growth factor I (Igf-1) and type 1 IGF receptor (Igf1r). Cell. 1993;75(1):59–72.
366. Kloting N et al. Autocrine IGF-1 action in adipocytes controls systemic IGF-1 concentrations and growth. Diabetes. 2008;57(8):2074–82.
367. Holzenberger M et al. Cre-mediated germline mosaicism: a method allowing rapid generation of several alleles of a target gene. Nucleic Acids Res. 2000;28(21):E92.
368. Xuan S et al. Defective insulin secretion in pancreatic beta cells lacking type 1 IGF receptor. J Clin Invest. 2002;110(7):1011–9.
369. Delmas V et al. Cre-mediated recombination in the skin melanocyte lineage. Genesis. 2003;36(2):73–80.
370. Desbois-Mouthon C et al. Hepatocyte proliferation during liver regeneration is impaired in mice with liver-specific IGF-1R knockout. FASEB J. 2006;20(6):773–5.
371. Sutherland BW et al. Conditional deletion of insulin-like growth factor-I receptor in prostate epithelium. Cancer Res. 2008;68(9):3495–504.
372. Kappeler L et al. Brain IGF-1 receptors control mammalian growth and lifespan through a neuroendocrine mechanism. PLoS Biol. 2008;6(10):e254.
373. Klinakis A et al. Igf1r as a therapeutic target in a mouse model of basal-like breast cancer. Proc Natl Acad Sci U S A. 2009;106(7):2359–64.
374. Rowland KJ et al. Loss of glucagon-like peptide-2-induced proliferation following intestinal epithelial insulin-like growth factor-1-receptor deletion. Gastroenterology. 2011;141(6):2166–2175e7.
375. Abbas A et al. The insulin-like growth factor-1 receptor is a negative regulator of nitric oxide bioavailability and insulin sensitivity in the endothelium. Diabetes. 2011;60(8):2169–78.
376. Freude S et al. Neuronal IGF-1 resistance reduces Abeta accumulation and protects against premature death in a model of Alzheimer's disease. FASEB J. 2009;23(10):3315–24.
377. Li P et al. IGF signaling directs ventricular cardiomyocyte proliferation during embryonic heart development. Development. 2011;138(9):1795–805.
378. Romero CJ et al. Targeted deletion of somatotroph insulin-like growth factor-I signaling in a cell-specific knockout mouse model. Mol Endocrinol. 2010;24(5):1077–89.
379. Moellendorf S et al. IGF-IR signaling attenuates the age-related decline of diastolic cardiac function. Am J Physiol Endocrinol Metab. 2012;303(2):E213–22.
380. McMullen JR et al. The insulin-like growth factor 1 receptor induces physiological heart growth via the phosphoinositide 3-kinase(p110alpha) pathway. J Biol Chem. 2004;279(6):4782–93.
381. Fernandez AM et al. Functional inactivation of the IGF-I and insulin receptors in skeletal muscle causes type 2 diabetes. Genes Dev. 2001;15(15):1926–34.
382. Fernandez AM et al. Muscle-specific inactivation of the IGF-I receptor induces compensatory hyperplasia in skeletal muscle. J Clin Invest. 2002;109(3):347–55.
383. Bridgewater DJ et al. The role of the type I insulin-like growth factor receptor (IGF-IR) in glomerular integrity. Growth Horm IGF Res. 2008;18(1):26–37.
384. Clement S et al. Low TSH requirement and goiter in transgenic mice overexpressing IGF-I and IGF-Ir receptor in the thyroid gland. Endocrinology. 2001;142(12):5131–9.
385. Carboni JM et al. Tumor development by transgenic expression of a constitutively active insulin-like growth factor I receptor. Cancer Res. 2005;65(9):3781–7.
386. Jones RA et al. Transgenic overexpression of IGF-IR disrupts mammary ductal morphogenesis and induces tumor formation. Oncogene. 2007;26(11):1636–44.
387. Wang ZQ et al. Regulation of embryonic growth and lysosomal targeting by the imprinted Igf2/Mpr gene. Nature. 1994;372(6505):464–7.
388. Lau MM et al. Loss of the imprinted IGF2/cation-independent mannose 6-phosphate receptor results in fetal overgrowth and perinatal lethality. Genes Dev. 1994;8(24):2953–63.

389. Ludwig T et al. Mouse mutants lacking the type 2 IGF receptor (IGF2R) are rescued from perinatal lethality in Igf2 and Igf1r null backgrounds. Dev Biol. 1996;177(2):517–35.
390. Wutz A et al. Non-imprinted Igf2r expression decreases growth and rescues the Tme mutation in mice. Development. 2001;128(10):1881–7.
391. Hughes J et al. Maternal transmission of a humanised igf2r allele results in an igf2 dependent hypomorphic and non-viable growth phenotype. PLoS One. 2013;8(2):e57270.
392. Wylie AA et al. Tissue-specific inactivation of murine M6P/IGF2R. Am J Pathol. 2003;162(1):321–8.
393. Pravtcheva DD, Wise TL. Igf2r improves the survival and transmission ratio of Igf2 transgenic mice. Mol Reprod Dev. 2008;75(11):1678–87.
394. Wise TL, Pravtcheva DD. Delayed onset of Igf2-induced mammary tumors in Igf2r transgenic mice. Cancer Res. 2006;66(3):1327–36.
395. Zaina S et al. Local reduction of organ size in transgenic mice expressing a soluble insulin-like growth factor II/mannose-6-phosphate receptor. Endocrinology. 1998;139(9):3886–95.
396. Harper J et al. Soluble IGF2 receptor rescues Apc(Min/+) intestinal adenoma progression induced by Igf2 loss of imprinting. Cancer Res. 2006;66(4):1940–8.
397. Kitamura T et al. Preserved pancreatic beta-cell development and function in mice lacking the insulin receptor-related receptor. Mol Cell Biol. 2001;21(16):5624–30.
398. Deyev IE et al. Deficient Response to Experimentally Induced Alkalosis in Mice with the Inactivated insrr Gene. Acta Naturae. 2011;3(4):114–7.
399. Park M et al. Sequence of MET protooncogene cDNA has features characteristic of the tyrosine kinase family of growth-factor receptors. Proc Natl Acad Sci U S A. 1987;84(18):6379–83.
400. Cooper CS et al. Molecular cloning of a new transforming gene from a chemically transformed human cell line. Nature. 1984;311(5981):29–33.
401. Giordano S et al. Tyrosine kinase receptor indistinguishable from the c-met protein. Nature. 1989;339(6220):155–6.
402. Ronsin C et al. A novel putative receptor protein tyrosine kinase of the met family. Oncogene. 1993;8(5):1195–202.
403. Huff JL et al. The protooncogene c-sea encodes a transmembrane protein-tyrosine kinase related to the Met/hepatocyte growth factor/scatter factor receptor. Proc Natl Acad Sci U S A. 1993;90(13):6140–4.
404. Bottaro DP et al. Identification of the hepatocyte growth factor receptor as the c-met proto-oncogene product. Science. 1991;251(4995):802–4.
405. Stoker M et al. Scatter factor is a fibroblast-derived modulator of epithelial cell mobility. Nature. 1987;327(6119):239–42.
406. Gherardi E et al. Purification of scatter factor, a fibroblast-derived basic protein that modulates epithelial interactions and movement. Proc Natl Acad Sci U S A. 1989;86(15):5844–8.
407. Di Renzo MF et al. Expression of the Met/HGF receptor in normal and neoplastic human tissues. Oncogene. 1991;6(11):1997–2003.
408. Bussolino F et al. Hepatocyte growth factor is a potent angiogenic factor which stimulates endothelial cell motility and growth. J Cell Biol. 1992;119(3):629–41.
409. Grant DS et al. Scatter factor induces blood vessel formation in vivo. Proc Natl Acad Sci U S A. 1993;90(5):1937–41.
410. Anastasi S et al. A natural hepatocyte growth factor/scatter factor autocrine loop in myoblast cells and the effect of the constitutive Met kinase activation on myogenic differentiation. J Cell Biol. 1997;137(5):1057–68.
411. Ebens A et al. Hepatocyte growth factor/scatter factor is an axonal chemoattractant and a neurotrophic factor for spinal motor neurons. Neuron. 1996;17(6):1157–72.
412. Galimi F et al. Hepatocyte growth factor induces proliferation and differentiation of multipotent and erythroid hemopoietic progenitors. J Cell Biol. 1994;127(6 Pt 1):1743–54.
413. Nishino T et al. Hepatocyte growth factor as a hematopoietic regulator. Blood. 1995; 85(11):3093–100.
414. Gherardi E, Stoker M. Hepatocyte growth factor–scatter factor: mitogen, motogen, and met. Cancer Cells. 1991;3(6):227–32.

415. Schmidt C et al. Scatter factor/hepatocyte growth factor is essential for liver development. Nature. 1995;373(6516):699–702.
416. Maina F et al. Uncoupling of Grb2 from the Met receptor in vivo reveals complex roles in muscle development. Cell. 1996;87(3):531–42.
417. Nakamura T, Teramoto H, Ichihara A. Purification and characterization of a growth factor from rat platelets for mature parenchymal hepatocytes in primary cultures. Proc Natl Acad Sci U S A. 1986;83(17):6489–93.
418. Michalopoulos GK. Liver regeneration: molecular mechanisms of growth control. FASEB J. 1990;4(2):176–87.
419. Rubin JS et al. A broad-spectrum human lung fibroblast-derived mitogen is a variant of hepatocyte growth factor. Proc Natl Acad Sci U S A. 1991;88(2):415–9.
420. Montesano R et al. Identification of a fibroblast-derived epithelial morphogen as hepatocyte growth factor. Cell. 1991;67(5):901–8.
421. Stoker M, Perryman M. An epithelial scatter factor released by embryo fibroblasts. J Cell Sci. 1985;77:209–23.
422. Wang MH et al. Identification of the ron gene product as the receptor for the human macrophage stimulating protein. Science. 1994;266(5182):117–9.
423. Iwama A et al. Terminal differentiation of murine resident peritoneal macrophages is characterized by expression of the STK protein tyrosine kinase, a receptor for macrophage-stimulating protein. Blood. 1995;86(9):3394–403.
424. Wang MH et al. Requirement of phosphatidylinositol-3 kinase for epithelial cell migration activated by human macrophage stimulating protein. Oncogene. 1996;13(10):2167–75.
425. Kurihara N et al. Macrophage-stimulating protein activates STK receptor tyrosine kinase on osteoclasts and facilitates bone resorption by osteoclast-like cells. Blood. 1996;87(9):3704–10.
426. Iwama A, Yamaguchi N, Suda T. STK/RON receptor tyrosine kinase mediates both apoptotic and growth signals via the multifunctional docking site conserved among the HGF receptor family. EMBO J. 1996;15(21):5866–75.
427. Banu N et al. Modulation of megakaryocytopoiesis by human macrophage-stimulating protein, the ligand for the RON receptor. J Immunol. 1996;156(8):2933–40.
428. Danilkovitch A, Leonard EJ. Kinases involved in MSP/RON signaling. J Leukoc Biol. 1999;65(3):345–8.
429. Danilkovitch A et al. Two independent signaling pathways mediate the antiapoptotic action of macrophage-stimulating protein on epithelial cells. Mol Cell Biol. 2000;20(6):2218–27.
430. Bladt F et al. Essential role for the c-met receptor in the migration of myogenic precursor cells into the limb bud. Nature. 1995;376(6543):768–71.
431. Chan AM et al. Primary structure of the met protein tyrosine kinase domain. Oncogene. 1987;1(2):229–33.
432. Low MJ et al. Tissue-specific posttranslational processing of pre-prosomatostatin encoded by a metallothionein-somatostatin fusion gene in transgenic mice. Cell. 1985;41(1):211–9.
433. Liang TJ et al. Transgenic expression of tpr-met oncogene leads to development of mammary hyperplasia and tumors. J Clin Invest. 1996;97(12):2872–7.
434. Palmiter RD et al. Distal regulatory elements from the mouse metallothionein locus stimulate gene expression in transgenic mice. Mol Cell Biol. 1993;13(9):5266–75.
435. Hogan B, Costantini F, Lacy E. Manipulating the mouse embryo : a laboratory manual. Cold Spring Harbor, NY: Cold Spring Harbor Laboratory; 1986. ix, 332 p.
436. Jeffers M et al. The mutationally activated Met receptor mediates motility and metastasis. Proc Natl Acad Sci U S A. 1998;95(24):14417–22.
437. Hogan B. Manipulating the mouse embryo : a laboratory manual. 2nd ed. Plainview, NY: Cold Spring Harbor Laboratory Press; 1994. xvii, 497 p.
438. Kistner A et al. Doxycycline-mediated quantitative and tissue-specific control of gene expression in transgenic mice. Proc Natl Acad Sci U S A. 1996;93(20):10933–8.
439. Wang R et al. Activation of the Met receptor by cell attachment induces and sustains hepatocellular carcinomas in transgenic mice. J Cell Biol. 2001;153(5):1023–34.

440. Moshitch-Moshkovitz S et al. In vivo direct molecular imaging of early tumorigenesis and malignant progression induced by transgenic expression of GFP-Met. Neoplasia. 2006;8(5):353–63.
441. Ponzo MG et al. Met induces mammary tumors with diverse histologies and is associated with poor outcome and human basal breast cancer. Proc Natl Acad Sci U S A. 2009;106(31):12903–8.
442. Williams-Simons L, Westphal H. EIIaCre – utility of a general deleter strain. Transgenic Res. 1999;8(4):53–4.
443. Postic C, Magnuson MA. DNA excision in liver by an albumin-Cre transgene occurs progressively with age. Genesis. 2000;26(2):149–50.
444. Huh CG et al. Hepatocyte growth factor/c-met signaling pathway is required for efficient liver regeneration and repair. Proc Natl Acad Sci U S A. 2004;101(13):4477–82.
445. Roccisana J et al. Targeted inactivation of hepatocyte growth factor receptor c-met in beta-cells leads to defective insulin secretion and GLUT-2 downregulation without alteration of beta-cell mass. Diabetes. 2005;54(7):2090–102.
446. Yu J, Carroll TJ, McMahon AP. Sonic hedgehog regulates proliferation and differentiation of mesenchymal cells in the mouse metanephric kidney. Development. 2002;129(22):5301–12.
447. Ishibe S et al. Met and the epidermal growth factor receptor act cooperatively to regulate final nephron number and maintain collecting duct morphology. Development. 2009;136(2):337–45.
448. Borowiak M et al. Met provides essential signals for liver regeneration. Proc Natl Acad Sci U S A. 2004;101(29):10608–13.
449. Uehara Y et al. Placental defect and embryonic lethality in mice lacking hepatocyte growth factor/scatter factor. Nature. 1995;373(6516):702–5.
450. Huelsken J et al. beta-Catenin controls hair follicle morphogenesis and stem cell differentiation in the skin. Cell. 2001;105(4):533–45.
451. Chmielowiec J et al. c-Met is essential for wound healing in the skin. J Cell Biol. 2007;177(1):151–62.
452. Graveel C et al. Activating Met mutations produce unique tumor profiles in mice with selective duplication of the mutant allele. Proc Natl Acad Sci U S A. 2004;101(49):17198–203.
453. Graveel CR et al. Met induces diverse mammary carcinomas in mice and is associated with human basal breast cancer. Proc Natl Acad Sci U S A. 2009;106(31):12909–14.
454. Gaudino G et al. RON is a heterodimeric tyrosine kinase receptor activated by the HGF homologue MSP. EMBO J. 1994;13(15):3524–32.
455. Correll PH et al. Deregulated inflammatory response in mice lacking the STK/RON receptor tyrosine kinase. Genes Funct. 1997;1(1):69–83.
456. Muraoka RS et al. The Ron/STK receptor tyrosine kinase is essential for peri-implantation development in the mouse. J Clin Invest. 1999;103(9):1277–85.
457. Monaghan AP et al. Postimplantation expression patterns indicate a role for the mouse forkhead/HNF-3 alpha, beta and gamma genes in determination of the definitive endoderm, chordamesoderm and neuroectoderm. Development. 1993;119(3):567–78.
458. Kaestner KH et al. The HNF-3 gene family of transcription factors in mice: gene structure, cDNA sequence, and mRNA distribution. Genomics. 1994;20(3):377–85.
459. Kaestner KH et al. Inactivation of the winged helix transcription factor HNF3alpha affects glucose homeostasis and islet glucagon gene expression in vivo. Genes Dev. 1999;13(4):495–504.
460. Waltz SE et al. Ron-mediated cytoplasmic signaling is dispensable for viability but is required to limit inflammatory responses. J Clin Invest. 2001;108(4):567–76.
461. Peace BE et al. Ron receptor signaling augments mammary tumor formation and metastasis in a murine model of breast cancer. Cancer Res. 2005;65(4):1285–93.
462. Chan EL et al. Ron tyrosine kinase receptor regulates papilloma growth and malignant conversion in a murine model of skin carcinogenesis. Oncogene. 2005;24(3):479–88.
463. Greenberg NM et al. Prostate cancer in a transgenic mouse. Proc Natl Acad Sci U S A. 1995;92(8):3439–43.

464. Thobe MN et al. The Ron receptor promotes prostate tumor growth in the TRAMP mouse model. Oncogene. 2011;30(50):4990–8.
465. Chen YQ et al. Multiple pulmonary adenomas in the lung of transgenic mice overexpressing the RON receptor tyrosine kinase. Carcinogenesis. 2002;23(11):1811–9.
466. Zinser GM et al. Mammary-specific Ron receptor overexpression induces highly metastatic mammary tumors associated with beta-catenin activation. Cancer Res. 2006;66(24):11967–74.
467. Babicky M. RON overexpression accelerates tumorigenesis and induces metastasis in a KRAS mutant mouse model of pancreatic cancer. Surgical Oncology I. 2011: S131. ISSN 1072-7515/11/$36.00 Published by Elsevier Inc. doi:10.1016/j.jamcollsurg.2011.06.314.
468. Gray JK et al. Ron receptor overexpression in the murine prostate induces prostate intraepithelial neoplasia. Cancer Lett. 2012;314(1):92–101.
469. DeChiara TM et al. The receptor tyrosine kinase MuSK is required for neuromuscular junction formation in vivo. Cell. 1996;85(4):501–12.
470. Glass DJ et al. Agrin acts via a MuSK receptor complex. Cell. 1996;85(4):513–23.
471. Jones G et al. Induction by agrin of ectopic and functional postsynaptic-like membrane in innervated muscle. Proc Natl Acad Sci U S A. 1997;94(6):2654–9.
472. Sander A et al. Gene transfer into individual muscle fibers and conditional gene expression in living animals. Cell Tissue Res. 2000;301(3):397–403.
473. Sander A, Hesser BA, Witzemann V. MuSK induces in vivo acetylcholine receptor clusters in a ligand-independent manner. J Cell Biol. 2001;155(7):1287–96.
474. Hesser BA, Henschel O, Witzemann V. Synapse disassembly and formation of new synapses in postnatal muscle upon conditional inactivation of MuSK. Mol Cell Neurosci. 2006;31(3):470–80.
475. Herbst R, Avetisova E, Burden SJ. Restoration of synapse formation in Musk mutant mice expressing a Musk/Trk chimeric receptor. Development. 2002;129(23):5449–60.
476. Herbst R, Burden SJ. The juxtamembrane region of MuSK has a critical role in agrin-mediated signaling. EMBO J. 2000;19(1):67–77.
477. Arber S et al. Requirement for the homeobox gene Hb9 in the consolidation of motor neuron identity. Neuron. 1999;23(4):659–74.
478. Yang X et al. Patterning of muscle acetylcholine receptor gene expression in the absence of motor innervation. Neuron. 2001;30(2):399–410.
479. Kim N, Burden SJ. MuSK controls where motor axons grow and form synapses. Nat Neurosci. 2008;11(1):19–27.
480. Chevessier F et al. A mouse model for congenital myasthenic syndrome due to MuSK mutations reveals defects in structure and function of neuromuscular junctions. Hum Mol Genet. 2008;17(22):3577–95.
481. Moll J et al. An agrin minigene rescues dystrophic symptoms in a mouse model for congenital muscular dystrophy. Nature. 2001;413(6853):302–7.
482. Punga AR et al. MuSK levels differ between adult skeletal muscles and influence postsynaptic plasticity. Eur J Neurosci. 2011;33(5):890–8.
483. Ben Ammar A et al. A mutation causes MuSK reduced sensitivity to agrin and congenital myasthenia. PLoS One. 2013;8(1):e53826.
484. Fredriksson L, Li H, Eriksson U. The PDGF family: four gene products form five dimeric isoforms. Cytokine Growth Factor Rev. 2004;15(4):197–204.
485. Levitzki A. PDGF receptor kinase inhibitors for the treatment of PDGF driven diseases. Cytokine Growth Factor Rev. 2004;15(4):229–35.
486. Hoch RV, Soriano P. Roles of PDGF in animal development. Development. 2003;130(20):4769–84.
487. Soriano P. The PDGF alpha receptor is required for neural crest cell development and for normal patterning of the somites. Development. 1997;124(14):2691–700.
488. Brennan J, Tilmann C, Capel B. Pdgfr-alpha mediates testis cord organization and fetal Leydig cell development in the XY gonad. Genes Dev. 2003;17(6):800–10.

489. Bostrom H, Gritli-Linde A, Betsholtz C. PDGF-A/PDGF alpha-receptor signaling is required for lung growth and the formation of alveoli but not for early lung branching morphogenesis. Dev Dyn. 2002;223(1):155–62.
490. Stephenson DA et al. Platelet-derived growth factor receptor alpha-subunit gene (Pdgfra) is deleted in the mouse patch (Ph) mutation. Proc Natl Acad Sci U S A. 1991;88(1):6–10.
491. Zhang FX, Hutchins JB. Expression of platelet-derived growth factor (PDGF) receptor alpha-subunit in mouse brain: comparison of Patch mutants and normal littermates. Cell Mol Neurobiol. 1996;16(4):479–87.
492. Morrison-Graham K et al. A PDGF receptor mutation in the mouse (Patch) perturbs the development of a non-neuronal subset of neural crest-derived cells. Development. 1992; 115(1):133–42.
493. Soriano P. Abnormal kidney development and hematological disorders in PDGF beta-receptor mutant mice. Genes Dev. 1994;8(16):1888–96.
494. Lindahl P et al. Pericyte loss and microaneurysm formation in PDGF-B-deficient mice. Science. 1997;277(5323):242–5.
495. Lindahl P et al. Paracrine PDGF-B/PDGF-Rbeta signaling controls mesangial cell development in kidney glomeruli. Development. 1998;125(17):3313–22.
496. Mellgren AM et al. Platelet-derived growth factor receptor beta signaling is required for efficient epicardial cell migration and development of two distinct coronary vascular smooth muscle cell populations. Circ Res. 2008;103(12):1393–401.
497. Tallquist MD, Soriano P. Cell autonomous requirement for PDGFRalpha in populations of cranial and cardiac neural crest cells. Development. 2003;130(3):507–18.
498. Jackson EL et al. PDGFR alpha-positive B cells are neural stem cells in the adult SVZ that form glioma-like growths in response to increased PDGF signaling. Neuron. 2006;51(2): 187–99.
499. Bleyl SB et al. Dysregulation of the PDGFRA gene causes inflow tract anomalies including TAPVR: integrating evidence from human genetics and model organisms. Hum Mol Genet. 2010;19(7):1286–301.
500. Chen H et al. PDGF signalling controls age-dependent proliferation in pancreatic beta-cells. Nature. 2011;478(7369):349–55.
501. Nakamura Y et al. Chondrocyte-specific microRNA-140 regulates endochondral bone development and targets Dnpep to modulate bone morphogenetic protein signaling. Mol Cell Biol. 2011;31(14):3019–28.
502. French WJ, Creemers EE, Tallquist MD. Platelet-derived growth factor receptors direct vascular development independent of vascular smooth muscle cell function. Mol Cell Biol. 2008;28(18):5646–57.
503. Schmahl J, Rizzolo K, Soriano P. The PDGF signaling pathway controls multiple steroid-producing lineages. Genes Dev. 2008;22(23):3255–67.
504. Pickett EA, Olsen GS, Tallquist MD. Disruption of PDGFRalpha-initiated PI3K activation and migration of somite derivatives leads to spina bifida. Development. 2008;135(3): 589–98.
505. Gao Z et al. Deletion of the PDGFR-beta gene affects key fibroblast functions important for wound healing. J Biol Chem. 2005;280(10):9375–89.
506. Ishii Y et al. Mouse brains deficient in neuronal PDGF receptor-beta develop normally but are vulnerable to injury. J Neurochem. 2006;98(2):588–600.
507. Chintalgattu V et al. Cardiomyocyte PDGFR-beta signaling is an essential component of the mouse cardiac response to load-induced stress. J Clin Invest. 2010;120(2):472–84.
508. Gromley A et al. Transient expression of the Arf tumor suppressor during male germ cell and eye development in Arf-Cre reporter mice. Proc Natl Acad Sci U S A. 2009;106(15): 6285–90.
509. Shen J et al. PDGFR-beta as a positive regulator of tissue repair in a mouse model of focal cerebral ischemia. J Cereb Blood Flow Metab. 2012;32(2):353–67.
510. Nguyen PT et al. Cognitive and socio-emotional deficits in platelet-derived growth factor receptor-beta gene knockout mice. PLoS One. 2011;6(3):e18004.

511. Nakagawa T et al. Roles of PDGF receptor-beta in the structure and function of postnatal kidney glomerulus. Nephrol Dial Transplant. 2011;26(2):458–68.
512. Kang J et al. PDGF-A as an epicardial mitogen during heart development. Dev Dyn. 2008;237(3):692–701.
513. Smith CL et al. Epicardial-derived cell epithelial-to-mesenchymal transition and fate specification require PDGF receptor signaling. Circ Res. 2011;108(12):e15–26.
514. Richarte AM, Mead HB, Tallquist MD. Cooperation between the PDGF receptors in cardiac neural crest cell migration. Dev Biol. 2007;306(2):785–96.
515. Klinghoffer RA et al. An allelic series at the PDGFalphaR locus indicates unequal contributions of distinct signaling pathways during development. Dev Cell. 2002;2(1):103–13.
516. Klinghoffer RA et al. The two PDGF receptors maintain conserved signaling in vivo despite divergent embryological functions. Mol Cell. 2001;7(2):343–54.
517. Olson LE, Soriano P. Increased PDGFRalpha activation disrupts connective tissue development and drives systemic fibrosis. Dev Cell. 2009;16(2):303–13.
518. Hamilton TG et al. Evolutionary divergence of platelet-derived growth factor alpha receptor signaling mechanisms. Mol Cell Biol. 2003;23(11):4013–25.
519. Moenning A et al. Sustained platelet-derived growth factor receptor alpha signaling in osteoblasts results in craniosynostosis by overactivating the phospholipase C-gamma pathway. Mol Cell Biol. 2009;29(3):881–91.
520. Kurth P et al. An activating mutation in the PDGF receptor alpha results in embryonic lethality caused by malformation of the vascular system. Dev Dyn. 2009;238(5):1064–72.
521. Chiara F et al. A gain of function mutation in the activation loop of platelet-derived growth factor beta-receptor deregulates its kinase activity. J Biol Chem. 2004;279(41):42516–27.
522. Krampert M, Heldin CH, Heuchel RL. A gain-of-function mutation in the PDGFR-beta alters the kinetics of injury response in liver and skin. Lab Invest. 2008;88(11):1204–14.
523. Looman C et al. An activating mutation in the PDGF receptor-beta causes abnormal morphology in the mouse placenta. Int J Dev Biol. 2007;51(5):361–70.
524. Tallquist MD, French WJ, Soriano P. Additive effects of PDGF receptor beta signaling pathways in vascular smooth muscle cell development. PLoS Biol. 2003;1(2):E52.
525. Heuchel R et al. Platelet-derived growth factor beta receptor regulates interstitial fluid homeostasis through phosphatidylinositol-3' kinase signaling. Proc Natl Acad Sci U S A. 1999;96(20):11410–5.
526. Tallquist MD et al. Retention of PDGFR-beta function in mice in the absence of phosphatidylinositol 3'-kinase and phospholipase Cgamma signaling pathways. Genes Dev. 2000; 14(24):3179–90.
527. Olson LE, Soriano P. PDGFRbeta signaling regulates mural cell plasticity and inhibits fat development. Dev Cell. 2011;20(6):815–26.
528. Sun T et al. A human YAC transgene rescues craniofacial and neural tube development in PDGFRalpha knockout mice and uncovers a role for PDGFRalpha in prenatal lung growth. Development. 2000;127(21):4519–29.
529. Reinertsen KK et al. Temporal and spatial specificity of PDGF alpha receptor promoter in transgenic mice. Gene Expr. 1997;6(5):301–14.
530. Liao X, Escobedo JA, Williams LT. Viability of transgenic mice expressing a platelet derived growth factor (PDGF) antagonist in plasma. J Investig Med. 1996;44(4):139–43.
531. Ritchie KA et al. The Tel-PDGFRbeta fusion gene produces a chronic myeloproliferative syndrome in transgenic mice. Leukemia. 1999;13(11):1790–803.
532. Yuasa T et al. Platelet-derived growth factor stimulates glucose transport in skeletal muscles of transgenic mice specifically expressing platelet-derived growth factor receptor in the muscle, but it does not affect blood glucose levels. Diabetes. 2004;53(11):2776–86.
533. Roesch K et al. The transcriptome of retinal Muller glial cells. J Comp Neurol. 2008; 509(2):225–38.
534. Carter CS et al. Abnormal development of NG2+PDGFR-alpha+neural progenitor cells leads to neonatal hydrocephalus in a ciliopathy mouse model. Nat Med. 2012;18(12): 1797–804.

535. Berry R, Rodeheffer MS. Characterization of the adipocyte cellular lineage in vivo. Nat Cell Biol. 2013;15(3):302–8.
536. Kang SH et al. NG2+ CNS glial progenitors remain committed to the oligodendrocyte lineage in postnatal life and following neurodegeneration. Neuron. 2010;68(4):668–81.
537. De Biase LM et al. NMDA receptor signaling in oligodendrocyte progenitors is not required for oligodendrogenesis and myelination. J Neurosci. 2011;31(35):12650–62.
538. Cuttler AS et al. Characterization of Pdgfrb-Cre transgenic mice reveals reduction of ROSA26 reporter activity in remodeling arteries. Genesis. 2011;49(8):673–80.
539. Lu X et al. PTK7/CCK-4 is a novel regulator of planar cell polarity in vertebrates. Nature. 2004;430(6995):93–8.
540. Caddy J et al. Epidermal wound repair is regulated by the planar cell polarity signaling pathway. Dev Cell. 2010;19(1):138–47.
541. Paudyal A et al. The novel mouse mutant, chuzhoi, has disruption of Ptk7 protein and exhibits defects in neural tube, heart and lung development and abnormal planar cell polarity in the ear. BMC Dev Biol. 2010;10:87.
542. Yen WW et al. PTK7 is essential for polarized cell motility and convergent extension during mouse gastrulation. Development. 2009;136(12):2039–48.
543. Takahashi M, Ritz J, Cooper GM. Activation of a novel human transforming gene, ret, by DNA rearrangement. Cell. 1985;42:581–8.
544. Schuchardt A et al. Defects in the kidney and enteric nervous system of mice lacking the tyrosine kinase receptor Ret. Nature. 1994;367:380–3.
545. Burton MD et al. RET proto-oncogene is important for the development of respiratory CO2 sensitivity. J Auton Nerv Syst. 1997;63:137–43.
546. Jijiwa M et al. A targeting mutation of tyrosine 1062 in Ret causes a marked decrease of enteric neurons and renal hypoplasia. Mol Cell Biol. 2004;24(18):8026–36.
547. Smith-Hicks CL et al. C-cell hyperplasia, pheochromocytoma and sympathoadrenal malformation in a mouse model of multiple endocrine neoplasia type 2B. EMBO J. 2000;19: 612–22.
548. Asai N et al. Targeted mutation of serine 697 in the Ret tyrosine kinase causes migration defect of enteric neural crest cells. Development. 2006;133:4507–16.
549. Carniti C et al. The Ret(C620R) mutation affects renal and enteric development in a mouse model of Hirschsprung's disease. Am J Pathol. 2006;168:1262–75.
550. Yin L et al. C620R mutation of the murine ret proto-oncogene: loss of function effect in homozygotes and possible gain of function effect in heterozygotes. Int J Cancer. 2007;121:292–300.
551. Gould TW et al. The neurotrophic effects of glial cell line-derived neurotrophic factor on spinal motoneurons are restricted to fusimotor subtypes. J Neurosci. 2008;28:2131–46.
552. Uesaka T et al. Diminished Ret expression compromises neuronal survival in the colon and causes intestinal aganglionosis in mice. J Clin Invest. 2008;118:1890–8.
553. Luo W et al. Molecular identification of rapidly adapting mechanoreceptors and their developmental dependence on ret signaling. Neuron. 2009;64:841–56.
554. Hippenmeyer S et al. A developmental switch in the response of DRG neurons to ETS transcription factor signaling. PLoS Biol. 2005;3(5):e159.
555. Jain S et al. Critical and distinct roles for key RET tyrosine docking sites in renal development. Genes Dev. 2006;20:321–33.
556. Encinas M et al. Tyrosine 981, a novel ret autophosphorylation site, binds c-Src to mediate neuronal survival. J Biol Chem. 2004;279(18):18262–9.
557. de Graaff E et al. Differential activities of the RET tyrosine kinase receptor isoforms during mammalian embryogenesis. Genes Dev. 2001;15:2433–44.
558. Wong A et al. Phosphotyrosine 1062 is critical for the in vivo activity of the Ret9 receptor tyrosine kinase isoform. Mol Cell Biol. 2005;25(21):9661–73.
559. Tronche F et al. Disruption of the glucocorticoid receptor gene in the nervous system results in reduced anxiety. Nat Genet. 1999;23(1):99–103.

560. Kramer ER et al. Cooperation between GDNF/Ret and ephrinA/EphA4 signals for motor-axon pathway selection in the limb. Neuron. 2006;50:35–47.
561. Zhuang X et al. Targeted gene expression in dopamine and serotonin neurons of the mouse brain. J Neurosci Methods. 2005;143(1):27–32.
562. Kramer ER et al. Absence of Ret signaling in mice causes progressive and late degeneration of the nigrostriatal system. PLoS Biol. 2007;5:e39.
563. Jain S et al. RET is dispensable for maintenance of midbrain dopaminergic neurons in adult mice. J Neurosci. 2006;26:11230–8.
564. Stirling LC et al. Nociceptor-specific gene deletion using heterozygous NaV1.8-Cre recombinase mice. Pain. 2005;113(1–2):27–36.
565. Golden JP et al. RET signaling is required for survival and normal function of nonpeptidergic nociceptors. J Neurosci. 2010;30:3983–94.
566. Enomoto H et al. RET signaling is essential for migration, axonal growth and axon guidance of developing sympathetic neurons. Development. 2001;128:3963–74.
567. Yan H et al. Neural cells in the esophagus respond to glial cell line-derived neurotrophic factor and neurturin, and are RET-dependent. Dev Biol. 2004;272:118–33.
568. Almeida ARM et al. RET/GFRα signals are dispensable for thymic T cell development in vivo. PLoS One. 2012;7:e52949.
569. de Boer J et al. Transgenic mice with hematopoietic and lymphoid specific expression of Cre. Eur J Immunol. 2003;33(2):314–25.
570. Baudet C et al. Retrograde signaling onto Ret during motor nerve terminal maturation. J Neurosci. 2008;28:963–75.
571. Vallstedt A et al. Different levels of repressor activity assign redundant and specific roles to Nkx6 genes in motor neuron and interneuron specification. Neuron. 2001;31(5):743–55.
572. Jain S et al. Mice expressing a dominant-negative Ret mutation phenocopy human Hirschsprung disease and delineate a direct role of Ret in spermatogenesis. Development. 2004;131:5503–13.
573. Hayashi H et al. A novel RFP-RET transgenic mouse model with abundant eumelanin in the cochlea. Hear Res. 2004;195:35–40.
574. Iwamoto T et al. Aberrant melanogenesis and melanocytic tumour development in transgenic mice that carry a metallothionein/ret fusion gene. EMBO J. 1991;10:3167–75.
575. Kato M et al. Transgenic mouse model for skin malignant melanoma. Oncogene. 1998;17:1885–8.
576. Kato M et al. RET tyrosine kinase enhances hair growth in association with promotion of melanogenesis. Oncogene. 2001;20:7536–41.
577. Kato M et al. Novel hairless RET-transgenic mouse line with melanocytic nevi and anagen hair follicles. J Invest Dermatol. 2006;126:2547–50.
578. Santoro M et al. Development of thyroid papillary carcinomas secondary to tissue-specific expression of the RET/PTC1 oncogene in transgenic mice. Oncogene. 1996;12:1821–6.
579. Cho JY et al. Early cellular abnormalities induced by RET/PTC1 oncogene in thyroid-targeted transgenic mice. Oncogene. 1999;18:3659–65.
580. Jhiang SM. Targeted expression of the ret/PTC1 oncogene induces papillary thyroid carcinomas. Endocrinology. 1996;137:375–8.
581. Iwamoto T et al. Oncogenicity of the ret transforming gene in MMTV/ret transgenic mice. Oncogene. 1990;5:535–42.
582. Sweetser DA et al. Ganglioneuromas and renal anomalies are induced by activated RET(MEN2B) in transgenic mice. Oncogene. 1999;18:877–86.
583. Kawai K et al. Tissue-specific carcinogenesis in transgenic mice expressing the RET proto-oncogene with a multiple endocrine neoplasia type 2A mutation. Cancer Res. 2000;60:5254–60.
584. Acton DS et al. Multiple endocrine neoplasia type 2B mutation in human RET oncogene induces medullary thyroid carcinoma in transgenic mice. Oncogene. 2000;19:3121–5.
585. Reynolds L et al. C-cell and thyroid epithelial tumours and altered follicular development in transgenic mice expressing the long isoform of MEN 2A RET. Oncogene. 2001;20:3986–94.

586. Michiels FM et al. Development of medullary thyroid carcinoma in transgenic mice expressing the RET protooncogene altered by a multiple endocrine neoplasia type 2A mutation. Proc Natl Acad Sci U S A. 1997;94:3330–5.
587. Srinivas S et al. Dominant effects of RET receptor misexpression and ligand-independent RET signaling on ureteric bud development. Development. 1999;126:1375–86.
588. Batourina E et al. Vitamin A controls epithelial/mesenchymal interactions through Ret expression. Nat Genet. 2001;27:74–8.
589. Green JL, Kuntz SG, Sternberg PW. Ror receptor tyrosine kinases: orphans no more. Trends Cell Biol. 2008;18(11):536–44.
590. Oishi I et al. The receptor tyrosine kinase Ror2 is involved in non-canonical Wnt5a/JNK signalling pathway. Genes Cells. 2003;8(7):645–54.
591. He F et al. Wnt5a regulates directional cell migration and cell proliferation via Ror2-mediated noncanonical pathway in mammalian palate development. Development. 2008;135(23):3871–9.
592. O'Connell MP et al. The orphan tyrosine kinase receptor, ROR2, mediates Wnt5A signaling in metastatic melanoma. Oncogene. 2010;29(1):34–44.
593. Wright TM et al. Ror2, a developmentally regulated kinase, promotes tumor growth potential in renal cell carcinoma. Oncogene. 2009;28(27):2513–23.
594. Nomi M et al. Loss of mRor1 enhances the heart and skeletal abnormalities in mRor2-deficient mice: redundant and pleiotropic functions of mRor1 and mRor2 receptor tyrosine kinases. Mol Cell Biol. 2001;21(24):8329–35.
595. Ho HY et al. Wnt5a-Ror-Dishevelled signaling constitutes a core developmental pathway that controls tissue morphogenesis. Proc Natl Acad Sci U S A. 2012;109(11):4044–51.
596. Lyashenko N et al. Mice lacking the orphan receptor ror1 have distinct skeletal abnormalities and are growth retarded. Dev Dyn. 2010;239(8):2266–77.
597. Yoda A, Oishi I, Minami Y. Expression and function of the Ror-family receptor tyrosine kinases during development: lessons from genetic analyses of nematodes, mice, and humans. J Recept Signal Transduct Res. 2003;23(1):1–15.
598. DeChiara TM et al. Ror2, encoding a receptor-like tyrosine kinase, is required for cartilage and growth plate development. Nat Genet. 2000;24(3):271–4.
599. Takeuchi S et al. Mouse Ror2 receptor tyrosine kinase is required for the heart development and limb formation. Genes Cells. 2000;5(1):71–8.
600. Valenzuela DM et al. High-throughput engineering of the mouse genome coupled with high-resolution expression analysis. Nat Biotechnol. 2003;21(6):652–9.
601. Raz R et al. The mutation ROR2W749X, linked to human BDB, is a recessive mutation in the mouse, causing brachydactyly, mediating patterning of joints and modeling recessive Robinow syndrome. Development. 2008;135(9):1713–23.
602. Matsushime H, Wang LH, Shibuya M. Human c-ros-1 gene homologous to the v-ros sequence of UR2 sarcoma virus encodes for a transmembrane receptorlike molecule. Mol Cell Biol. 1986;6(8):3000–4.
603. Neckameyer WS et al. Proto-oncogene c-ros codes for a molecule with structural features common to those of growth factor receptors and displays tissue specific and developmentally regulated expression. Mol Cell Biol. 1986;6(5):1478–86.
604. Chen JM et al. The proto-oncogene c-ros codes for a transmembrane tyrosine protein kinase sharing sequence and structural homology with sevenless protein of Drosophila melanogaster. Oncogene. 1991;6(2):257–64.
605. Tessarollo L, Nagarajan L, Parada LF. c-ros: the vertebrate homolog of the sevenless tyrosine kinase receptor is tightly regulated during organogenesis in mouse embryonic development. Development. 1992;115(1):11–20.
606. El-Deeb IM, Yoo KH, Lee SH. ROS receptor tyrosine kinase: a new potential target for anti-cancer drugs. Med Res Rev. 2011;31(5):794-818. doi: 10.1002/med.20206. Epub 2010 Aug 4.
607. Takeuchi K et al. RET, ROS1 and ALK fusions in lung cancer. Nat Med. 2012;18(3):378–81.
608. Lee J et al. Identification of ROS1 rearrangement in gastric adenocarcinoma. Cancer. 2013;119(9):1627–35.

609. Sonnenberg-Riethmacher E et al. The c-ros tyrosine kinase receptor controls regionalization and differentiation of epithelial cells in the epididymis. Genes Dev. 1996;10(10):1184–93.
610. Wagenfeld A et al. Lack of glutamate transporter EAAC1 in the epididymis of infertile c-ros receptor tyrosine-kinase deficient mice. J Androl. 2002;23(6):772–82.
611. Yeung CH, Sonnenberg-Riethmacher E, Cooper TG. Infertile spermatozoa of c-ros tyrosine kinase receptor knockout mice show flagellar angulation and maturational defects in cell volume regulatory mechanisms. Biol Reprod. 1999;61(4):1062–9.
612. Halford MM et al. Ryk-deficient mice exhibit craniofacial defects associated with perturbed Eph receptor crosstalk. Nat Genet. 2000;25(4):414–8.
613. Keeble TR et al. The Wnt receptor Ryk is required for Wnt5a-mediated axon guidance on the contralateral side of the corpus callosum. J Neurosci. 2006;26(21):5840–8.
614. Lu W et al. Mammalian Ryk is a Wnt coreceptor required for stimulation of neurite outgrowth. Cell. 2004;119(1):97–108.
615. Augustin HG et al. Control of vascular morphogenesis and homeostasis through the angiopoietin-Tie system. Nat Rev Mol Cell Biol. 2009;10(3):165–77.
616. Rodewald HR, Sato TN. Tie1, a receptor tyrosine kinase essential for vascular endothelial cell integrity, is not critical for the development of hematopoietic cells. Oncogene. 1996;12(2):397–404.
617. Girling JE, Rogers PA. Regulation of endometrial vascular remodelling: role of the vascular endothelial growth factor family and the angiopoietin-TIE signalling system. Reproduction. 2009;138(6):883–93.
618. Saharinen P, Bry M, Alitalo K. How do angiopoietins Tie in with vascular endothelial growth factors? Curr Opin Hematol. 2010;17(3):198–205.
619. Arai F et al. Tie2/angiopoietin-1 signaling regulates hematopoietic stem cell quiescence in the bone marrow niche. Cell. 2004;118(2):149–61.
620. Takakura N et al. Critical role of the TIE2 endothelial cell receptor in the development of definitive hematopoiesis. Immunity. 1998;9(5):677–86.
621. David S et al. Mending leaky blood vessels: the angiopoietin-Tie2 pathway in sepsis. J Pharmacol Exp Ther. 2013;345(1):2–6.
622. Holash J et al. Vessel cooption, regression, and growth in tumors mediated by angiopoietins and VEGF. Science. 1999;284(5422):1994–8.
623. Saharinen P et al. Multiple angiopoietin recombinant proteins activate the Tie1 receptor tyrosine kinase and promote its interaction with Tie2. J Cell Biol. 2005;169(2):239–43.
624. Winderlich M et al. VE-PTP controls blood vessel development by balancing Tie-2 activity. J Cell Biol. 2009;185(4):657–71.
625. Yuan HT et al. Activation of the orphan endothelial receptor Tie1 modifies Tie2-mediated intracellular signaling and cell survival. FASEB J. 2007;21(12):3171–83.
626. Eklund L, Olsen BR. Tie receptors and their angiopoietin ligands are context-dependent regulators of vascular remodeling. Exp Cell Res. 2006;312(5):630–41.
627. Wong AL et al. Tie2 expression and phosphorylation in angiogenic and quiescent adult tissues. Circ Res. 1997;81(4):567–74.
628. Puri MC et al. The receptor tyrosine kinase TIE is required for integrity and survival of vascular endothelial cells. EMBO J. 1995;14(23):5884–91.
629. Woo KV et al. Tie1 attenuation reduces murine atherosclerosis in a dose-dependent and shear stress-specific manner. J Clin Invest. 2011;121(4):1624–35.
630. Sato TN et al. Distinct roles of the receptor tyrosine kinases Tie-1 and Tie-2 in blood vessel formation. Nature. 1995;376(6535):70–4.
631. Patan S. TIE1 and TIE2 receptor tyrosine kinases inversely regulate embryonic angiogenesis by the mechanism of intussusceptive microvascular growth. Microvasc Res. 1998;56(1): 1–21.
632. Qu X et al. Abnormal embryonic lymphatic vessel development in Tie1 hypomorphic mice. Development. 2010;137(8):1285–95.
633. D'Amico G et al. Loss of endothelial Tie1 receptor impairs lymphatic vessel development-brief report. Arterioscler Thromb Vasc Biol. 2010;30(2):207–9.

634. Dumont DJ et al. Dominant-negative and targeted null mutations in the endothelial receptor tyrosine kinase, tek, reveal a critical role in vasculogenesis of the embryo. Genes Dev. 1994;8(16):1897–909.
635. Tachibana K et al. Selective role of a distinct tyrosine residue on Tie2 in heart development and early hematopoiesis. Mol Cell Biol. 2005;25(11):4693–702.
636. Puri MC et al. Interaction of the TEK and TIE receptor tyrosine kinases during cardiovascular development. Development. 1999;126(20):4569–80.
637. Sarao R, Dumont DJ. Conditional transgene expression in endothelial cells. Transgenic Res. 1998;7(6):421–7.
638. Jones N et al. Rescue of the early vascular defects in Tek/Tie2 null mice reveals an essential survival function. EMBO Rep. 2001;2(5):438–45.
639. Voskas D et al. A cyclosporine-sensitive psoriasis-like disease produced in Tie2 transgenic mice. Am J Pathol. 2005;166(3):843–55.
640. Diamond I et al. Conditional gene expression in the epidermis of transgenic mice using the tetracycline-regulated transactivators tTA and rTA linked to the keratin 5 promoter. J Invest Dermatol. 2000;115(5):788–94.
641. Wolfram JA et al. Keratinocyte but not endothelial cell-specific overexpression of Tie2 leads to the development of psoriasis. Am J Pathol. 2009;174(4):1443–58.
642. Hantzopoulos PA et al. The low affinity NGF receptor, p75, can collaborate with each of the Trks to potentiate functional responses to the neurotrophins. Neuron. 1994;13(1):187–201.
643. Nakagawara A et al. Cloning and chromosomal localization of the human TRK-B tyrosine kinase receptor gene (NTRK2). Genomics. 1995;25(2):538–46.
644. Kaplan DR, Miller FD. Signal transduction by the neurotrophin receptors. Curr Opin Cell Biol. 1997;9(2):213–21.
645. Dixon JE, McKinnon D. Expression of the trk gene family of neurotrophin receptors in prevertebral sympathetic ganglia. Brain Res Dev Brain Res. 1994;77(2):177–82.
646. Schober A et al. Reduced acetylcholinesterase (AChE) activity in adrenal medulla and loss of sympathetic preganglionic neurons in TrkA-deficient, but not TrkB-deficient, mice. J Neurosci. 1997;17(3):891–903.
647. Huang EJ, Reichardt LF. Trk receptors: roles in neuronal signal transduction. Annu Rev Biochem. 2003;72:609–42.
648. Klein R et al. Disruption of the neurotrophin-3 receptor gene trkC eliminates la muscle afferents and results in abnormal movements. Nature. 1994;368(6468):249–51.
649. Klein R et al. Targeted disruption of the trkB neurotrophin receptor gene results in nervous system lesions and neonatal death. Cell. 1993;75(1):113–22.
650. Smeyne RJ et al. Severe sensory and sympathetic neuropathies in mice carrying a disrupted Trk/NGF receptor gene. Nature. 1994;368(6468):246–9.
651. Barbacid M. Structural and functional properties of the TRK family of neurotrophin receptors. Ann N Y Acad Sci. 1995;766:442–58.
652. Fagan AM et al. TrkA, but not TrkC, receptors are essential for survival of sympathetic neurons in vivo. J Neurosci. 1996;16(19):6208–18.
653. Liebl DJ et al. Loss of brain-derived neurotrophic factor-dependent neural crest-derived sensory neurons in neurotrophin-4 mutant mice. Proc Natl Acad Sci U S A. 2000;97(5): 2297–302.
654. Chen X et al. A chemical-genetic approach to studying neurotrophin signaling. Neuron. 2005;46(1):13–21.
655. Coppola V et al. Ablation of TrkA function in the immune system causes B cell abnormalities. Development. 2004;131(20):5185–95.
656. Osborn L et al. Tissue-specific and insulin-dependent expression of a pancreatic amylase gene in transgenic mice. Mol Cell Biol. 1987;7(1):326–34.
657. Dorsey SG et al. In vivo restoration of physiological levels of truncated TrkB.T1 receptor rescues neuronal cell death in a trisomic mouse model. Neuron. 2006;51(1):21–8.
658. Lai KO et al. TrkB phosphorylation by Cdk5 is required for activity-dependent structural plasticity and spatial memory. Nat Neurosci. 2012;15(11):1506–15.

659. Luikart BW et al. In vivo role of truncated trkb receptors during sensory ganglion neurogenesis. Neuroscience. 2003;117(4):847–58.
660. Medina DL et al. TrkB regulates neocortex formation through the Shc/PLCgamma-mediated control of neuronal migration. EMBO J. 2004;23(19):3803–14.
661. Minichiello L et al. Mechanism of TrkB-mediated hippocampal long-term potentiation. Neuron. 2002;36(1):121–37.
662. Minichiello L et al. Point mutation in trkB causes loss of NT4-dependent neurons without major effects on diverse BDNF responses. Neuron. 1998;21(2):335–45.
663. Rohrer B et al. Role of neurotrophin receptor TrkB in the maturation of rod photoreceptors and establishment of synaptic transmission to the inner retina. J Neurosci. 1999; 19(20):8919–30.
664. Xu B et al. Brain-derived neurotrophic factor regulates energy balance downstream of melanocortin-4 receptor. Nat Neurosci. 2003;6(7):736–42.
665. Rohrer B et al. Neurotrophin receptor TrkB activation is not required for the postnatal survival of retinal ganglion cells in vivo. Exp Neurol. 2001;172(1):81–91.
666. Paredes A et al. TrkB receptors are required for follicular growth and oocyte survival in the mammalian ovary. Dev Biol. 2004;267(2):430–49.
667. He XP et al. Conditional deletion of TrkB but not BDNF prevents epileptogenesis in the kindling model. Neuron. 2004;43(1):31–42.
668. Liu X et al. Brain-derived neurotrophic factor and TrkB modulate visual experience-dependent refinement of neuronal pathways in retina. J Neurosci. 2007;27(27):7256–67.
669. Minichiello L et al. Essential role for TrkB receptors in hippocampus-mediated learning. Neuron. 1999;24(2):401–14.
670. Bergami M et al. Deletion of TrkB in adult progenitors alters newborn neuron integration into hippocampal circuits and increases anxiety-like behavior. Proc Natl Acad Sci U S A. 2008;105(40):15570–5.
671. Chakravarthy S et al. Postsynaptic TrkB signaling has distinct roles in spine maintenance in adult visual cortex and hippocampus. Proc Natl Acad Sci U S A. 2006;103(4):1071–6.
672. Tessarollo L et al. Targeted deletion of all isoforms of the trkC gene suggests the use of alternate receptors by its ligand neurotrophin-3 in neuronal development and implicates trkC in normal cardiogenesis. Proc Natl Acad Sci U S A. 1997;94(26):14776–81.
673. Postigo A et al. Distinct requirements for TrkB and TrkC signaling in target innervation by sensory neurons. Genes Dev. 2002;16(5):633–45.
674. Bai Y et al. In glaucoma the upregulated truncated TrkC.T1 receptor isoform in glia causes increased TNF-alpha production, leading to retinal ganglion cell death. Invest Ophthalmol Vis Sci. 2010;51(12):6639–51.
675. Moqrich A et al. Expressing TrkC from the TrkA locus causes a subset of dorsal root ganglia neurons to switch fate. Nat Neurosci. 2004;7(8):812–8.
676. Koch S, Claesson-Welsh L. Signal transduction by vascular endothelial growth factor receptors. Cold Spring Harb Perspect Med. 2012;2(7):a006502.
677. Olsson AK et al. VEGF receptor signalling - in control of vascular function. Nat Rev Mol Cell Biol. 2006;7(5):359–71.
678. Shinkai A et al. Mapping of the sites involved in ligand association and dissociation at the extracellular domain of the kinase insert domain-containing receptor for vascular endothelial growth factor. J Biol Chem. 1998;273(47):31283–8.
679. Fuh G et al. Requirements for binding and signaling of the kinase domain receptor for vascular endothelial growth factor. J Biol Chem. 1998;273(18):11197–204.
680. Albuquerque RJ et al. Alternatively spliced vascular endothelial growth factor receptor-2 is an essential endogenous inhibitor of lymphatic vessel growth. Nat Med. 2009;15(9):1023–30.
681. Fujisawa H et al. Roles of a neuronal cell-surface molecule, neuropilin, in nerve fiber fasciculation and guidance. Cell Tissue Res. 1997;290(2):465–70.
682. He Z, Tessier-Lavigne M. Neuropilin is a receptor for the axonal chemorepellent Semaphorin III. Cell. 1997;90(4):739–51.
683. Nakamura F et al. Neuropilin-1 extracellular domains mediate semaphorin D/III-induced growth cone collapse. Neuron. 1998;21(5):1093–100.

684. Kitsukawa T et al. Neuropilin-semaphorin III/D-mediated chemorepulsive signals play a crucial role in peripheral nerve projection in mice. Neuron. 1997;19(5):995–1005.
685. Chen H et al. Neuropilin-2, a novel member of the neuropilin family, is a high affinity receptor for the semaphorins Sema E and Sema IV but not Sema III. Neuron. 1997;19(3):547–59.
686. Kitsukawa T et al. Overexpression of a membrane protein, neuropilin, in chimeric mice causes anomalies in the cardiovascular system, nervous system and limbs. Development. 1995;121(12):4309–18.
687. Favier B et al. Neuropilin-2 interacts with VEGFR-2 and VEGFR-3 and promotes human endothelial cell survival and migration. Blood. 2006;108(4):1243–50.
688. Soker S et al. VEGF165 mediates formation of complexes containing VEGFR-2 and neuropilin-1 that enhance VEGF165-receptor binding. J Cell Biochem. 2002;85(2):357–68.
689. Fong GH et al. Role of the Flt-1 receptor tyrosine kinase in regulating the assembly of vascular endothelium. Nature. 1995;376(6535):66–70.
690. Hiratsuka S et al. Flt-1 lacking the tyrosine kinase domain is sufficient for normal development and angiogenesis in mice. Proc Natl Acad Sci U S A. 1998;95(16):9349–54.
691. Hiratsuka S et al. Involvement of Flt-1 tyrosine kinase (vascular endothelial growth factor receptor-1) in pathological angiogenesis. Cancer Res. 2001;61(3):1207–13.
692. Hiratsuka S et al. Membrane fixation of vascular endothelial growth factor receptor 1 ligand-binding domain is important for vasculogenesis and angiogenesis in mice. Mol Cell Biol. 2005;25(1):346–54.
693. Ku CH et al. Inducible overexpression of sFlt-1 in podocytes ameliorates glomerulopathy in diabetic mice. Diabetes. 2008;57(10):2824–33.
694. Ambati BK et al. Corneal avascularity is due to soluble VEGF receptor-1. Nature. 2006;443(7114):993–7.
695. Shalaby F et al. Failure of blood-island formation and vasculogenesis in Flk-1-deficient mice. Nature. 1995;376(6535):62–6.
696. Sakurai Y et al. Essential role of Flk-1 (VEGF receptor 2) tyrosine residue 1173 in vasculogenesis in mice. Proc Natl Acad Sci U S A. 2005;102(4):1076–81.
697. Ema M, Takahashi S, Rossant J. Deletion of the selection cassette, but not cis-acting elements, in targeted Flk1-lacZ allele reveals Flk1 expression in multipotent mesodermal progenitors. Blood. 2006;107(1):111–7.
698. Licht AH et al. Endothelium-specific Cre recombinase activity in flk-1-Cre transgenic mice. Dev Dyn. 2004;229(2):312–8.
699. Dumont DJ et al. Cardiovascular failure in mouse embryos deficient in VEGF receptor-3. Science. 1998;282(5390):946–9.
700. Karkkainen MJ et al. A model for gene therapy of human hereditary lymphedema. Proc Natl Acad Sci U S A. 2001;98(22):12677–82.
701. Ichise T, Yoshida N, Ichise H. H-, N- and Kras cooperatively regulate lymphatic vessel growth by modulating VEGFR3 expression in lymphatic endothelial cells in mice. Development. 2010;137(6):1003–13.
702. Haiko P et al. Deletion of vascular endothelial growth factor C (VEGF-C) and VEGF-D is not equivalent to VEGF receptor 3 deletion in mouse embryos. Mol Cell Biol. 2008;28(15): 4843–50.
703. Zhang L et al. VEGFR-3 ligand-binding and kinase activity are required for lymphangiogenesis but not for angiogenesis. Cell Res. 2010;20(12):1319–31.
704. Martinez-Corral I et al. In vivo imaging of lymphatic vessels in development, wound healing, inflammation, and tumor metastasis. Proc Natl Acad Sci U S A. 2012;109(16):6223–8.
705. Makinen T et al. Inhibition of lymphangiogenesis with resulting lymphedema in transgenic mice expressing soluble VEGF receptor-3. Nat Med. 2001;7(2):199–205.
706. Kawasaki T et al. A requirement for neuropilin-1 in embryonic vessel formation. Development. 1999;126(21):4895–902.
707. Giger RJ et al. Neuropilin-2 is required in vivo for selective axon guidance responses to secreted semaphorins. Neuron. 2000;25(1):29–41.
708. Yuan L et al. Abnormal lymphatic vessel development in neuropilin 2 mutant mice. Development. 2002;129(20):4797–806.

709. Takashima S et al. Targeting of both mouse neuropilin-1 and neuropilin-2 genes severely impairs developmental yolk sac and embryonic angiogenesis. Proc Natl Acad Sci U S A. 2002;99(6):3657–62.
710. Gu C et al. Neuropilin-1 conveys semaphorin and VEGF signaling during neural and cardiovascular development. Dev Cell. 2003;5(1):45–57.
711. Goel HL et al. Neuropilin-2 promotes branching morphogenesis in the mouse mammary gland. Development. 2011;138(14):2969–76.
712. Walz A, Rodriguez I, Mombaerts P. Aberrant sensory innervation of the olfactory bulb in neuropilin-2 mutant mice. J Neurosci. 2002;22(10):4025–35.
713. Chen H et al. Neuropilin-2 regulates the development of selective cranial and sensory nerves and hippocampal mossy fiber projections. Neuron. 2000;25(1):43–56.
714. Fong GH et al. Increased hemangioblast commitment, not vascular disorganization, is the primary defect in flt-1 knock-out mice. Development. 1999;126(13):3015–25.
715. Eremina V et al. Glomerular-specific gene excision in vivo. J Am Soc Nephrol. 2002;13(3): 788–93.
716. Jin J et al. Soluble FLT1 binds lipid microdomains in podocytes to control cell morphology and glomerular barrier function. Cell. 2012;151(2):384–99.
717. Partanen TA et al. Neuropilin-2 and vascular endothelial growth factor receptor-3 are up-regulated in human vascular malformations. Angiogenesis. 2013;16(1):137–46.
718. Cursiefen C et al. Nonvascular VEGF receptor 3 expression by corneal epithelium maintains avascularity and vision. Proc Natl Acad Sci U S A. 2006;103(30):11405–10.
719. Vasioukhin V et al. The magical touch: genome targeting in epidermal stem cells induced by tamoxifen application to mouse skin. Proc Natl Acad Sci U S A. 1999;96(15):8551–6.
720. Ashery-Padan R et al. Pax6 activity in the lens primordium is required for lens formation and for correct placement of a single retina in the eye. Genes Dev. 2000;14(21):2701–11.
721. Lallemand Y et al. Maternally expressed PGK-Cre transgene as a tool for early and uniform activation of the Cre site-specific recombinase. Transgenic Res. 1998;7(2):105–12.
722. Means AL et al. Keratin 19 gene drives Cre recombinase expression throughout the early postimplantation mouse embryo. Genesis. 2005;42(1):23–7.
723. Claxton S et al. Efficient, inducible Cre-recombinase activation in vascular endothelium. Genesis. 2008;46(2):74–80.
724. Tammela T et al. VEGFR-3 controls tip to stalk conversion at vessel fusion sites by reinforcing Notch signalling. Nat Cell Biol. 2011;13(10):1202–13.
725. Mao X, Fujiwara Y, Orkin SH. Improved reporter strain for monitoring Cre recombinase-mediated DNA excisions in mice. Proc Natl Acad Sci U S A. 1999;96(9):5037–42.
726. Rodriguez CI et al. High-efficiency deleter mice show that FLPe is an alternative to Cre-loxP. Nat Genet. 2000;25(2):139–40.
727. Kisanuki YY et al. Tie2-Cre transgenic mice: a new model for endothelial cell-lineage analysis in vivo. Dev Biol. 2001;230(2):230–42.
728. Suemori H et al. A mouse embryonic stem cell line showing pluripotency of differentiation in early embryos and ubiquitous beta-galactosidase expression. Cell Differ Dev. 1990; 29(3):181–6.
729. Dymecki SM. Flp recombinase promotes site-specific DNA recombination in embryonic stem cells and transgenic mice. Proc Natl Acad Sci U S A. 1996;93(12):6191–6.
730. Skarnes WC et al. Capturing genes encoding membrane and secreted proteins important for mouse development. Proc Natl Acad Sci U S A. 1995;92(14):6592–6.
731. Lanaya H et al. EGFR has a tumour-promoting role in liver macrophages during hepatocellular carcinoma formation. Nat Cell Biol. 2014;16(10):972–81. doi:10.1038/ncb3031. Epub 2014 Aug 31.
732. Wang Y et al. Cre/lox recombination in the lower urinary tract. Genesis. 2009;47(6):409–13.
733. Jin C et al. Transgenic mouse with high Cre recombinase activity in all prostate lobes, seminal vesicle, and ductus deferens. Prostate. 2003;57:160–164.

Index

D.L. Wheeler, Y. Yarden (eds.), *Receptor Tyrosine Kinases: Structure, Functions and Role in Human Disease*, DOI 10.1007/978-1-4939-2053-2